The Pauson-Khand Reaction

The Pauson-Khand Reaction

Scope, Variations and Applications

Edited by

RAMON RIOS TORRES

Department of Organic Chemistry, University of Barcelona, Barcelona, Spain and Catalan Institute of Research and Advanced Studies (ICREA), Spain

WILEY

A John Wiley & Sons, Ltd., Publication

This edition first published 2012

Registered office
John Wiley & Sons Ltd, The Atrium, Southern Gate, Chichester, West Sussex, PO19 8SQ, United Kingdom

For details of our global editorial offices, for customer services and for information about how to apply for permission to reuse the copyright material in this book please see our website at www.wiley.com.

Library of Congress Cataloging-in-Publication Data

The Pauson-Khand reaction : scope, variations, and applications / edited by Ramon Rios Torres.
p. cm.
Includes bibliographical references and index.
ISBN 978-0-470-97076-8 (hardback)
1. Pauson-Khand reaction. I. Rios Torres, Ramon.
QD281.R5P38 2012
547′.412–dc23

2011051669

A catalogue record for this book is available from the British Library.

ISBN: 9780470970768 (13 digits)

Set in 10/12pt Times by Aptara Inc., New Delhi, India.
Printed and bound in Malaysia by Vivar Printing Sdn Bhd

1 2012

Contents

List of Contributors

Xacobe C. Cambeiro Institute of Chemical Research of Catalonia (ICIQ), Tarragona, Spain

Young Keun Chung Department of Chemistry, College of Natural Sciences, Seoul National University, Seoul, Korea

James M. Cook Department of Chemistry, University of Wisconsin-Milwaukee, Milwaukee, Wisconsin, USA

Sarah Hoerner Department of Natural Science, Concordia University Wisconsin, Mequon, Wisconsin, USA

Martin Kamlar Department of Organic and Nuclear Chemistry, Charles University in Prague, Prague, Czech Republic

William J. Kerr Department of Pure and Applied Chemistry, WestCHEM, Glasgow, Scotland

Fuk Yee Kwong State Key Laboratory of Chirosciences and Department of Applied Biology and Chemical Technology, The Hong Kong Polytechnic University, Hung Hom, Kowloon, Hong Kong

Fuk Loi Lam State Key Laboratory of Chirosciences and Department of Applied Biology and Chemical Technology, The Hong Kong Polytechnic University, Hung Hom, Kowloon, Hong Kong

Hang Wai Lee State Key Laboratory of Chirosciences and Department of Applied Biology and Chemical Technology, The Hong Kong Polytechnic University, Hung Hom, Kowloon, Hong Kong

Agustí Lledó Institute for Research in Biomedicine (IRB) Barcelona, Barcelona, Spain

Albert Moyano Department of Organic Chemistry, University of Barcelona, Barcelona, Spain

Miquel A. Pericàs Institute of Chemical Research of Catalonia (ICIQ), Tarragona, Spain and Department of Organic Chemistry, University of Barcelona, Barcelona, Spain

Antoni Riera Institute for Research in Biomedicine (IRB) Barcelona, and Department of Organic Chemistry, University of Barcelona, Barcelona, Spain

Ramon Rios Torres Department of Organic Chemistry, University of Barcelona, Spain, and Catalan Institute of Research and Advanced Studies (ICREA), Spain

Takanori Shibata Department of Chemistry and Biochemistry, Waseda University, Okubo, Shinjuku, Tokyo, Japan

Jun Wang State Key Laboratory of Chirosciences and Department of Applied Biology and Chemical Technology, The Hong Kong Polytechnic University, Hong Kong

Scott G. Van Ornum Department of Natural Science, Concordia University Wisconsin, Mequon, Wisconsin, USA

Xavier Verdaguer Institute for Research in Biomedicine (IRB) Barcelona, and Department of Organic Chemistry, University of Barcelona, Barcelona, Spain

Jan Vesely Department of Organic and Nuclear Chemistry, Charles University in Prague, Prague, Czech Republic

Foreword

In 1973, Pauson and Khand reported a [2+2+1] reaction between an alkyne, an alkene, and CO, thereby furnishing a cyclopentenone in a highly convergent manner. Because five-membered carbocycles are common sub-units in organic molecules, the potential utility of this new process, the Pauson-Khand reaction, was rapidly appreciated by synthetic chemists.

Since its discovery, the Pauson-Khand reaction has been most frequently applied in the context of reactions between alkenes and pre-formed cobalt–alkyne complexes. However, it is worth noting that, in their original 1973 report, Pauson and Khand recognized the potential advantages of a catalytic variant, and indeed described successful cyclizations that employed $Co_2(CO)_8$ as a catalyst.

The power of the Pauson-Khand reaction has been beautifully highlighted in the syntheses of a wide range of bioactive compounds, including pentalenene, hirsutene, epoxydictymene, hydroxymethylacylfulvene, dendrobine, and axinellamines A and B. In general, these applications have either generated the target molecules as racemates, or they have relied upon pre-existing stereogenic centers to guide diastereoselective reactions. The development of catalytic, enantioselective Pauson-Khand reactions is still in its infancy, and therefore remains one of the important frontiers for future investigation.

This monograph provides an excellent summary of the current state-of-the-art for the Pauson-Khand reaction. It will no doubt stimulate further advances in the development of this powerful method for five-membered ring formation.

Professor Gregory C. Fu
Massachussets Institute of Technology
December 2011

Preface

"Where observation is concerned, chance favors only the prepared mind."
—Louis Pasteur

This book on the Pauson-Khand reaction, edited for Wiley in 2012, is intended to provide an overview of the Pauson-Khand reaction from its discovery in 1973 up until the present day. No book on the topic is to be found at the time of writing. And yet there is a real need for a work that covers the issue.

The sections included in the present book deal with different aspects of the reaction under study, ranging from the mechanism to the later enantioselective methodologies; from the original cobalt catalyzed reaction to the use of different transition metal catalysts that improve the outcome of the reaction.

It was a pleasure to be the editor of this compendia, I had the chance to revise my PhD work on the topic and to honor the work of so many chemists.

I would like to thank all the distinguished scientists and their coauthors for their rewarding and timely contributions. I acknowledge the great work done by the Wiley editorial staff, in particular that of Sarah Tilley, her help was invaluable.

I also want to thank Professors G.C. Fu, P. Walsh, B. List, and A. Cordova, who introduced me into the world of chemistry. Their example and knowledge has always led me throughout my career.

Finally, I want to thank my parents for all their support and help. Without them I could never have carried out this project.

Ramon Rios Torres, Barcelona
October 2011

1

The Pauson-Khand Reaction – an Introduction

William J. Kerr

1.1 The Discovery and Early Evolution of the Khand Reaction

Over the period from the 1950s, metal-mediated methods have changed the profile and escalated the preparative potential of organic synthesis well beyond that which was previously possible or imaginable. New organic cyclisation techniques have been at the heart

The Pauson-Khand Reaction: Scope, Variations and Applications, First Edition. Edited by Ramon Rios Torres.

of the developments in this domain. In relation to this, the preparative interconversion now known as the Pauson-Khand reaction is a cyclopentenone-forming [2+2+1] annulation process involving an alkyne, an alkene, and a unit of carbon monoxide, originally supplied as a ligand in the starting hexacarbonylalkynedicobalt complex, as illustrated in Scheme 1.1. Despite many citations to this process having been divulged initially in 1973, the first account of this cobalt-mediated cyclisation was noted within a very short *Chemical Communications* article from Pauson's laboratory, submitted in November, 1970 and appearing in press in 1971, with the following line, "... the reaction of norbornadiene with complexes (I) [hexacarbonylalkynedicobalt complexes] yields hydrocarbon and ketonic products derived from norbornadiene, acetylene, and carbon monoxide."[1] Pauson and co-workers went on to describe this process in appreciably more detail in 1973.[2]

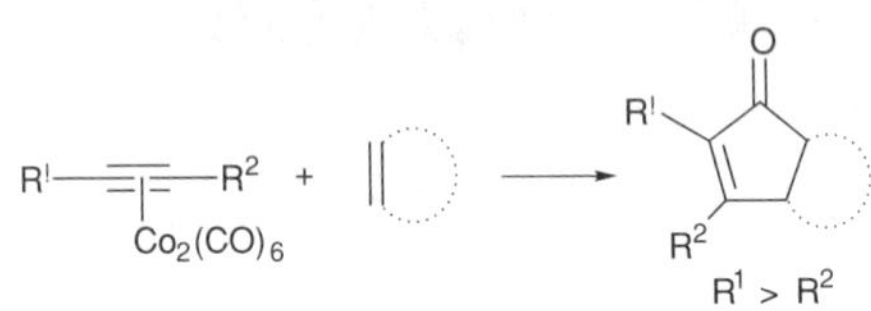

Scheme 1.1

As detailed by Pauson and Khand, the formation of cyclopentenones in a single step was a serendipitous discovery.[3] Within Pauson's laboratories at the University of Strathclyde, a programme of work had been aiming to establish the preparation of a series of cobalt-containing compounds from accessible cobalt complexes and alkynes, as a method by which to probe, for example, the reaction pathway involved in alkyne trimerisation processes. Believing that sufficiently reactive alkenes could be employed in a similar fashion to alkynes, norbornadiene was reacted with hexacarbonylalkynedicobalt complexes. Surprisingly, the main organic products from these reactions were ketonic in nature and, more specifically, possessed the general cyclopentenone structure, as shown in Scheme 1.2.[2a] This seminal discovery established the foundations for the study and enhancement of the synthetic transformation that is the subject of this monograph.

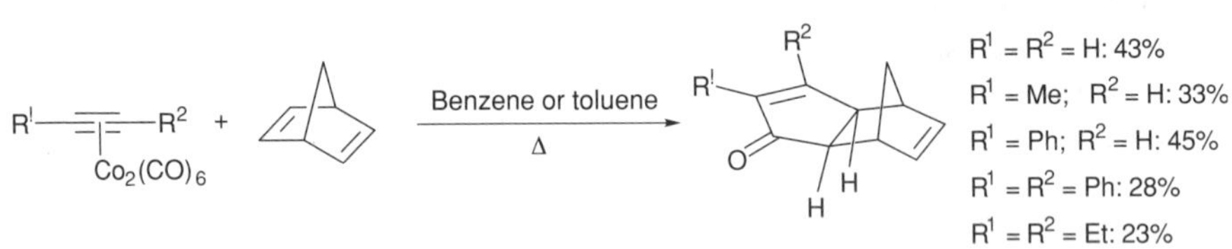

Scheme 1.2

Throughout the remainder of the 1970s, the initial scope and generality of this cyclopentannulation process was established through an extensive series of studies by Pauson and his co-workers. In relation to this, these reactions were conveniently performed by typically heating a mixture of the alkyne Co-complex and the alkene in hydrocarbon (or ethereal) solvents. Additionally, even from the initial series of reactions, for example those shown in Scheme 1.2, good levels of regio- and stereoselectivity were achievable within these cyclisation processes, with the larger alkyne substituent being installed in the position α to the cyclopentenone carbonyl unit and the *exo*-product predominating in all instances. Despite this, it is worth noting that other workers did not engage in research incorporating this preparative process, in any appreciable sense, until into the 1980s. The reasons for this somewhat delayed engagement by the preparative community may well have their foundations in the only low to moderate cyclopentenone yields delivered in many of the preliminary transformations, coupled with the variety of organometallic by-products formed within a typical reaction process and from which the desired organic compound had to be removed and purified. Having stated all of this and following a series of subsequent enhancements to this general preparative method (*vide infra*), the development and use of the Pauson-Khand reaction escalated considerably. By way of illustration, a simple Web of Science^{SM} search (using "Khand" as the topic keyword), conducted during the production of this manuscript,[4] delivered some 1,339 publications with this cyclopentenone forming technique cited in the paper abstract or keywords. Notably, over 1,300 of these publications have appeared since 1991 and, to illustrate the level of active and on-going employment of this cyclisation process, almost 6% of all these cited publications appeared in the year 2010 alone. In addition, this key annulation reaction has been the subject of a number of review articles.[3,5]

In relation to the use of the term "Khand" in the search process mentioned above and in a more general sense, it is worth discussing the name of this now widely employed transformation. As noted above, the cyclopentenone-forming transformation was discovered by Peter L. Pauson and co-workers at the University of Strathclyde in Glasgow. One of these researchers was Ihsan U. Khand, at this stage a postdoctoral research associate in Pauson's laboratory. Khand was born in India in 1935 and had moved to Glasgow in 1965, before initiating PhD studies in the Pauson laboratory in 1966. Khand's PhD thesis described the preparation and use of cationic iron sandwich complexes, and was completed in 1969, with the degree of PhD award being made in 1970. Thereafter and following a short period undergoing secondary school teacher training in Glasgow, Khand convinced Pauson to allow him to return to the laboratory for a one year period as a postdoctoral associate. It was during this period that the reaction of alkyne-cobalt complexes with alkenes to form cyclopentenones was discovered. Typical of his generosity and remarkable modesty, Peter Pauson only ever referred to this annulation process in spoken or written word as "The Khand Reaction". This can be seen quite clearly in Pauson's 1985 *Tetrahedron* review of this annulation protocol,[5a] as well as within all of his publications describing the development and use of this process. Indeed, it was quite a frustration to Pauson that other workers had amended the name of this, now world-renowned, transformation to include his own name. This was most vividly highlighted during the reviewing and acceptance of Pauson's final paper, in collaboration with our own laboratory; during this process the reviewers and editors had stipulated that the name of the reaction be changed to that which was more universally accepted. Accordingly, Pauson insisted that, if we were to do this, a footnote be inserted. This was duly done and read, "Hitherto one of us has preferred to use the initially designated name of 'Khand' reaction for this synthetic process; in light of what has now become common usage and following comments from the referees, we have subsequently employed the more widely accepted term 'Pauson-Khand' throughout this paper."[6] Based on all of this and returning to the search term initially utilised, if one was to employ "Pauson-Khand", almost one hundred papers would be missed.

Following on from the establishment of ferrocene preparation and reactivity, throughout the 1950s, 1960s, and 1970s Pauson became an internationally-leading contributor to preparative organometallic chemistry and the discovery of the "Khand Reaction" further enhanced his impact on the world stage, which continued well beyond his formal retirement. As for Dr Khand, despite returning to secondary school and college teaching employment from 1971, on an entirely voluntary basis he would appear in the Pauson laboratory most evenings, as well as at weekends and during school holidays, and throughout these periods established much of the original scope of this cycloaddition reaction. Based on this level of drive and dedication, perhaps one can see some of the reasons behind Peter Pauson's desire to have the reaction named as he did. Sadly, at the age of 44, Ihsan Khand passed away in January 1980.

1.2 The Intermolecular Pauson-Khand Reaction

The intermolecular Pauson-Khand (P–K) process has an appreciably wide tolerance in relation to substrate structure and functionality. In relation to the alkyne component, acetylene and terminal alkynes operate most effectively, whilst internal alkynes tend to deliver generally lower yields.[5c] With regards to the alkene partner, strained cyclic species tend to give good yields of cyclopentenone products, whereas increasing the sterics around the olefin reduces annulation effectiveness. In relation to these points relating to alkene reactivity, it is noteworthy that ethylene has been shown to be an effective and widely used PK cyclisation component, although elevated temperatures (80–160 °C) and pressures of ethylene (50-120 atm) were traditionally required.[7] Associated with the alkene strain considerations, a theoretical study by Milet, Gimbert, and co-workers has rationalised that the reactivity of the alkene component within P–K reactions is correlated to the back-donation of electrons from the metal centre to the π^* orbital (LUMO) of the olefin.[8] Furthermore, these authors have noted that a relationship exists between the C=C–C bond angle and the energy of the LUMO: the smaller the angle, the lower the LUMO energy. This is in direct accord with the reactivity pattern previously displayed across an olefin series: norbornene (107 °) > cyclopentene (112 °) ≫ cyclohexene (128 °).[2,9]

One distinct limitation in relation to alkene applicability relates to olefins with conjugated electron-withdrawing units, which react to deliver conjugated dienes, where the new C–C bond has formed between the two most accessible carbon atoms of both the alkyne and alkene components. In relation to this, styrene substrates display intermediate behaviour. In the example shown in Scheme 1.3, following reflux in toluene, the phenylacetylene complex **1** couples with styrene **2** to give the conjugated diene **3** as the predominant product (39%), along with the cyclopentenone **4** in a more moderate 12% yield.[9] This outcome is in contrast with the more recent developments from Wender and co-workers,

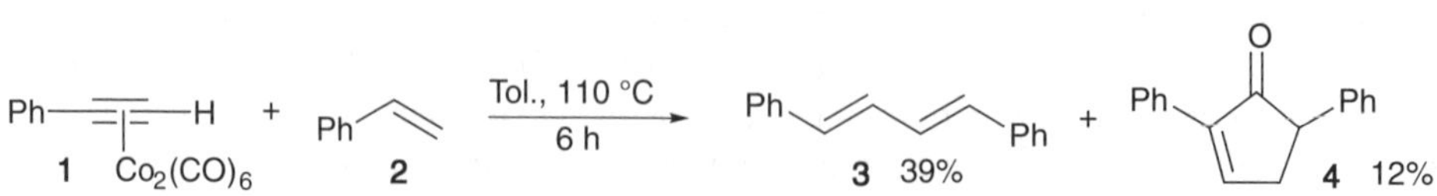

Scheme 1.3

where remarkable increases in efficiency and, indeed, selectivity for the cyclopentenone products have been achieved using rhodium-catalysed techniques.[10]

1.2.1 Regioselectivity of Alkyne Insertion

As detailed previously in relation to Pauson's original findings, alkyne regiochemistry is most often based on substituent sterics, with the larger of the C≡C triple bond components residing ultimately in the 2-position of the PK cyclopentenone product. The reasons for this appear to be embedded in key steps within the perceived reaction pathway (as will be detailed in Chapter 2). In most instances, the regioselectivity of alkyne insertion is complete;[5c] that shown in Scheme 1.4 effectively illustrates this point, as well as providing an example of the typical conditions traditionally employed for the use of a gaseous olefin such as ethylene (*vide supra*).[7d]

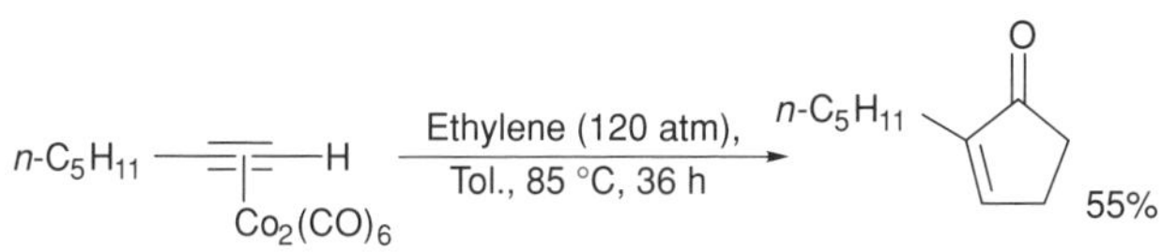

Scheme 1.4

In relation to internal alkynes where the substituents are similar in size, mixtures of products can result. One such example from Billington and Pauson, where both alkyne insertion regioisomers are observed, is provided in Scheme 1.5;[11] despite a mixture of products resulting, an 8:1 ratio of cyclopentenones (**5**:**6**) is obtained, with the less prevalent regioisomer **6** being formed in only 3% yield.

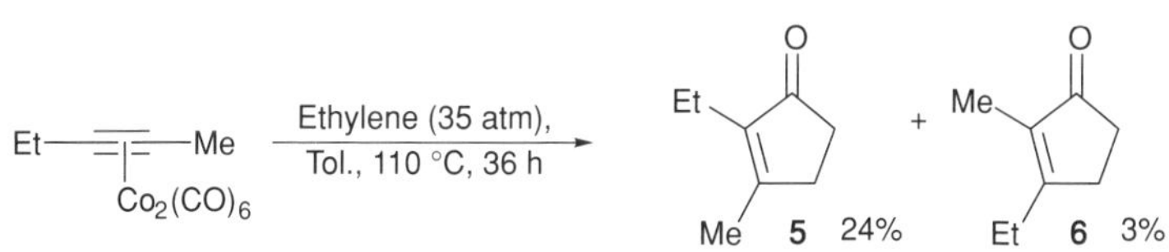

Scheme 1.5

Further examples have appeared in the literature where the mode of alkyne insertion cannot be explained by steric factors. In this regard, it has been shown by Krafft that, in the case of internal alkynes bearing electron-withdrawing components, electronic considerations are required to be invoked in attempts to explain the regiochemical outcome of the P–K process. More specifically, with alkynes conjugated with an electron-withdrawing group (EWG), the cyclisation product formed is that with the EWG installed at the cyclopentenone 3-position.[12] This alkyne polarisation effect is further illustrated by Gimbert and Greene, with the example of an internal alkyne such as **7** possessing electronically distinct yet sterically rather similar substituents. In this instance, the sole cyclopentenone product **8** is that with the more electron-deficient arene positioned in the β-position (Scheme 1.6).[13] In attempts to explain these observations, in relation to the anticipated cyclisation mechanism, Krafft has proposed an argument based on preferential alkene insertion into one Co–C bond within a polarised alkyne complex intermediate. In contrast, following theoretical studies, Gimbert and Greene have used electronic differences between the alkyne termini in order to support the loss of a specific CO ligand (via a *trans*-effect exerted by the alkyne), in turn, leading to selective olefin insertion and, ultimately, delivering the cyclopentenone regiochemistry observed.

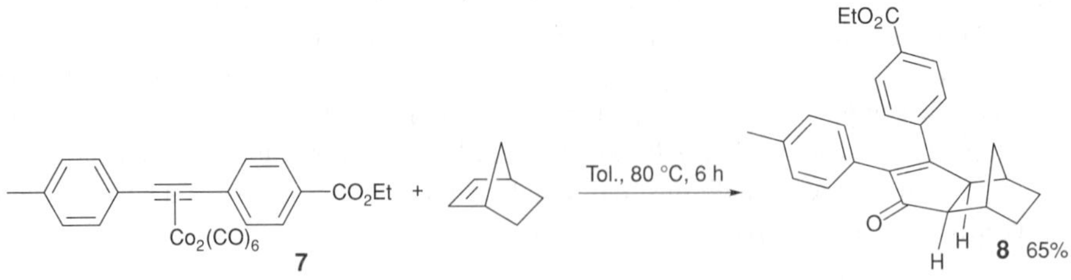

Scheme 1.6

1.2.2 Regioselectivity of Alkene Insertion

From the broad array of examples performed as the scope of the PK annulation process was being established, it became clear that the selectivity associated with the alkene reaction partner was significantly less predictable.[5c,9] In particular, as well as delivering only modest yields of cyclopentenones, simple acyclic alkene substrates are incorporated with no regioselection when reacted with terminal alkynes (although the elevated levels of selectivity associated with the alkyne component are retained). This is illustrated in Scheme 1.7 by the reaction of 1-octene **9** with the phenylacetylene complex **1** to give a 1:1 mixture of the PK cyclisation products.[14] In contrast, when internal alkynes (e.g. **10**) were employed with the same class of aliphatic alkene substrates (e.g. **9**) significantly enhanced regioselectivity in relation to alkene insertion is achieved, albeit at the cost of appreciably lowered chemical yields in most instances.[14]

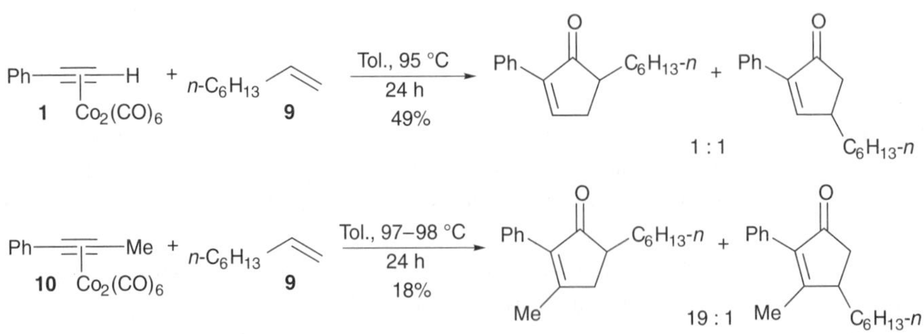

Scheme 1.7

It is worth noting that incorporation of aryl-substituted and heteroatom-containing alkenes can offer good to excellent levels of regioselectivity with terminal alkyne partners.[5c] Indeed, the same can be said, in a general sense, for unsymmetrically-substituted cyclic alkenes. Returning to the issue of employing more simple alkenes with terminal alkynes and in a drive to enhance the selectivity of the alkene incorporation, Krafft turned to the use of heteroatoms tethered to the olefin in attempts to control annulation regiochemistry. Using homallylic (or bishomoallylic) sulfides and amines, as functional units capable of acting as soft donor ligands, both the efficiency and regioselectivity of these PK processes were shown to be appreciable, and certainly when compared to the previously established and normally much less effective intermolecular reactions (Scheme 1.8).[15] These outcomes are believed to result from the tethered heteroatom acting as a ligand for cobalt along the reaction pathway, in turn acting to restrict the conformational flexibility of the complexed alkene, and leading to the more favourable formation of the 5-substituted products.

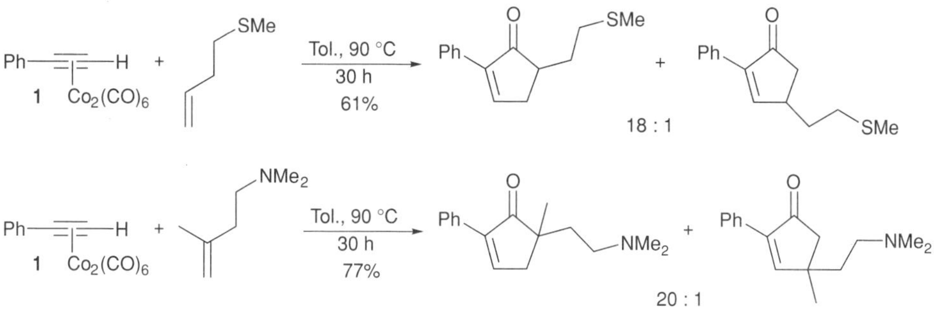

Scheme 1.8

Following on from this pioneering work by Krafft and co-workers in the area of directed intermolecular PK reactions, studies within our own laboratory have led to the development of an analogous system with allylphosphonates. Optimised conditions with phenylacetylene complex **1** and diethyl allylphosphonate **11** led to the formation of cyclopentenones **12** in high yield and good levels of regioselectivity.[16] In due course, it was demonstrated that an extended class of alkenes possessing a tethered phosphonate ester unit could also be employed within P–K processes, again with good levels of olefin regiocontrol (Scheme 1.9), to deliver an additional class of phosphorus-containing cyclopentenones of further synthetic potential.[17]

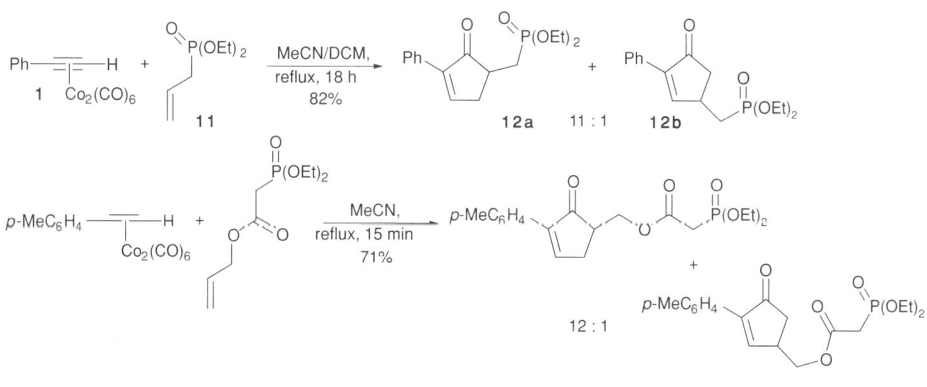

Scheme 1.9

1.3 The Intramolecular Pauson-Khand Reaction

The establishment of the intramolecular variant of the PK reaction by Schore in 1981 provided a crucial advance associated with this methodology, which has allowed this metal-mediated cyclisation technique to grow considerably in importance and utility within preparative organic chemistry.[18] As illustrated in Scheme 1.10, it was shown within the original landmark publication that both hept-6-en-1-ynes and oct-7-en-1-ynes perform in the intramolecular PK process to deliver the bicyclic enones shown. In addition to this and by extension, others went on to show that similar oxygen- and nitrogen-tethered enynes could also take part in similar cyclisations to deliver the equivalent bicyclopentenone products.[19]

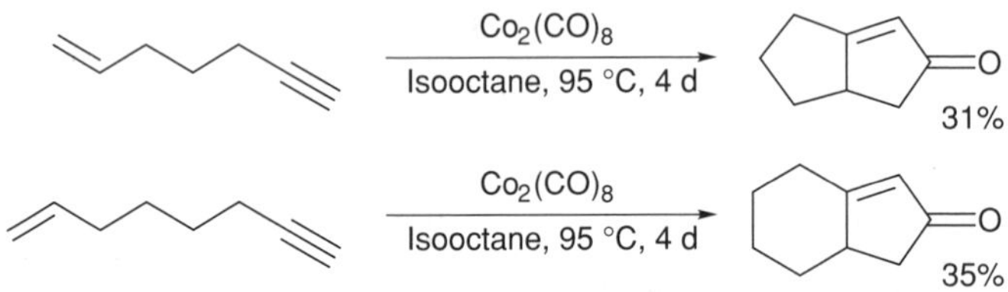

Scheme 1.10

A short period after Schore's initial publications in this area, Magnus further enhanced the efficacy of this method by showing that appreciable stereoselectivity could be embedded within intramolecular processes, a feature that was to prove key to the application of the PK reaction in natural product synthesis. As illustrated in Scheme 1.11, substituents at the propargylic (C3) and allylic (C5) positions have a preference to become situated on the *exo*-face of the resulting bicyclic PK products, and *syn* to the ring fusion hydrogen.[20] Furthermore, the level of stereoselectivity achieved is enhanced by bulky substitution at the terminal alkyne carbon. With the stereochemical reaction outcomes established for a number of substrate substitution patterns, Magnus and co-workers went on to utilise these methods within the synthesis of precursors to the linearly-fused triquinane natural products, coriolin and hirsutic acid.[20a,c] In addition, the employment of a similar stereoselective

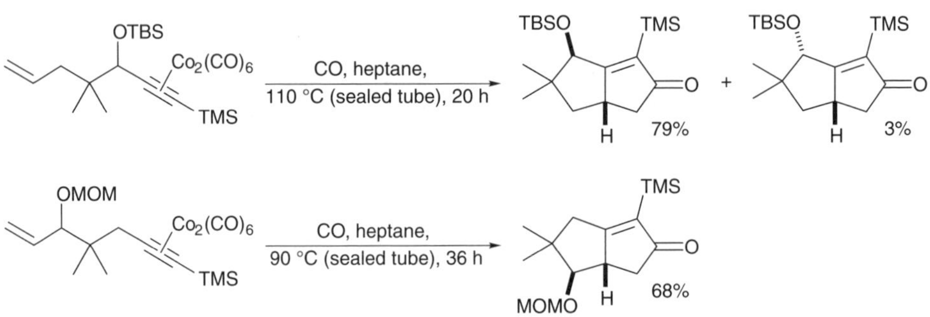

Scheme 1.11

intramolecular PK strategy allowed the same research group to deliver an elegant approach to the cytotoxic sesquiterpene, quadrone, as well as the related biologically-active natural product, terrecyclic acid.[21]

Following on from these seminal examples produced during the early- to mid-1980s, the intramolecular version of the PK annulation has continued to expand in terms of substrate scope and reaction selectivity, as well as general utility and effectiveness. In particular, this growth in PK potential has been driven by a range of emerging procedures and techniques, which have appreciably enhanced the overall efficiency of the cyclisation method, and which will be discussed through the following sections of this introductory chapter.

1.4 Enhancing the Pauson-Khand Annulation by Reaction Promotion

Traditionally, and as detailed in many of the examples cited to this stage, PK reactions were performed at elevated temperatures (and pressures) in aromatic or hydrocarbon solvents.[2,3,5a–d] Moreover, isolation and purification of the cyclopentenone products in the face of a plethora of organometallic by-products often presented further practical difficulties. Additionally and at least as importantly, the PK cyclisation product yields were more usually low to moderate. Accordingly, from the mid-1980s considerable efforts were applied to the enhancement of PK techniques by a number of researchers worldwide.

1.4.1 Dry State Adsorption

In 1985–86, Smit and co-workers reported on an appreciable practical advance in PK methodology.[22] By adsorbing the complexed alkyne onto a support such as silica or alumina and by gently heating in the dry state, reaction times were markedly reduced and yields dramatically enhanced. To illustrate this, an example that compares a more traditional thermal process with the dry state adsorption technique is shown in Scheme 1.12. These authors proposed that the donor sites on the solid surface were acting to enhance decarbonylative ligand exchange. Additionally, the adsorption of the substrate onto the (silica) support may well also act to limit conformational movement of the substrate, akin to the Restricted Rotamer Effect, and promote cyclisation. Through the remainder of the 1980s, Smit, Caple, and co-workers, as well as others, went on expanding the range of substrates utilised in both intra-[19b,23] and intermolecular[24] processes with this dry state method.

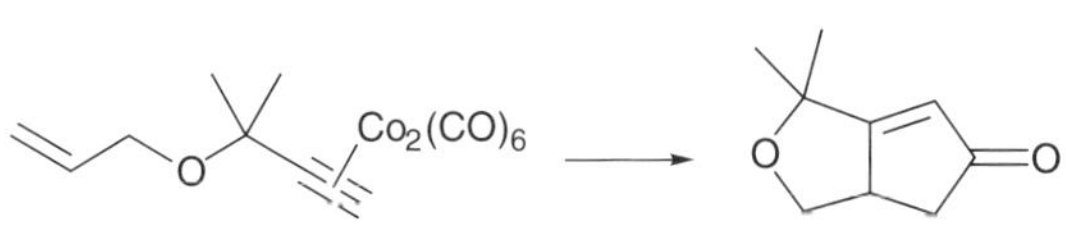

Conditions
Thermal: isooctane, CO, 60 °C, 24 h: 29%[19a]
Dry State Adsorption: SiO_2, O_2, 45 °C, 30 min: 75%[22]

Scheme 1.12

1.4.2 Ultrasound Techniques

Metal-carbonyl bond cleavage, which is believed to be the requisite first step in the P–K annulation, is promoted by ultrasonication.[25] In relation to this, the use of low powered sonication (e.g. an ultrasound cleaning bath) has been applied to both inter-[7d] and intramolecular[26] PK processes. Other than notably decreasing the required reaction times, these methods have had, at best, little general effect on the product yields. In contrast, studies from our own laboratory have shown that the use of high intensity ultrasound, in combination with other reaction promoters, can appreciably improve cyclisation efficiencies.[27] As shown in Scheme 1.13, the use of a titanium ultrasonic horn (Vibracell VC 50) alone reduces the reaction time for the reaction of complex **1** and norbornene from 4 h to 10 min, along with an uplift in reaction yield. When this sonication method is combined with an amine *N*-oxide promoter (*vide infra*) the intermolecular cyclisation is further enhanced in terms of both reaction time and cyclopentenone yield.

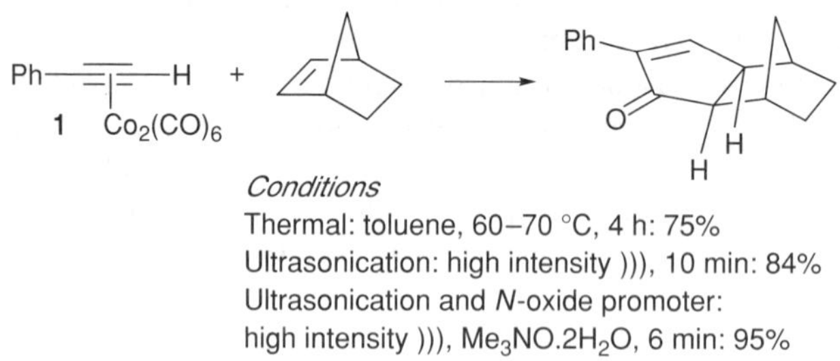

Scheme 1.13

1.4.3 Microwave Promotion

Over recent years and as an alternative to thermal heating, organic transformations have been subjected to irradiation within tuned microwave reactors, which has delivered generally elevated reaction efficiencies, particularly in relation to appreciably reduced reaction times.[28] In 2002, Evans reported the application of microwave (MW) techniques to a series of PK reactions, with the cyclopentenone products being delivered in good to excellent yields, as illustrated in Scheme 1.14.[29] In the same year, Groth *et al*. reported on a MW-promoted catalytic protocol, with cyclohexylamine as a reaction additive;[30] in general and despite the introduction of no additional CO gas being required, when compared to other methods this technique provided relatively moderate efficiencies with a series of normally reactive substrates.

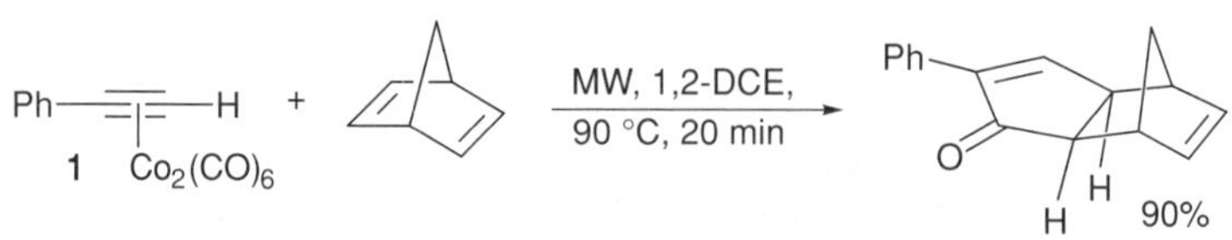

Scheme 1.14

1.4.4 Amine *N*-Oxide Additives

Perhaps the most significant of all advances in the promotion of PK cyclisations, and in turn the elevation of this method as a widely utilisable tool in organic synthesis, came with the application of amine *N*-oxide additives, as first described by Schreiber *et al.* in 1990.[31] It had been well-known that amine *N*-oxides could be employed to remove carbon monoxide ligands from transition metal complexes by oxidation to carbon dioxide.[32] By extension and when applied within the PK manifold, this would render the first step of this process irreversible *and* potentially allow the use of significantly more mild reaction conditions. As shown in Scheme 1.15, the use of *N*-methylmorpholine *N*-oxide (NMO) as the reaction promoter allowed cyclisations to be performed under the remarkably mild conditions of stirring at room temperature;[31] by way of further illustration of the effectiveness of this *N*-oxide method, the outcome of the annulation with the same substrate using the previously optimal dry state absorption method is provided for comparison.[22] Following this publication by Schreiber, Jeong *et al.* divulged similarly mild and effective techniques with the equivalent trimethylamine *N*-oxide (TMANO).[33]

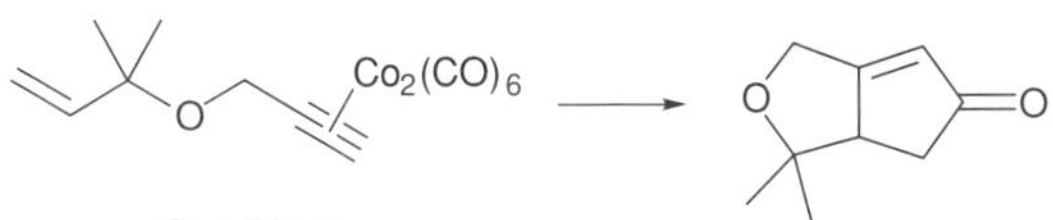

Conditions

Dry state adsorption: SiO_2, O_2, 55 °C, 1.5 h: 59%[22]

N-oxide promotion: NMO, CH_2Cl_2, r.t., ~12 h: 92%[31]

Scheme 1.15

In relation to these amine *N*-oxide promoted methods, it is important to note that the PK reactions with such additives appear to be significantly more clean, with the final cyclopentenone products normally requiring a trivial filtration or short silica column purification. Having stated this, one issue that is often overlooked or dismissed within publications in this area relates to the hydration state of the *N*-oxide additive used. More specifically, we[6,34] and others[12a] have found that the monohydrate of NMO and the dihydrate of TMANO offer optimal yields in most instances. The equivalent anhydrous *N*-oxides tend to lead to varying degrees of alkyne decomplexation or decomposition, unless there is a requirement for such, more active, *N*-oxides, for example, with less reactive alkyne complexes.[35]

Additional techniques that have emerged around this area include the formulation and use of recyclable polymer-supported amine *N*-oxides,[36] which provide the added advantage of sequestering cobalt by-products to deliver an even cleaner reaction process, and the use of molecular sieves in combination with *N*-oxide promoters.[37]

As stated above, the emergence of *N*-oxide promoted PK methods has significantly enhanced the preparative applicability of this annulation process. This point is beautifully illustrated within Schreiber's total synthesis of the diterpene natural product, (+)-epoxydictymene.[26,31,38] In this particular example, the central PK cyclisation proceeds under extremely mild *N*-oxide conditions to deliver the desired cyclopentenone product with elevated levels of selectivity over that obtained under the more standard thermal protocol (Scheme 1.16).

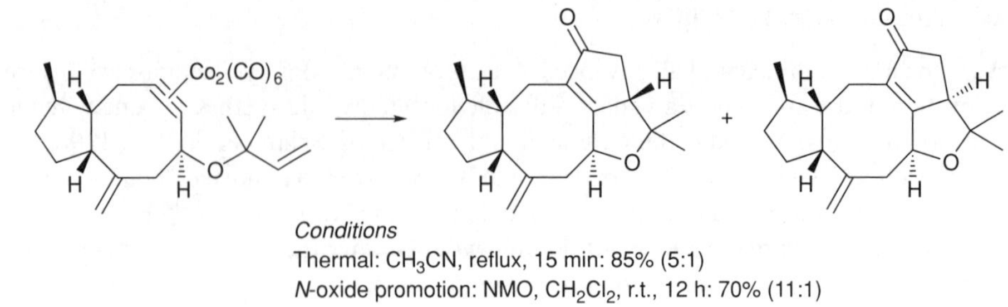

Scheme 1.16

As discussed previously, conditions for use of gaseous olefins in the PK annulation require more forcing conditions and generally lead to only low to moderate yields of cyclopentenone products.[7] In relation to this, work in our own laboratory has formulated optimised conditions with *N*-oxide promotion for the use of ethylene as the alkene reaction partner under appreciably more moderate temperatures and pressures.[34a–c] Indeed, whilst mild autoclave pressure (25–30 atm) and temperatures (40 °C) provided optimum yields in most instances, it was also found that atmospheric pressures of ethylene at room temperature also delivered the targeted cyclopentenones. It is also worth noting that both techniques result in considerably cleaner product reaction mixtures than those obtained under the more traditional thermal protocols. As shown in Scheme 1.17, these developments allowed us to efficiently complete a total synthesis of the sesquiterpene, (+)-taylorione **13**;[34a,c] the requisite chiral alkyne complex **14** was prepared in seven steps from the inexpensive and readily available starting material, (+)-2-carene **15**, and, following successful application of the *N*-oxide promoted PK conditions with ethylene, (+)-**13** was obtained in short order and with high optical purity.

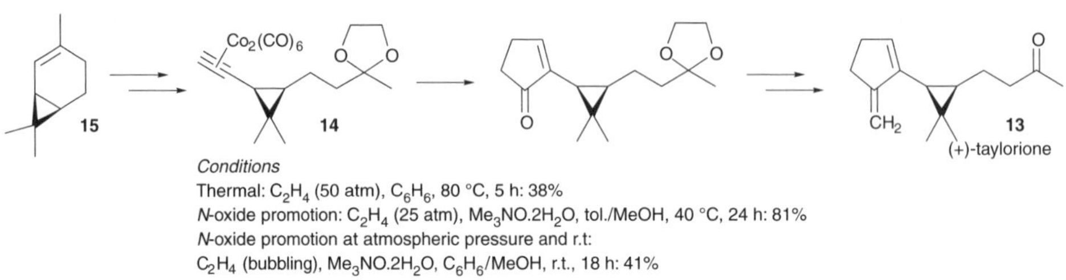

Scheme 1.17

In continuing efforts to further develop the PK reaction towards delivering cyclopentenones from ethylene incorporation, when vinyl esters, such as vinyl benzoate **16**, were employed with *N*-oxide promotion, the products which resulted were those where the expected ester unit had been cleaved, presumably via an *in situ* reduction process with low valent cobalt.[34d–f] This outcome clearly establishes an extremely mild and utilisable protocol for employment of the vinyl ester as an ethylene equivalent, without the requirement for handling of gaseous reagents. Moreover and as illustrated in Scheme 1.18, the new vinyl benzoate technique is at least comparable with the optimum ethylene methods (*vide supra*) and, in most instances, significant enhancement in the yield of cyclopentenone can be realised. It should also be noted that similar reductive outcomes have been observed in PK reactions with vinyl halide substrates.[9,39]

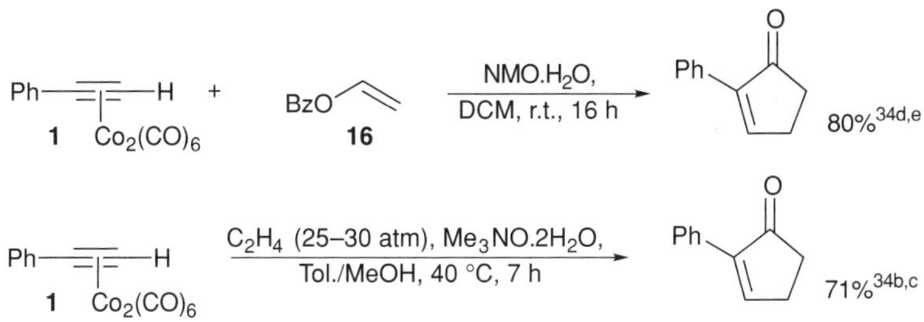

Scheme 1.18

1.4.5 Sulfide Promoters

In 1997, Sugihara and Yamaguchi introduced the concept of employing Lewis basic additives to promote the PK reaction, and demonstrated this with the use of, in particular, cyclohexylamine to enhance a series of both intra- and intermolecular examples,[40] albeit under considerably greater forcing temperatures than those used with amine *N*-oxides. Despite the apparent success of the amine additive method, less reactive alkenes did not perform well under the developed conditions and cleavage of carbon-heteroatom bonds was also observed. Accordingly, further advances by the same researchers using sulfide-based promoters have become significantly more widely employed by the preparative community.[41] An example using the recommended *n*-butyl methyl sulfide promoter compared with the cyclohexylamine additive in reaction with the allyl propargyl amine is shown in Scheme 1.19.

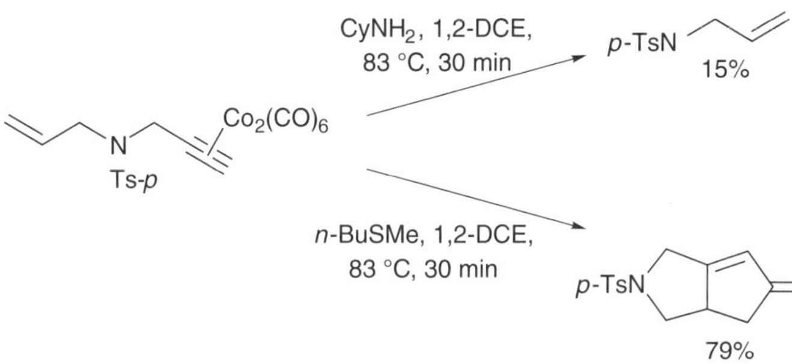

Scheme 1.19

Despite the more elevated temperatures that are required, this sulfide-based method appears to be complementary with the amine *N*-oxide techniques. Having stated this, some appreciable practical drawbacks are encountered when employing this low molecular weight sulfide, which possesses a rather unpleasant smell and is lachrymatory in nature. Accordingly and in seeking to establish a more user-friendly protocol, work from our own laboratory has shown that the inexpensive and stench-free alternative, *n*-dodecyl methyl sulfide (DodSMe) is a highly effective promoter of both intra- and intermolecular PK annulations (Scheme 1.20).[42] Additionally, in every case investigated, the reaction efficiencies and yields obtained when utilising DodSMe were at least comparable with those from the use of the less practically convenient sulfide, *n*-BuSMe.

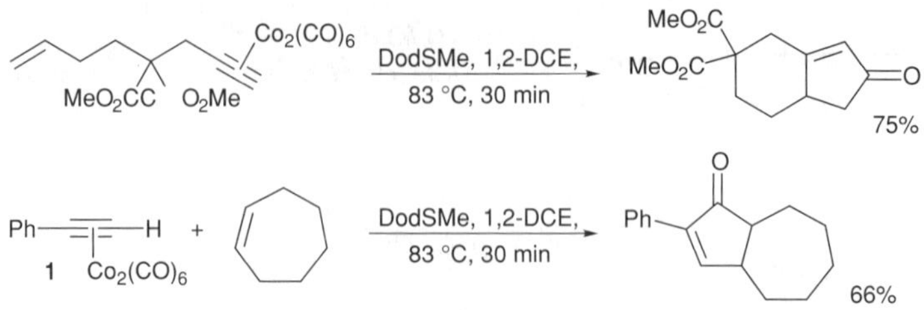

Scheme 1.20

In a similar fashion to that described for amine *N*-oxide promoters (*vide supra*), work from our own laboratory has shown how the sulfur additive can be attached to a solid support. These polymer-bound and odourless materials display PK promoting abilities in line with the solution-phase additives, with the further advantage that they can be reused extremely effectively over a number of cycles.[43] Further practical benefit arises from the reaction work-up now merely involving filtration and minimal cyclopentenone purification. Finally in this section, and by way of illustrating how these sulfide-promoted methods can provide further techniques for use within natural product synthesis (when compared to other available protocols), Scheme 1.21 illustrates the key PK cyclisations used within our

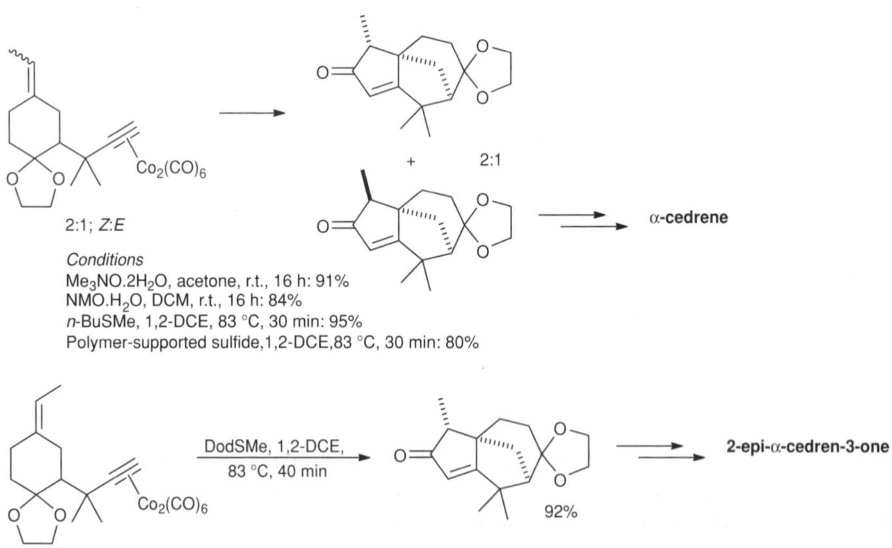

Scheme 1.21

laboratory in routes towards the tricyclic sesquiterpenes, α-cedrene and 2-epi-α-cedren-3-one.[6,34g,42]

In relation to how these Lewis base additives operate to enhance the efficiency of PK reactions, it had been assumed that such species accelerate the decarbonylative-Lewis base ligand exchange process and, in turn, promote the loss of a further CO ligand to provide a vacant coordination site for olefin complexation. Despite these assumptions, a more recent theoretical study by Milet and Gimbert has provided quite a different scenario, and leading to the proposal that the Lewis base promoters achieve their PK annulation enhancing effects by making the olefin insertion step of the reaction pathway an irreversible process.[44]

1.5 Catalytic Pauson-Khand Protocols

Moving the cobalt-based PK reaction into the catalytic domain would allow this already widely utilised transformation to assume magnified importance not only within industrial settings but within the synthetic community as a whole. In this regard, attempts were made to employ sub-stoichiometric quantities of octacarbonyldicobalt in some of the earliest studies on the PK annulation process,[2a,21] with efficient cyclisations only being achieved at relatively high catalyst loadings and when the most reactive strained alkenes were used. In 1990 an appreciably more productive catalytic system was divulged by Rautenstrauch and co-workers.[45] Despite the employment of only 0.22 mol% catalyst leading to a high turnover number of 220 for $Co_2(CO)_8$, the reaction conditions for this PK annulation between 1-heptyne and ethylene required partial pressures at room temperature of 100 bar CO and 40 bar C_2H_4, and heating to 150 °C (to deliver a total pressure of 310–360 bar) over 16 h reaction time, leading to a 49% yield of the desired cyclopentenone product. From this stage and over recent years a number of significantly more effective Co-catalysed processes have been formulated by a number of research groups.[5j] Since the protocols using cobalt-based catalysts will be discussed in more detail elsewhere (Chapter 3), only a selection of methods will be discussed here to provided a flavour of the advances made to date.

In 1994 Jeong and co-workers divulged the first practically efficient catalytic PK system.[46] As illustrated in Scheme 1.22, with the key introduction of a phosphite ligand, low levels of $Co_2(CO)_8$ could be employed combined with relatively modest CO pressures. Despite the obvious successes of this approach, this method appeared to be limited to intermolecular examples, coupled with a requirement for 120 °C reaction temperatures. Following on from this, Krafft has also formulated a series of equally effective catalytic $Co_2(CO)_8$-based protocols (at lower temperatures and CO pressures) with cyclohexylamine as the reaction additive.[47]

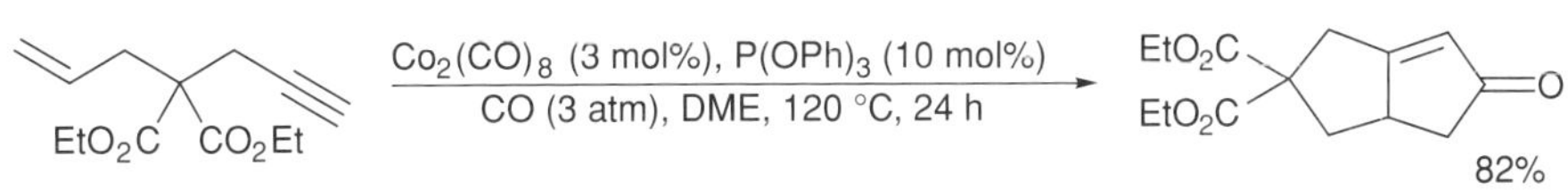

Scheme 1.22

In a further example with $Co_2(CO)_8$ as catalyst, Hashimoto and Saigo have shown that the introduction of tributylphosphine sulfide allows the use of only atmospheric pressure CO to deliver efficient cyclisations in a series of intramolecular PK reactions (Scheme 1.23).[48] Notably with this method, an intermolecular reaction was also reported, albeit with the strained and reactive olefin, norbornene.

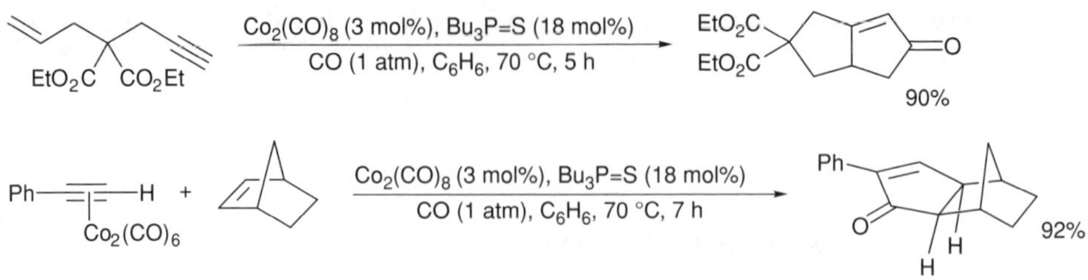

Scheme 1.23

Heterogenous cobalt-based catalysis has also been applied, in particular with the use of Co loaded onto a range of supports, or by the employment of recyclable colloidal Co nanoparticles, as divulged in an extensive series of reports by Chung, Hyeon, and co-workers.[49] As illustrated in Scheme 1.24, cobalt on charcoal can be employed effectively in intramolecular PK annulations, albeit at relatively elevated temperature and CO pressure. Interestingly, the Co on C catalyst can be recycled extensively without loss in efficiency and with <0.1 ppm of Co being bled from the support. This specific technique has also been shown to be successful in intermolecular examples, again, with the reactive alkene, norbornadiene.

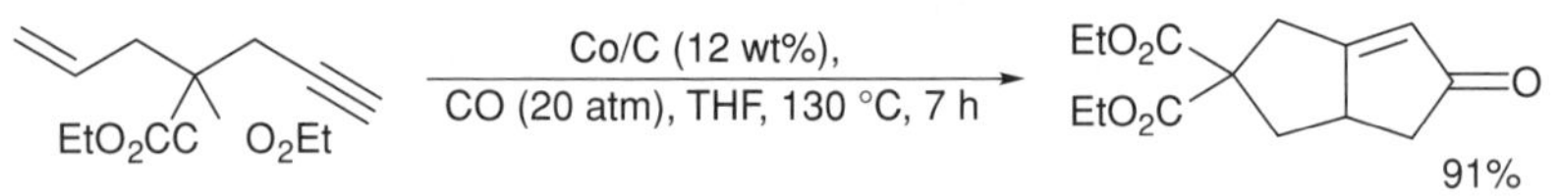

Scheme 1.24

From this brief summary of cobalt-based catalytic methods, it is clear that systems have yet to be developed that would allow less reactive alkenes to be applied in an intermolecular sense and, more generally, the employment of milder reaction conditions to be adopted. Accordingly, catalytic PK protocols continue to be a focus of the preparative chemistry community. In relation to this, a series of additional and complementary catalytic techniques using other metal species are adding extensively to the growing repertoire of PK methods, and these will be detailed in Chapter 7.

1.6 Concluding Remarks

From relatively modest beginnings at the University of Strathclyde, due to the outstanding contributions of an extensive spectrum of international researchers, the Pauson-Khand reaction has grown to become a synthetic method of immense value to the global chemistry community. As discussed above, a series of methodological advances now allow this annulation process to be used with preparative confidence, to deliver generally very good levels of efficiency and selectivity in both an intra- and intermolecular sense. Accordingly and as illustrated with some examples above, the PK technique is finding continual use in targeted natural product synthesis (see Chapter 8). Furthermore and as will be detailed in Chapters 4, 5, and 6, an array of enantioselective methods have also now been developed.

This monograph will describe the development and application of the Pauson-Khand reaction and related processes over the 40-or-so years since the initial discovery of the original cobalt-mediated annulations. The development of this method forms just one, albeit significant, part of Peter Pauson's seminal contributions to organometallic methods and organic synthesis. As for Dr Khand, one feels that he would be immensely pleased to observe where his serendipitous discovery had been elevated to, as well as the impact that this cyclisation process has had on preparative organic chemistry.

Acknowledgements

I am honoured and indebted to have had Peter Pauson as a mentor and friend for the past 28 years and I offer my sincerest gratitude for his unstinting guidance and support. I also thank my own students and postdoctoral associates in the area of Pauson-Khand chemistry for their dedication and inspiration: Steve Christie, Udo Lange, Gavin Forsyth, Tony Murray, Craig Johnstone, Hans Donkervoort, George Kirk, Alison Gordon, Alastair Hay, Jamie Scott, Gair Ford, John Caldwell, Sarah Robertson, Dave Lindsay, Elaine Rankin, Dave Carbery, Mark McLaughlin, Angus Morrison, Iain Cameron, Paul Ratcliffe, Elaine Campbell, James Crawford, Emma Carswell, Colin Pearson, Claire Russell, Graeme Thurston, Jack Brown, Tomasz Janecki, Stephanie Irvine, Allan McPherson, Jacek Kędzia, Laura Paterson (née McEvoy), Alison Cochrane, Malcolm Gordon, and Calum Forsyth.

References

[1] I. U. Khand, G. R. Knox, P. L. Pauson, and W. E. Watts, *J. Chem. Soc., Chem. Commun.*, 36 (1971).

[2] (a) I. U. Khand, G. R. Knox, P. L. Pauson, W. E. Watts, and M. I. Foreman, *J. Chem. Soc., Perkin Trans. 1*, 977–981 (1973); (b) I. U. Khand, G. R. Knox, P. L. Pauson, and W. E. Watts, *J. Chem. Soc., Perkin Trans. 1*, 975–977 (1973).

[3] P. L. Pauson and I. U. Khand, *Ann. N. Y. Acad. Sci.*, **295**, 2–14 (1977).

[4] Web of Science^{SM} conducted on 26 September, 2011.

[5] (a) P. L. Pauson, *Tetrahedron*, **41**, 5855–5860 (1985); (b) P. L. Pauson, In *Organometallics in Organic Synthesis. Aspects of a Modern Interdisciplinary Field*, A. de Meijere and H. tom Dieck (Eds), Springer, Berlin, 1988, pp 233–246; (c) N. E. Schore, *Org. React.*, **40**, 1–90 (1991); (d) N. E. Schore, In *Comprehensive Organic Synthesis*, B. M. Trost and I. Fleming (Eds), Pergamon, Oxford, 1991, vol. 5, pp 1037–1064; (e) O. Geis and H.-G. Schmalz, *Angew. Chem. Int. Ed.*, **37**, 911–914 (1998); (f) Y. K. Chung, *Coord. Chem. Rev.*, **188**, 297–341 (1999); (g) K. M. Brummond and J. L. Kent, *Tetrahedron*, **56**, 3263–3283 (2000); (h) J. Fletcher and S. D. R.

Christie, *J. Chem. Soc., Perkin Trans. 1*, 1–13 (2001); (i) M. R. Rivero, J. Adrio, and J. C. Carretero, *Eur. J. Org. Chem.*, 2881–2889 (2002); (j) S. E. Gibson (née Thomas) and A. Stevenazzi, *Angew. Chem. Int. Ed.*, **42** 1800–1810 (2003); (k) J. Blanco-Urgoiti, L. Añorbe, L. Pérez-Serrano, G. Dominguez, and J. Pérez-Castells, *Chem. Soc. Rev.*, **33**, 32–42 (2004); (l) I. Nakamura and Y. Yamamoto, *Chem. Rev.*, **104**, 2127–2198 (2004); (m) S. E. Gibson and N. Mainolfi, *Angew. Chem. Int. Ed.*, **44**, 3022–3037 (2005); (n) S. Laschat, A. Becheanu, T. Bell, and A. Baro, *Synlett*, 2547–2570 (2005); (o) T. Shibata, *Adv. Synth. Catal.*, **348**, 2328–2336 (2006); (p) I. Omae, *App. Organomet. Chem.*, **21**, 318–344 (2007); (q) H.-W. Lee and F.-Y. Kwong, *Eur. J. Org. Chem.* 7[illegible] 811 (2010).

[6] J. J. Crawford, W. J. Kerr, M. McLaughlin, A. J. Morrison, P. L. Pauson, and G. J. Thurston, *Tetrahedron*, **62**, 11360–11370 (2006).

[7] (a) R. F. Newton, P. L. Pauson, and R. G. Taylor, *J. Chem. Res. (M)*, 3501–3522 (1980); *J. Chem. Res. (S)*, 277 (1980); (b) H. J. Jaffer and P. L. Pauson, *J. Chem. Res. (M)*, 2201–2218 (1983); *J. Chem. Res. (S)*, 244 (1983); (c) L. Daalman, R. F. Newton, P. L. Pauson, and A. Wadsworth, *J. Chem. Res. (M)*, 3150–3164 (1984); *J. Chem. Res. (S)*, 346–347 (1984); (d) D. C. Billington, I. M. Helps, P. L. Pauson, W. Thomson, and D. Willison, *J. Organomet. Chem.*, **354**, 233–242 (1988).

[8] T. J. M. de Bruin, A. Milet, A. E. Greene, and Y. Gimbert, *J. Org. Chem.*, **69**, 1075–1080 (2004).

[9] I. U. Khand and P. L. Pauson, *J. Chem. Res. (M)*, 168–187 (1977); *J. Chem. Res. (S)*, 9 (1977).

[10] (a) P. A. Wender, N. M. Deschamps, and T. J. Williams, *Angew. Chem. Int. Ed.*, **43**, 3076–3079 (2004); (b) P. A. Wender, N. M. Deschamps, and G. G. Gamber, *Angew. Chem. Int. Ed.*, **42**, 1853–1857 (2003); (c) M. P. Croatt and P. A. Wender, *Eur. J. Org. Chem.*, 19–32 (2010); see also (d) P. A. Wender, M. P. Croatt, and N. M. Deschamps, *J. Am. Chem. Soc.*, **126**, 5948–5949 (2004); (e) P. A. Wender, M. P. Croatt, and N. M. Deschamps, *Angew. Chem. Int. Ed.*, **45**, 2459–2462 (2006).

[11] D. C. Billington, P. Bladon, I. M. Helps, P. L. Pauson, W. Thomson, and D. Willison, *J. Chem. Res. (M)*, 2601–2622 (1988); *J. Chem. Res. (S)*, 326–327 (1988).

[12] (a) M. E. Krafft, R. H. Romero, and I. L. Scott, *J. Org. Chem.*, **57**, 5277–5278 (1992); see also (b) M. E. Krafft, R. H. Romero, and I. L. Scott, *Synlett*, 577–578 (1995); (c) T. R. Hoye and J. A. Suriano, *J. Org. Chem.*, **58**, 1659–1660 (1993).

[13] F. Robert, A. Milet, Y. Gimbert, D. Konya, and A. E. Greene, *J. Am. Chem. Soc.*, **123**, 5396–5400 (2001).

[14] M. E. Krafft, *Tetrahedron Lett.*, **29**, 999–1002 (1988).

[15] (a) M. E. Krafft, *J. Am. Chem. Soc.*, **110**, 968–970 (1988); (b) M. E. Krafft, C. A. Juliano, I. L. Scott, C. Wright, and M. E. McEachin, *J. Am. Chem. Soc.*, **113**, 1693–1703 (1991); (c) M. E. Krafft, I. L. Scott, R. H. Romero, S. Feibelmann, and C. E. Van Pelt, *J. Am. Chem. Soc.*, **115**, 7199–7207 (1993).

[16] J. A. Brown, T. Janecki, and W. J. Kerr, *Synlett*, 2023–2026 (2005).

[17] J. L. Kędzia, W. J. Kerr, and A. R. McPherson, *Synlett*, 649–653 (2010).

[18] N. E. Schore and M. C. Croudace, *J. Org. Chem.*, **46**, 5436–5438 (1981).

[19] (a) D. C. Billington and D. Willison, *Tetrahedron Lett.*, **25**, 4041–4044 (1984); (b) S. W. Brown and P. L. Pauson, *J. Chem. Soc., Perkin Trans. 1*, 1205–1209 (1990).

[20] (a) C. Exon and P. Magnus, *J. Am. Chem. Soc.*, **105**, 2477–2478 (1983); (b) P. Magnus and L. M. Principe, *Tetrahedron Lett.*, **26**, 4851–4854 (1985); (c) P. Magnus, C. Exon, and P. Albaugh-Robertson, *Tetrahedron*, **41**, 5861–5869 (1985).

[21] P. Magnus, L. M. Principe, and M. J. Slater, *J. Org. Chem.*, **52**, 1483–1486 (1987).

[22] (a) V. A. Smit, A. S. Gybin, S. O. Simonyan, A. S. Shashkov, V. A. Tarasov, and I. I. Ibragimov, *Izv. Akad. Nauk SSSR, Ser. Khim.*, 2650–2651 (1985); *Bull. Acad. Sci. USSR, Div. Chem. Sci.*, **34**, 2455–2456 (1985); (b) S. O. Simonian, W. A. Smit, A. S. Gybin, A. S. Shashkov, G. S. Mikaelian, V. A. Tarasov, I. I. Ibragimov, R. Caple, and D. E. Froen, *Tetrahedron Lett.*, **27**, 1245–1248 (1986); (c) W. A. Smit, A. S. Gybin, A. S. Shashkov, Y. T. Strychkov, L. G. Kyz'mina, G. S. Mikaelian, R. Caple, and E. D. Swanson, *Tetrahedron Lett.*, **27**, 1241–1244 (1986); (d) W. A. Smit, S. O. Simonyan, V. A. Tarasov, G. S. Mikaelian, A. S. Gybin, I. I. Ibragimov, R. Caple, D. E. Froen, and A. Kreager, *Synthesis*, 472–476 (1989); (e) V. A. Smit, S. O. Simonyan, A. S. Shashkov, S. S. Mamyan, V. A. Tarasov, and I. I. Ibragimov, *Izv. Akad. Nauk SSSR, Ser. Khim.*, 234–235 (1987); *Bull. Acad. Sci. USSR, Div. Chem. Sci.*, **36**, 213–214 (1987); (f) V. A. Smit, S. O. Simonyan, V. A. Tarasov, A. S. Gybin, G. S. Mikaélyan, A. S. Shashkov, S. S. Mamyan, I. I. Ibragimov, and R. Kéipl, *Izv. Akad. Nauk SSSR, Ser. Khim.*, 2802–2812 (1988); *Bull. Acad. Sci. USSR, Div. Chem. Sci.*, **37**, 2526–2535 (1988).

[23] (a) V. A. Smit, A. S. Gybin, S. O. Veretenov, and A. S. Shashkov, *Izv. Akad. Nauk SSSR, Ser. Khim.*, 232–233 (1987); *Bull. Acad. Sci. USSR, Div. Chem. Sci.*, **36**, 211–212 (1987); (b) V. A. Smit, S. O. Simonyan, V. A. Tarasov, A. S. Shashkov, S. S. Mamyan, A. S. Gybin, and I. I. Ibragimov, *Izv. Akad. Nauk SSSR, Ser. Khim.*, 2796–2802 (1988); *Bull. Acad. Sci. USSR, Div. Chem. Sci.*, **37**, 2521–2526 (1988); (c) V. A. Smit, S. M. Bukhanyuk, A. S. Shashkov, Yu. T. Struchkov, and A. I. Yanovskii, *Izv. Akad. Nauk SSSR, Ser. Khim.*, 2876 (1988); *Bull. Acad. Sci. USSR, Div. Chem. Sci.*, **37**, 2597–2598 (1988).

[24] (a) V. A. Smit, S. L. Kireev, and O. M. Nefedov, *Izv. Akad. Nauk SSSR, Ser. Khim.*, 2637–2638 (1987); *Bull. Acad. Sci. USSR, Div. Chem. Sci.*, **36**, 2452 (1987); (b) V. A. Smit, V. A. Tarasov, E. D. Daeva, and I. I. Ibragimov, *Izv. Akad. Nauk SSSR, Ser. Khim.*, 2870–2871 (1987); *Bull. Acad. Sci. USSR, Div. Chem. Sci.*, **36**, 2669–2670 (1987); (c) W. A. Smit, S. L. Kireev, O. M. Nefedov, and V. A. Tarasov, *Tetrahedron Lett.*, **30**, 4021–4024 (1989).

[25] (a) K. S. Suslick, *Adv. Organomet. Chem.*, **25**, 73–119 (1986); (b) K. S. Suslick, J. W. Goodale, P. F. Schubert, and H. H. Wang, *J. Am. Chem. Soc.*, **105**, 5781–5785 (1983); (c) S. V. Ley and C. M. R. Low, *Ultrasound in Synthesis*, Springer, Berlin, 1989.

[26] T. F. Jamison, S. Shambayati, W. E. Crowe, and S. L. Schreiber, *J. Am. Chem. Soc.*, **119**, 4353–4363 (1997).

[27] J. G. Ford, W. J. Kerr, G. G. Kirk, D. M. Lindsay, and D. Middlemiss, *Synlett*, 1415–1418 (2000).

[28] C. O. Kappe, *Angew. Chem. Int. Ed.*, **43**, 625–6284 (2004).

[29] M. Iqbal, N. Vyse, J. Dauvergne, and P. Evans, *Tetrahedron Lett.*, **43**, 7859–7862 (2002).

[30] S. Fischer, U. Groth, M. Jung, and A. Schneider, *Synlett*, 2023–2026 (2002).

[31] S. Shambayati, W. E. Crowe, and S. L. Schreiber, *Tetrahedron Lett.*, **31**, 5289–5292 (1990).

[32] (a) H. Alper and J. T. Edward, *Can. J. Chem.*, **48**, 1543–1549 (1970); (b) Y. Shvo and E. Hazum, *J. Chem. Soc., Chem. Commun.*, 336–337 (1974); (c) J.-K. Shen, Y.-C. Gao, Q.-Z. Shi, and F. Basolo, *Organometallics*, **8**, 2144–2147 (1989).

[33] N. Jeong, Y. K. Chung, B. Y. Lee, S. H. Lee, and S.-E. Yoo, *Synlett*, 204–206 (1991).

[34] (a) C. Johnstone, W. J. Kerr, and U. Lange, *J. Chem. Soc., Chem. Commun.*, 457–458 (1995); (b) A. R. Gordon, C. Johnstone, and W. J. Kerr, *Synlett*, 1083–1084 (1995); (c) J. G. Donkervoort, A. R. Gordon. C. Johnstone, W. J. Kerr, and U. Lange, *Tetrahedron*, **52**, 7391–7420 (1996); (d) W. J. Kerr, M. McLaughlin, P. L. Pauson, and S. M. Robertson, *Chem. Commun.*, 2171–2172 (1999); (e) W. J. Kerr, M. McLaughlin, P. L. Pauson, and S. M. Robertson, *J. Organomet. Chem.*, **630**, 104–117 (2001); (f) W. J. Kerr, M. McLaughlin, and P. L. Pauson, *J. Organomet. Chem.*, **630**, 118–124 (2001); (g) W. J. Kerr, M. McLaughlin, A. J. Morrison, and P. L. Pauson, *Org. Lett.*, **3**, 2945–2948 (2001).

[35] A. M. Hay, W. J. Kerr, G. G. Kirk, and D. Middlemiss, *Organometallics*, **14**, 4986–4988 (1995).

[36] (a) W. J. Kerr, D. M. Lindsay, and S. P. Watson, *Chem. Commun.*, 2551–2552 (1999); (b) D. S. Brown, E. Campbell, W. J. Kerr, D. M. Lindsay, A. J. Morrison, K. G. Pike, and S. P. Watson, *Synlett*, 1573–1576 (2000).

[37] (a) L. Pérez–Serrano, L. Casarrubios, G. Domínguez, and J. Pérez-Castells, *Org. Lett.*, **1**, 1187-1188 (1999); (b) L. Pérez-Serrano, J. Blanco-Urgoiti, L. Casarrubios, G. Domínguez, and J. Pérez-Castells, *J. Org. Chem.*, **65**, 3513–3519 (2000).

[38] T. Jamison, S. Shambayati, W. E. Crowe, and S. L. Schreiber, *J. Am. Chem. Soc.*, **116**, 5505–5506 (1994).

[39] (a) O. Arjona, A. G. Csákÿ, R. Medel, and J. Plumet, *Tetrahedron Lett.*, **42**, 3805–3807 (2001); (b) O. Arjona, A. G. Csákÿ, M. C. Murcia, and J. Plumet, *J. Org. Chem.*, **64**, 7338–7341 (1999).

[40] (a) T. Sugihara, M. Yamada, H. Ban, M. Yamaguchi, and C. Kaneko, *Angew. Chem. Int. Ed.*, **36**, 2801–2804 (1997); see also (b) T. Sugihara, M. Yamaguchi, and M. Nishizawa, *Chem. Eur. J.*, **7**, 1589–1595 (2001).

[41] T. Sugihara, M. Yamada, H. Ban, M. Yamaguchi, and M. Nishizawa, *Synlett*, 771–773 (1999).

[42] J. A. Brown, S. Irvine, W. J. Kerr, and C. M. Pearson, *Org. Biomol. Chem.*, **3**, 2396–2398 (2005).

[43] W. J. Kerr, D. M. Lindsay, M. McLaughlin, and P. L. Pauson, *Chem. Commun.*, 1467–1468 (2000).

[44] C. Perez del Valle, A. Milet, Y. Gimbert, and A. E. Greene, *Angew. Chem. Int. Ed.*, **44**, 5717–5719 (2005).

[45] V. Rautenstrauch, P. Mégard, J. Conesa, and W. Küster, *Angew. Chem. Int. Ed.*, **29**, 1413–1416 (1990).

[46] N. Jeong, S. H. Hwang, Y. Lee, and Y. K. Chung, *J. Am. Chem. Soc.*, **116**, 3159–3160 (1994).

[47] (a) M. E. Krafft, L. V. R. Bonaga, and C. Hirosawa, *Tetrahedron Lett.*, **40**, 9171–9175 (1999); (b) M. E. Krafft, C. Hirosawa, and L. V. R. Bonaga, *Tetrahedron Lett.*, **40**, 9177–9181 (1999); (c) M. E. Krafft and L. V. R. Boñaga, *Synlett*, 959–962 (2000); for

a detailed overview, see (d) M. E. Krafft, L. V. R. Boñaga, and C. Hirosawa, *J. Org. Chem.*, **66**, 3004–3020 (2001).

[48] M. Hayashi, Y. Hashimoto, Y. Yamamoto, J. Usuki, and K. Saigo, *Angew. Chem. Int. Ed.*, **39**, 631–633 (2000).

[49] (a) S. U. Son, S. I. Lee, and Y. K. Chung, *Angew. Chem. Int. Ed.*, **39**, 4158–4160 (2000); (b) S.-W. Kim, S. U. Son, S. I. Lee, T. Hyeon, and Y. K. Chung, *J. Am. Chem. Soc.*, **122**, 1550–1551 (2000); (c) K. H. Park and Y. K. Chung, *Synlett*, 545–559 (2005).

2

The Mechanism of the Pauson-Khand Reaction: Hypothesis, Experimental Facts, and Theoretical Investigations

Xacobe C. Cambeiro and Miquel A. Pericàs

2.1 Introduction

The prototypal Pauson-Khand (PK) synthesis of cyclopentenones, first introduced in 1973,[1] combines in a single step an alkyne, an alkene, and carbon monoxide (coming from octacarbonyl dicobalt) in a regioselective and stereoselective manner (Scheme 2.1). For this

The Pauson-Khand Reaction: Scope, Variations and Applications, First Edition. Edited by Ramon Rios Torres.

reason, it has evolved over the years into a powerful and well-established procedure for the assembly of five-membered rings. In a parallel manner, PK-type reactions involving transition metals other than cobalt have also importantly developed.[2]

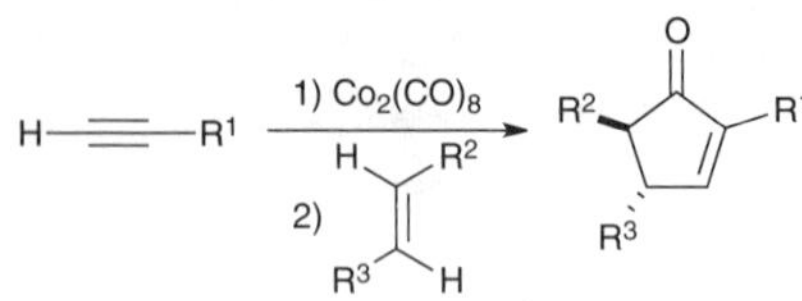

Scheme 2.1 *A prototypic Pauson-Khand reaction.*

From the perspective of reaction mechanisms, the Pauson-Khand reaction (PKR) is a complex process (three fragments are combined and three covalent bonds are built in a single step), taking place essentially as a black box. Thus, in stoichiometric reactions the starting materials are normally the dicobalt hexacarbonyl complex of the reacting alkyne and the counterpart olefin, and cyclopentenone formation takes place without detection of any intermediate reaction.

A mechanistic sequence for the reaction was proposed by Magnus in 1985.[3] It was based on general knowledge of organometallic chemistry, with the single purpose of facilitating the rationalization of the observed diastereoselectivity of a particular set of examples of the reaction (Scheme 2.2). This proposal, however, had the virtue of accommodating the known experimental facts of the reaction and has gained universal acceptation as the *Magnus mechanism* of the PKR.

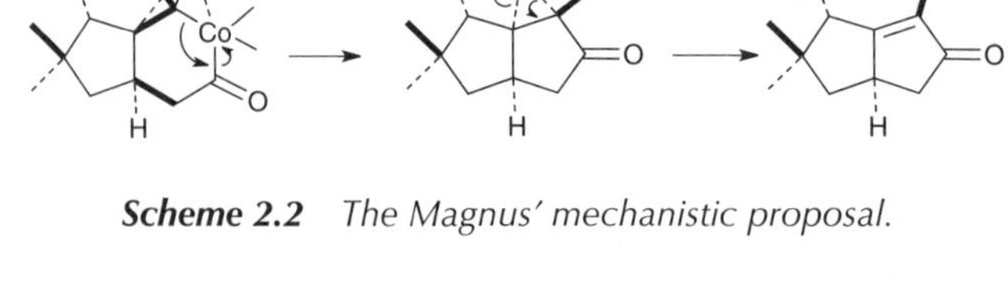

Scheme 2.2 *The Magnus' mechanistic proposal.*

In fact, considerations based on the Magnus mechanism have led to the rationalization of regiochemical and stereochemical outputs of the PKR, and theoretical calculations (see below) have shown the energetic viability of the proposed pathway. A more general formulation of this mechanistic proposal for the PKR can be found in Scheme 2.3.

Thus, four main events are involved in the pathway connecting the starting alkyne dicobalt hexacarbonyl complex with the product cyclopentenone: i) Ligand substitution, where the reacting alkene becomes associated with the initial complex, ii) cobaltacycle formation, of outmost importance since the regiochemistry and the stereochemistry of the

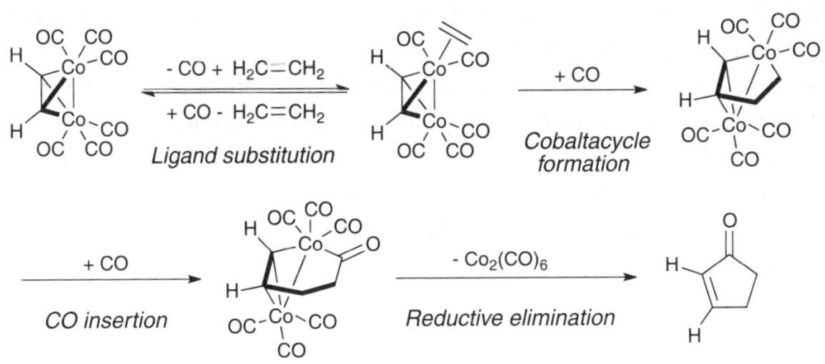

Scheme 2.3 General formulation of the Magnus mechanism of PKR.

final adduct become determined at this point, iii) CO insertion, and iv) reductive elimination and decomplexation of a weakly bonded cyclopentenone-$Co_2(CO)_6$ complex.

We will discuss in the next section the available knowledge of these steps in stoichiometrical PKR.

2.2 Stoichiometric Pauson-Khand Reaction

2.2.1 The Ligand Substitution Steps

It can be reasonably assumed that the ligand substitution process in stoichiometric PKR takes place through a dissociative mechanism whose first step leads to a coordinatively unsaturated dicobalt pentacarbonyl species (Scheme 2.4). Thus, operation under a CO overpressure leads to a lowering in reaction rate (through prevention of CO dissociation), while solvents with Lewis base character[4] (or Lewis base additives)[5] accelerate the reaction likewise through stabilization of the coordinatively unsaturated intermediate and subsequent labilization of CO ligands in the Lewis base-substituted complex.[6]

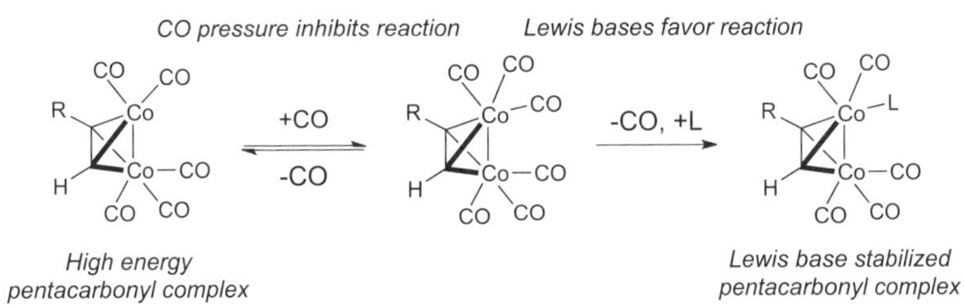

Scheme 2.4 Mechanisms for CO dissociation.

An important aspect of the CO dissociation from the initial hexacarbonyl complex refers to the enthalpy characteristics of this step. For PKR performed in the absence of nucleophilic additives, under purely thermal conditions, this is with high probability the rate-determining step of the reaction. In the search for milder reaction conditions and increased selectivity control, efforts have been devoted to developing chemical methodologies that could render this initial dissociative step irreversible.

The most successful approach in this direction has been based on the use of amine oxides as chemical promoters of CO dissociation. Following early work on oxygen atom transfer to metal carbonyls,[7] Schreiber[8] and Jeong[9] independently reported very significant accelerations of both intramolecular and intermolecular PKR through the use of amine oxides as *N*-methylmorpholine oxide (NMO) and trimethylamine *N*-oxide (TMAO) as promoters for the reaction (Scheme 2.5).

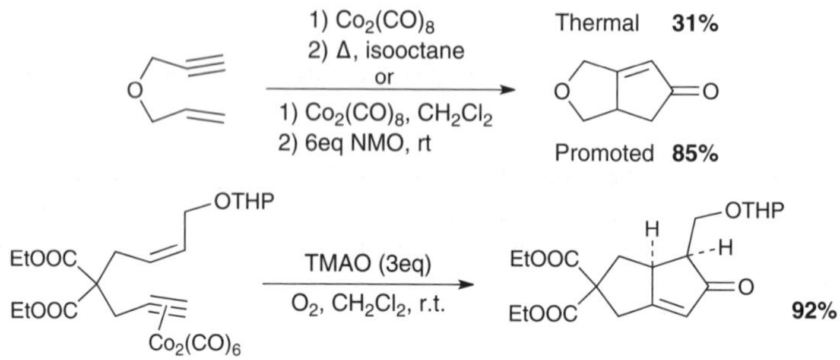

Scheme 2.5 *Use of* N*-oxides as promoters for the PKR.*

In these cases, the mechanism of activation by the *N*-oxide would be that proposed by Alper for the *N*-oxide activation of metal carbonyls:[7b] a nucleophilic attack of the *N*-oxide on one of the carbonyl carbons to generate the goods, leaving group CO_2 and a coordinatively unsaturated metal carbonyl (Scheme 2.6). Probably depending on the nature of the reacting alkene, the *N*-oxide promoter and the employed reaction conditions, this coordinatively unsaturated species reacts preferentially with the excess *N*-oxide, the tertiary amine generated in its deoxygenation, or directly enters the reaction pathway through coordination of the alkene.

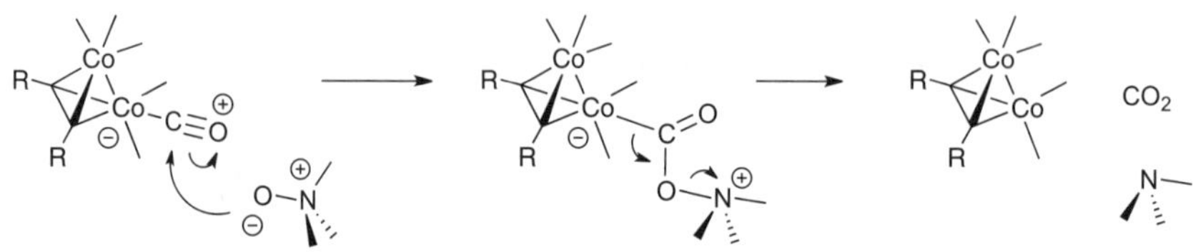

Scheme 2.6 *Mechanism for the activation of PK substrates by* N*-oxides (formal charges made explicit for clarity).*

From a practical perspective, the promotion of the PKR by *N*-oxides has become a well established and much used methodology. The possibility of selectively inducing decarbonylation at one of the enantiotopic cobalt atoms of a dicobalt hexacarbonyl process in view of controlling the stereochemical course of the reaction with the use of enantiopure brucine *N*-oxide (Figure 2.1) has been explored by Kerr and co-workers.[10]

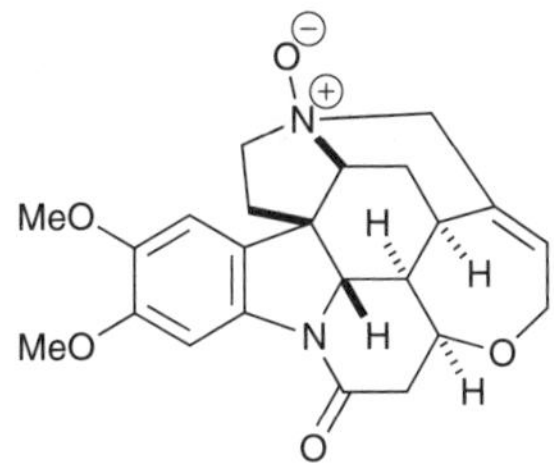

Figure 2.1 *Structure of brucine* N*-oxide (BNO).*

Other approaches to the promotion of the decarbonylation of dicobalt hexacarbonyl complexes have been explored. Photochemical promotion,[11] ultrasound promotion,[12] and the use of TEMPO to promote single-electron activation of cobalt-carbonyl bonds[13] have been studied, and even mechanistic proposals have been formulated for some of these approaches.[11b,13] However, synthetic use of these activation methods has been up to now very limited and will not be discussed in detail here.

The pentacarbonyl complex resulting from CO loss of the initial $Co_2(CO)_6$(alkyne) complex is a high energy species that can only be observed under special conditions. Thus, photolysis[14] of $Co_2(CO)_6$(alkyne) complexes, involving alkynes such as acetylene, simple symmetrically disubstituted alkynes or phenylacetylene, in frozen Nujol at ca. 90 K results in loss of CO to give the coordinatively unsaturated $Co_2(CO)_5$(alkyne) complexes, where CO loss appears to have occurred from the axial position. However, complete conversion to a second isomer, presumably having the coordination vacancy in an equatorial position, is observed for all compounds upon annealing the matrices to ca. 140 K (Scheme 2.7).

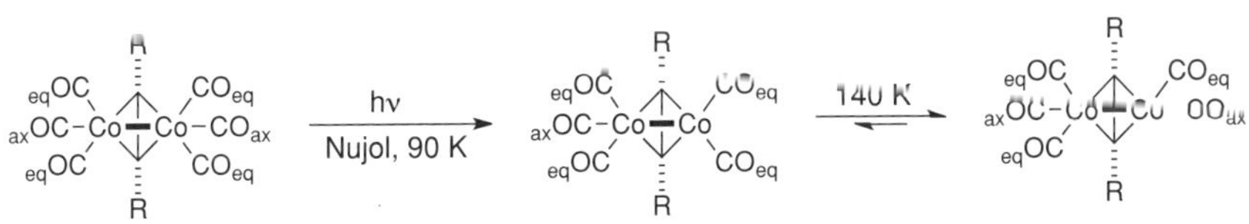

Scheme 2.7 *Detection of the coordinatively unsaturated dicobalt pentacarbonyl species.*

In addition, $Co_2(CO)_6$(alkyne) complexes where the alkyne bears properly placed thioether[15b] or thioacetal[15a] substituents afford, upon decarbonylation, thermally stable, yet reactive in PKR, pentacarbonyl complexes involving the thioether function as a hemilabile ligand[15] (Scheme 2.8).

Coordination of the reactant alkene to the coordinatively unsaturated $Co_2(CO)_5$(alkyne) intermediates is an energetically favorable process taking place with very low (if any) activation energy. Product stereochemistry in PKR involving alkenes with diastereotopic faces (for instance, norbornene and norbornadiene derivatives) indicates that coordination takes place involving the less hindered face of the olefin (Scheme 2.9). According to theoretical studies[16] (see below), even with poorly reacting olefins the energy barrier associated to the next step in the PKR mechanistic pathway (cobaltacycle formation)

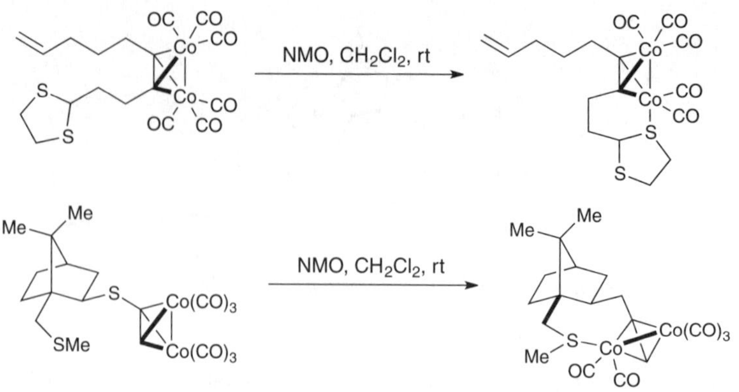

Scheme 2.8 *Chelation-aided decarbonylation of PK substrates.*

has an activation energy lower than that of olefin decomplexation. For this reason, olefin complexation is a kinetically controlled process and the resulting $Co_2(CO)_5$(alkyne)(alkene) complexes cannot, in general, be detected as intermediates.

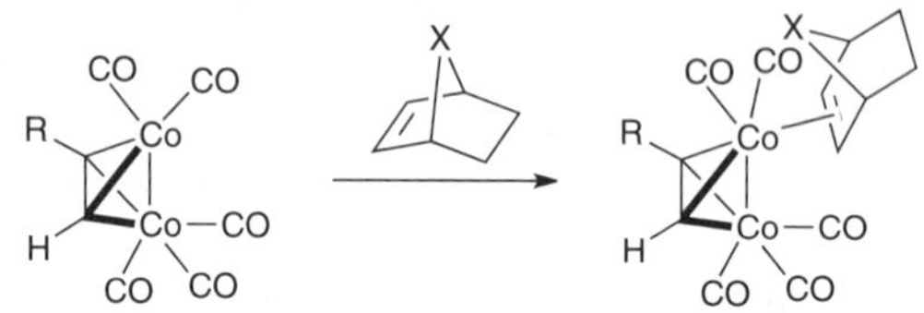

Scheme 2.9 *Olefin complexation.*

There are, however, some known exceptions to this behavior. Evans and McGlinchey isolated and characterized by X-ray diffraction $Co_2(CO)_5$(enyne) complexes (Figure 2.2, left), where an olefin is intramolecularly coordinated to a cobalt atom.[17] It is noteworthy that these complexes do not evolve to the corresponding PKR products, and this behavior is attributed to two factors: the severe strain in the possible reaction product, and the pseudoequatorial coordination of the alkene in the isolated complexes. More recently, an alkyne tetracarbonyl dicobalt complex with a chelated phosphine-alkene ligand, in which the phosphorus atom and the alkene moieties of the ligand are attached to the same cobalt atom, has been characterized by X-ray diffraction by Riera, Verdaguer, and co-workers[18] (Figure 2.2, center). Very interestingly, although the alkene fragment is located in this complex in an equatorial coordination site with an appropriate orientation and, therefore, should undergo insertion into a C(sp)-Co bond leading to cobaltacycle formation, it fails to provide the PKR product upon either thermal or *N*-methylmorpholine *N*-oxide activation. On the other hand, a similar $Co_2(CO)_5$(phosphine) complex containing a terminal alkene as one of the phosphine substituents, failed to afford any isolatable chelated phosphine-alkene complex (Figure 2.2, right), and readily provided the corresponding PKR product. This change in reactivity was attributed by the authors to the different ability of the olefins involved to undergo 1,2-insertion.

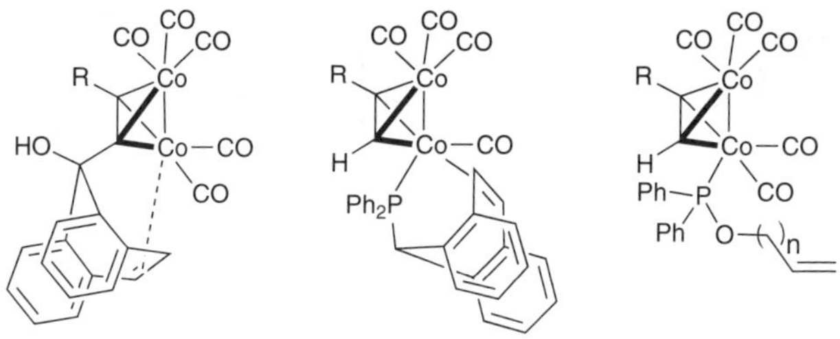

Figure 2.2 Studies on the isolation of the intermediates bearing the coordinated alkene.

2.2.2 Cobaltacycle Formation

Cobaltacycle formation is the most important step in the mechanistic pathway of the PKR because it is in this step where the regiochemistry and the stereochemistry of the final cyclopentenone adduct are established. The available experimental data indicate that, in the absence of geometrical restrictions, the alkene-substituted complex $Co_2(CO)_5$(alkyne)(alkene) readily evolves into a coordinatively unsaturated cobaltacycle. The stereoelectronic requirement for this step to proceed is the achievement of a synperiplanar arrangement between the carbon–cobalt bond and the carbon–carbon double bond participating in the metathesis-like process (Scheme 2.10).

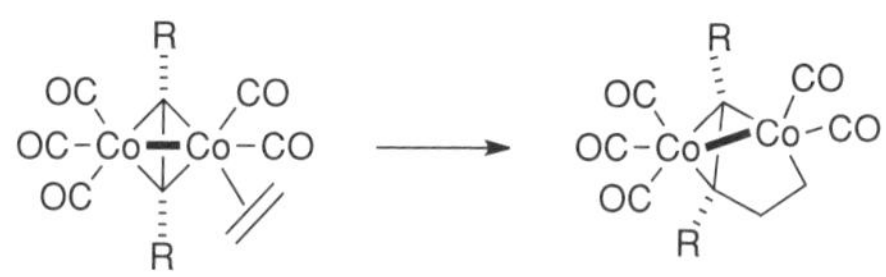

Scheme 2.10 The alkene insertion step.

As we have already mentioned, the olefin coordination process is likewise kinetically controlled. Thus, for olefins where the double bond depicts diastereotopic faces, coordination involving the less hindered face will be favored. Even if this step is irreversible, the alkene moiety of a $Co_2(CO)_5$(alkyne)(alkene) complex can experience two types of low-energy process before cobaltacycle formation: i) Rotation of the alkene ligand with respect to the complex, and ii) fluxional processes of the $Co(CO)_2$(alkene) moiety leading to interconversion between pseudoaxial and pseudoequatorial positions.[19]

Rotation of the alkene ligand (Scheme 2.11) is a highly relevant process in connection with the regioselectivity of the PKR (whenever non-symmetrically substituted alkenes are involved in the process) or with its stereoselectivity (when non-symmetrically substituted alkynes and alkenes depicting diastereotopic faces are simultaneously involved in the process).

Fluxional processes at a single $Co(CO)_2$(alkene) moiety have, in turn, no direct influence on the regio- or the stereochemistry of the PKR product.

Although considerations based on the conformational analysis of the $Co_2(CO)_5$(alkyne)(alkene) complexes are normally used to predict or to rationalize

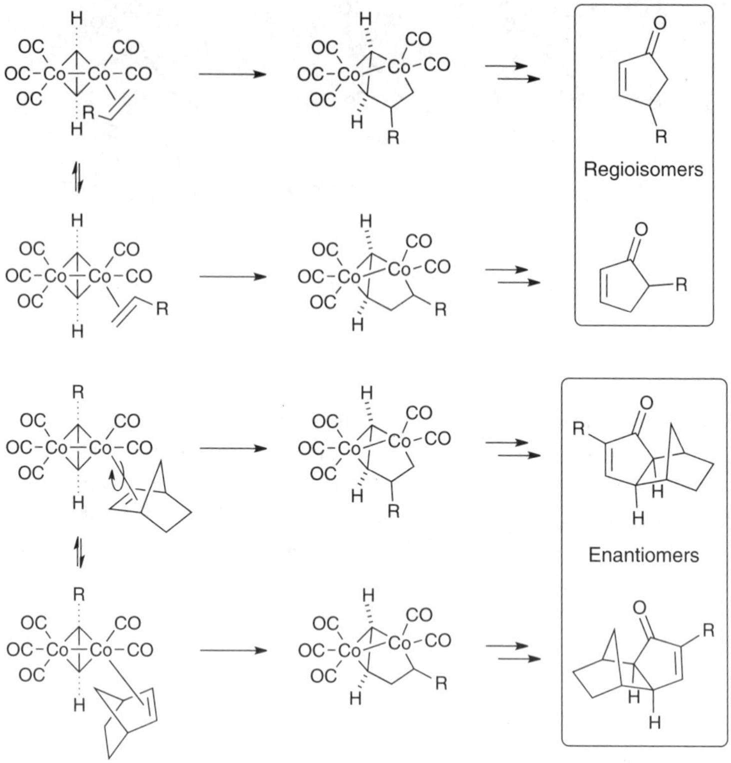

Scheme 2.11 Importance of alkene rotation processes.

the course of cobaltacycle formation (and of the whole PKR), the situation may well not be this simple. Since the energetic cost of olefin rotation or $Co(CO)_2$(alkene) pseudorotation should be small in comparison with cobaltacycle formation, this last process must be governed (assuming kinetic control) by the Curtin–Hammett principle, so that only the relative energies of the different transition states connecting the conformers of the $Co_2(CO)_5$(alkyne)(alkene) complex with the product cobaltacycle(s) determine the course of the reaction. Thus, irrespective of the nature (pseudoaxial or pseudoequatorial) of the vacant coordination site generated by CO loss, of the initially formed olefin complex, and of the nature (pseudoaxial or pseudoequatorial) of the most stable olefin complex, cobaltacycle formation is governed by the kinetics of the metathesis step.

Simple thermochemical considerations based on bond energies, confirmed by the available theoretical studies (see below) indicate that formation of a coordinatively unsaturated cobaltacycle, like those depicted in Schemes 2.10 and 2.11, should be essentially thermoneutral and this can raise a question over the possible reversibility of this step. It is important to consider, however, that the coordinatively unsaturated cobaltacycle intermediate will react without any energy barrier with CO or with Lewis bases present in the reaction medium to afford a much more stable species (Scheme 2.12). In this manner, the overall cobaltacycle formation step should become irreversible. This prediction, fully confirmed by theoretical studies, is in agreement with the fact that no equilibration of PK adducts has ever been observed.

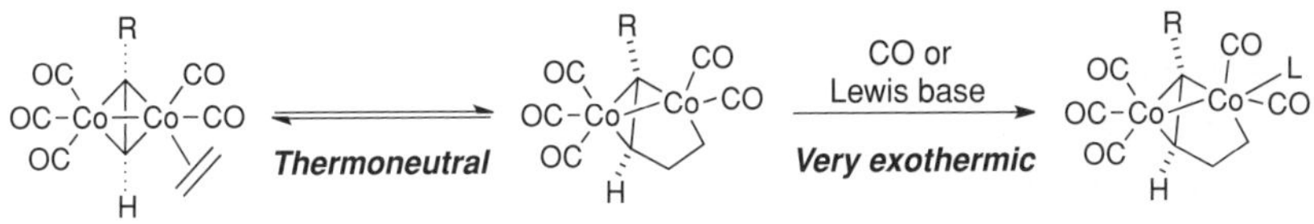

Scheme 2.12 *Irreversibility of the cobaltacycle formation process.*

2.2.3 CO Insertion

The coordinatively saturated cobaltacycles resulting from the previous step in the PKR mechanism have never been detected from an experimental point of view. According to the Magnus mechanism and to subsequent theoretical studies, one of the CO ligands in the coordination sphere of the active Co atom is next inserted into one of the carbon–cobalt bonds, installing in this manner the future carbonyl group of the cyclopentenone product. According to precedents with geometrically unconstrained systems, this insertion step should involve very low activation energy, since the transition state geometry is achieved by simple modification of a rather flexible C–Co–CO angle.

Although CO insertion could take place in any of the two carbon–cobalt sigma bonds in the cobaltacycle (Scheme 2.13), it has always been considered that insertion takes place in the $C(sp^3)$–Co bond. Yamanaka and Nakamura[16a] studied this problem from the theoretical point of view (see below) and found that while this process is in fact more favorable, the alternative insertion in the C(sp)–Co bond involves a transition state only $4\,kcal{\cdot}mol^{-1}$ higher in energy. In any case, the nature of the bond experiencing CO insertion should have no effect on the nature of the reaction product since both acylcobaltacycles should lead upon reductive elimination to the same cyclopentenone.

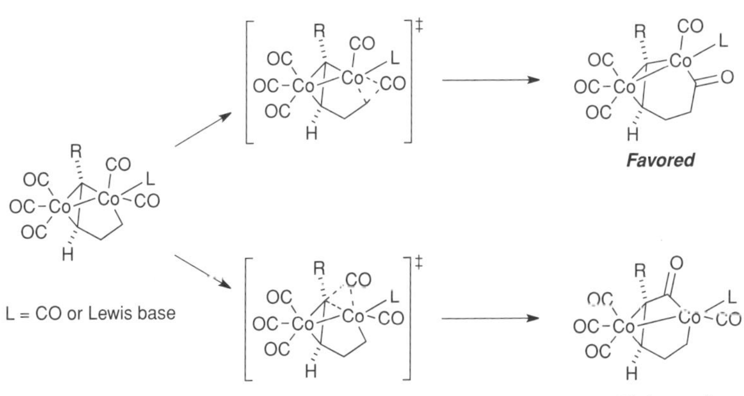

Scheme 2.13 *Possible routes for the CO insertion.*

As already discussed for the cobaltacycle formation step, the enthalpic balance of the CO insertion step should be made more favorable by coordinative saturation of the active cobalt atom in the initially formed species with carbon monoxide or with some other Lewis base present in the reaction medium.

2.2.4 Reductive Elimination

As with cobaltacyles, acylcobaltacycles have never been detected as intermediates in the PKR. It is assumed that, as soon as they are formed, they will evolve by reductive elimination into weakly bonded cyclopentenone dicobalt complexes which will readily decompose to afford the cyclopentenone product and unstable cobalt residues (Scheme 2.14).

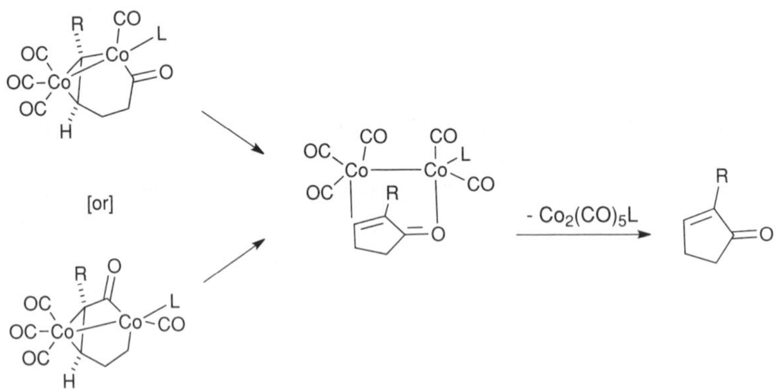

Scheme 2.14 *Reductive elimination and release of the cyclopentenone.*

While this step is irrelevant to the course of stoichiometric PKR, it becomes of critical importance for the success of catalytic versions of the reaction. In this type of process (see below), the survival of the catalytic cycle depends on the successful transfer of the dicobalt carbonyl fragment from the cyclopentenone complexes to new alkyne and alkene (or enyne) molecules entering the reaction cycle.

2.3 Catalytic Pauson-Khand Reaction

Sustainability concerns have in recent times fostered a shift in interest from stoichiometric to cobalt-catalytic versions of the PKR.[20] Different classes of additives were reported to improve catalytic PKR,[21] but no mechanistic study was reported until quite recently. Even the effect of the pressure of CO used in the process on reaction rate remained unclear. Thus, while initial reports on catalytic PKR involved the use of high temperatures and CO pressures,[22] catalytic PKR at atmospheric CO pressure was reported later.[23] The only thorough kinetic study on the PKR reported so far is from Verdaguer, Riera, and co-workers,[24] who studied by in situ FTIR the reaction of trimethylsilylacetylene with norbornadiene in the presence of a catalytic amount of $Co_2(CO)_8$, and applied Blackmond's reaction progress kinetic analysis[25] to the study of the mechanism of the reaction (Scheme 2.15).

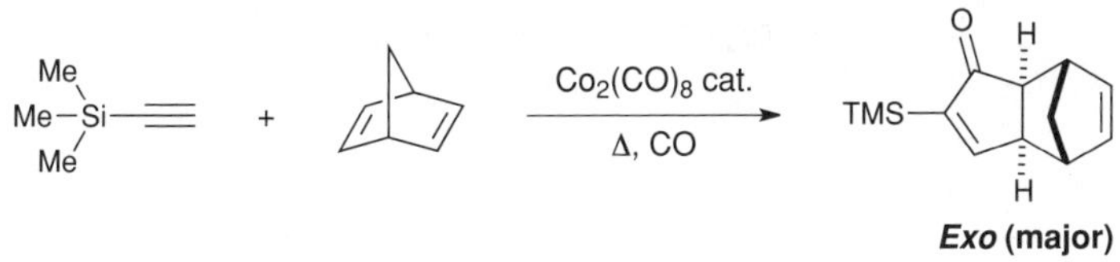

Scheme 2.15 *Catalytic PKR between trimethylsilylacetylene and norbornadiene.*

From the experimental data obtained in this study, the rate equation 2.1 for the catalytic intermolecular PKR between trimethylsilylacetylene and norbornadiene (NBD) was deduced.

$$v = k[Co_2(CO)_8]^{1.3}[NBD]^{0.3-1.2}[CO]^{-1.9} \quad (2.1)$$

The reaction turned out to be zero order with respect to alkyne, 1.3 order with respect to the catalyst, between 0.3 and 1.2 order with respect to NBD, and −1.9 order with respect to CO. On the basis of this rate equation, the authors suggested that a complex reaction mechanism was in operation in the catalytic intermolecular PKR of norbornadiene. However, the positive dependence on NBD and the negative dependence on CO, together with the observation that the corresponding dicobalt alkyne complex **I** (Scheme 2.16), clearly identified by IR spectroscopy, was acting as a major resting state, meant they could infer that the CO-alkene ligand exchange process was an energetically demanding equilibrium preceding the rate-limiting step. In this manner, the following step in the sequence, consisting of alkene insertion to give cobaltacycle **III**, would be the rate-limiting step in the catalytic process. In this scenario, a higher concentration of NBD and lower concentration of CO are beneficial for the kinetics of the reaction, since they favor formation of the key intermediate **II**.

A very important conclusion is that these experimental data are consistent with the overall catalytic cycle shown in Scheme 2.16, which is in agreement with the original Magnus proposal and with the theoretical knowledge accumulated on the reaction (see next section).

2.4 Theoretical Studies

2.4.1 General Approach to the Mechanism

Despite the great interest in the Pauson-Khand reaction, only in the last decade, 30 years after its discovery, has significant progress been made in computational approaches to elucidate and understand its mechanism. Only limited attempts to explain the diastereoselectivity observed when using chiral auxiliaries, by semi-empirical and molecular mechanics methods, were published in the late 1990s[19c,26] (see below).

It was not until 2001 that the first comprehensive study on the full reaction mechanism was published by Nakamura and Yamanaka.[16a] In this study, performed at the DFT level with simple substrates, the authors confirmed the feasibility of the Magnus mechanism, while shedding some light on the particular issues affecting the reaction outcome in different

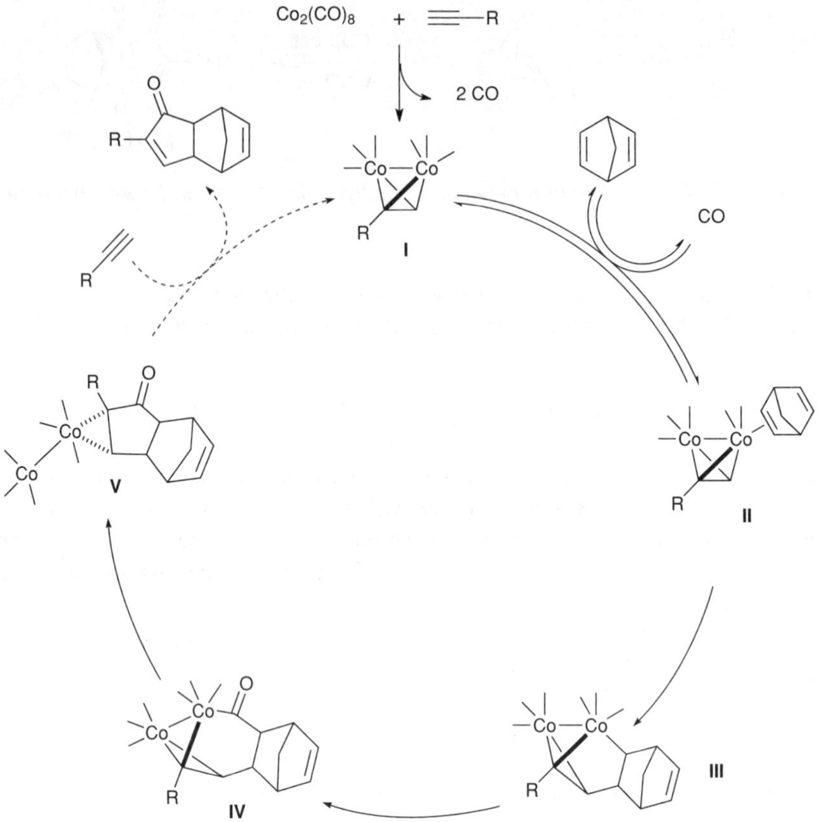

Scheme 2.16 *Catalytic cycle for the PKR as proposed by Verdaguer, Riera* et al.[24]

aspects, as regio- and stereoselectivity. This pioneering work has thus turned into a reference for any computational approach to the Pauson-Khand reaction.

In Scheme 2.17 and Figure 2.3, the reaction pathway as determined by Nakamura and coworkers is shown. Following their results, the mechanism can be clearly divided in two parts, the first (from the starting materials to intermediate **C**, before the alkene insertion) being endothermic and the second (from the alkene insertion to the ending products) being strongly exothermic and presumably irreversible.

Although the authors do not delve deeply into the exact mechanism of the carbonyl substitution by olefin, they assume a dissociative process has taken place, through the intermediate unsaturated species **B**. As shown by calculations, formation of **B** would be the most energetically demanding step of the whole process, which would explain the difficulties encountered up to that moment to experimentally isolate or even detect any of the reaction intermediates. Even if the reaction was not to proceed through **B** but through a lower energy associative transition state, the significant (ca. 15 kcal·mol^{-1}) difference between **A** and **C** would make the latter at best hard to isolate or even observe by experimental means.

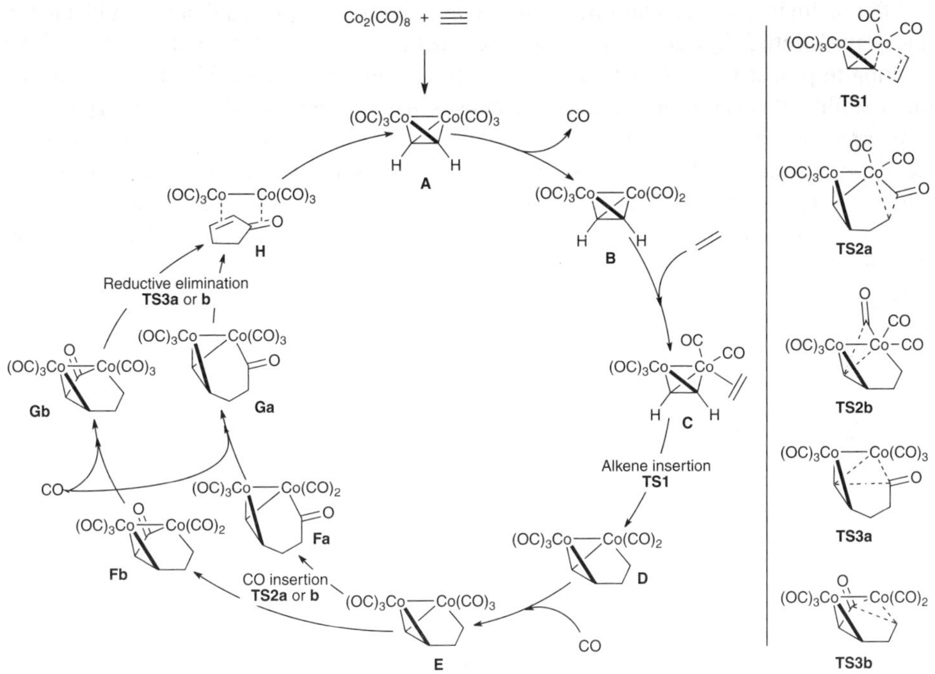

Scheme 2.17 *Detailed catalytic cycle of the PKR as defined by Nakamura and Yamanaka.[16a] The same nomenclature will be used throughout this section.*

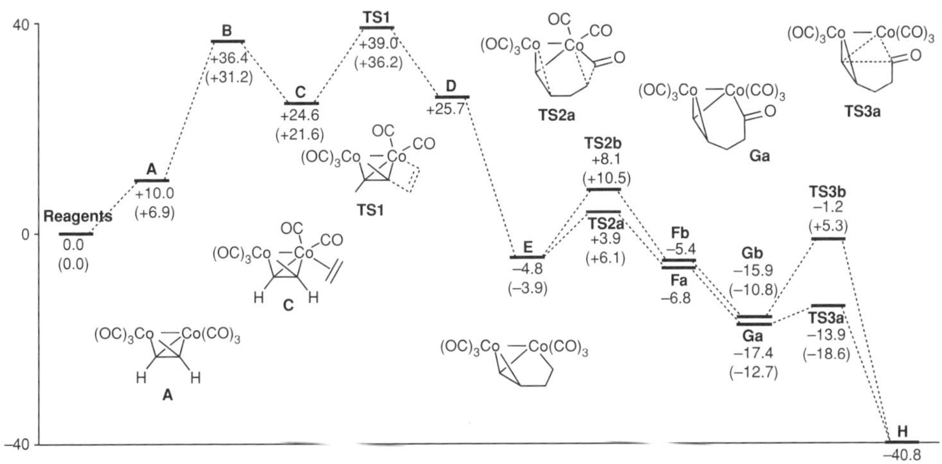

Figure 2.3 *Reaction profile of the PK reaction at the B3LYP/631LAN level. The values in parentheses are for B3LYP/6311SDD//B3LYP/631LAN.*

After olefin insertion, on the other hand, coordination of a carbonyl ligand would quickly transform **D** into **E** in a strongly exothermic process (by over 30 kcal·mol^{-1}) allowing the reaction to proceed to completion with no other significant barrier, this also explains the impossibility of isolating any of the intermediates in the second part of the reaction pathway.

Intermediate **E**, thus, would experience insertion of one of the carbonyl ligands from Co(1) giving place to the acylcobalt species **F**, further stabilized again by coordination of a new CO ligand to obtain the corresponding coordinatively saturated species **G**. In this sequence, two possibilities were considered, namely the carbonyl insertion taking place at the Co–C(1) (alkenyl) or at the Co–C(4) (alkyl) bonds. The calculations showed the later to be preferred, as had been silently assumed before, although the difference was relatively small (ca. 4 kcal·mol^{-1}).

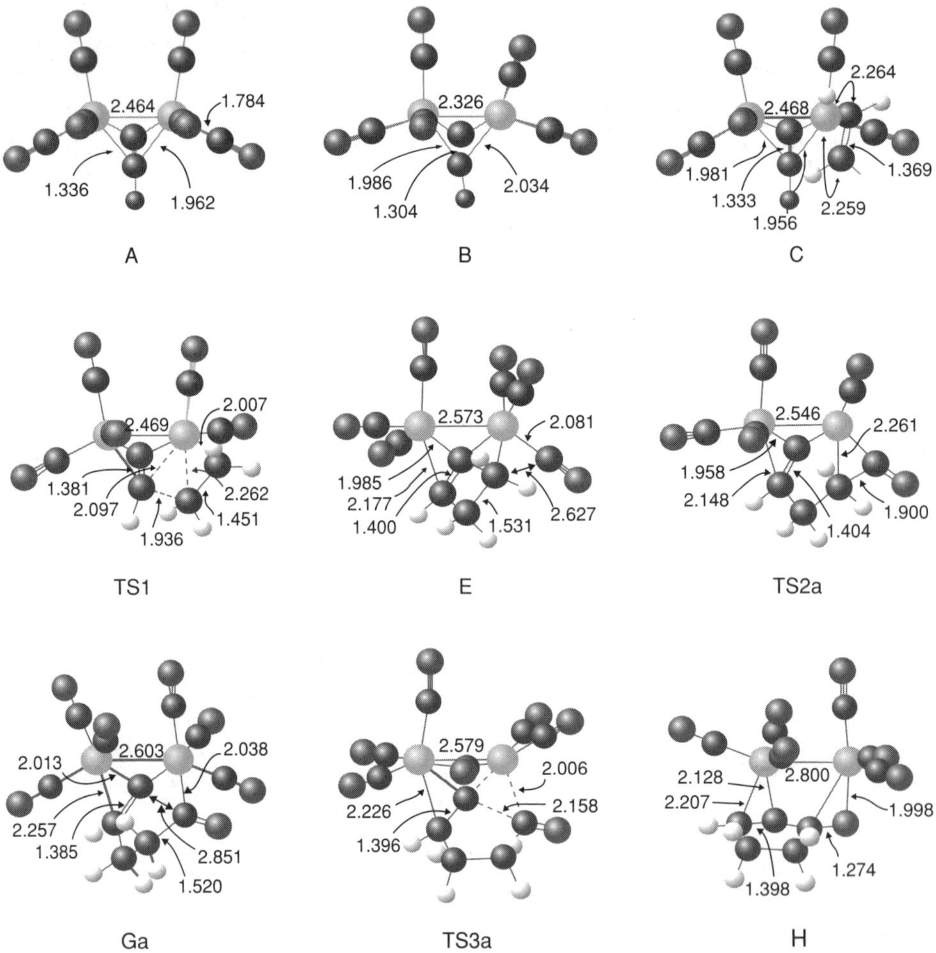

Figure 2.4 *Relevant minima and transition states for the Pauson-Khand reaction optimized at B3LYP/631LAN level. Numbers indicate bond lengths and distances in Å.*

From both acylated species **Ga** and **Gb**, reductive elimination gives way to the same coordinated cyclopentenone product **H** which can be released by decomposition of the complex (stoichiometric systems) or, under the adequate conditions, substituted by a new alkyne unit to regenerate complex **A** (catalytic systems).

This high level theoretical study has had a very strong impact on further investigation dealing with different aspects of the PKR from a mechanistic perspective. The key aspects of its importance are: i) the theoretical analysis provided further support to the Magnus mechanism (accepted as the working paradigm for research on this topic), while complementing it with an effective rational of previous experimental observations and a theoretical frame for designing new experiments; ii) the determination of the particular structures of the intermediates and transition states involved in this mechanism (Figure 2.4) allowed careful study of their most relevant features in both geometrical and electronic aspects.

Regarding the geometrical aspects, it is worth noting that the essential backbone of the complex was maintained throughout the process, with the Co–Co bond being retained (2.326–2.695 Å) and the Co^2 atom firmly attached to the acetylene moiety (Co^2–C^1 and Co^2–C^2 bond lengths of 1.952–2.257 Å). The acetylene unit in **A** had a C–C distance of 1.336 Å and a C–C–H angle of 142.9 °, this clearly indicating strong backbonding from the Co atoms. On the other hand, the olefin in **C** presented a nearly planar geometry and long Co–C distances, this showing a weak coordination with no significant backbonding.

For the electronic aspects, natural population calculations were used to determine the charge distribution in each step. Comparison of the evolution of the charge on both Co atoms showed that Co^2 not only acted as a spectator, structural moiety but as a reservoir of electrons for Co^1. On the other hand, the acetylene moiety was negatively charged in the first stages of the process (as a result of the above-mentioned strong back donation from the Co atoms) while the ethylene moiety was essentially neutral, this situation changing at the insertion stage.

These fundamental considerations provide the frame for further research on more particular issues as the regio- and stereoselectivity of the PKR, not only from computational but also experimental approaches.

2.4.2 Regioselectivity. Early Steps of the Pauson-Khand Reaction

The above-mentioned analysis of the electronic characteristics of the intermediates and transition states served as a basis for a preliminary theoretical rational of the experimentally observed regioselectivity.[16a] Particularly, the lack of polarization at the alkyne, or the alkene observed in the transition state for the insertion step, (**TS1**, see above) suggested that the regioselectivity had to be based on steric effects.

Analysis of this **TS1** structure for the model reaction with acetylene and ethylene showed that, while for the alkyne the C^1 position was clearly less hindered than C^2, for the olefin there was no apparent difference between both carbon atoms, both C^3 and C^4 occupying relatively unhindered positions. This was consistent with the experimentally observed lack of selectivity regarding the alkene and high selectivity regarding the alkyne counterpart in the reaction between terminal alkenes and alkynes (Scheme 2.18).

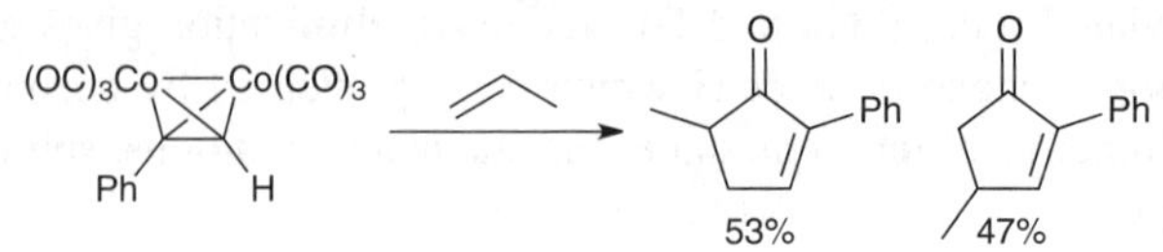

Scheme 2.18 *Regioselectivity of the PK reaction between phenylacetylene and propene.*

With olefins bearing a functional group capable of coordinating a Co atom, this total lack of selectivity in the olefin part completely disappears experimentally. In the computational approach, 4-aminobutene was used as a model for this kind of system and the four possible structures of **TS1** were considered. A partial optimization approach was used, keeping the core reaction center frozen (Co^1–Co^2–C^1–C^2–C^3–C^4) in the geometry of **TS1** for the unsubstituted model and optimizing the substituents (Figure 2.5). A significant preference for **TS1a**, in which the chelating substituent was linked to C^4 and coordinating Co^1, was observed. This result was consistent with the experimental observation of the chelating substituents occupying preferentially the C^4 position in the final product (Scheme 2.19).

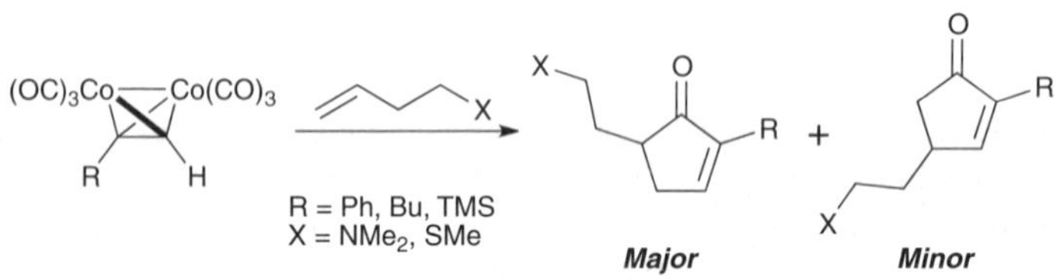

Scheme 2.19 *P–K reaction of chelating olefin substrates with terminal alkynes.*

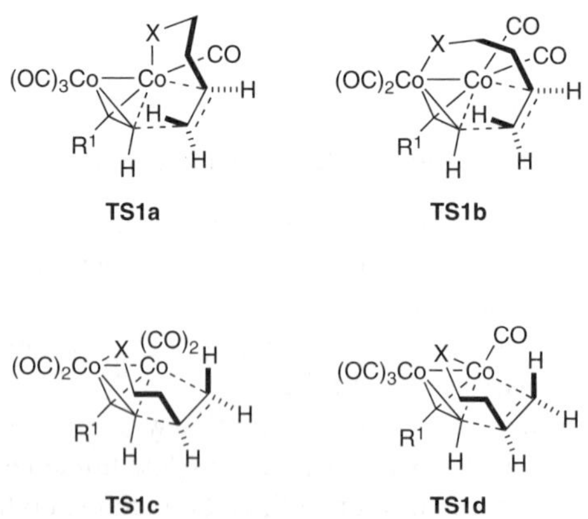

Figure 2.5 *Schematic representation of the transition state models for the stereodetermining alkene insertion step with chelating olefins (X = NH_2).*

With alkynes bearing two sterically similar but electronically different substituents, however, polarization on the alkyne turned out to exert a strong influence on the regiochemical outcome of the reaction (Scheme 2.20).

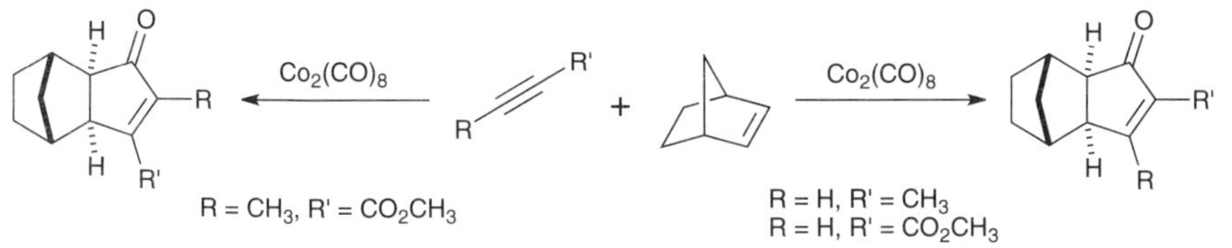

Scheme 2.20 *Regioselectivity in the PK reaction with differently polarized alkynes.*

This issue was addressed by a theoretical study on the basis of a *trans* effect claimed to affect the preferred olefin coordination position and, in consequence, the position for olefin coordination and insertion[27] (Table 2.1).

In this work the authors argued that, with polarized alkynes such as methyl 2-butynoate (Table 2.1, $R = CH_3$, $R' = CO_2CH_3$), the carbonyl ligand in eq_{cis} position with respect to the negatively charged C^2 carbon atom was more labile than the others, and this causes the reaction to proceed through olefin coordination and insertion from this point. This study correctly reproduced the experimental observations for alkynes bearing sterically similar but electronically different substituents, although it had the flaw of obviating the possibility of a Curtin–Hammett situation with the coordination of the olefin not being the regioselectivity-determining step (see Section 2.2).

A similar approach, based on the polarization of the acetylene unit in the $Co_2(CO)_6$(alkyne) complex was attempted for the case of *S*-alkyl-substituted alkynes,

Table 2.1 *Atomic charges (e^-) based on natural population analysis of the acetylenic carbons in complexes* $\boldsymbol{A_{1\text{-}3}}$ *and first carbonyl dissociation energies (kcal·mol^{-1}) for the formation of the different isomeric dicobalt pentacarbonyl alkyne complexes* ***B****.*

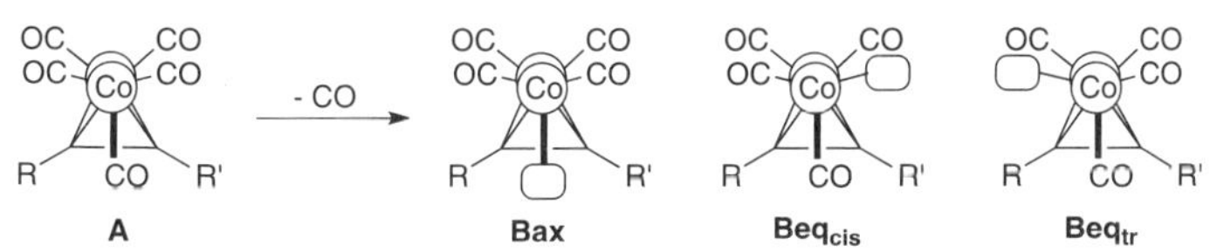

						$\Delta E_{A\rightarrow B}$		
Starting complex	R	R′	$q(C_R)$	$q(C_{R'})$	Δq	**Bax**	**Beq$_{cis}$**	**Beq$_{tr}$**
A_1	H	CH_3	−0.31	−0.18	−0.13	–[a]	–[b]	24.2
A_2	H	CO_2CH_3	−0.36	−0.33	−0.03	37.1	26.4	26.7
A_3	CH_3	CO_2CH_3	−0.14	−0.34	+0.20	37.9	24.5	–[c]

[a]Optimized structure exhibited a small imaginary frequency and reoptimization yielded $\mathbf{B_1eq_{tr}}$. [b]Optimization yielded $\mathbf{B_1eq_{tr}}$. [c]Optimization yielded $\mathbf{B_3eq_{cis}}$.

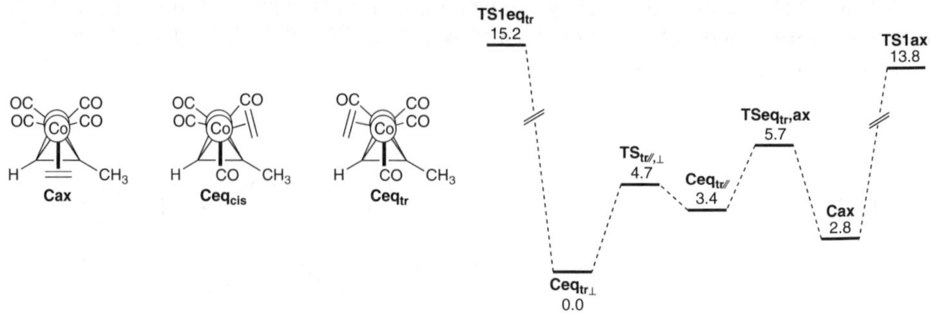

Figure 2.6 Computational analysis of the interconversion between the different olefin-coordinated complexes **C** for the PKR between propyne and ethylene.

although with no success, this making apparent that a more complete theoretical analysis was necessary for obtaining a good model for the regioselectivity with polarized alkynes.[28]

In a more detailed study, focused on the olefin insertion step,[29] the possible modes of olefin coordination in a $Co_2(CO)_5$(propyne)(ethylene) complex were analyzed. After demonstrating that the perpendicular conformation of the olefin is preferred, the barriers for olefin rotation and for the $Co(CO)_2$(alkene) unit pseudorotation were determined. The low barriers found demonstrated that the initial coordination position is not the key issue for determining the regioselectivity of the reaction (Figure 2.6).

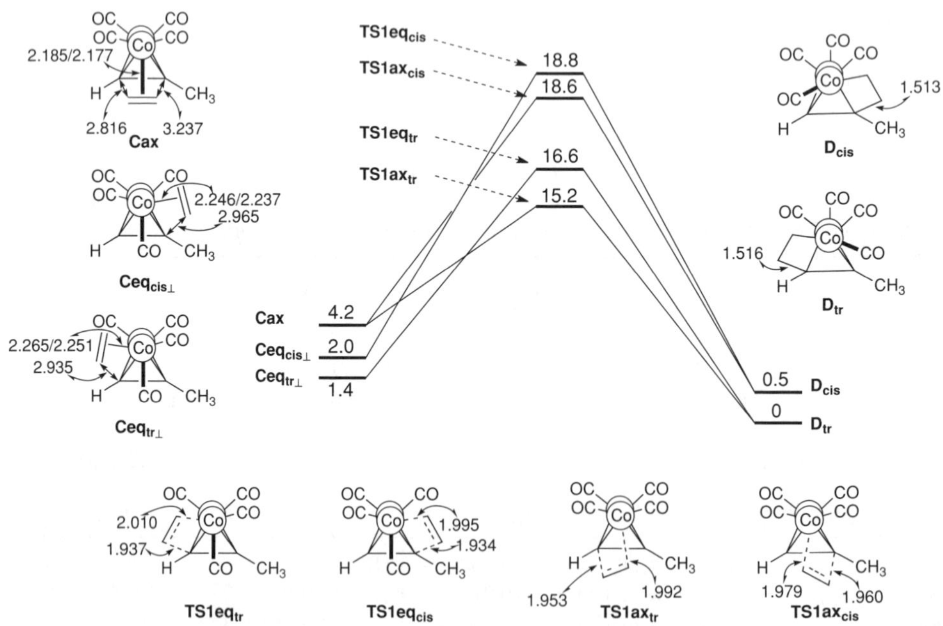

Figure 2.7 Analysis of the transition states for the olefin insertion step (***TS1***) as a function of the different isomers of intermediate **C**, in the PKR of propyne and ethylene.

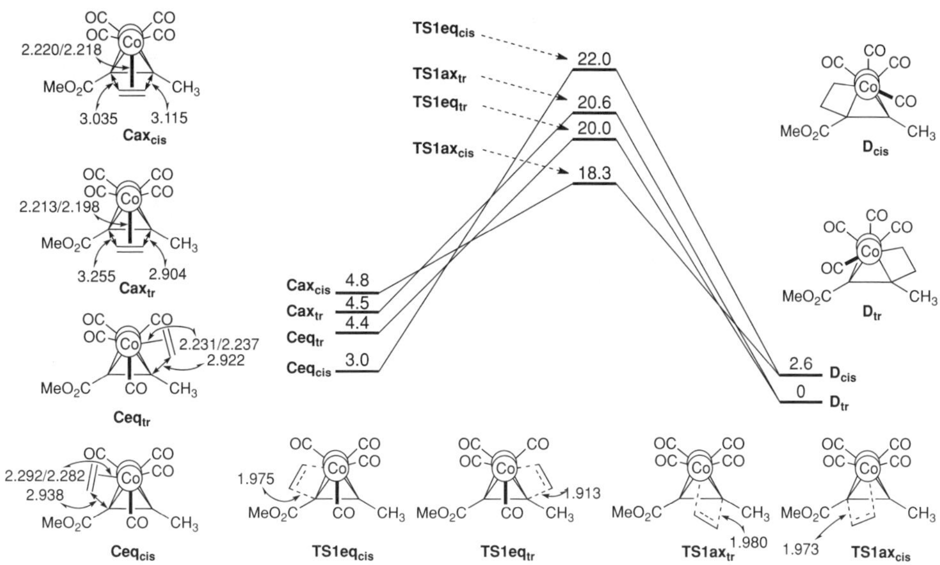

Figure 2.8 *Energy diagram for the alkene insertion step in the PK reaction of methyl butynoate with ethylene.*

On this basis, the theoretical analysis was focused on calculating the transition states for the olefin insertion starting from the three favored isomers (**Ceq**$_{\text{cis},\perp}$, **Ceq**$_{\text{trans},\perp}$ and **Cax**, Figure 2.6), observing interestingly that the system did indeed show a Curtin–Hammett behavior, the axial coordination giving place to the lowest transition state for the insertion (**TS1ax**$_{\text{tr}}$) despite the olefin complex (**Cax**) itself lying significantly higher in energy than the equatorial analogues. Having in consideration the two possible TS's for insertion of the olefin from the axial position (**TS1ax**$_{\text{cis}}$ and **TS1ax**$_{\text{tr}}$), as well as the ones for insertion from both equatorial positions (**TS1eq**$_{\text{cis}}$, **TS1eq**$_{\text{tr}}$), the experimentally observed regioselectivity for these kind of complexes was correctly reproduced (Figure 2.7).

This approach has been extended to explain the regioselectivity of three differently polarized alkynes as propyne, methyl propiolate and methyl butynoate[30] (Figures 2.8 and 2.9), thus integrating the former, polarization-based analysis (see above). The results show that, although in all cases the insertion step seems to take place from the complex with the olefin in pseudoaxial position (**Cax**), the different polarizations of the acetylene unit in these complexes played a fundamental role in determining which end of it was to be involved in the C–C bond formation event (**TS1ax**$_{\text{cis}}$ or **TS1ax**$_{\text{tr}}$).

2.4.3 Stereoselectivity

Despite the important advances made in the last years in stereoselective Pauson-Khand reactions, only a few attempts to computationally model this kind of reactions have been made. Pioneering studies dealing with diastereoselective, stoichiometric versions (with chiral substrates or chiral auxiliaries) were published prior to the complete DFT study of the mechanism, modeling the stereo-determining step by simple MM or semi-empirical methods.

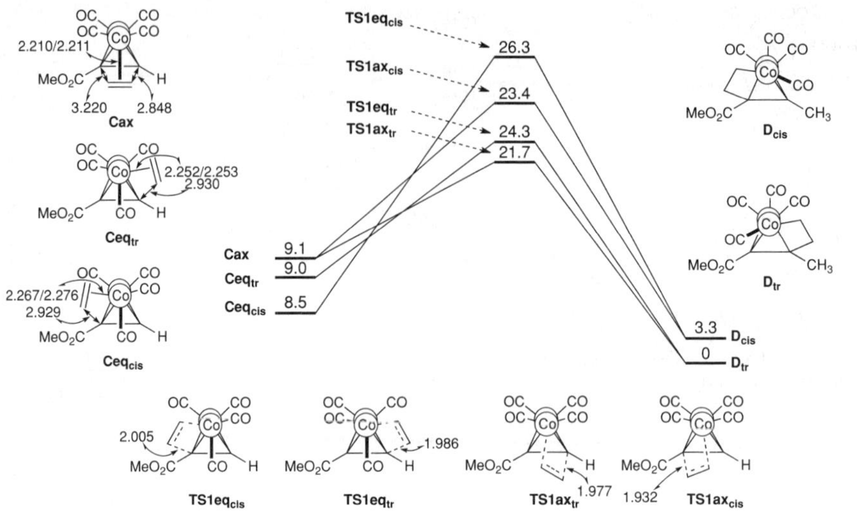

Figure 2.9 *Energy diagram for the alkene insertion step in the PK reaction of methyl propynoate with ethylene.*

Table 2.2 *Correlation of the diastereoselectivity of intramolecular PK reactions and the MM-computed energies of the corresponding intermediate metallacycles (**D**).*

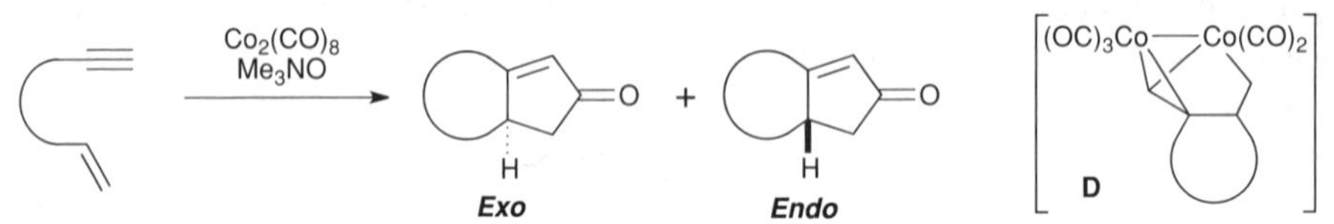

Reaction	*exo:endo* (exp.)	$\Delta H_{endo} - \Delta H_{exo}$ (metallacycles)	$\Delta H_{endo} - \Delta H_{exo}$ (products)
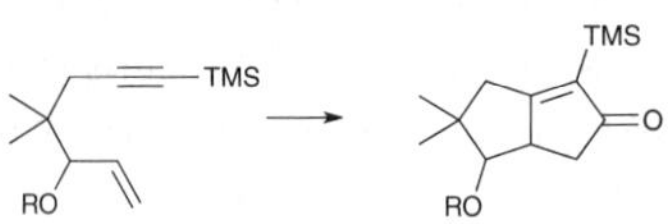	(R = H) 72:28 (R = MOM) 100:0	−0.3 −1.5	0.1 1.8
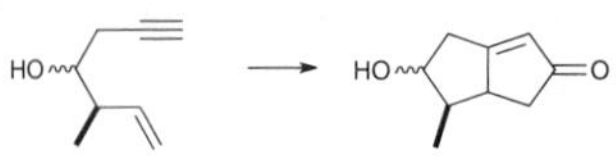	(cis) 92:8 (trans) 75:25	−1.9 −2.1	2.5 1.7
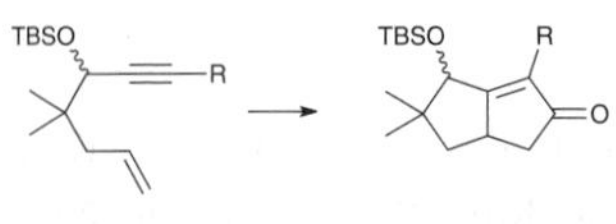	(R = H) 39:61 (R = Me) 77:23 (R = TMS) 96:4	−1.1 −1.2 −3.1	0.6 1.2 4.8
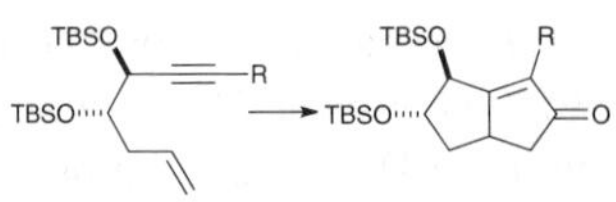	(R = H) 50:50 (R = TMS) 100:0	−0.2 −3.7	0.9 1.7

Intramolecular PK reactions on chiral dienynes giving place to bicyclo-[3.3.0]-octenones have been rationalized on the basis of the relative energies of the intermediate metallacycles (**D**) obtained upon olefin insertion, calculated at a MM level. This method, although extremely simple, allows to correctly reproduce some of the experimental results obtained with this kind of substrates[26] (Table 2.2).

Also, computational studies were used to explain the selectivity obtained in the intermolecular reaction with alkynes bearing camphor-derived chelating chiral auxiliaries. Both for substituted thioisoborneol[19c] (Figure 2.10) and for Oppolzer's camphorsultam

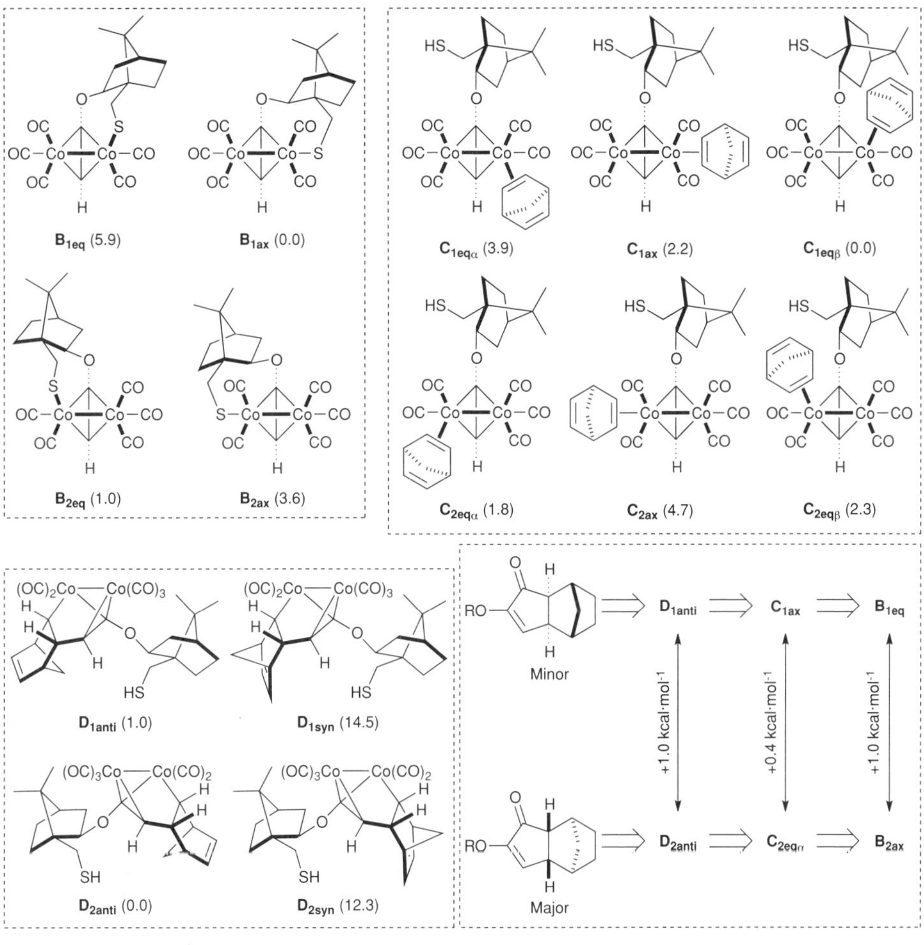

Figure 2.10 *Study of the stereoselective PK reaction of O-alkynyl-(2R)-10-mercaptoisoborneol with norbornadiene. Top, left: Dicobaltpentacarbonyl alkyne complex **B**, with the thiol group coordinating one of the Co atoms. Top, right: Alkene complex **C**, after substitution of the thiol by norbornadiene. Bottom, left: Cyclometalated complexes **D**, after alkene insertion into the acetylenic C–Co bond. Bottom, right: Genealogy of the two observed products, accounting for the observed selectivity. Numbers in brackets are relative energies in kcal·mol^{-1}, obtained from single point calculations at DFT level on PM3(tm)-optimized structures.*

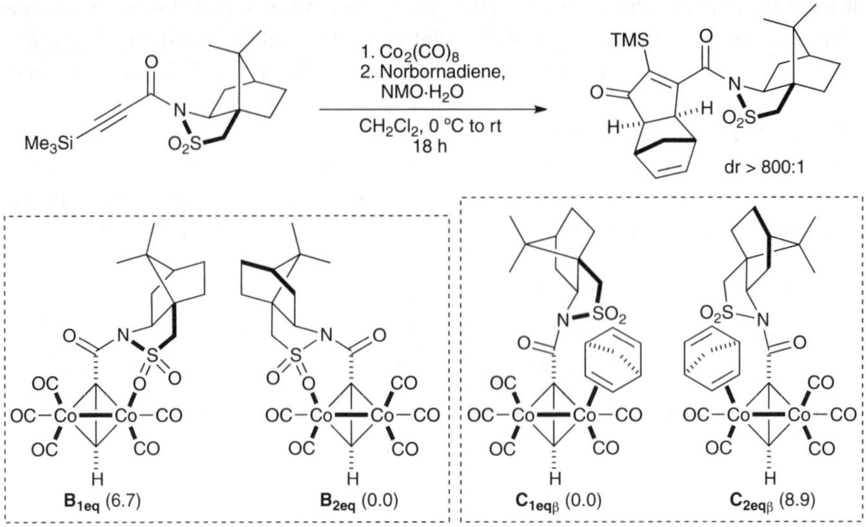

Figure 2.11 *Highly diastereoselective PK reaction of* (1R,5S,7S)-N-*(3-trimethylsilylpropiolyl)-bornane-2,10-sultam and norbornadiene (top) and explanation of the selectivity by the chelated intermediates* ***B*** *(bottom, left) and the olefin complexes* ***C*** *(bottom, right). Relative energies in kcal·mol^{-1} are shown between brackets, determined by MP3(tm) (structures* ***B****) or DFT (structures* ***C****).*

derivatives[31] (Figure 2.11), analysis of the reaction intermediates at the semi-empirical/DFT level allowed interpretation of the stereoselectivity based on the preference of the sulfur-containing chelating group to displace one particular carbonyl ligand, thus directing the olefin coordination to that position.

Very recently, studies on the enantioselective catalytic PKR at DFT level have been published. On the one hand, a QM/MM ONIOM approach has been employed for studying the

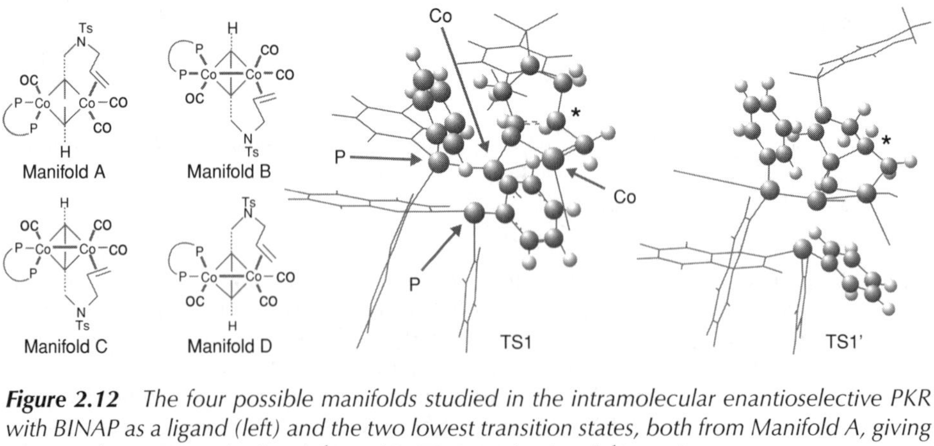

Figure 2.12 *The four possible manifolds studied in the intramolecular enantioselective PKR with BINAP as a ligand (left) and the two lowest transition states, both from Manifold A, giving place to the* R *(**TS1**, 0 kcal·mol^{-1}) and* S *(**TS1'**, 9.0 kcal·mol^{-1}) products.*

intramolecular PKR with BINAP as a ligand.[32] Modeling of the possible transition states for the alkene insertion step resulted in accurate reproduction of the experimental observations, thus allowing rationalization of how the chiral induction takes place (Figure 2.12).

In a similar approach, combining QM/MM ONIOM with full DFT calculations, the use of chiral *N*-oxides as promoters for the enantioselective PKR was studied.[33] In this case, the asymmetric induction was obtained from the selective oxidation of one CO ligand by the chiral, enantiopure *N*-oxide (Figure 2.13, top). After this initial selection, the barrier for racemization of the pentacarbonyl complex was demonstrated to be significantly higher than the barrier for alkene insertion, this being responsible for the transfer of chirality to the final product (Figure 2.13, bottom).

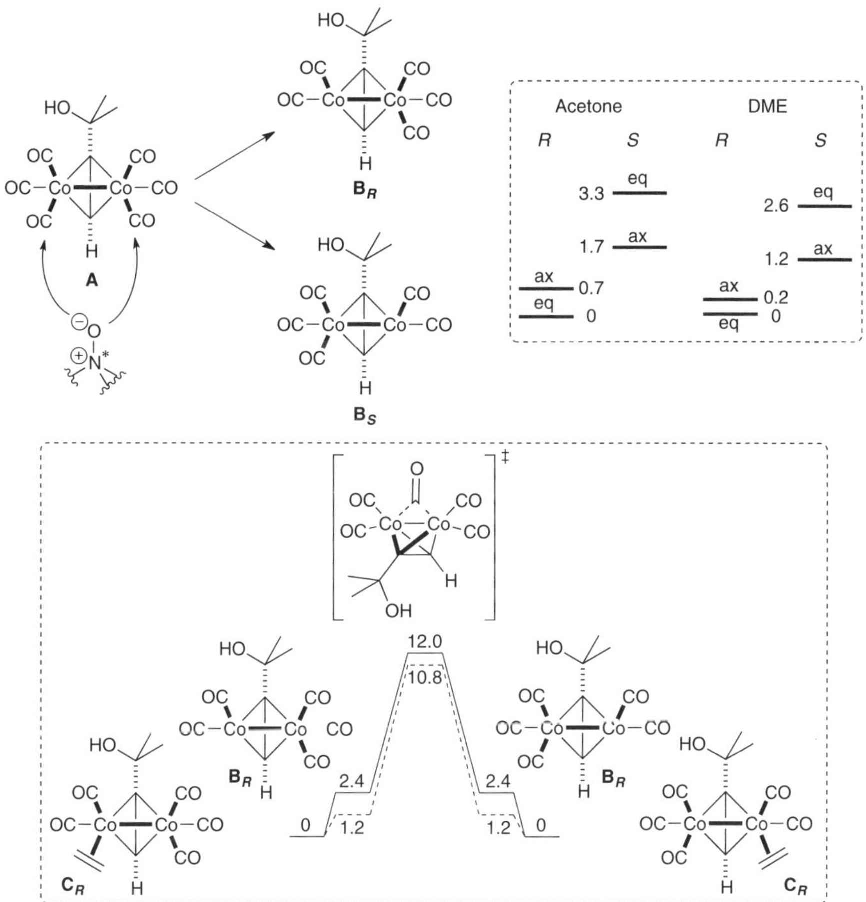

Figure 2.13 *Top: BNO-promoted selective loss of CO from $Co_2(CO)_6$(alkyne) complex and energies (in kcal·mol^{-1}) of the four lower transition states. Bottom: Energetic barrier for the interconversion between enantiomers of the alkene complexes $C_{R/S}$.*

2.5 Conclusions

The Magnus proposal for the mechanism of the PKR, put forward more than 25 years ago on the basis of common knowledge of organometallic chemistry, has guided over the years progress in knowledge about this reaction. Although experimental confirmation of the original proposal has remained elusive, high-level theoretical calculations have fully confirmed the viability of the different steps in the sequence.

In addition to the many examples of *normal* Pauson-Khand reactions leading to the expected cyclopentenones, the reaction behaves rather capriciously in some instances, leading to a plethora of alternative reaction products. These variations of PKR have been thoroughly reviewed by Boñaga and Krafft.[34] In any case, it is interesting to recall that all of these variations can be accommodated in the Magnus mechanism,[3] resulting from mechanism diversions taking place at the level of some of its different intermediates.

References

[1] I. U. Khand, G. R. Knox, P. L. Pauson, W. E. Watts, M. I. Foreman *J. Chem. Soc., Perkin Trans.* **1**, 977–981 (1973).

[2] For reviews, see: (a) N. E. Schore *Org. React.* **40**, 1–90 (1991); (b) O. Geis, H.-G. Schmalz *Angew. Chem. Int. Ed.* **37**, 911–914 (1998); (c) K. M. Brummond, J. L. Kent *Tetrahedron* **56**, 3263–3283 (2000); (d) S. E. Gibson, N. Mainolfi *Angew. Chem. Int. Ed.* **44**, 3022–3037 (2005).

[3] (a) P. C. Magnus, C. Exon, P. Albaugh-Robertson *Tetrahedron* **41**, 5861–5869 (1985); (b) P. C. Magnus, L. M. Principe. *Tetrahedron Lett.* **26**, 4851–4854 (1985).

[4] Y. K. Chung, B. Y. Lee, N. Jeong, M. Hudecek, P. L. Pauson *Organometallics* **12**, 220 (1993).

[5] Amines: (a) T. Sugihara, M. Yamada, H. Ban, M. Yamaguchi, C. Kaneko *Angew. Chem. Int. Ed. Engl.* **36**, 2801 (1997); see also: (b) T. Sugihara, M. Yamaguchi, M. Nishizawa *Chem. Eur. J.* **7**, 1589 (2001); thioethers: (c) T. Sugihara, M. Yamada, M. Yamaguchi, M. Nishizawa *Synlett* **1999**, 771; phosphines: (d) D. C. Billington, I. M. Helps, P. L. Pauson, W. Thomson, D. Willison *J. Organomet. Chem.* **354**, 233 (1988).

[6] (a) R. J. Dennenberg, D. J. Darensbourg, *Inorg. Chem.* **11**, 72 (1977); (b) J. D. Atwood, T. L. Brown, *J. Am. Chem. Soc.* **98**, 3160 (1976); (c) D. J. Darensbourg, R. L. Kump, *Inorg. Chem.* **17**, 2680 (1976).

[7] (a) W. Hieber, A. Lipp *Chem. Ber.* **92**, 2085 (1959); (b) H. Alper, J. T. Edward *Can. J. Chem.* **48**, 1543 (1970); (c) Y. Shvo, E. Hazum *J. Chem. Soc., Chem. Commun.* **1975**, 829; (d) Y. L. Shi, Y. C. Gao, Q. Z. Shi, F. Basolo *Organometallics* **6**, 1528 (1987); (e) J. K. Shen, Y. C. Gao, Q. Z. Shi, F. Basolo *Organometallics* **8**, 2144 (1989).

[8] S. Shambayati, W. E. Crowe, S. L. Schreiber, *Tetrahedron Lett.* **31**, 5289 (1990).

[9] N. Jeong, Y. K. Chung, B. Y. Lee, S. H. Lee, S.-E. Yoo, *Synlett*, 1991, 204.

[10] (a) W. J. Kerr, G. G. Kirk, D. Middlemiss, *Synlett* 1995, 1085; (b) D. R. Carbery, W. J. Kerr, D. M. Lindsay, J. S. Scott, S. P. Watson, *Tetrahedron Lett.* **41**, 3235–3239 (2000); (c) A. R. Kennedy, W. J. Kerr, D. M. Lindsay, J. S. Scott, S. P.

Watson, *J. Chem. Soc. Perkin Trans.* **1**, 4366–4372 (2000). (d) W. J. Kerr, D. M. Lindsay, E. M. Rankin, J. S. Scott, S. P. Watson, *Tetrahedron Lett.* **41**, 3229–3233 (2000).

[11] (a) B. L. Pagenkopf, T. Livinghouse *J. Am. Chem. Soc.* **118**, 2285–2286 (1996); (b) C. M. Gordon, M. Kiszka, I. R. Dunkin, W. J. Kerr, J. S. Scott, J. Gebicki *J. Organomet. Chem.* **554**, 147–154 (1998); (c) A. C. Coleman, C. Long, A. Meetsma, B. L. Feringa, W. R. Browne, M. T. Pryce, *Dalton Trans.* **38**, 7885–7887 (2009).

[12] J. G. Ford, W. J. Kerr, G. G. Kirk, D. M. Lindsay, D. Middlemiss *Synlett* 2000, 1415–1418.

[13] A. Lagunas, A. Mairata i Payeras, C. Jimeno, M. A. Pericàs *Org. Lett.* **7**, 3033–3036 (2005).

[14] T. E. Bitterwolf, W. B. Scallorn, C. A. Weiss *J. Organomet. Chem.* **605**, 7–14 (2000).

[15] a) M. E. Krafft, I. L. Scott, R. H. Romero, S. Feibelmann, C. E. van Pelt *J. Am. Chem. Soc.* **115**, 7199–7207 (1993); b) X. Verdaguer, A. Moyano, M. A. Pericàs, A. Riera, V. Bernardes, A. E. Greene, A. Alvarez-Larena, J. F. Piniella *J. Am. Chem. Soc.* **116**, 2153–2154 (1994).

[16] (a) M. Yamanaka, E. Nakamura *J. Am. Chem. Soc.* **123**, 1703–1708 (2001); (b) M. A. Pericàs, J. Balsells, J. Castro, I. Marchueta, A. Moyano, A. Riera, J. Vázquez, X. Verdaguer *Pure Appl. Chem.* **74**, 167–174 (2002).

[17] (a) E. V. Banide, H. Mueller-Bunz, A. R. Manning, P. Evans, M. J. McGlinchey *Angew. Chem. Int. Ed.* **46**, 2907–2910 (2007); (b) S. A. Brusey, E. V. Banide, S. Dorrich, P. O. Donohue, Y. Ortin, H. Muller-Bunz, C. Long, P.Evans, M. J. McGlinchey *Organometallics* **28**, 6308–6319 (2009).

[18] C. Ferrer, J. Benet-Buchholz, A. Riera, X. Verdaguer *Chem. Eur. J.* **16**, 8340–8346 (2010).

[19] (a) S. Aime, L. Milone, R. Rossetti, P. L. Stanghellini *Inorg. Chim. Acta* **22**, 135–139 (1977); (b) E. Band, E. L. Muetterties *Chem. Rev.* **78**, 639–658 (1978); (c) X. Verdaguer, J. Vazquez, G. Fuster, V. Bernardes-Genisson, A. E. Greene, A. Moyano, M. A. Pericàs, A. Riera *J. Org. Chem.* **63**, 7037–7052 (1998).

[20] (a) S. L. Buchwald, F. A. Hicks in *Comprehensive Asymmetric Catalysis I-III*; Springer-Verlag: Berlin, 1999; 491–510; (b) S. E. Gibson, A. Stevenazzi *Angew. Chem., Int. Ed.* **42**, 1800–1810 (2003); (c) S. Laschat, A. Becheanu, T. Bell, A. Baro *Synlett* **2005**, 2547–2570.

[21] T. Sugihara, M. Yamaguchi, M. Nishizawa *Chem. Eur. J.* **7**, 1589–1595 (2001).

[22] V. Rautenstrauch, P. Megard, J. Conesa, W. Kuester *Angew. Chem., Int. Ed. Engl.* **29**, 1413–1416 (1999).

[23] (a) B. L. Pagenkopf, T. Livinghouse *J. Am. Chem. Soc.* **118**, 2285–2286 (1996); (b) D. B. Belanger, D. J. R. O'Mahony, T. Livinghouse *Tetrahedron Lett.* **39**, 7637–7640 (1998).

[24] R. Cabot, A. Lledó, M. Revés, A. Riera, X. Verdaguer *Organometallics* **26**, 1134–1142 (2007).

[25] D. G. Blackmond *Angew. Chem. Int. Ed.* **44**, 4302–4320 (2005).

[26] P. M. Breczinski, A. Stumpf, H. Hope, M. E. Krafft, J. A. Casalnuovo, N. E. Schore, *Tetrahedron* **55**, 6797–6812 (1999).

[27] F. Robert, A. Milet, Y. Gimbert, D. Konya, A. E. Greene, *J. Am. Chem. Soc.* **123**, 5396–5400 (2001).

[28] J. H. Schulte, R. Gleiter, F. Rominger, *Org. Lett.* **4**, 3301–3304 (2002).
[29] T. J. M. de Bruin, A. Milet, F. Robert, Y. Gimbert, A. E. Greene, *J. Am. Chem. Soc.* **123**, 7184–7185 (2001).
[30] T. J. M. de Bruin, C. Michel, K. Vekey, A. E. Greene, Y. Gimbert, A. Milet, *J. Organometal. Chem.* **691**, 4281–4288 (2006).
[31] J. Vázquez, S. Fonquerna, A. Moyano, M. A. Pericàs, A. Riera, *Tetrahedron: Asym.* **12**, 1837–1850 (2001).
[32] T. Fjermestad, M. A. Pericàs, F. Maseras, *J. Mol. Catal. A* **324**, 127–132 (2010).
[33] T. Fjermestad, M. A. Pericàs, F. Maseras*Chem. Eur. J.* **17**, 10050–10057 (2011).
[34] L. V. R. Boñaga, M. E. Maseras, Chem. Krafft, *Tetrahedron* **60**, 9795–9833 (2004).

3

Non Chiral Pauson-Khand Reaction

Takanori Shibata

3.1 History of Co-Mediated Pauson-Khand Reaction

In 1954,I. Wender reported the facile complexation of dicobalt octacarbonyl with alkynes along with the generation of carbon monoxide.[1] But it took about 20 years for the Pauson-Khand reaction to be published in a journal. In 1973, P. L. Pauson and I. U. Khand published the reaction of alkyne-dicobalt hexacarbonyl complex with norbornadiene at 60–70 °C in 1,2-dimethoxyethane (DME), which provided carbonylated tricyclic products (Scheme 3.1).[2]

The whole protocol can be categorized as [2+2+1] cycloaddition of alkyne, alkene and carbon monoxide. In the early stages, the intermolecular reactions of various alkynes and alkenes were examined, and synthetically useful cyclic enones were provided. But

The Pauson-Khand Reaction: Scope, Variations and Applications, First Edition. Edited by Ramon Rios Torres.

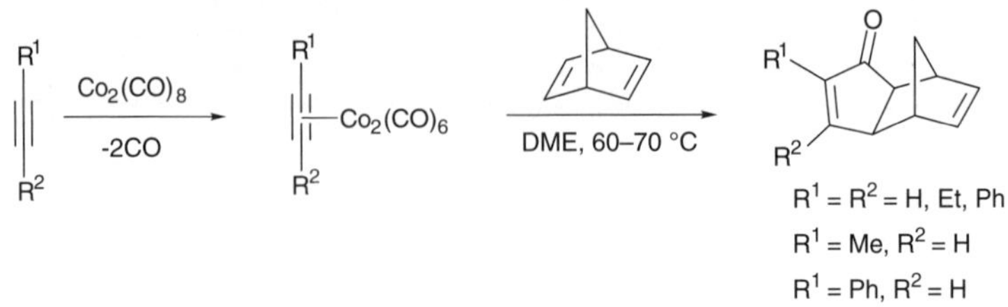

Scheme 3.1

the problem was the formation of possible regioisomers in the reaction of unsymmetrical alkynes and alkenes (Scheme 3.2), therefore, symmetrical and/or active alkenes, such as ethylene and norbornene, were conventional substrates, and excess amounts of the alkenes were used for the high yield.

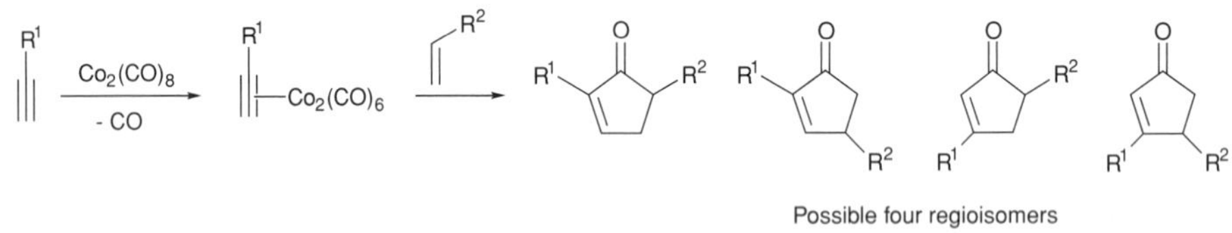

Scheme 3.2

In contrast, in the intramolecular reaction of enynes with relatively short tethers, the orientation of alkyne and alkene are uniquely determined, and bicyclic cyclopentenones can be obtained (Scheme 3.3). Moreover, intramolecular reactions are generally more efficient than intermolecular ones: when the tether between alkyne and alkene includes three or four atoms ($n = 1$ or 2), the yield is generally high because of the formation of a five- or six-membered ring.

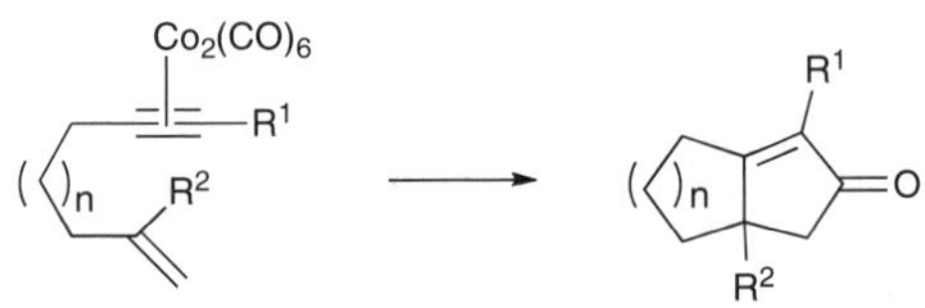

Scheme 3.3

Actually, the Pauson-Khand Reaction gained prominence because it was used as a key reaction in natural product syntheses. Magnus was a pioneer and reported the pioneering work for the synthesis of the multicyclic compound, *dl*-coriolin (Scheme 3.4).[3]

3.2 Mechanism of the Pauson-Khand Reaction

Magnus first proposed the mechanism of this novel carbonylative coupling, and it has been well accepted (Scheme 3.5).[4] The Alkyne-dicobalt hexacarbonyl complex possesses a

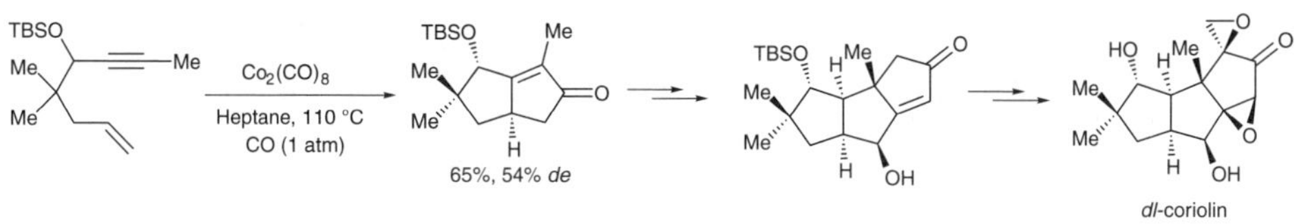

Scheme 3.4

trigonal pyramidal bimetallic structure. Coordination of the alkene proceeds to give a π-complex **A** along with dissociation of one of the CO ligands. The coordination of the generated carbon monoxide induces alkene insertion, which provides complex **B**. The following migratory insertion of carbon monoxide gives the carbonyl complex **C**. Reductive elimination affords the cyclopentenone, along with the generation of the dicobalt hexacarbonyl complex. In the course of various mechanistic studies, complex **A** has been isolated and characterized, but the structures of complexes **B** and **C** have not been fully characterized yet.

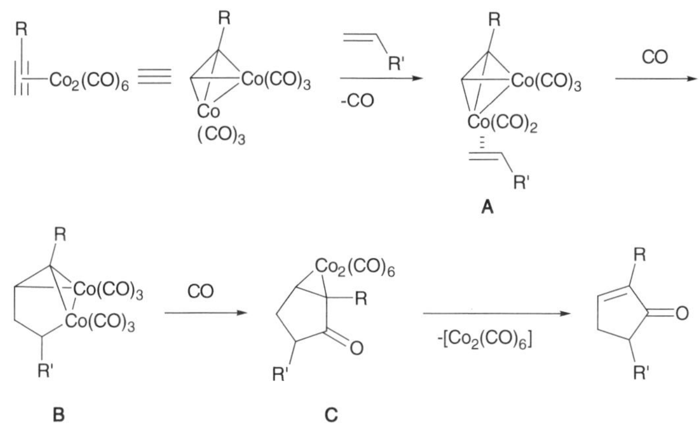

Scheme 3.5

Based upon the above mechanism, the dicobalt carbonyl complex is regenerated, and the catalytic reaction can be achieved by the supply of carbon monoxide. However, the development of the catalytic Pauson-Khand reaction has been difficult to achieve, probably because the coordinatively unsaturated $Co_2(CO)_6$ complex is unstable and is readily transformed into a more stable complex by oligomerization.

3.3 An Early Example of Catalytic Reaction

In the first paper on the Pauson-Khand reaction, the catalytic reaction was already disclosed. Acetylene-dicobalt hexacarbonyl was used as a catalyst precursor, and the reaction of norbornene was examined under the mixed gas of acetylene and carbon monoxide (Scheme 3.6).[2]

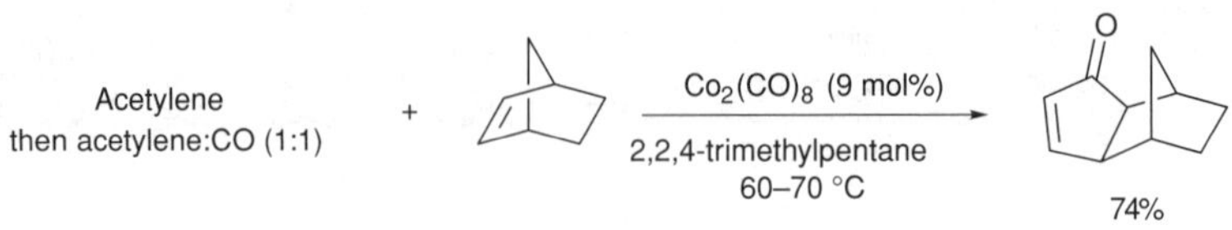

Scheme 3.6

Another example required more harsh reaction conditions (Scheme 3.7). In the presence of a catalytic amount of $Co_2(CO)_8$, alkyl acetylene reacted under the highly pressurized condition of ethylene and carbon monoxide at a high temperature.[5]

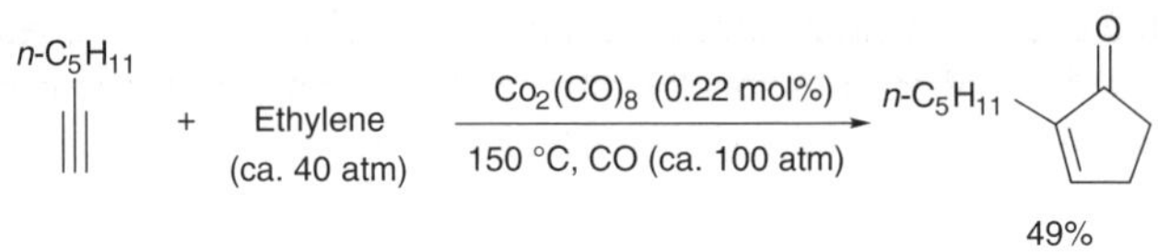

Scheme 3.7

The substrate scope of the above two examples was narrow, and the reactions were practically premature, but they showed the possibility of catalytic reaction.

3.4 Catalytic Reactions by Aid of Additives

The "appropriate" stabilization of $Co_2(CO)_6$, which possibly forms in the reaction pathway, was considered important for the achievement of a catalytic reaction. Displacement of the CO ligand with triarylphosphine generally stabilizes the cobalt carbonyl complex, but $Co_2(CO)_6(PPh_3)_2$ is too stable and has little catalytic activity.

The avenue for the catalytic Pauson-Khand reaction was brought about by another reaction using $Co_2(CO)_8$: Iwasawa reported the $Co_2(CO)_8$-mediated transformation of alkynylcyclopropanols into cyclopentenones.[6] He further examined the catalytic version of this rearrangement and found that bulky phosphite was effective as an additive (Scheme 3.8).[7]

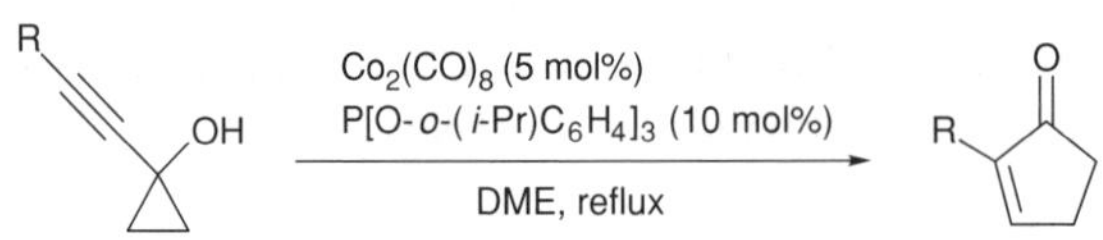

Scheme 3.8

He speculated on the reaction mechanism as follows: the carbon–carbon bond cleavage of the cyclopropane ring is an initial step, which would be induced by the electron-donating effect of the hydroxy group (Scheme 3.9).[8] The mechanism of the second half resembles that of the Pauson-Khand reaction (Scheme 3.5).

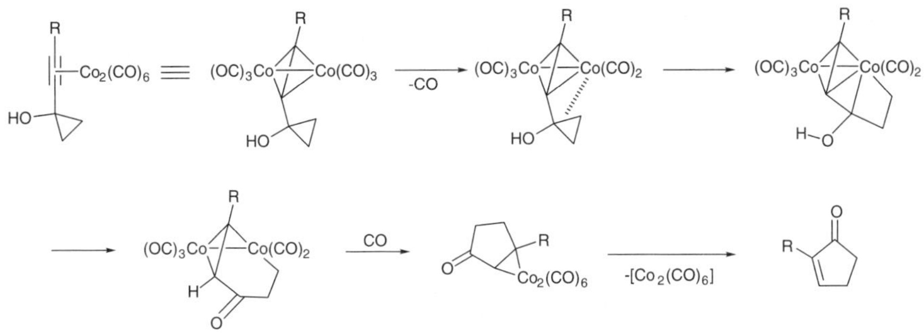

Scheme 3.9

Soon after this publication, Jeong used triphenylphosphite as an additive for $Co_2(CO)_8$ and achieved the first practical and catalytic Pauson-Khand reaction.[9] Various 1,6-enynes were transformed into the corresponding bicyclic cyclopentenones at 120 °C under a slightly pressurized condition of carbon monoxide (Scheme 3.10).

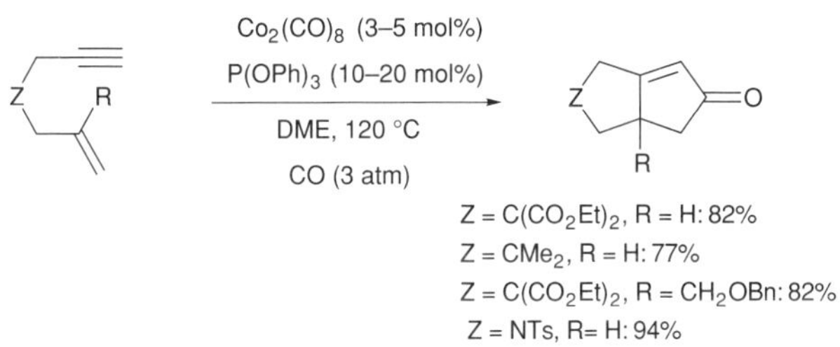

Scheme 3.10

Later, triphenylphosphine was found to be an effective additive, but the ratio against $Co_2(CO)_8$ was very important (Scheme 3.11).[10,11] The heptacarbonyl (triarylphosphine)dicobalt complex, where only one of CO ligands was displaced with triphenylphosphine, was an effective catalyst, but the hexacarbonyl bis(triarylphosphine)dicobalt complex had almost no catalytic activity, as previously mentioned above.

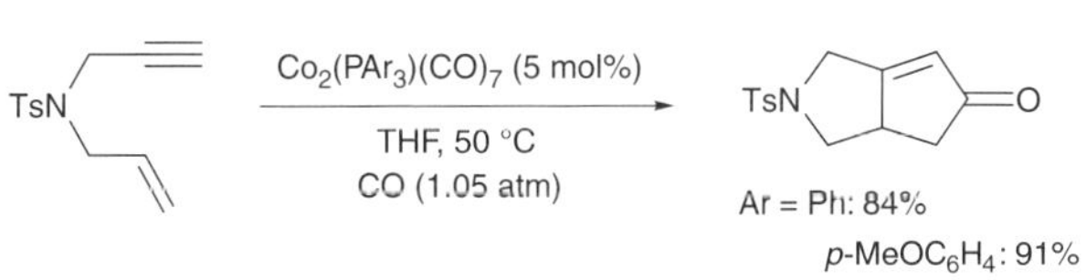

Scheme 3.11

Hashimoto and Saigo reported tributyl phosphine sulfide as an efficient ligand for the catalytic reaction (Scheme 3.12). The effect of sulfur is unclear, but triphenylphosphine oxide was less effective. The carbonylative reaction of various enynes proceeded at lower temperature (70 °C) under ambient pressure of carbon monoxide to give bicyclic enones in a good to high yield.[12] An intermolecular reaction also proceeded under the same reaction conditions to give a tricyclic enone in high yield.

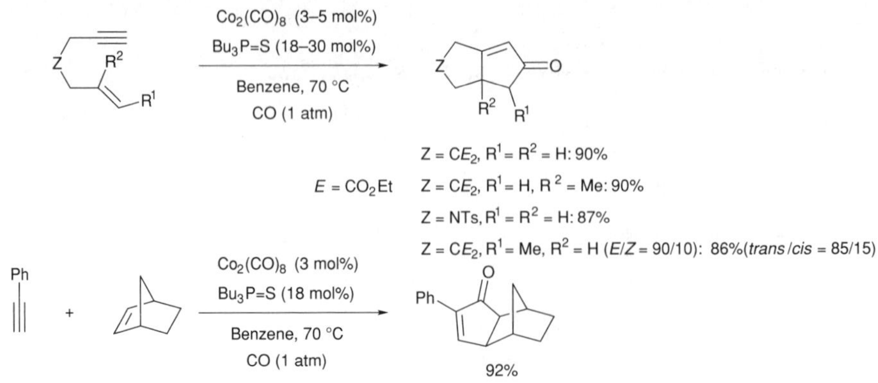

Scheme 3.12

Chen and Yang reported thiourea as an additive (Scheme 3.13). The $Co_2(CO)_8$-catalyzed reaction proceeded smoothly in the presence of tetramethylthiourea (TMTU) under ambient pressure of carbon monoxide.[13] It is noteworthy that 1,7-enynes (n = 2) were also good substrates, and a 5,6-bicyclic system could be constructed. An ester-tethered enyne was transformed into a bicyclic lactone. An example of the intermolecular reaction of phenylacetylene with norbornene was also listed.

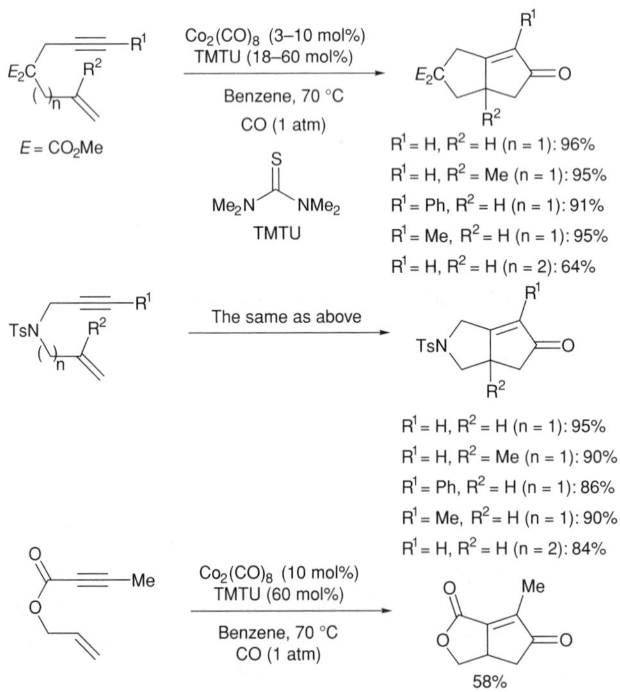

Scheme 3.13

The structures of the true catalysts were unclear, but the above-mentioned additives could work as ligands for the cobalt complex by displacement of the CO ligand. Sugihara and Yamaguchi disclosed a different type of additive: in the presence of a catalytic amount of 1,2-dimethoxyethane or water, the Pauson-Khand reaction proceeded (Scheme 3.14).[14] A pressurized condition of carbon monoxide (7 atm) was required, but 2–3 mol % of $Co_2(CO)_8$ realized a good to high yield. As for the intermolecular reaction, mono-substituted acetylenes were good substrates.

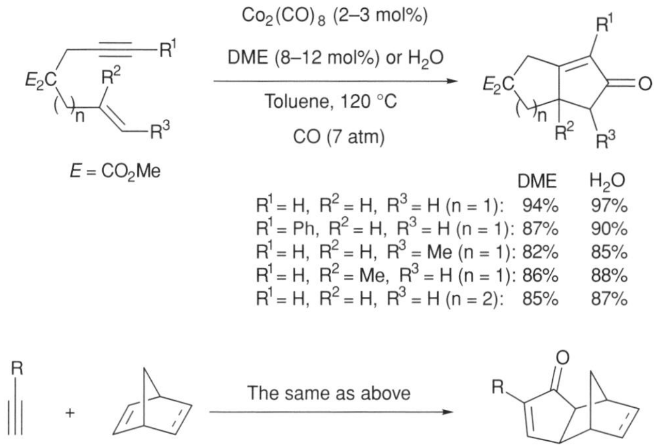

	DME	H₂O
R^1 = H, R^2 = H, R^3 = H (n = 1):	94%	97%
R^1 = Ph, R^2 = H, R^3 = H (n = 1):	87%	90%
R^1 = H, R^2 = H, R^3 = Me (n = 1):	82%	85%
R^1 = H, R^2 = Me, R^3 = H (n = 1):	86%	88%
R^1 = H, R^2 = H, R^3 = H (n = 2):	85%	87%

		DME	H_2O
With norbornene:	R = Ph:	98%	97%
	R = n-C_8H_{17}:	100%	98%
With norbornadiene	R = Ph:	92%	90%

Scheme 3.14

Pérez-Castells realized a catalytic reaction in the presence of molecular sieves (zeolite) without CO gas: molecular sieve 4A was pretreated under CO atmosphere at 200 °C for the absorption of carbon monoxide, and was used as a CO source for the $Co_2(CO)_8$-catalyzed reaction (Scheme 3.15).[15] The stereospecific reaction proceeded, and the enyne with Z-olefin moiety provided carbonylated cycloadduct, where the geometry of the methoxy group and hydrogen at the ring-fused position was *trans*. *Ortho*-phenylene tethered 1,7-enyne with a hydroxy group was also a good substrate.

3.5 Catalytic Reaction Using *in-situ* Generated Low-Valent Cobalt Complex

The catalytic Pauson-Khand reaction is generally difficult to achieve, probably because of the instability of the low-valent cobalt carbonyl complex. Actually, several catalytic examples were reported, where low-valent cobalt carbonyl complexes were prepared *in-situ* from relatively stable cobalt complexes.

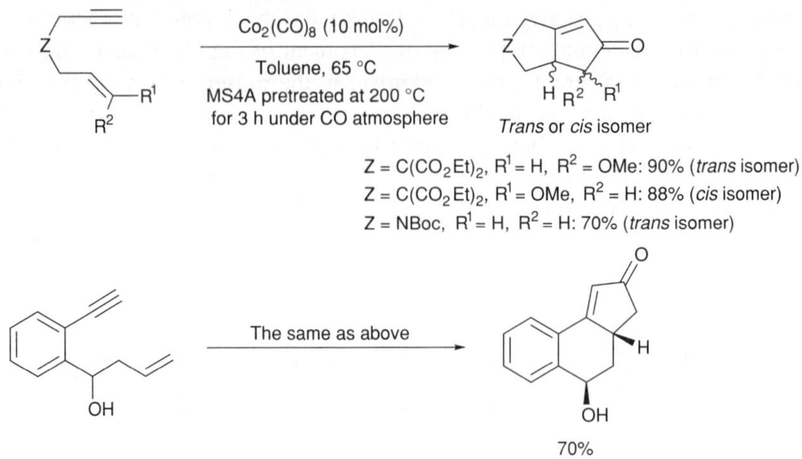

Scheme 3.15

Chung reported two protocols using non-carbonyl cobalt complexes as carbonyl complex precursors. (Indenyl)Co(octa-1,5-diene) was used as relatively stable Co(I) complex.[16] Monosubstituted acetylenes even with a hydroxy group were good substrates and just 1 mol% cobalt catalyst realized an excellent yield. As for disubstituted acetylenes, the yield was moderate when using 5 mol% cobalt catalyst (Scheme 3.16).

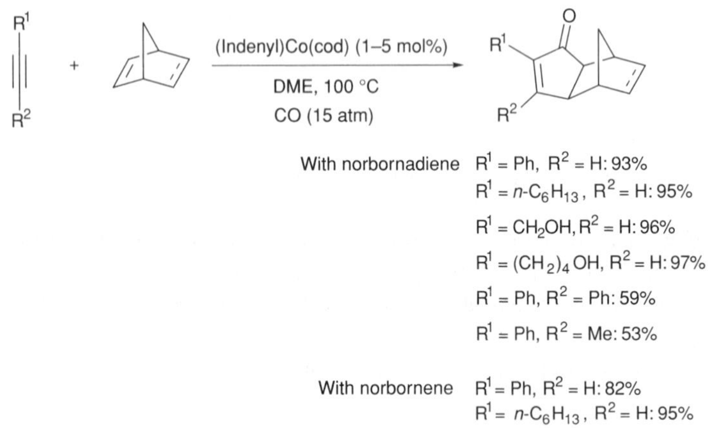

Scheme 3.16

Another approach was the use of the more stable Co(II): $Co(acac)_2$ was reduced by $NaBH_4$ for the preparation of an active catalyst.[17] Compared with $Co_2(CO)_8$, the cobalt carbonyl complex precursors are easy to handle, but a pressurized condition of CO is required. Various monosubstituted acetylenes were submitted, and ethyl propiolate also reacted with norbornadiene, yet in a low yield (Scheme 3.17). In the reaction of norbornene, a catalytic reaction of phenylacetylene quantitatively proceeded without $NaBH_4$.

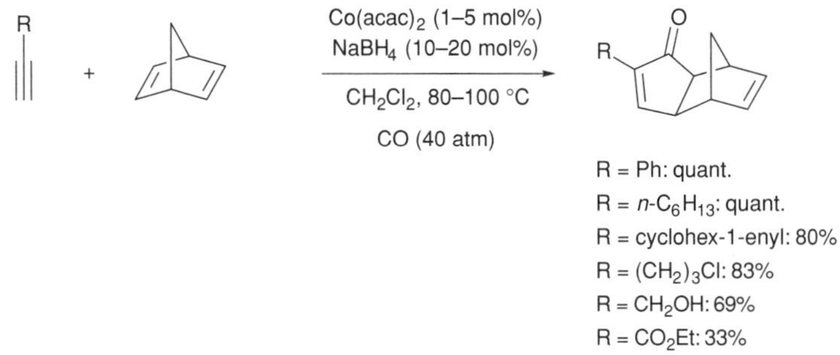

Scheme 3.17

In the above two reports, the intermolecular reaction of alkyne with norbornadiene or norbornene was mainly stated, but the intramolecular reaction of 1,6-enynes with an unsubstituted alkyne terminus also proceeded.

Livinghouse considered that impurities in $Co_2(CO)_8$ would induce the formation of oligomers as an inactive complex, and paid attention to the pure grade of $Co_2(CO)_8$. He prepared highly pure $Co_2(CO)_8$ by recrystallization using HPLC grade hexane or sublimation at room temperature under reduced pressure (50 mmHg), and submitted it to the reaction. In his first report,[18] a pure $Co_2(CO)_8$-catalyzed reaction was achieved by aid of photo-irradiation at 50–55 °C (Scheme 3.18).

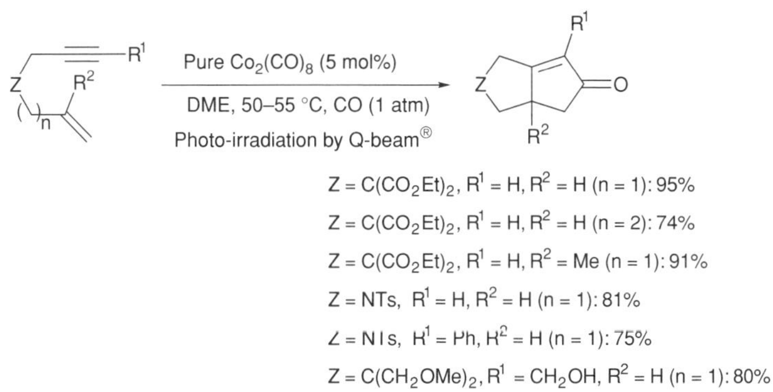

Scheme 3.18

In the following report,[19] he realized the catalytic reaction by the strict control of the reaction temperature (60 °C) without the aid of photo-irradiation (Scheme 3.19). The stereospecific transformation of an enyne with *Z*-olefinic moiety to the corresponding cycloadduct could also be possible.

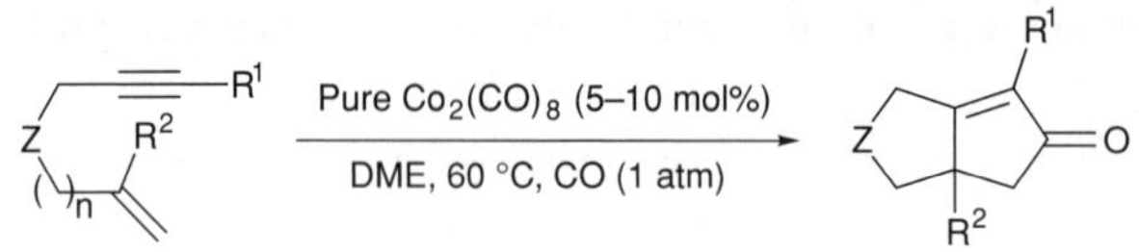

Z = $C(CO_2Et)_2$, R^1 = H, R^2 = H (n = 1): 83%
Z = $C(CO_2Et)_2$, R^1 = H, R^2 = H (n = 2): 77%
Z = $C(CO_2Et)_2$, R^1 = H, R^2 = Me (n = 1): 85%
Z = NTs, R^1 = H, R^2 = H (n = 1): 86%
Z = NTs, R^1 = Me, R^2 = H (n = 1): 78%

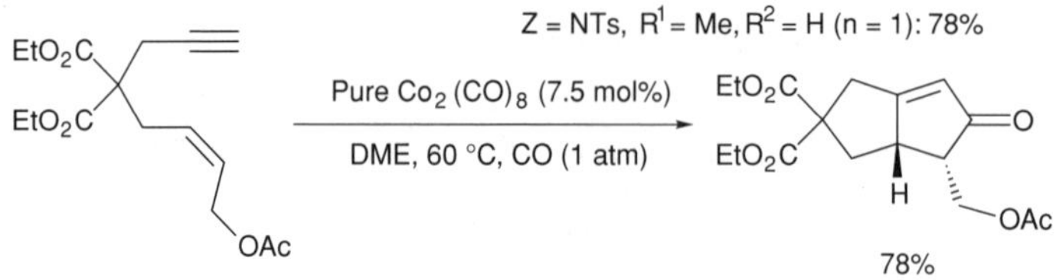

Scheme 3.19

Compared with $Co_2(CO)_8$, the $Co_2(CO)_6$-alkyne complex is generally more stable, and can be purified even by silica-gel chromatography. Livinghouse used $Co_2(CO)_6$-2-methylbut-3-yn-2-ol as a pre-catalyst (Scheme 3.20). The complex was treated with triethylsilane and cyclohexylamine for the dissociation of the alkyne, and was submitted to the reaction. The reaction proceeded smoothly under the ambient pressure of CO atmosphere.[20] Also in this case, a stereospecific reaction of an enyne with *Z*-olefinic moiety was listed.

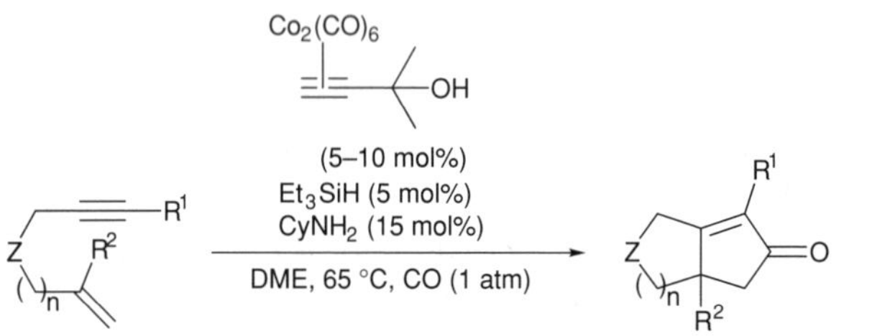

Z = $C(CO_2Et)_2$, R^1 = H, R^2 = H (n = 1): 92%
Z = $C(CO_2Me)_2$, R^1 = H, R^2 = H (n = 2): 86%
Z = $C(CO_2Et)_2$, R^1 = H, R^2 = Me (n = 1): 95%
Z = NTs, R^1 = H, R^2 = H (n = 1): 77%
Z = NTs, R^1 = Me, R^2 = H (n = 1): 92%

Scheme 3.20

Based on the same concept, Krafft used the $Co_2(CO)_6$-enyne complex. The transfer of $Co_2(CO)_6$ to other enynes as substrates readily occurred without the aid of silane, and could operate as a catalyst (Scheme 3.21). In some cases, amine was not needed.[21]

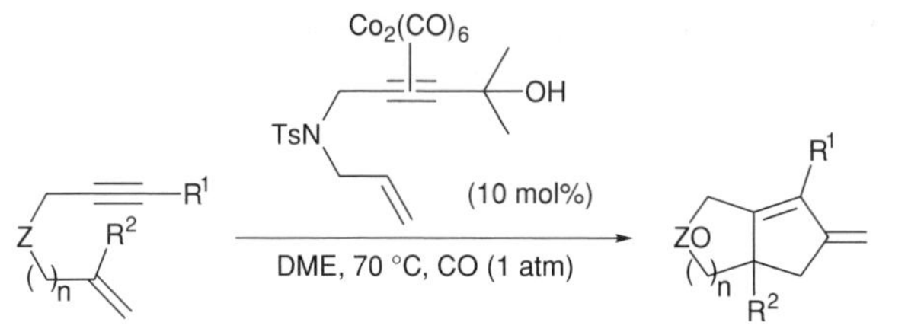

Without $CyNH_2$ Z = $C(CO_2Et)_2$, R^1 = H, R^2 = H (n = 1): 78%
Z = $C(CO_2Me)_2$, R^1 = H, R^2 = Me (n = 1): 84%
With $CyNH_2$ (20 mol%) Z = $C(CO_2Et)_2$, R^1 = H, R^2 = H (n = 2): 81%
Z = NTs, R^1 = H, R^2 = H (n = 1): 94%
Z = $CHCH_2OTBS$, R^1 = H, R^2 = H (n = 1): 81%

Scheme 3.21

3.6 Catalytic Reaction Using Multinuclear Cobalt Carbonyl Catalysts

The problem for catalytic reactions using $Co_2(CO)_8$ is the formation of a stable and inactive oligomeric complex. Chung used stable tetrametallic complex $Co_4(CO)_{12}$ as a catalyst, which would *in situ* be transformed to active dimetallic complex (Scheme 3.22).[22] Although the harsh reaction conditions of the high temperature (150 °C) and high pressure of carbon monoxide (10 atm) were required, just 0.5–1 mol% o the catalyst realized a high yield both in inter- and intramolecular reactions.

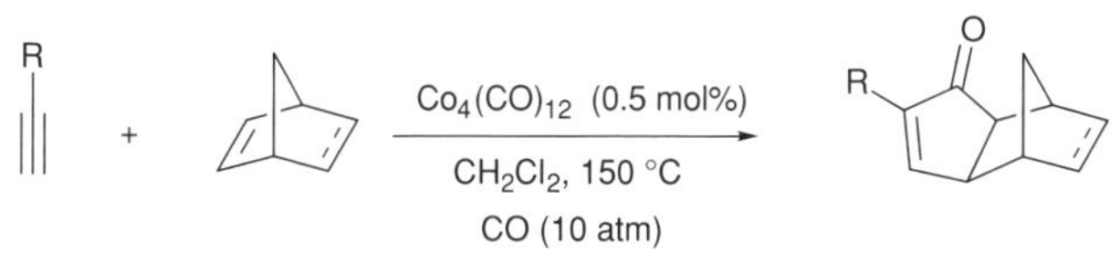

With norbornadiene R = Ph: 97%
R = n-C_4H_9: 98%
R = $(CH_2)_3Cl$: 71%
R = cyclohex-1-enyl: 75%
R = CH_2OH: 80%

With norbornene R = Ph: 93%
R = n-C_4H_9: 100%

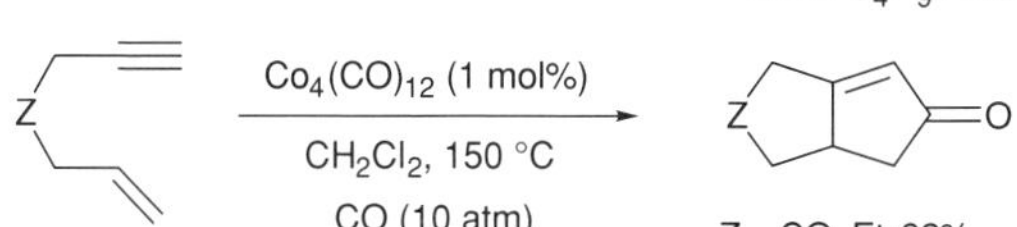

Z = CO_2Et: 92%
Z = NTs: 87%

Scheme 3.22

Sugihara used tri-nuclear $Co_3(CO)_9(\mu^3\text{-CH})$ as a stable catalyst, which can be recognized as the carbon-analogue of $Co_4(CO)_{12}$.[23] The reaction conditions were a little milder than those of the $Co_4(CO)_{12}$ -catalyzed reaction, and intra- and intermolecular reactions proceeded to give the corresponding cyclopentenones (Scheme 3.23).

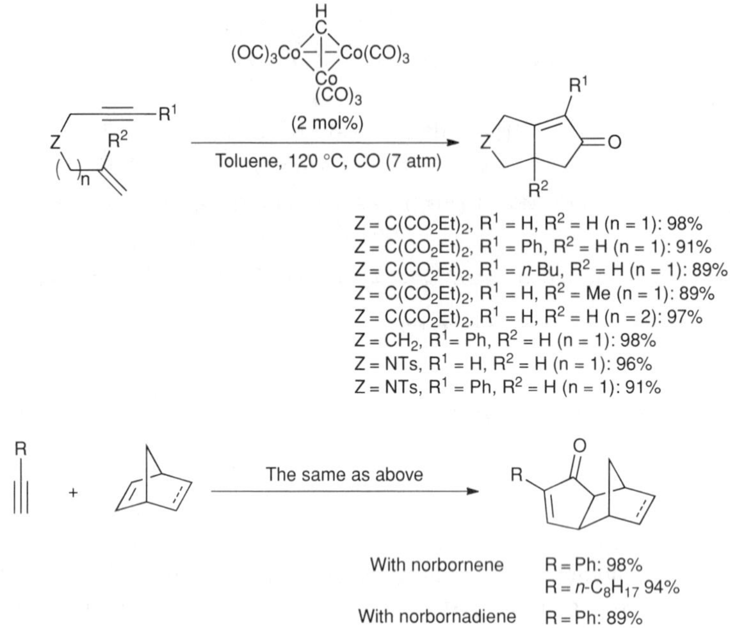

Scheme 3.23

Later, Krafft reported a $Co_4(CO)_{12}$–catalyzed reaction under much milder reaction conditions (70 °C, CO (1 atm)) by using excess amounts of cyclohexylamine (Scheme 3.24).[24] 1,6-enynes were transformed into bicyclic enones in a good yield.

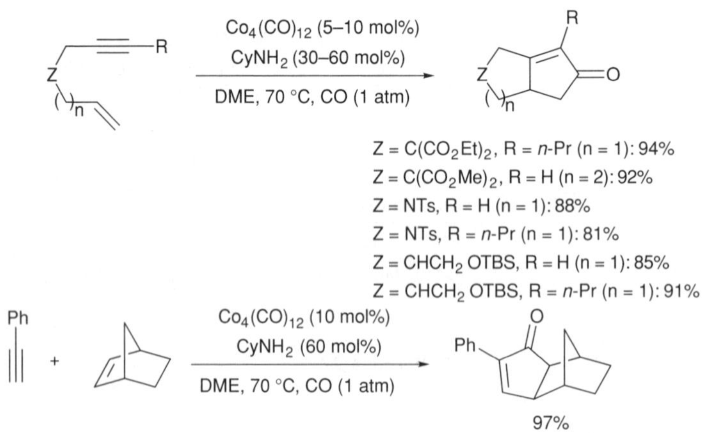

Scheme 3.24

Groth used a microwave in place of heating.[25] The reaction concluded within 5 min at 100 °C in the presence of relatively large amounts of $Co_2(CO)_8$ along with a stoichiometric amount of cyclohexylamine (Scheme 3.25). The intermolecular reaction of several alkynes gave acceptable yields, but only an example of an intramolecular reaction with a moderate yield (46%) was listed.

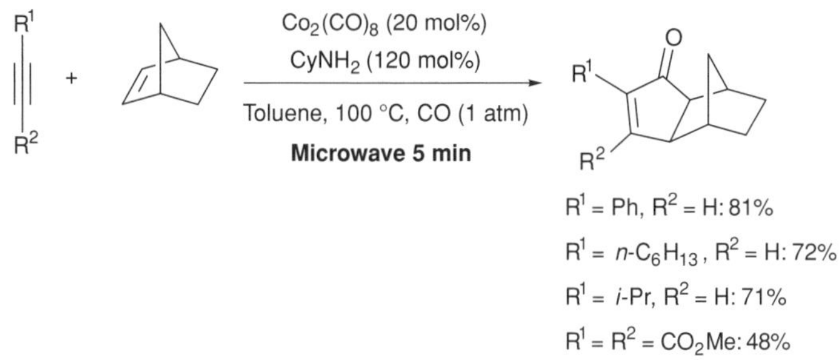

Scheme 3.25

3.7 Catalytic Reaction Using Heterogeneous Catalysts

Several examples of co-catalyzed Pauson-Khand reactions in heterogeneous conditions are listed in this section. Chung disclosed the pioneering work by using metallic cobalt on mesoporous silica (SBA-15) with 9–10 wt.% cobalt content.[26] A high reaction temperature (130 °C) and high pressure of carbon monoxide (20 atm) were required, but an intramolecular reaction of enynes proceeded smoothly to give carbonylative cycloadducts in a high to excellent yield (Scheme 3.26). Moreover, three-times recycling of the heterogeneous catalyst was ascertained without loss of catalytic activity. This catalyst was less effective for the intermolecular reaction of norbornene derivatives with phenylacetylene (11–36% yield).

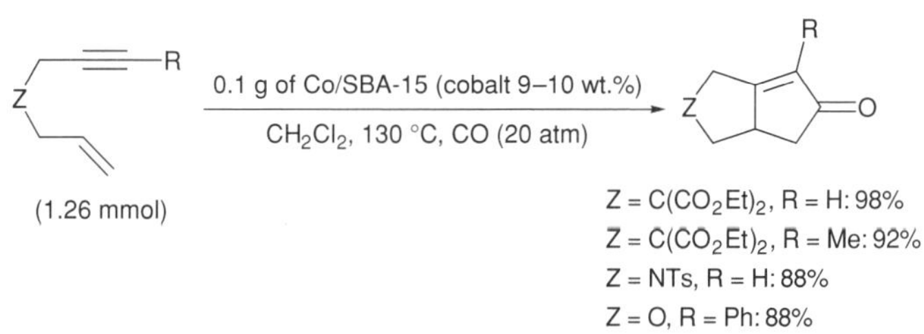

Scheme 3.26

Chung further realized a more convenient and less expensive heterogeneous system using cobalt on charcoal.[27] The reaction conditions were still harsh, but nine-times recycling without loss of catalytic activity could be possible, moreover, this catalyst could also operate in the intermolecular reaction of norbornadiene with terminal alkynes (Scheme 3.27).

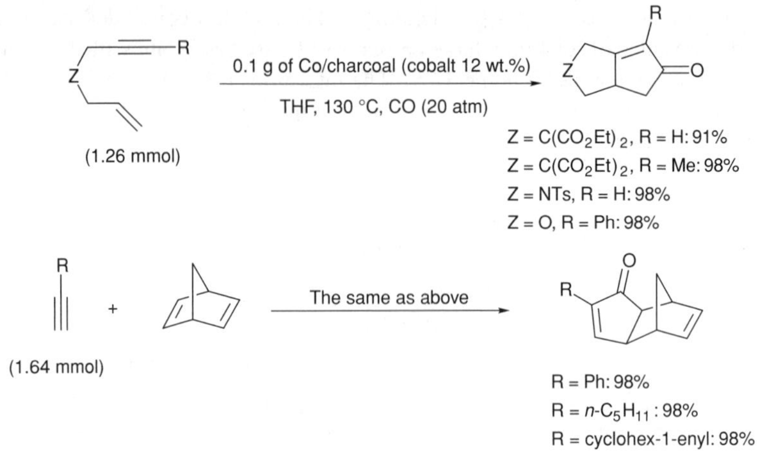

Scheme 3.27

Chung disclosed colloidal cobalt nanoparticles and the cobalt nanoparticles on charcoal as other examples of reusable heterogeneous catalysts. The former catalyst was prepared by stirring $Co_2(CO)_8$, oleic acid, trioctylphosphine and dioctyl ether under the reflux condition, and was precipitated by adding ethanol to the cooled reaction mixture.[27] Refluxing the prepared colloidal cobalt nanoparticles with charcoal provided the latter catalyst.[28] Intra- and intermolecular reaction proceeded under a lower CO pressure of 5 atm but at a high temperature (130 °C) by using these heterogeneous catalysts.

As for other choices of heterogeneous catalyst, Leitner used Raney cobalt for the catalytic reaction (Scheme 3.28).[29] It is commercially available and readily accessible, but requires a high pressure of carbon monoxide (23 atm) and high reaction temperature (130 °C), and catalytic activity was significantly decreased by twice recycling.

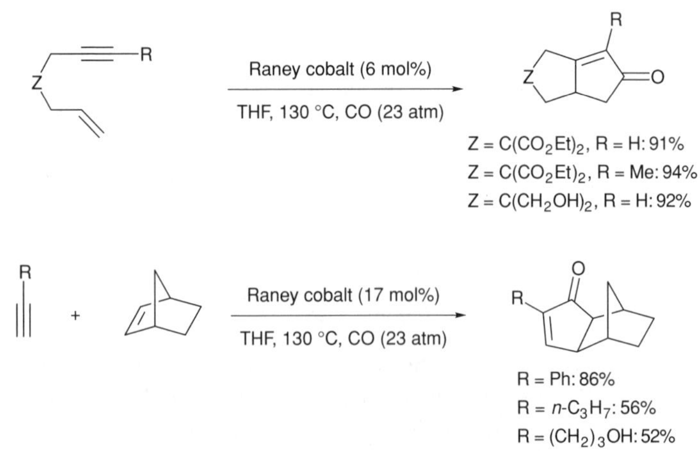

Scheme 3.28

Leitner reported cobalt nanoparticles stabilized on poly(ethylene glycol) (PEG).[30] They were prepared by thermal decomposition of $Co_2(CO)_8$ in PEG. The heterogeneous catalyst operated as an efficient catalyst for intra- and intramolecular reactions under the harsh conditions (Scheme 3.29). Also in this case, a significant decrease of catalytic activity was observed by twice recycling.

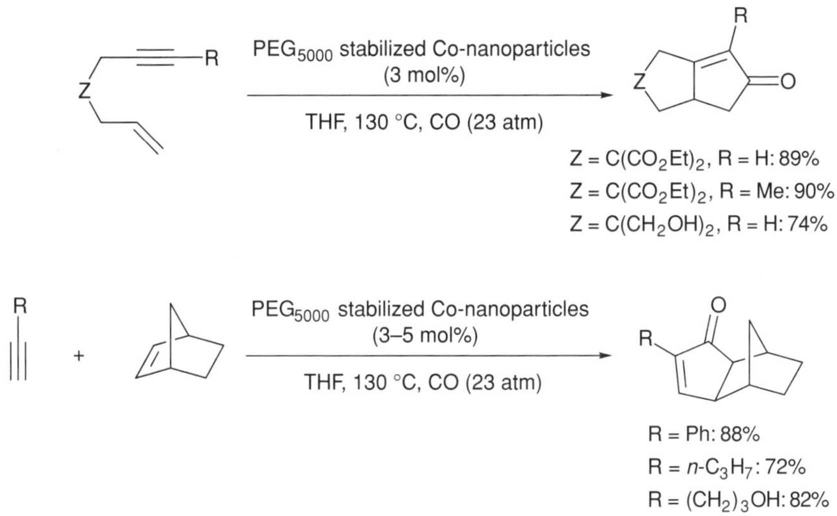

Scheme 3.29

The polymer-bound triphenylphosphines could immobilize $Co_2(CO)_8$ to give heterogeneous catalysts. For example, polystyrene cross-linked with 1% divinylbenzene[31] and polystyrene functionalized by second or third generation dendron[32] were reported as carriers, and the heterogeneous catalysts were used in the Pauson-Khand reaction. The reaction surely proceeded, but was far from practically useful, because the yield was generally low to moderate and the substrate scope was narrow.

3.8 Catalytic Reaction in Other Than Conventional Solvents

Hydrocarbons such as benzene and toluene, and etherate solvents such as THF and 1,2-dimethoxyethane are the usual solvents for the catalytic Pauson-Khand reaction. Jeong used a supercritical fluid for the catalytic Pauson-Khand reaction.[33] $Co_2(CO)_8$ was submitted under 30 atm pressure of CO (at 23 °C) in supercritical carbon dioxide (final pressure was 110–120 atm at 36–39 °C), and the reaction mixture was heated at 90–95 °C. Intra- and intermolecular reactions proceeded to give cyclopentenones (Scheme 3.30). The reaction of nitrogen-tethered enynes did not proceed, because they were not sufficiently soluble in supercritical carbon dioxide

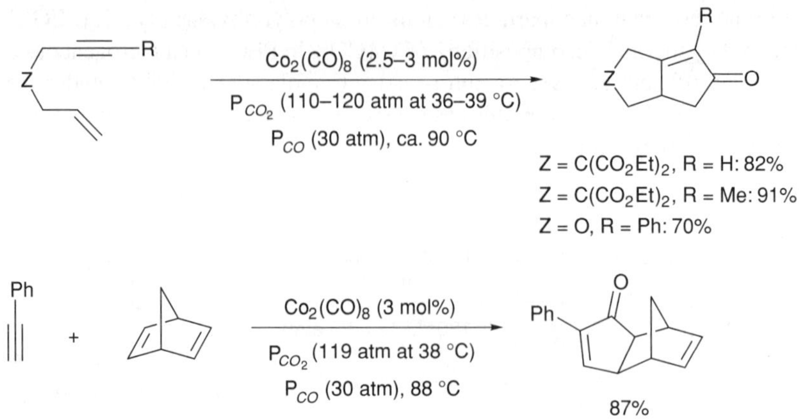

Scheme 3.30

Jeong further reported an intermolecular reaction in supercritical ethylene, which was used as both reaction media and reactant.[34] The tetranuclear complex $[Co_4(CO)_{11}\{P(OPh)_3\}]$ worked well, and various mono-substituted alkynes were submitted to the reaction with ethylene (Scheme 3.31).

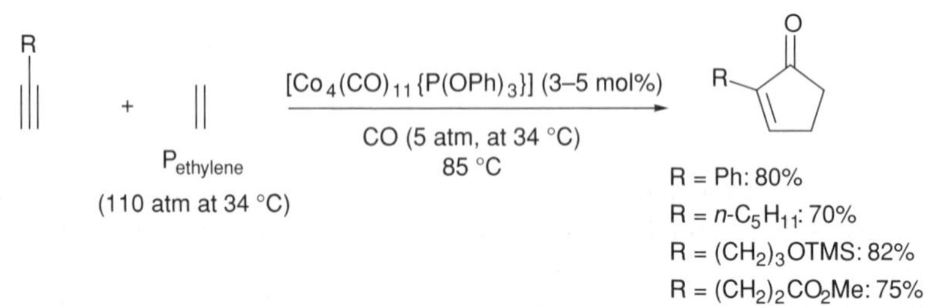

Scheme 3.31

The above mentioned colloidal cobalt nanoparticles[27] could operate as a catalyst even in water, though a higher pressure of CO (20 atm) and a high reaction temperature (130 °C) were required for a high yield.[35] In the same reaction conditions, $Co_2(CO)_8$ had little catalytic activity.

In contrast, $Co_4(CO)_{12}$ was an efficient catalyst in water under much milder reaction conditions.[36] The addition of surfactant (triton X) was the key to success, and the intramolecular reaction of various enynes proceeded even under the ambient pressure of CO at 70 °C (Scheme 3.32).

$Co_2(CO)_8$ acted as a catalyst even in an ionic liquid, such as 1-butyl-2,3-dimethylimidazolium salt, which is known to be a non-volatile and recyclable solvent.[37]

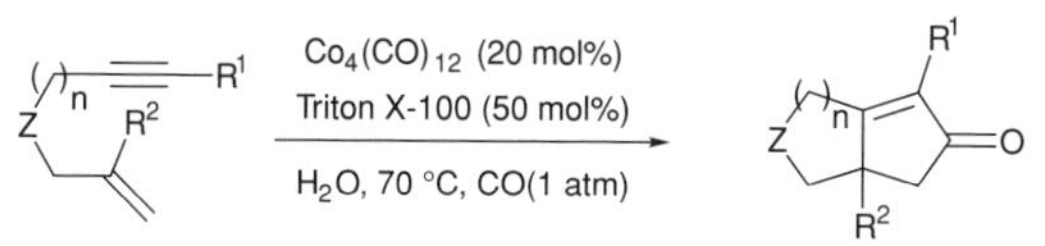

$Z = C(CO_2Et)_2$, R^1 = Me, R^2 = H (n = 1): 79%
$Z = C(CO_2Et)_2$, R^1 = Ph, R^2 = H (n = 1): 85%
$Z = C(CO_2Me)_2$, R^1 = Me, R^2 = Et (n = 1): 86%
$Z = C(CO_2Et)_2$, R^1 = Me, R^2 = H (n = 2): 76%
$Z = CH(CO_2Et)$, R^1 = Me, R^2 = H (n = 1): 78%
Z = NTs, R^1 = *n*-Pr, R^2 = H (n = 1): 84%
Z = O, R^1 = Ph, R^2 = H (n = 1): 70%
$Z = CH_2$, $R^1 = (CH_2)_2SEt$, R^2 = H (n = 1): 68%

Scheme 3.32

But a high pressure of CO was needed, substrate scope was narrow, and there was no comment on the recyclability of the reaction media.

3.9 Intramolecular Reaction of Carbodiimides with Alkynes

Among cumulated compounds, carbodiimide was used in a hetero Pauson-Khand reaction. Mukai disclosed an intramolecular reaction of *ortho*-phenylene tethered alkyne-carbodiimide for the synthesis of tricyclic heterocycles.[38] The above mentioned $Co_2(CO)_8$-TMTU system could operate as a catalyst (Scheme 3.33). The yield was generally moderate, but this protocol was used as a key reaction for the synthesis of indole alkaloid (±)-physostigmine.

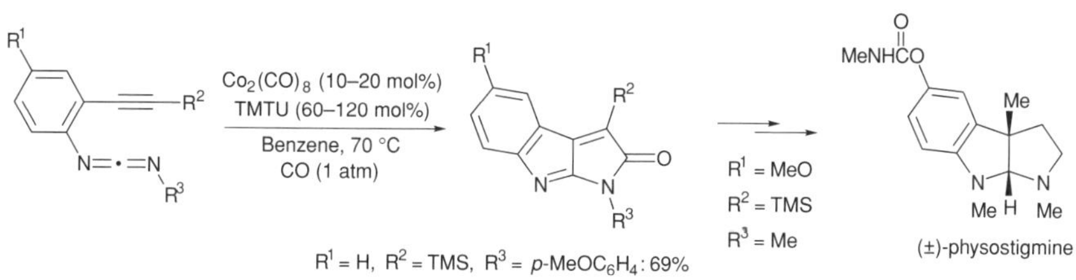

R^1 = H, R^2 = TMS, R^3 = *p*-$MeOC_6H_4$: 69%
R^1 = H, R^2 = *n*-Pr, R^3 = *p*-$MeOC_6H_4$: 66%
R^1 = H, $R^2 = (CH_2)_2OTBS$, R^3 = *p*-$MeOC_6H_4$: 48%
R^1 = Me, R^2 = TMS, R^3 = *p*-$MeOC_6H_4$: 54%
R^1 = MeO, R^2 = TMS, R^3 = *p*-$MeOC_6H_4$: 54%
R^1 = Cl, R^2 = TMS, R^3 = *p*-$MeOC_6H_4$: 52%
R^1 = MeO, R^2 = TMS, R^3 = Me: 55%

Scheme 3.33

References

[1] H. W. Sternberg, H. Greenfield, R. A. Friedel, J. Wotiz, R. Markby, and I. Wender, A new type of of metallo-organic complex derived from dicobalt octacarbonyl and acetylenes, *J. Am. Chem. Soc.*, **76**, 1457–1458 (1954).

[2] I. U. Khand, G. R. Knox, P. L. Pauson, and W. E. Watts, M. I. Foremann Reaction of acetylenehexacarbonyldicobalt complexes, $(R^1C_2R^2)Co_2(CO)_6$, with norbornene and its derivatives, *J. Chem. Soc., Perkin Trans. 1*, 977–981 (1973).

[3] C. Exon and P. Magnus, Stereoselectivity of intramolecular dicobalt octacarbonyl alkene-alkyne cyclizations: short synthesis of *dl*-coriolin, *J. Am. Chem. Soc.*, **105**, 2477–2478 (1983).

[4] P. Magnus and L. M. Principe, Origins of 1,2- and 1,3-stereoselectivity in dicobaltoctacarbonyl alkene-alkyne cyclizations for the synthesis of substituted bicyclo[3.3.0]octenones, *Tetrahedron Lett.*, **26**, 4851–4854 (1985).

[5] V. Rautenstrauch, P. Mégard, J. Conesa, and W. Küster, 2-Pentylcyclopent-2-en-1-one by catalytic Pauson-Khand reaction, *Angew. Chem. Int. Ed. Engl.*, **29**, 1413–1416 (1990).

[6] N. Iwasawa, A novel rearrangement of 1-alkynylcyclopropanol to 2-cyclopenten-1-one via dicobalt hexacarbonyl complex, *Chem. Lett.*, 21, 473–476 (1992).

[7] N. Iwasawa and T. Matsuo, Rearrangement of 1-alkynylcyclopropanol to 3-substituted 2-cyclopentenone promoted by a catalytic amount of octacarbonyldicobalt-triphenylphosphite, *Chem. Lett.*, **22**, 997–1000 (1993).

[8] N. Iwasawa, T. Matsuo, M. Iwamoto, and T. Ikeno, Rearrangement of 1-(1-alkynyl)cyclopropanols to 2-cyclopentenones via their hexacarbonyldicobalt complexes. A new use of alkyne$-Co_2(CO)_6$ complexes in organic synthesis, *J. Am. Chem. Soc.*, **120**, 3903–3914 (1998).

[9] N. Jeong, S. H. Hwang, Y. Lee, and Y. K. Chung, Catalytic version of the intramolecular Pauson-Khand reaction, *J. Am. Chem. Soc.*, **116**, 3159–3160 (1994).

[10] A. C. Comely, S. E. Gibson, A. Stevenazzi, and N. J. Hales, New stable catalysts of the Pauson-Khand annelation, *Tetrahedron Lett.*, **42**, 1183–1185 (2001).

[11] S. E. Gibson, C. Johnstone, and A. Stevenazzi, A stable catalyst of the Pauson-Khand annelation, *Tetrahedron*, **58**, 4937–4942 (2002).

[12] M. Hayashi, Y. Hashimoto, Y. Yamamoto, J. Usuki, and K. Saigo, Phosphane sulfide/octacarbonyldicobalt-catalyzed Pauson-Khand reaction under an atmospheric pressure of carbon monoxide, *Angew. Chem., Int. Ed.*, **39**, 631–633 (2000).

[13] Y. Tang, L. Deng, Y. Zhang, G. Dong, J. Chen, and Z. Yang, Tetramethyl thiourea/$Co_2(CO)_8$-catalyzed Pauson-Khand reaction under balloon pressure of CO, *Org. Lett.*, **7**, 593–595 (2005).

[14] T. Sugihara and M. Yamaguchi, The catalytic Pauson-Khand reaction promoted by a small amount of 1,2-dimethoxyethane or water, *Synlett*, 1384–1386 (1998).

[15] J. Blanco-Urgoiti, L. Casarrubios, G. Domínguez, and J. Pérez-Castells, New protocol for the catalytic Pauson-Khand reaction induced by molecular sieves, *Tetrahedron Lett.*, **43**, 5763–5765 (2002).

[16] B. Y. Lee, Y. K. Chung, N. Jeong, Y. Lee, and S. H. Hwang, (Indenyl)cobalt(I)-catalyzed cocyclization of alkyne, alkene, and carbon monoxide to cyclopentenones, *J. Am. Chem. Soc.*, **116**, 8793–8794 (1994).

[17] N. Y. Lee and Y. K. Chung, Synthesis of cyclopentenones: the new catalytic cocyclization reaction of alkyne, alkene, and carbon monoxide employing catalytic $Co(acac)_2$ and $NaBH_4$, *Tetrahedron Lett.*, **37**, 3145–3148 (1996).

[18] B. L. Pagenkopf and T. Livinghouse, Photochemical promotion of the intramolecular Pauson-Khand reaction. a new experimental protocol for cobalt-catalyzed [2 + 2 + 1] cycloadditions, *J. Am. Chem. Soc.*, **118**, 2285–2286 (1996).

[19] D. B. Belanger, D. J. R. O'Mahony, and T. Livinghouse, Thermal promotion of the cobalt catalyzed intramolecular Pauson-Khand reaction - an alternative experimental protocol for cyclopentenone synthesis, *Tetrahedron Lett.*, **39**, 7637–7640 (1998).

[20] D. B. Belanger and T. Livinghouse, Hexacarbonyldicobalt-alkyne complexes as convenient $Co_2(CO)_8$ surrogates in the catalytic Pauson-Khand reaction, *Tetrahedron Lett.*, **39**, 7641–7644 (1998).

[21] M. E. Krafft, C. Hirosawa, and L. V. R. Boñaga, Catalyst precursors for the catalytic Pauson-Khand reaction, *Tetrahedron Lett.*, **40**, 9177–9181 (1999).

[22] J. W. Kim and Y. K. Chung, Pauson-Khand reaction catalyzed by $Co_4(CO)_{12}$, *Synthesis*, 142–144 (1998).

[23] T. Sugihara and M. Yamaguchi, The Pauson-Khand Reaction catalyzed by the methylidynetricobalt nonacarbonyl cluster, *J. Am. Chem. Soc.*, **120**, 10782–10783 (1998).

[24] M. E. Krafft and L. V. R. Boñaga, Dodecacarbonyltetracobalt catalysis in the thermal Pauson-Khand reaction, *Angew. Chem. Int. Ed.*, **39**, 3676–3680 (2000).

[25] S. Fischer, U. Groth, M. Jung, and A. Schneider, $Co_2(CO)_8$ catalyzed Pauson-Khand reaction under microwave irradiation, *Synlett*, 2023–2026 (2002).

[26] S.-W. Kim, S. U. Son, S. I. Lee, T. Hyeon, and Y. K. Chung, Cobalt on mesoporous silica: the first heterogeneous Pauson-Khand catalyst, *J. Am. Chem. Soc.*, **122**, 1550–1551 (2000).

[27] S.-W. Kim, S. U. Son, S. S. Lee, T. Hyeon, and Y. K. Chung, Colloidal cobalt nanoparticles: a highly active and reusable Pauson-Khand catalyst, *Chem. Commun.*, 2212–2213 (2001).

[28] S. U. Son, K. H. Park, and Y. K. Chung, Cobalt nanoparticles on charcoal: a versatile catalyst in the Pauson-Khand reaction, hydrogenation, and the reductive Pauson-Khand reaction, *Org. Lett.*, **4**, 3983–3986 (2002).

[29] J.-L. Muller, A. Rickers, and W. Leitner, Raney cobalt: an effective and recyclable catalyst for the Pauson-Khand reaction, *Adv. Syn. Catal.*, **349**, 287–291 (2007).

[30] J.-L. Muller, J. Klankermayer, W. Leitner, Poly(ethylene glycol) stabilized Co nanoparticles as highly active and selective catalysts for the Pauson-Khand reaction. *Chem. Commun.*, 1939–1941 (2007).

[31] A. C. Comely, S. E. Gibson, and N. J. Hales, Polymer-supported cobalt carbonyl complexes as novel solid-phase catalysts of the Pauson-Khand reaction, *Chem. Commun.*, 305–306 (2000).

[32] A. Dahan and M. Portnoy, Dendritic effect in polymer-supported catalysis of the intramolecular Pauson-Khand reaction, *Chem. Commun.*, 2700–2701 (2002).

[33] N. Jeong, S. H. Hwang, Y. W. Lee, and J. S. Lim, Catalytic Pauson-Khand reaction in super critical fluids, *J. Am. Chem. Soc.*, **119**, 10549–10550 (1997).

[34] N. Jeong and S. H. Hwang, Catalytic intermolecular Pauson-Khand reactions in supercritical ethylene, *Angew. Chem. Int. Ed.*, **39**, 636–638 (2000).
[35] S. U. Son, S. I. Lee, Y. K. Chung, S.-W. Kim, and T. Hyeon, The first intramolecular Pauson-Khand reaction in water using aqueous colloidal cobalt nanoparticles as catalysts, *Org. Lett.*, **4**, 277–279 (2002).
[36] L. V. R. Boñaga, J. A. Wright, and M. E. Krafft, Catalytic activity of dodecacarbonyltetracobalt in aqueous media: a "greening" of the Pauson-Khand reaction, *Chem. Commun.*, 1746–1747 (2004).
[37] P. Mastrorilli, C. F. Nobile, R. Paolillo, and G. P. Suranna, Catalytic Pauson-Khand reaction in ionic liquids, *J. Mol. Catal. A: Chem.*, **214**, 103–106 (2004).
[38] C. Mukai, T. Yoshida, M. Sorimachi, and A. Odani, $Co_2(CO)_8$-catalyzed intramolecular hetero-Pauson-Khand reaction of alkynecarbodiimide: synthesis of (±)-physostigmine, *Org. Lett.*, **8**, 83–86 (2006).

4

Diastereoselective Pauson-Khand Reaction using Chiral Pool Techniques (Chiral Substrates)

Martin Kamlar, Jan Vesely, and Ramon Rios Torres

"Anyone who has never made a mistake has never tried anything new."

—Albert Einstein

4.1 Introduction and Background

The Pauson Khand reaction is a formal [2+2+1] cycloaddition reaction that involves an alkyne, an alkene, and carbon monoxide and is mediated by a hexacarbonyldicobalt complex to yield cyclopentanones in a single step. Since its discovery, many asymmetric Pauson-Khand reactions (PKRs) have been developed. In this chapter we will

The Pauson-Khand Reaction: Scope, Variations and Applications, First Edition. Edited by Ramon Rios Torres.

disclose several approaches to a chiral version of the Pauson-Khand reaction using chiral substrates.

4.2 Intramolecular Diastereoselective Pauson-Khand Reaction

The earliest example of diastereoselective PKR based on chiral pool methodology was published by Magnus *et al.* They described an elegant, enantiospecific synthesis of a 6a-carbocycline analogue, **4**, based on stereospecific dicobaltoctacarbonyl-mediated cyclization of acetonide **2** (Scheme 4.1), which is produced in a few steps from D-ribonolactone **1**.[1] Soon after, the same group published a study on the stereospecific synthesis of cytotoxic sesquiterpene (+/-)-quadrone based on the same concept, that is, a dicobaltoctacarbonyl-mediated enyne cyclization reaction.[2]

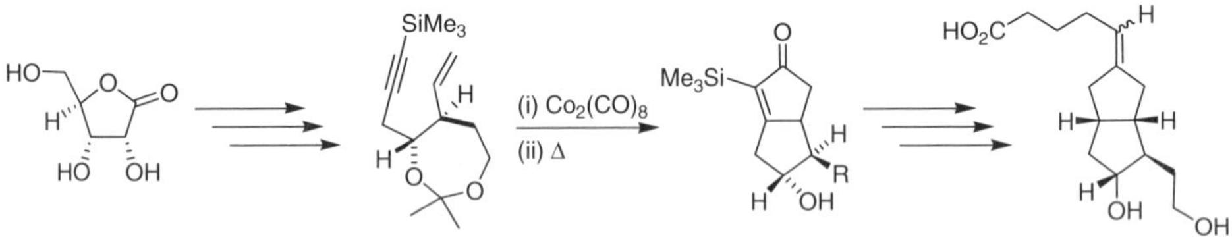

1 2 3 4

Scheme 4.1 *Synthesis of 6a-carbocycline analog* **4** *based on the stereospecific dicobaltoctacarbonyl mediated cyclization.*

Subsequently, Mulzer *et al.* independently investigated dicobaltoctacarbonyl-mediated cyclization of acyclic enynes **6a–d**, which are readily accessible from (*R*)- and (*S*)-2,3-*O*-isopropylidene-glyceraldehyde, **5**.[3] As an example, the treatment of enyne **6b** with $Co_2(CO)_8$ at room temperature followed by thermal decomposition afforded the enantiomerically pure, bicyclic compound **7** in good yield (43%). In contrast, derivatives **6c** and **6d** gave mixtures of diastereoisomers **8** and **9** in 3:1 and 1:1 ratios, respectively, via the same reaction (Scheme 4.2). The stereoselectivity of the cyclization of **6b–6d** is dictated by the requirement that C-1/C-6 and C-2/C-5 are sufficiently close to allow formation of the cyclopentenone ring. This is best fulfilled by a roughly parallel alignment of the olefin and cobalt-complexed alkyne moieties. Based on the stereochemical outcomes of two possible reactive geometries, that is, **10** and **11** (Scheme 4.3), which leads to the formation of **7** and its C-5 epimer, respectively, transition state **10** is highly favored. This is probably due to the anti-periplanar arrangement of R^2 with respect to the newly created C–C bond. For the cyclization of **6c** and **6d**, the same effect dominates; however, the preference for a geometry like **10** is not as high due to the smaller size of R^2 (OBz). Moreover, it is evident that OR^1 has a stereoregulating influence similar in magnitude to that of R^2 and directs cyclization to preferentially place OR^1 on the convex face of the bicyclic system.

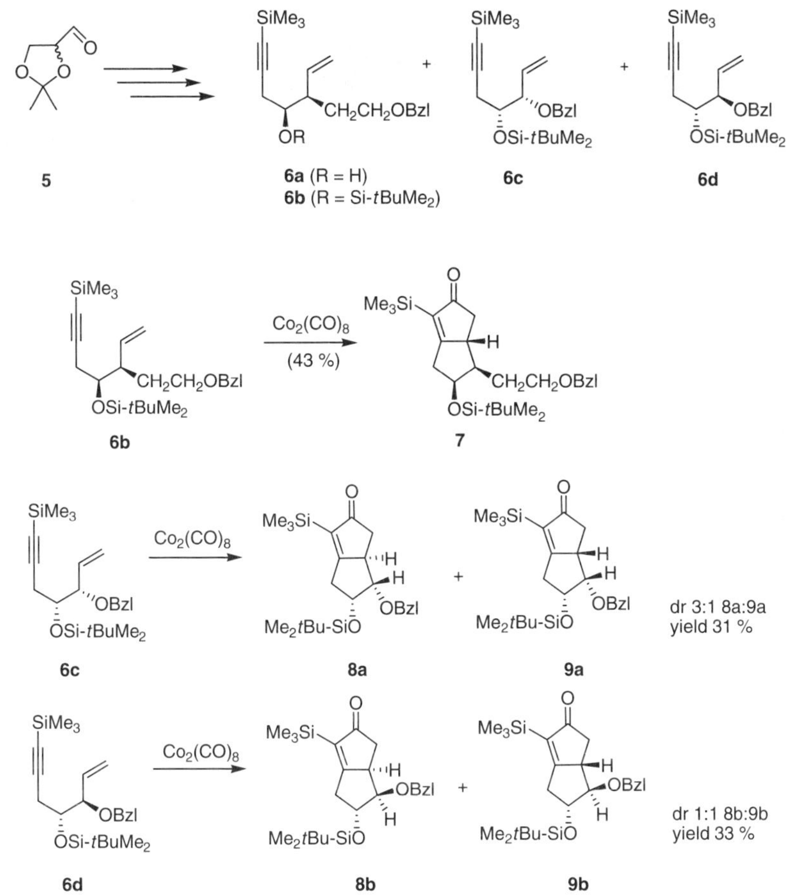

Scheme 4.2 *Diastereoselective dicobaltoctacarbonyl-mediated cyclization of the acyclic enynes* ***6a–d*** *described by Mulzer* et al.

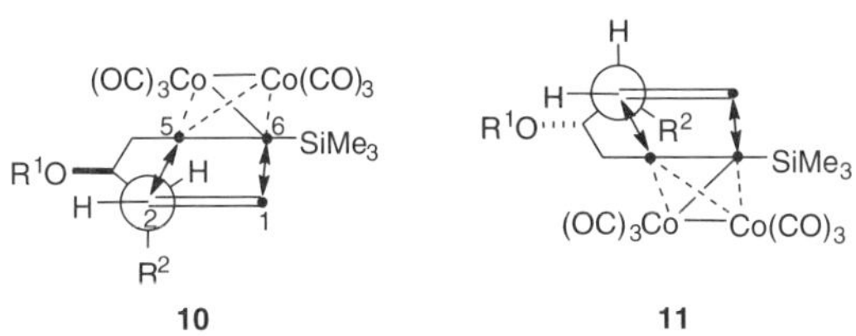

Scheme 4.3 *Transition state models* ***10*** *and* ***11****.*

In 1991, Roush *et al.* reported the enantioselective crotylboration of 3-decynal dicobalt hexacarbonyl **12**. This reaction provides a very convenient route to stereochemically defined 1,6-enyne dicobalt hexacarbonyl complexes **13** and **14**, which undergo a highly diastereoselective intramolecular PKR to form bicyclooctenones **15** and **16** (Scheme 4.4).[4]

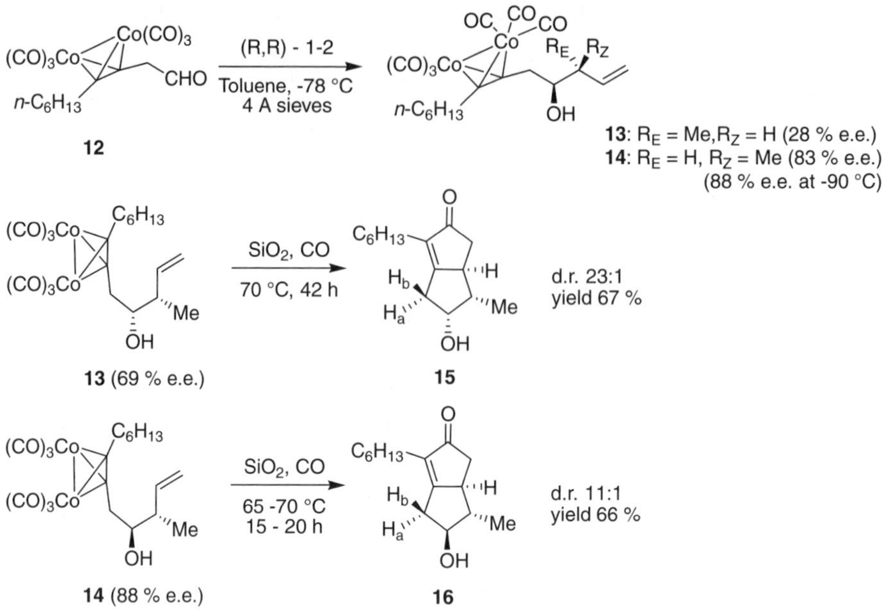

Scheme 4.4 *Enantioselective crotylboration of 3-decynal dicobalt hexacarbonyl.*

The Pauson-Khand reaction was also successfully applied to the enantioselective synthesis of (-)-kainic acid and its derivatives, which are known for their anthelmintic and neuroexcitatory activities. In the first example, Takano *et al.* described a new enantiospecific route to (-)-kainic acid from (*R*)-4-benzyloxy-l-butyn-3-ol **16** that employs the intramolecular Pauson-Khand reaction as the key step (Scheme 4.5).[5] Via a few steps, acetylene alcohol **16** was first transformed into corresponding tertiary carbamate **17** in a 76% overall yield. Subsequent treatment of **17** with dicobalt octacarbonyl furnished complex **18** in an 82% yield, which was reacted with excess *N*-methylmorpholine *N*-oxide (NMO) to yield an inseparable mixture of bicyclic enones **18a,b** in a 6:1 diastereomeric ratio and in an 85% yield. Reduction of the mixture using a complex prepared from lithium aluminum hydride and copper(I) iodide in THF and HMPA followed by removal of the THP group in acidic methanol furnished separable adducts **19a** and **19b**. Isomer **19a** was then used for the enantiospecific synthesis of natural (-)-kainic acid **20.**

A similar approach to the enantiospecific synthesis of (-)-kainic acid and its derivatives was independently developed by Yoo *et al.* The sources of chirality for this method are optically vinylglycine derivative **21** and L-glutamic acid derivative **27**, respectively.[6] When enyne **21** was subjected to the Pauson-Khand reaction using dicobalt octacarbonyl followed by treatment with trimethylamine *N*-oxide or MNO, the reaction proceeded smoothly to give an inseparable mixture of two diastereoisomers, **22** and **23**, in a 1.7:1 ratio and a 95% yield. Hydrogenation of the mixture of enones (i.e., **22** and **23**) resulted in a mixture of ketones **24** and **25**, which were separated using silica gel column chromatography. Treatment of the major isomer, **24**, using Holton's method regioselectively (97:3 ratio) produced thermodynamically more stable enolether **26**, which was used for the enantiospecific synthesis of (-)-kainic acid (Scheme 4.6).

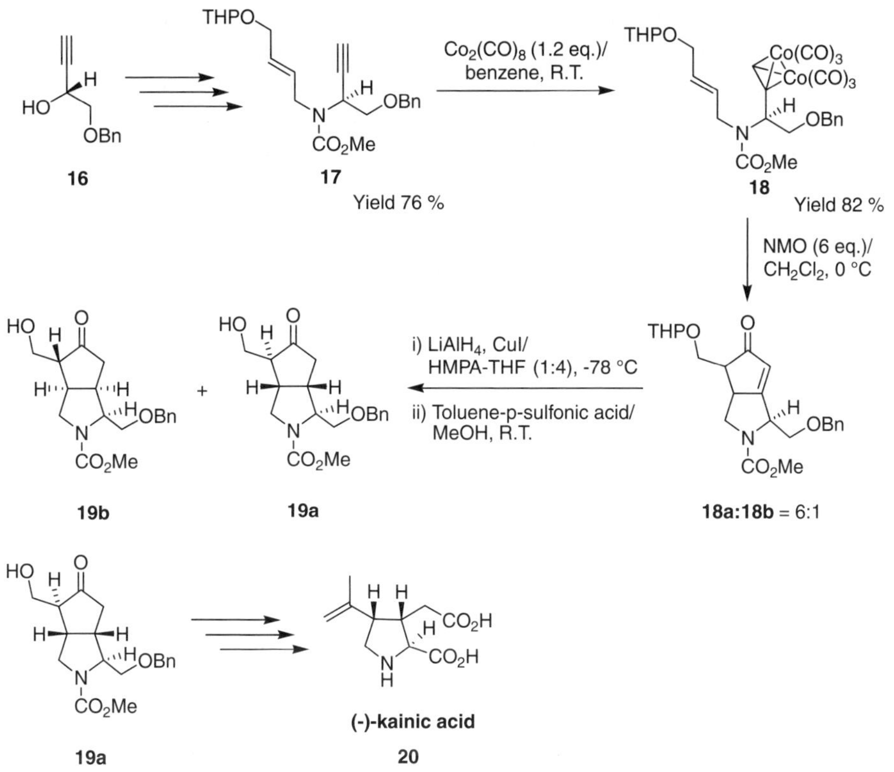

Scheme 4.5 Synthesis of (-)-kainic acid **20** utilizing the Pauson-Khand reaction.

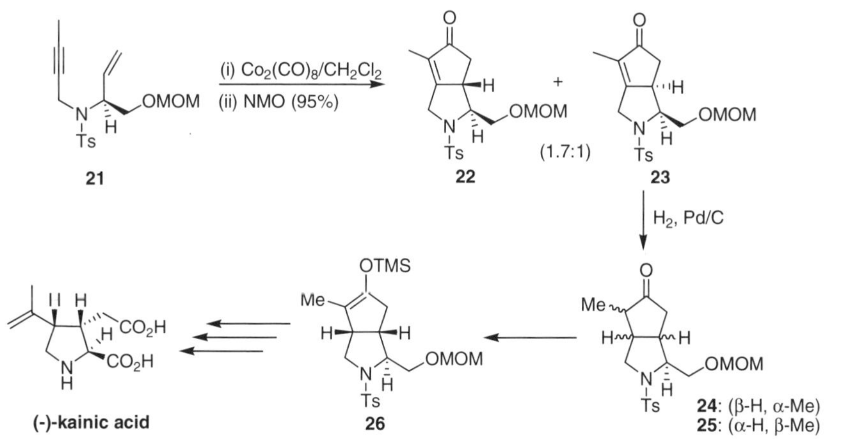

Scheme 4.6 The enantiospecific synthesis of (-)-kainic acid and its derivatives developed by Yoo.

In another approach,[6b] chiral oxazolidinone derivative **27**, which is available in a few steps from L-glutamic acid, was used as the substrate (Scheme 4.7). In this reaction, the built-in oxazolidinone ring serves as a rigid template for good diastereofacial selectivity in the cyclization step, which gives **28** in a 93% yield. Subsequent reduction afforded a separable mixture of isomers **29**; one isomer was then used to synthesize (-)-kainic acid.

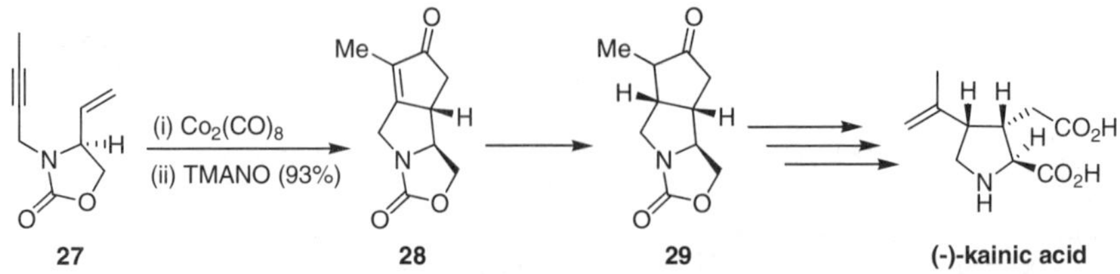

Scheme 4.7 *Another aproach to synthesize (-)-kainic acid via the Pauson-Khand reaction.*

In the same year, that is, 1994, several studies using carbohydrates as chiral substrates for PKR appeared. The first attempt at a PKR with carbohydrate derivatives was reported by Rettie *et al.*; they studied and characterized hexacarbonyldicobalt complexes derived from 2-propynyl and 3-butynyl 4,6-di-O-acetyl-2,3-dideoxy-α-D-erythro-hex-2-enopyranosides (Scheme 4.8).[8] 2,3-Dideoxyhexenepyranosides **30a–d** were prepared from tri-O-acetyl-D-glucal and an appropriate alcohol in the presence of boron trifluoride etherate. Then, hex-2-enopyranosides **30a,b** were treated with octacarbonyldicobalt in diethyl ether to produce hexacarbonyldicobalt complexes **31a,b** as analytically pure red oils in good yields. However, attempts to promote intramolecular Pauson-Khand reactions of **31a,b** (e.g., in solution under a CO atmosphere or on SiO_2) were unsuccessful.

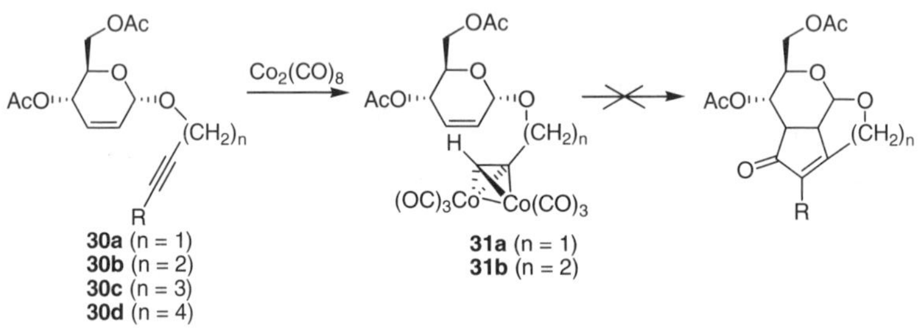

Scheme 4.8 *Carbohydrates as chiral substrates in the Pauson-Khand reaction.*

Soon after, the first successful PK cyclization of sugar-derived enynes (Scheme 4.9) was reported by Marco-Contelles *et al al.*[9] The readily available D-glucose-derived 1,6-enyne precursors, **30,32**, and D-galacto derivative, **33**, were converted into the corresponding cobalt complexes, which readily decomposed upon treatment with NMO to provide bis-heteroannulated pyranosides **34a–d** and **35** in good yields. The moderate overall yield of the PKR is compensated for by the efficiency of the one-pot process and the highly functionalized final products, which are difficult to obtain via other methods.[10]

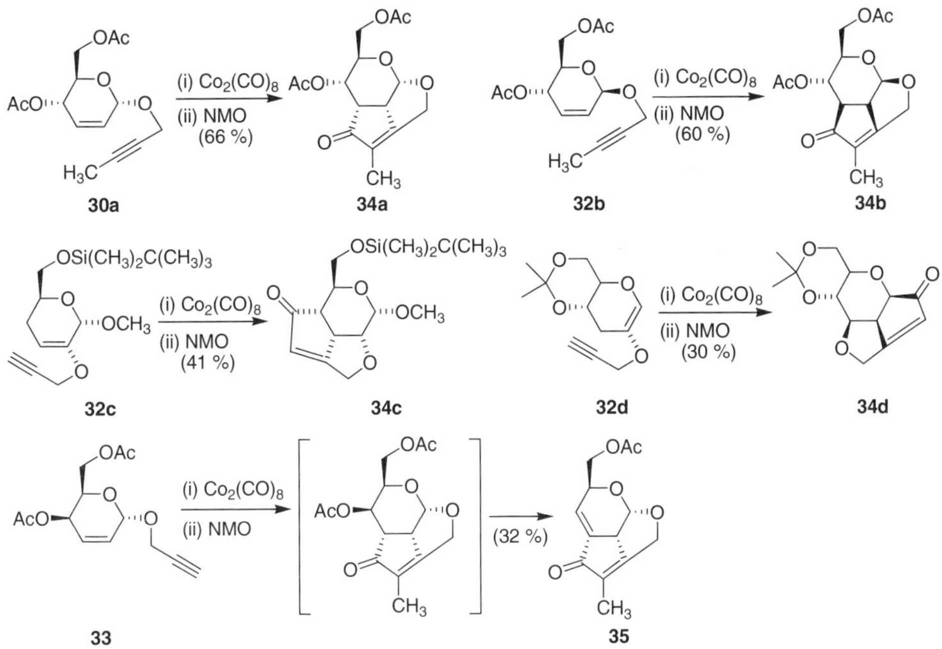

Scheme 4.9 *The first successful Pauson-Khand cyclization of sugar derived enynes.*

Independently, Voelter *et al.* published their closely related research on the synthesis of polyfunctionalized bis-annulated pyranosides (Scheme 4.10).[11] In these reports, intramolecular PKR was applied to propargylmalonate branched-chain sugars **36** and **37** using a catalytic amount of DMSO rather than NMO or other *N*-oxides for the complex decomposition step; this reaction resulted in bis-heteroannulated pyranosides **38** and **39**.

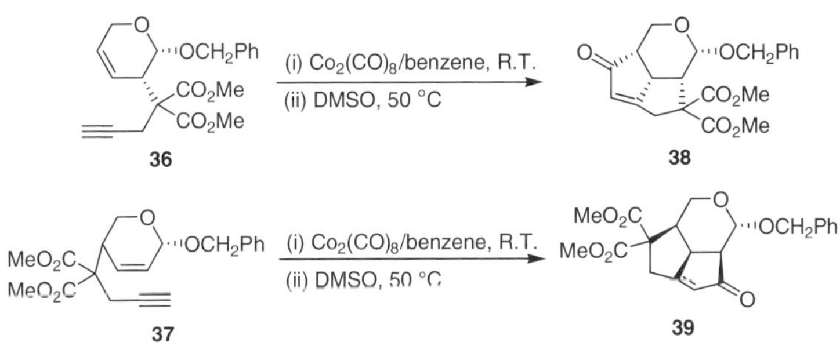

Scheme 4.10 *Synthesis of polyfunctionalized bis-annulated pyranosides via the Pauson-Khand reaction.*

In 1995, Mukai and Hanaoka *et al.* reported highly diastereoselective construction of optically active bicyclo[3.3.0]octenone derivatives **44** and **45** from chiral enynes **43** via the PKR (Scheme 4.11).[12] The required starting enynes, **43**, were easily prepared from L-ascorbic acid **40**, or dimethyl L-tartrate **41** via dioxolane intermediate **42**.

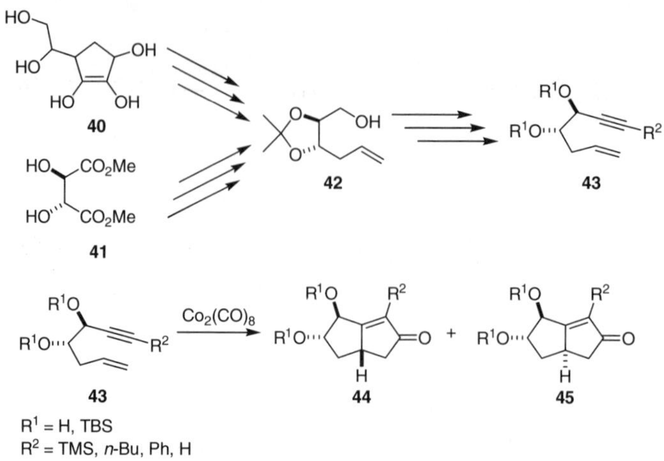

Scheme 4.11 *Highly diastereoselective construction of optically active bicyclo-[3.3.0]octenone derivatives.*

PKRs of enynes **43** were executed using both conventional thermal conditions and an alternative procedure involving a mild oxidant (TMANO). As a result, bicyclo[3.3.0]octenone derivatives **44** and **45** were obtained in high yields with good to excellent diastereoselectivity when the substrates contained bulkier R^2 substituents. In contrast, when the R^2 group was hydrogen, the diastereoselectivity dropped to a nearly 1:1 ratio

In 1996, Alcaide and Sierra *et al.* successfully applied the PKR for the synthesis of fused tricyclic β-lactams.[13] In this study, various enyne-2-azetidinones **46**, which were prepared using standard methodology, were used as substrates for the cyclization. The treatment of enyne-β-lactams **46a–c** with $CO_2(CO)_8$ followed by *in situ* thermal or TMANO decomposition of the resultant alkyne-$CO_2(CO)_6$ complexes gave tricyclic β-lactams **47a–c** as single stereoisomers in good to excellent yields (Scheme 4.12).

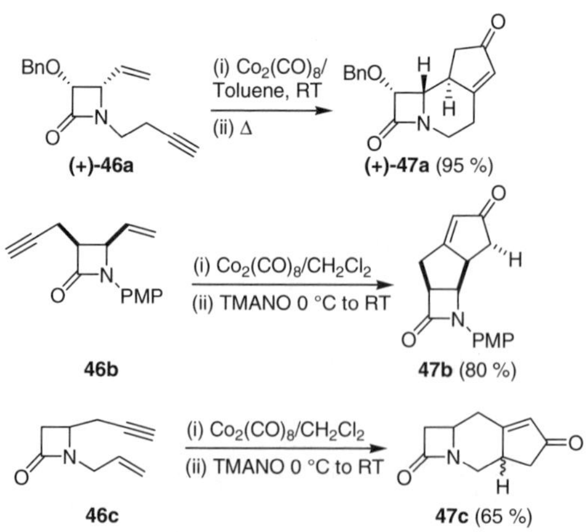

Scheme 4.12 *Synthesis of fused tricyclic β-lactams.*

Interesting results in the area of solid-phase synthesis were reported by Bolton in 1996.[14] He developed a method for stereocontrolled construction of highly functionalized fused bicyclic amino acid derivatives in which the key step involved an intramolecular PKR. Cyclization precursors **49** and **50** were easily obtained via sulfonylation and *N*-alkylation of either (*S*)-allyl or racemic propargyl glycine methyl esters **48a,b**, respectively. Treatment of tosyl derivatives **49** and **50** with $CO_2(CO)_8$ resulted in complete conversion to the cobalt complex. Subsequent NMO oxidation provided cyclopentanone derivatives **51** as a single enantiomer and diastereomer, respectively, in good yields (Scheme 4.13). As an extension of this methodology, a solid-phase synthetic strategy using commercial Wang resin improved the yield of **52** and simplified the purification steps.

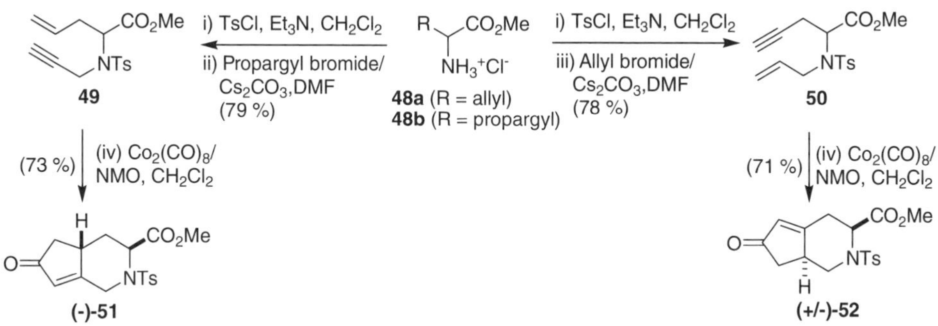

Scheme 4.13 *Stereocontrolled construction of highly functionalized fused bicyclic amino acid derivatives.*

In 2000, an interesting study on asymmetric PKR via optically pure alkyne-$Co_2(CO)_5PPh_3$ complexes was reported by Kerr *et al.*[15] They investigated the use of amine *N*-oxides for the formation of monophosphine complexes **55a** and **55b** from corresponding chiral hexacarbonyl compounds **54a** and **54b**, which were prepared from commercially available phenylprop-2-yn-1-ols **53a** and **53b** (Scheme 4.14).

At low temperatures (−60 °C) no diastereoselectivity was observed for the formation of complexes **55a**, **56a** and **55b**, **56b** from enantiomerically pure hexacarbonyl compounds **54a** and **54b**, respectively, with NMO. Surprisingly, chiral BNO gave a similar result for complex **54b**; however, complex **54a** consistently afforded a 7:3 ratio of diastereomers **55a**:**56a**. The stereoselectivity of the reaction with **54a** is believed to originate from the matching interaction of the chiral *N*-oxide with the propargylic stereocenter in combination with Co_2C_2 core of the complex. Stereopure complexes **55a–56b**, which were purified using a simple flash column, were converted into corresponding pure stereoisomers **57a–58b** using norbornene with anhydrous NMO at room temperature. By determining the absolute stereochemistry of optically pure alkyne-$Co_2(CO)_5PPh_3$ complexes **55a–56b** and cyclopentanones **57a–58b**, these studies explicitly demonstrated that decarbonylation and alkene complexation occurs exclusively at the $Co(CO)_3$ vertex in monophosphinated alkyne complexes in PK annulations.

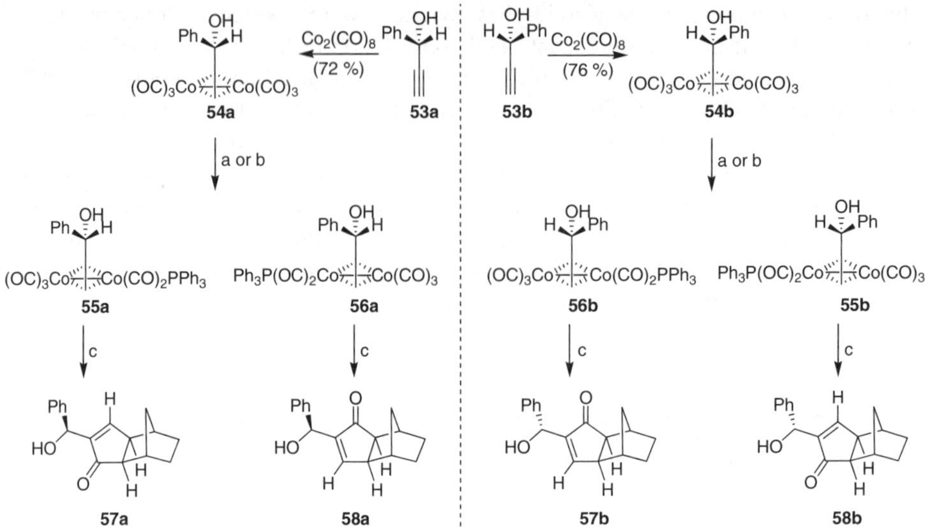

a) NMO, PPh$_3$, THF, -60 °C; 98 % (51:49 **3a**:**4a**) from complex **2a**; 95 % (50:50 **3b**:**4b**) from complex **2b**; b) BNO, PPh$_3$, -59 °C; 66 % (70:30 **3a**:**4a**), from complex **2a**; 77 % (49:51 **3b**:**4b**) from complex **2b**; c) norbornene, NMO, DCM, rt.; 77 % for **5a**, 40 % for **6a**, 69 % for **5b**, 58 % for **6b**.

Scheme 4.14 *Asymmetric Pauson-Khand reactions via optically pure alkyne-$Co_2(CO)_5PPh_3$ complexes.*

In 2002, Areces *et al.* reported a new synthesis of enantiomerically pure 3-azabicyclo[3.3.0]octen-7-one derivatives **61–63** via intramolecular PK cycloaddition using a sugar moiety as the chiral template (Scheme 4.15).[16] As starting materials, aza-enynes **60a–c** were prepared from aldehyde **59**, which is easily accessible from tri-O-actyl-D-glucal

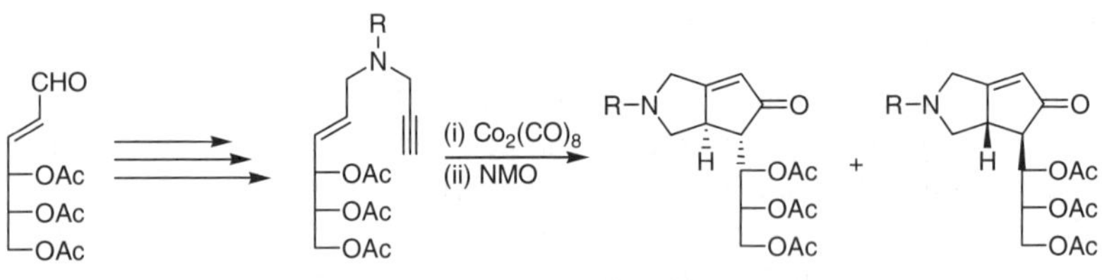

59

60a (R = Tosyl)
60b (R = COCH$_3$)
60c (R = Boc)

61a (R = Tosyl)
62a (R = COCH$_3$)
63a (R = Boc)

61b (R = Tosyl)
62b (R = COCH$_3$)
63b (R = Boc)

Scheme 4.15 *Synthesis of enantiomerically pure 3-azabicyclo[3.3.0]octen-7-one derivatives.*

The intramolecular PK reactions of **60a–c** afforded the corresponding 3-azabicyclo[3.3.0]octen-7-one derivatives, **61–63**, in good yields (47–60%) as 1:1 mixtures of their diastereomers, which were separated via column chromatography followed by crystallization. Three years later, a more detailed synthesis of enantiomerically pure 3-azabicyclo[3.3.0]octen-7-one derivatives via PKR was described by the same group.[17]

Also in 2002, Jiang *et al.* described a new strategy for the preparation of enantiopure, fused cyclic chiral amino acids **65**.[18] Optically pure enyne amino acid derivatives **64**, which were easily prepared by an alkenylboronic acid-mediated Mannich-type reaction, were successfully used for intramolecular PK reactions with a variety of catalytic systems. Optimal results featuring yields up to 82% were obtained using 10 mol% $Co_2(CO)_8$ and 1,6-enyne amino esters **64** in the presence of 60 mol% Bu_3PS co-ligand in benzene at 70 °C under 1 atm of carbon monoxide (Scheme 4.16).

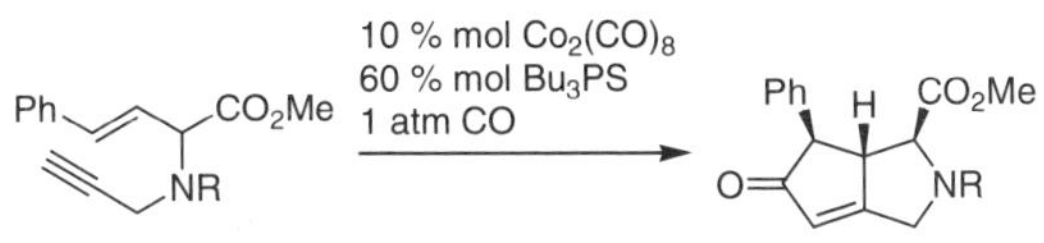

Scheme 4.16 *An intramolecular Pauson-Khand reaction reported by Jinag.*

In 2004, Saito *et al.* reported a novel diastereoselective total synthesis of isocarbacyclin **68** that features a fused bicyclic key intermediate obtained from L-ascorbic acid.[19] This intermediate was prepared in multigram quantities via the Pauson-Khand reaction of L-ascorbic acid-based (R)-4,4-diallyl-2,2-dimethyl-5-(trimethylsilyl)ethynyl-1,3-dioxolane **66**, by discriminating between the diastereotopic groups and faces of the geminal allyl substituents. The functional group assembly in **67** and the conventional synthetic reactions used for the conversion of **67** to **68** makes this approach attractive for flexible access to side-chain analogs of **68** (Scheme 4.17).

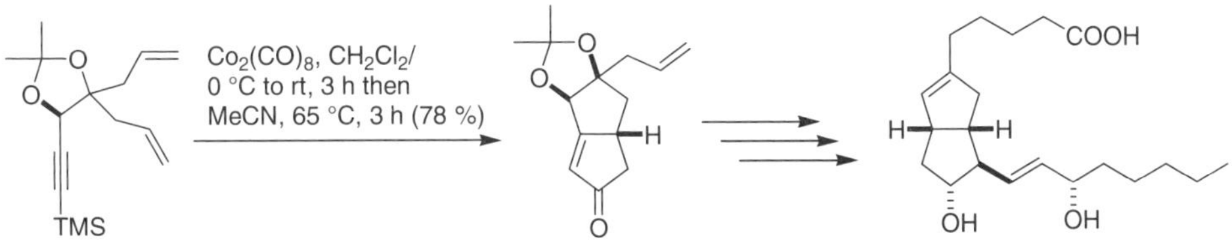

Scheme 4.17 *Construction of key intermediate **67** via Pauson-Khand reaction.*

An interesting use of PKR was reported by Jeong *et al.*, who studied the desymmetrization of *meso*-dienynes such as propargyl 1-vinylallyl *N*-tosylamides **69a–c** and propargyl 1-vinylallyl ethers **69d–e** using asymmetric Pauson-Khand-type catalysts (Scheme 4.18).[20] The corresponding vinyl-substituted optically active N- or O-containing bicyclo[3.3.0]octenes, **70** and **71**, were obtained with high diastereoselectivity (up to 75:1) and good enantioselectivity (up to 81% ee)

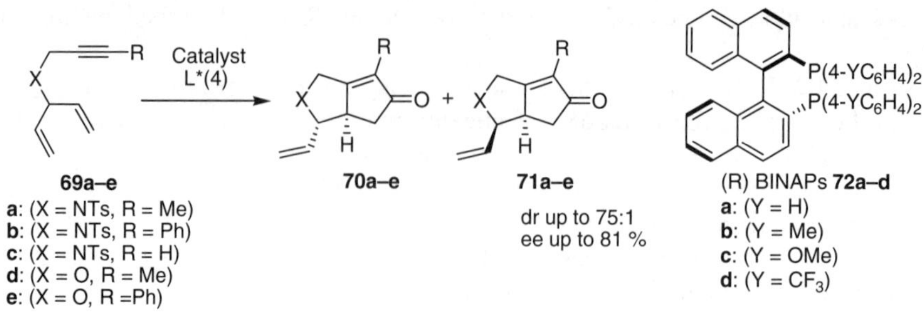

Scheme 4.18 *Preparation of vinyl substituted optically active* N- *or* O-*containing bicyclo[3.3.0]octenes.*

In 2004, Moriarty *et al.* published a general and novel solution to the synthesis of biologically important stable analogs of prostacyclin PGI2, that is, benzindene prostacyclins.[21] This work provided the first demonstration of the synthetic utility and reliability of the asymmetric PK cyclization route for the synthesis and subsequent manufacture of complex drug compounds on a multikilogram scale. The key step in the synthesis involves efficient stereoselection via the PK cyclization of benzoenyne **73** under the agency of the benzylic OTBDMS group, which serves as a temporary stereodirecting group and is conveniently removed via benzylic hydrogenolysis concomitantly with the catalytic hydrogenation of the enone product (Scheme 4.19). Thus, the benzylic chiral center dictates the stereochemistry of the stereogenic centers at three carbon atoms, that is, C3a, C9a, and C1.

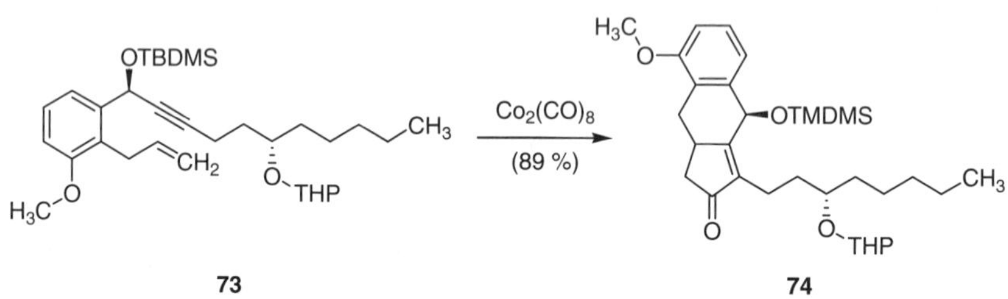

Scheme 4.19 *An asymmetric Pauson-Khand cyclization, formation of key intermediate* ***74*** *used in the synthesis of prostacyclin PGI_2 analogs.*

The PK approach was also successfully applied to the synthesis of three new carbocyclic nucleoside analogs with nucleobases attached to 3′-hydroxymethylcyclopent-2′-en-1′-yl scaffolds, as reported by Schmalz *et al.* in 2005.[22] A variety of symmetric dienynes (i.e., propargylic acetals **75**) were used as substrates in the cobalt-mediated PK reaction to yield racemic bicyclic cyclopentenone derivatives **76** with high diastereoselectivity (Scheme 4.20).

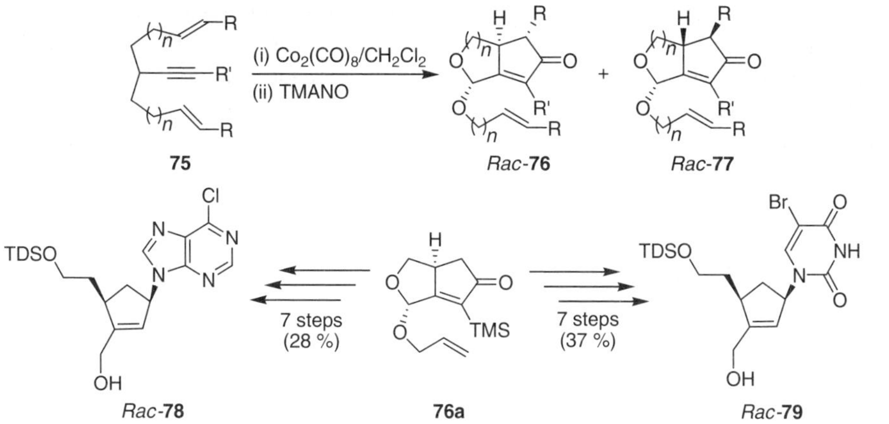

Scheme 4.20 *Diastereoselective Pauson-Khand construction of bicyclic cyclopentenone derivatives* ***76***.

These compounds are valuable building blocks for the synthesis of structurally diverse carbocyclic nucleoside analogs that have a high biological potential as apoptosis-inducing agents. Starting from PK product ***rac*-76**, the synthesis of 4′-trialkylsilyloxyethy l-substituted nucleosides ***rac*-78** and ***rac*-79** with 5-bromouracil and 6-chloropurine as nucleobases, respectively, was accomplished in seven steps (28 and 37% overall yields, respectively)

Another example of a diastereoselective PK reaction was reported by Honda *et al.* in 2007.[23] They established a diastereoselective formal synthesis of a monoterpene alkaloid, (-)-incarvilline **83**, that incorporates the intramolecular Pauson-Khand reaction of the corresponding enyne amide, that is, (*S*)-*N*-[(*E*)-2-butenyl]-*N*-(3-butynyl-2-methoxymethoxy)-p-toluenesulfonamide, **80**. In this synthesis, the stereochemistry at the 7 and 7a positions of the target compound was controlled through the reflection of the stereochemistry at the 4-position to providing the desired absolute configuration; this was achieved by employing E-olefin as the starting material. The optimal result was obtained when the reaction was carried out with 1.05 equivalents of $Co_2(CO)_8$ in refluxing DCE in the presence of 3.5 equivalents of *tert*-butyl methyl sulfide as a promoter for 2.5 h under an atmosphere of CO to furnish **81** and **82** in 73 and 8% yields, respectively (Scheme 4.21).

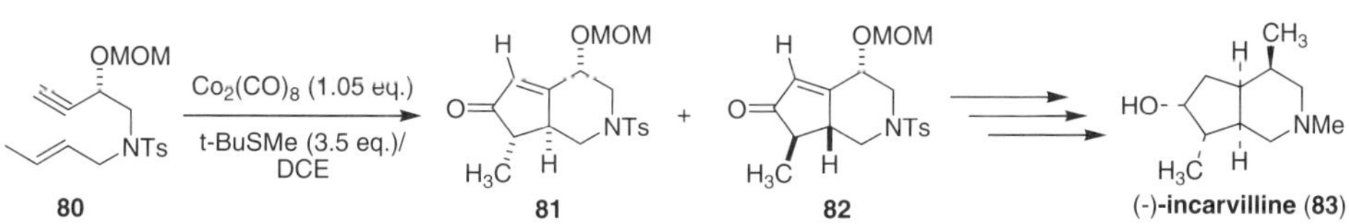

Scheme 4.21 *Pauson-Khand reaction used in diastereoselective synthesis of (-)-incarvilline.*

In the same year, the application of the PK reaction to a cyclopropene moiety was described by Fox *et al.*[24] In this study, the enantioselective synthesis of (-)-pentalenene **87** based on catalytic enantioselective cyclopropenation followed by an intramolecular PKR established the quaternary center. The transformation of cyclopropene **84** into tricyclic PK product **85** failed when N-oxides were used as promoters. Optimal reactivity was achieved using tetramethylthiourea (TMU); this promoter was found to be very effective even with substoichiometric amounts of $Co_2(CO)_8$ (Scheme 4.22).

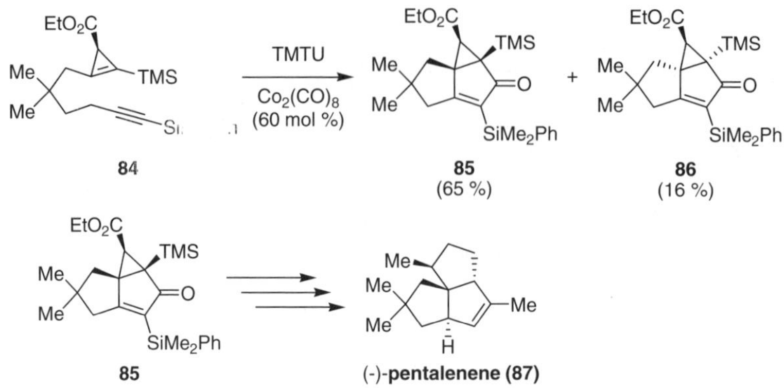

Scheme 4.22 *An intramolecular Pauson-Khand reaction used to establish the quaternary center.*

In the same year, the diastereoselective synthesis of **91**, which is a key intermediate of naturally occurring anti-HIV isolitseane B and its analogs, via diastereoselective PKR was reported by Coquerel and Rodriguez *et al.*[25] In this work, the introduction of a sulfur atom not only improved the yield and stereoselectivity of the PKR but also enabled the isolation of the final product **91**, in an enantiopure form by diastereomeric separation of an intermediate bearing a chiral, configurationally defined, sulfur atom (Scheme 4.23).

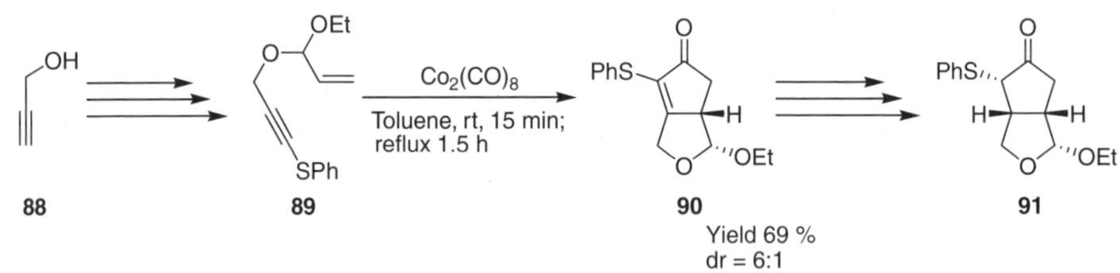

Scheme 4.23 *Diastereoselective synthesis of a key intermediate **91** of anti-HIV Isolitseane B and its analogs.*

In 2007, an interesting study on the intramolecular PKR of 1,8-enynes derived from salicylaldehyde derivatives was reported by Lovely *et al.*[26] Substrates derived from salicylaldehyde **92** itself reacted poorly in this reaction but related substrates containing *ortho-tert*-butyl substituents reacted effectively: in many cases the cyclizations proceeded with high levels of diastereoselectivity (dr up to 10:1) and good yields (up to 95%) (Scheme 4.24).

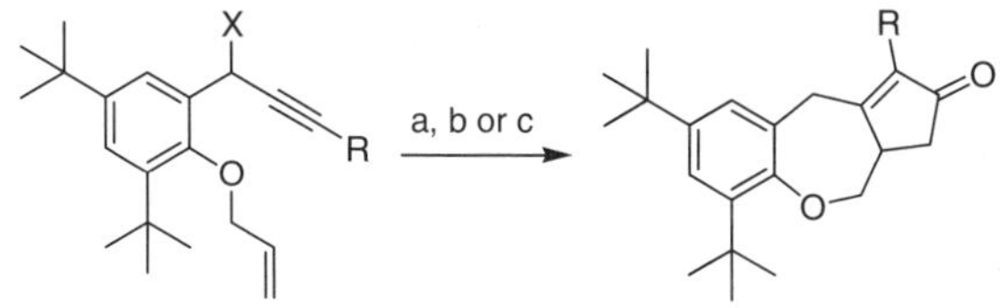

92 **93**

X = OH, H
R = H, TMS, Ph

a) $Co_2(CO)_8$, toluene, 70 °C; b) $Co_2(CO)_8$, CH_2Cl_2, NMO; c) (i) $Co_2(CO)_8$, CH_2Cl_2, (ii) $NaBH_4$, TFA, (iii) NMO.

Scheme 4.24 *Intramolecular Pauson-Khand reaction of 1,8-enynes derived from salicylaldehyde.*

Another example of the successful application of PKR to the total synthesis of natural products, that is, the fusicoccane family of diterpenes, was reported by Dake *et al.* in 2008.[27] In this work, the construction of the A-B ring system of the compounds was successfully achieved by diastereoselective PKR followed by Norrish Type 1 fragmentation. Corresponding enynes **94a–e** were synthesized using standard procedures, that is, silylation with bulky groups followed by ketalization, and subjected to PK conditions (i.e., $CO_2(CO)_8$-NMO) to produce bicyclic products **95a–e** and **96a–e**. When isopropyl acetonide was used as a protecting group for the diol, the PKR processed with a high yield and good diastereoselectivity (6:1) (Scheme 4.25).

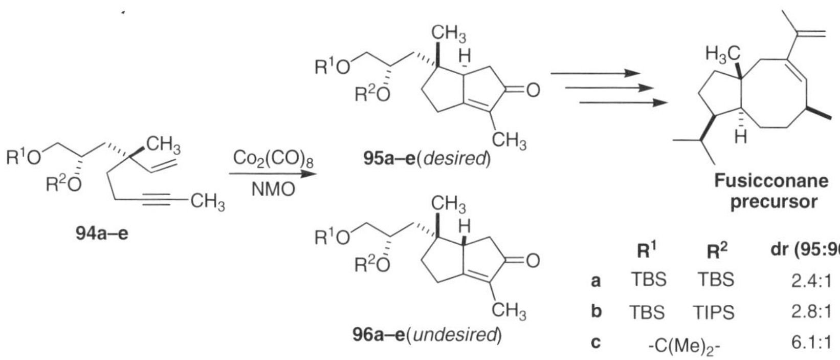

	R^1	R^2	dr (95:96)
a	TBS	TBS	2.4:1
b	TBS	TIPS	2.8:1
c	-C(Me)$_2$-		6.1:1
d	-C(Et)$_2$-		1.3:1
e	-C(C$_5$H$_{10}$)$_2$-		2.2:1

Scheme 4.25 *Successful application of Pauson-Khand reaction into total synthesis of fusicoccane family of diterpens.*

In 2008, Williams *et al.* described an asymmetric approach to the synthesis of a B-C ring system in tuberosteminol, which is from the Stemona alkaloids family, based on an intramolecular PKR stereocontrolled by readily available chiral glycinates.[28] Treatment of chiral glycinate derivatives **98a–d**, which were prepared from commercial glycinate **97** in three steps, with CO_2CO_8 followed by the addition of NMO to initiate cyclization, provided the corresponding cyclopentenone products **99a–d** at 10–98% yields. A detailed study of the PKR of various 1,7-enynes, **98**, showed that an increase in the electron-deficiency of **98** reduced the yield to 10% (Scheme 4.26).

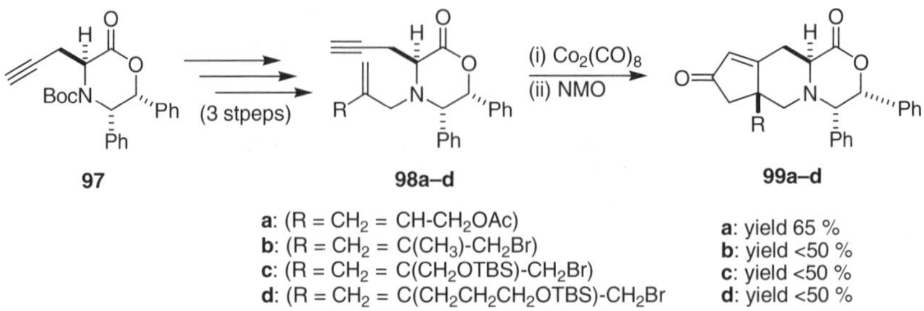

Scheme 4.26 *Intramolecular Pauson-Khand reaction stereocontrolled by easily available chiral glycinates.*

Also in 2008, Honda *et al.* established an alternative stereoselective chiral synthesis of (+)-α-skytanthine **102** that employed an intramolecular PKR of the corresponding enyne **100**.[29] In this study, it was determined that the configuration at the 4a position of the cyclized product was stereoselectively constructed by the reflection of the stereochemistry of the methyl group in the starting enyne. Moreover, it was found that the nitro group was reduced to an amino group by PKR even when hydrogen-free solvents were used. Although the mechanism for the reduction of the nitro group was not described, the highest yield (71%) of the desired product, **101**, was obtained when trimethylamine *N*-oxide was used as a promoter (Scheme 4.27).

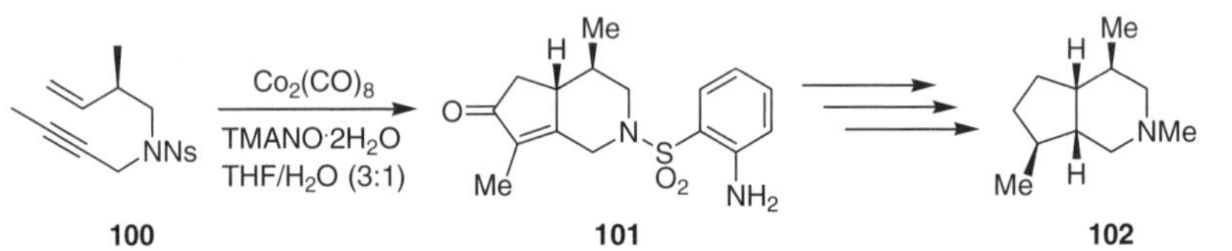

Scheme 4.27 *Stereoselective chiral synthesis of (+)-α-skytanthine **102** by employing an intramolecular Pauson-Khand reaction.*

In 2008, Tanyeli *et al.* reported a simple route to the synthesis of camphor-anchored achiral spiro cyclopentenone-pyran systems **106** and **107** with excellent stereoselectivities based on the intramolecular PKR of camphor-tethered enynes **104, 105**.[30] The camphor-tethered enynes, which were derived from homoallyl, homomethallyl, and homopropargyl alcohols **103a–c**, were easily prepared in two steps from commercially available (1*R*)-(+)camphor via the addition of alkylmagnesium bromide followed by consecutive O-propargylation, O-methanylation, and O-allylation. The addition of alkylmagnesium bromides resulted in excellent endo-face diastereoselectivity, as expected. The enantiomerically pure enynes, **104–105**, afforded their respective spirocyclic cyclopenta[c]pyran products, **106–107**, in high yields (82–86%) as the single, conformationally most stable diastereoisomer under standard PKR conditions (i.e., the Co_2CO_8-NMO system) (Scheme 4.28).

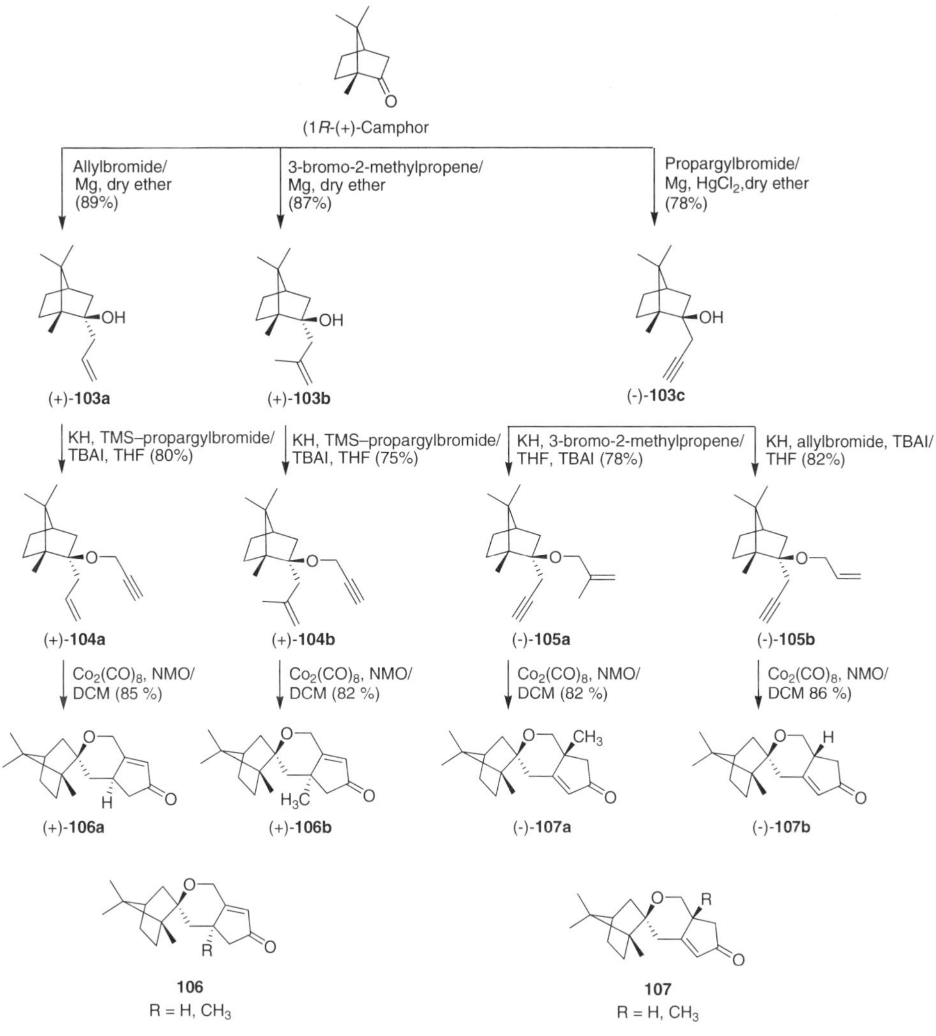

Scheme 4.28 *Simple route to synthesis camphor-anchored achiral spiro cyclopentenone-pyran systems.*

In 2008, Verdaguer and Riera *et al.* reported a new enantioselective approach to various carbanucleosides from chiral cyclopentenone **108** via Pauson-Khand reaction.[31] Multigram quantities of **108** were obtained in good yield (53%) with excellent enantiomeric purity (>99%) via a PKR involving trimethylsilylacetylene, $Co_2(CO)_8$, and norbornadiene with *N*-benzyl-*N*-diphenylphosphino-*tert*-butyl-sulfinamide as a chiral P,S ligand, as previously described by the same authors.[32] Next, a hydroxymethyl group was efficiently added to PK cycloadduct **108** via photochemical conjugate addition. Protection using a triisopropylsilyl ether followed by a retro-Diels–Alder reaction afforded the protected hydroxymethyl cyclopentenone, **110**. Diastereoselective reduction to the corresponding allyl alcohol and subsequent palladium-catalyzed substitution afforded an advanced intermediate, **111**, that was converted into enantiomerically pure (-)-Carbovir and (-)-Abacavir (Scheme 4.29).

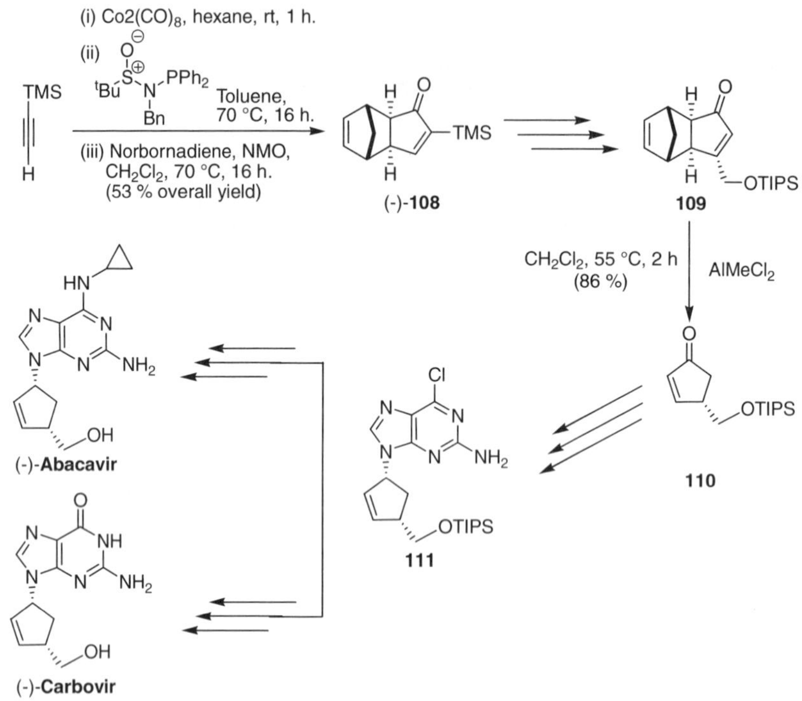

Scheme 4.29 Enantioselective approach to various carbanucleosides from chiral cyclopentenone ***108*** via Pauson-Khand reaction.

In 2009, Shea *et al.* described an interesting study of a tandem intramolecular Nicholas/PKR strategy for the synthesis of tricyclic oxygen- and nitrogen-containing heterocycles.[33] This methodology enables the conversion of simple acyclic starting materials, **112**, into a series of previously unknown heterocycles, **114**. Tricyclic ethers (Z = O) with [5,7,5]- and [5,8,5]-systems and tricyclic [5,7,5]- and [5,8,5]-amine-containing (Z = NTs) heterocycles were successfully prepared via the Nicholas/PKR reaction (Scheme 4.30) .

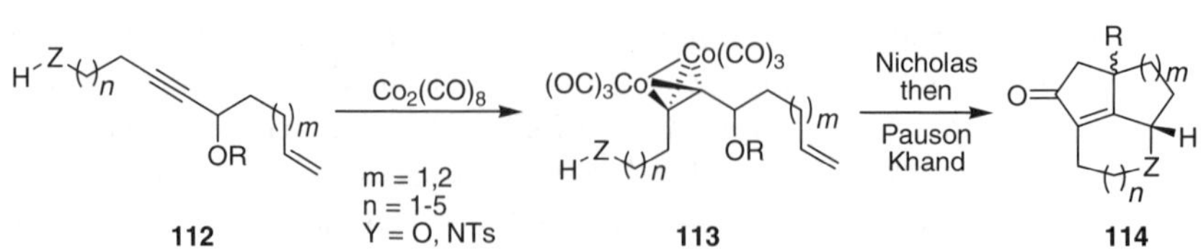

Scheme 4.30 Tandem intramolecular Nicholas/Pauson-Khand reaction strategy leading to tricyclic oxygen- and nitrogen-containing heterocycles.

Attempts to make larger ring systems or prepare lactones via Nicholas reactions with carboxylic acid nucleophiles resulted in decomposition or dimerization of the starting materials. Nevertheless, the latter process enables the formation of 14-, 16-, and 18-membered ring diolides. The use of chiral amine promoters in the Pauson-Khand step was also investigated but did not result in asymmetric induction.

In 2010, Tanyeli *et al.* developed a stereoselective synthesis of an optically active cyclopenta[c]pyran and cyclopenta[c]furan ring system by intramolecular PKR.[34] 2-Heteroaryl-substituted homoallyl **115**, homopropargyl **116**, and allyl alcohols **117** were efficiently resolved in high ee (91–99%) and with known stereochemistry through enzymatic resolution. Each enantiomerically enriched enyne derived from the homoallyl and homopropargyl alcohols afforded the conformationally most stable diastereomeric cyclopenta[c]pyran ring system as the sole product. In contrast, enantiomerically enriched enynes derived from the allyl alcohols gave a diastereomeric cis:trans mixture of their corresponding cyclopenta[c]furan ring system **123** and **126** (Scheme 4.31).

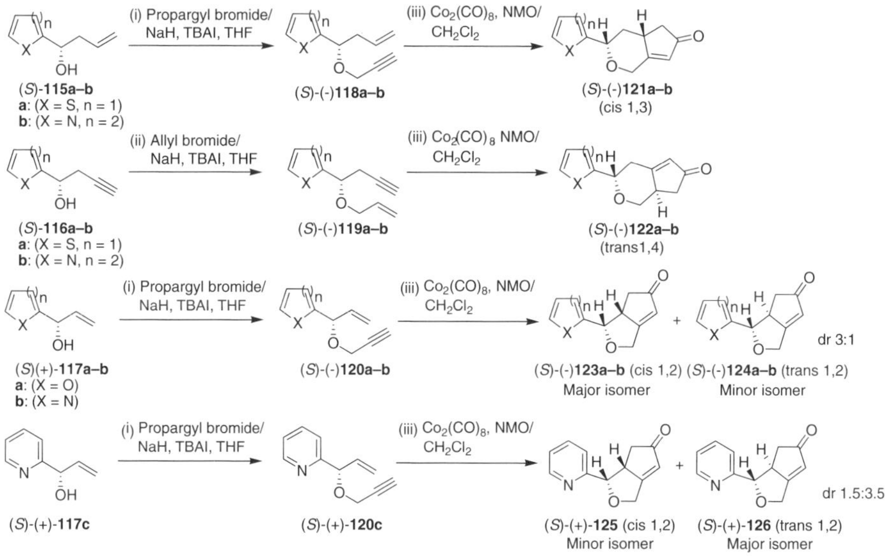

Scheme 4.31 *Stereoselective synthesis of optically active cyclopenta[c]pyrans and cyclopenta[c]furans.*

4.3 Intermolecular Diastereoselective Pauson-Khand Reaction

In contrast to the plethora of papers on intramolecular diastereoselective PKRs, only a few studies on the intermolecular version of this transformation have been reported. The first example of intermolecular PK cycloaddition was described by Hsung *et al.* in 2003.[35] In this study, chiral ynamides **127** and **128** were successfully used as starting materials to obtain the corresponding adducts with excellent regioselectivity, albeit with modest stereoselectivity. When norbornadiene and substituted ynamides were used, an unusual endo addition occurred leading to cycloadducts **129–131** that were not observed in previous studies using ynamines or ynamides (Scheme 4.32).

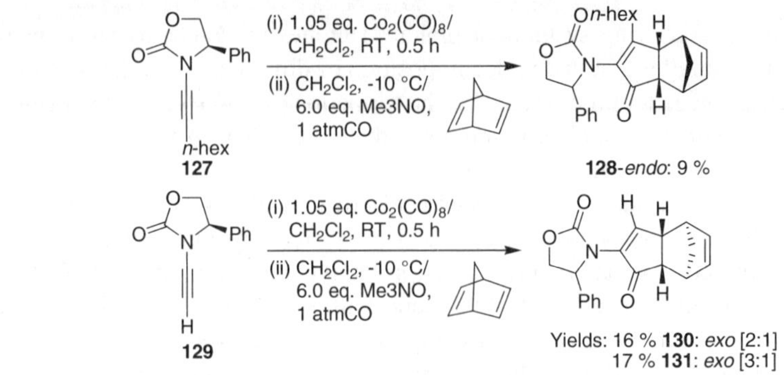

Scheme 4.32 *First example of intermolecular Pauson-Khand cycloaddition.*

In 2005, Fox *et al.* demonstrated that the unusual reactivity of cyclopropenes increased the scope and utility of intermolecular Pauson-Khand reactions.[36] It was shown that the well-defined chiral environment of cyclopropenes, **132**, has a significant influence on the diastereoselectivity of the PK reactions and leads to the production of cyclopentenones **133** as single diastereomers. Exo-diastereoselectivity confirmed via X-ray crystallography and NOE studies for derivative **133b** (Scheme 4.33).

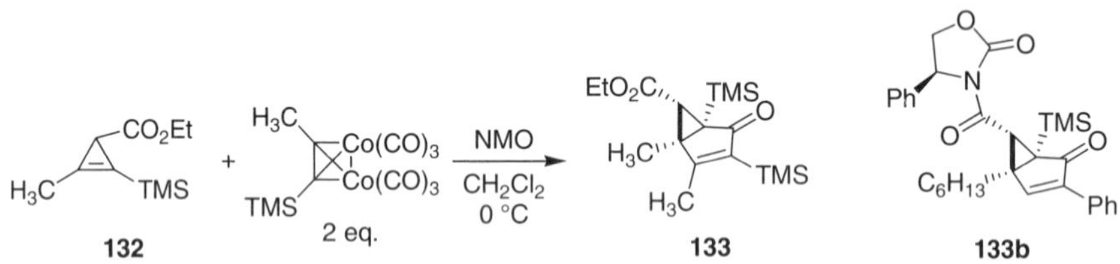

Scheme 4.33 *Intermolecular Pauson-Khand reactions of cyclopropenes.*

Moreover, the three-membered ring can subsequently be cleaved under mild conditions to afford cyclopentanones **135**, **136**, which feature quaternary carbon centers (Scheme 4.34).

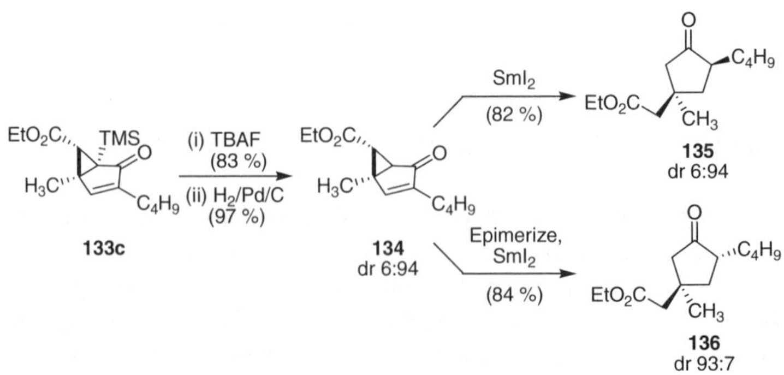

Scheme 4.34 *Construction of cyclopentanones **135**, **136** bearing quaternary carbon centers.*

An interesting example of an intermolecular PK reaction of sugar-derived azaenynes was described by Areces *et al.* in 2006.[37] Sugar azaenynes **137a,b**, which are readily obtained from D-glucal and D-galactal, respectively, react with norbornene and norbornadiene under Pauson-Khand conditions to give intermolecular adducts **138** and **139** as the major or exclusive product (Scheme 4.35). In all cases, only one diastereomer was obtained in a good yield (up to 80%). This reaction represents an unprecedented example of a completely diastereoselective intermolecular Pauson-Khand reaction using a sugar moiety as the chiral auxiliary

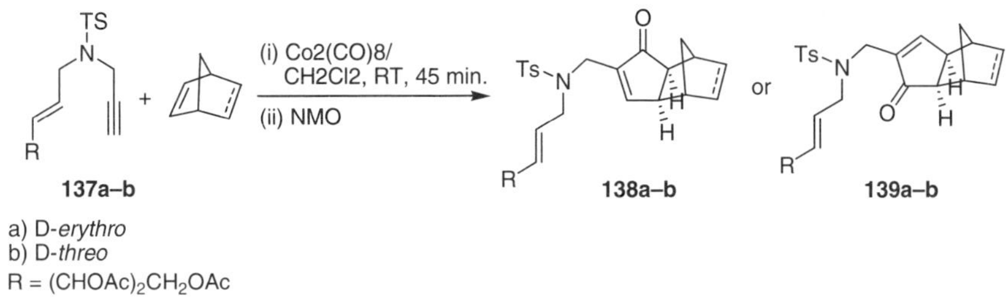

Scheme 4.35 *Diastereoselective intermolecular Pauson-Khand reaction using a sugar moiety as chiral auxiliary.*

In 2010, Cazes *et al.* demonstrated that the PKR of allenic hydrocarbons **142**, gives 4-alkylidenecyclopentenones, **143**, with high regio- and stereo-selectivities (E/Z$\geq$70:30) along with minor amounts of regioisomeric cyclopentenones **144** and **145**.[38] The regio- and stereo-selectivities of isomeric cyclopentenones **143–145** depend mainly on the substitution pattern of both the alkyne, **140**, and the allenic moieties, **142** (Scheme 4.36).

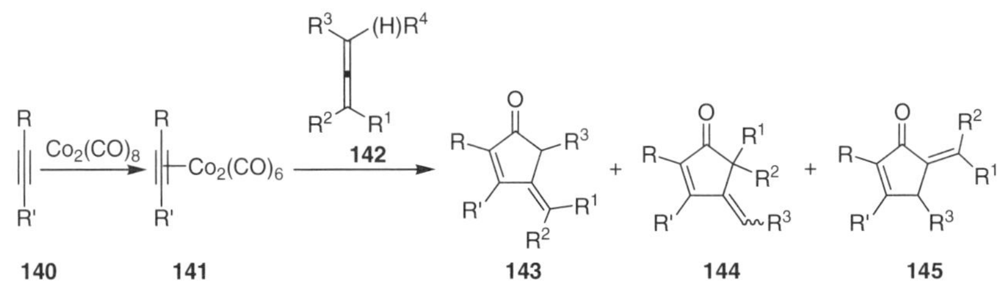

Scheme 4.36 *Intermolecular Pauson-Khand reactions of allenic hydrocarbons.*

By studying the relationship between the change in selectivity of these isomeric cyclopentenones, **143–145**, and the substitution patterns of both the acetylenic and allenic components, competitive mechanistic pathways were established via several allene–dicobalt π complexes. DFT calculations were carried out on the most relevant of these intermediate π complexes; the results provided evidence for the first time that the PKR may involve both pseudoequatorial and pseudoaxial coordination of a double bond to one of the cobalt atoms leading to two isomeric cyclopentenones.

In the same year, Pu *et al.* developed catalytic asymmetric alkyne addition methodology that provides rapid access to a variety of chiral propargylic alcohol-based enynes, **147,**

148 with high enantiomeric purity.[39] In contrast to other studies concerning the PKR of enantiomerically enriched propargylic alcohols, which featured additional chiral centers in the PK substrates in order to isolate the enantiomers and contribute to the stereocontrol of the reaction, two types of chiral enynes **147, 148**, were synthesized and observed to undergo highly diastereoselective intramolecular PK cycloaddition with retention of their enantiomeric purity. Corresponding optically active 5,5- and 5,6-fused bicyclic products **149, 150** were obtained in good to high yields (up to 94%) with high diastereoselectivity (dr up to 99:1). In the major product, the propargylic substituent and the bridgehead hydrogen are located cis to each other on the fused bicyclic systems (Scheme 4.37).

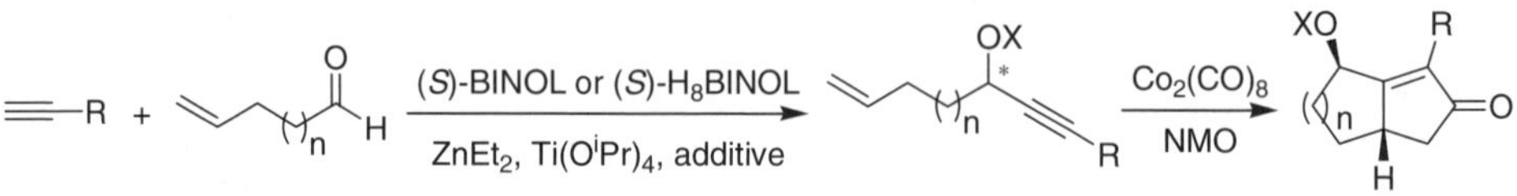

147 (n = 1)
148 (n = 2)
X = H, Ac, Me

149 (n = 1)
150 (n = 2)
dr up to 99:1
ee up to 95 %

R = Ph, -CH_2Ph, -CH_2CH_2Ph, -$CH_2CH_2CH_2CH_3$, CO_2Me, TMS

Scheme 4.37 *Highly enantio- and diastereoselective Pauson-Khand reactions of enantiomerically enriched propargylic alcohols.*

Similar results were obtained with allylic ethers of chiral propargylic alcohols **151**: they also underwent a highly diastereoselective PKR and retained their high enantiomeric purity (Scheme 4.38). This study demonstrated that the sizes of the substituents at the propargylic position and on the alkyne are important for diastereoselectivity, as more bulky substituents resulted in higher diastereoselectivity.

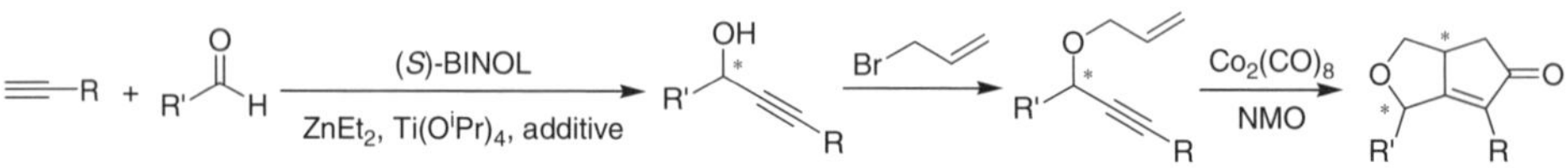

151 **152** **153**

dr up to 99:1
ee up to 95 %

R = Ph, -CH_2CH_2Ph, -$CH_2CH_2CH_2CH_3$, TMS
R' = -$CH_2CH_2CH_2CH_3$, cyclohexyl

Scheme 4.38 *Highly diastereoselective Pauson-Khand reactions of the chiral propargylic alcohols **151**.*

4.4 Conclusion

In this chapter we disclose alternatives to the diastereoselective Pauson-Khand reaction for the synthesis of various cyclopentenone derivatives. Both methodologies, the intra- and intermolecular versions, seem very attractive and offer in some cases a great degree of regioselectivity. In the majority of the examples, the transference of the chiral information is high obtaining the valuable chiral cyclopentenones in excellent diastereoselectivities.

References

[1] P. Magnus, D. P. Becker, "Stereospecific Dicobalt Octacarbonyl Mediated Enyne Cyclization for the Enantiospecific Synthesis of a 6a-Carbocycline Analog" *J. Am. Chem. Soc.* 1987, **109** (24), 7495–7498.

[2] P. Magnus, L. M. Principe, J. M. Slater, "Stereospecific Dicobalt Octacarbonyl Mediated Enyne Cyclization for the Synthesis of the Cytotoxic Sesquiterpene (+/-)-Quadrone" *J.Org. Chem.* 1987, **52** (8), 1483–1486.

[3] J. Mulzer, K. D. Graske, B. Kirste, "Asymmetric synthesis of carbacyclin precursors by Pauson-Khand cyclization" *Liebigs Ann. Chem.* 1988, (9), 891–897.

[4] W. R. Roush, J. C. Park, "Asymmetric Allylborations of 3-Decynal Dicobalt Hexacarbonyl - Cobalt Induced Reversal of Enantioselectivity and Applications to Highly Diastereoselective Intramolecular Pauson-Khand Reactions" *Tetrahedron Lett.* 1991, **32** (44), 6285–6288.

[5] S. Takano, S, K. Inomata, K. Ogasawara, "A New Enantiospecific Route to (-)-Kainic Acid Via the Intramolecular Pauson-Khand Reaction" *Chem. Commun.* 1992, (2), 169–170.

[6] (a) S. E. Yoo, S. H. Lee, N. Jeong, I. Cho, "A Total Synthesis of (-)-Alpha-Kainic Acid by the Pauson-Khand Reaction" *Tetrahedron Lett.* 1993, **34** (21), 3435–3438; (b) S. E. Yoo, S. H. Lee, "A Total Synthesis of (-)-Alpha-Kainic Acid Involving a Pauson-Khand Reaction as the Key Step" *J. Org. Chem.* 1994, **59** (23), 6968–6972.

[7] M. E. Kraft, R. A. Holton, "Preparation of a New, Highly Reactive Form of Iron(0) and Its Use in Deprotonation of Carbonyl-Compounds" *J. Org. Chem.* 1984, **49** (19), 3669–3670.

[8] W. E. Lindsell, P. N. Preston, A. B. Rettie, " Synthesis and Characterization of Hexacarbonyldicobalt Complexes Derived from 2-Propynyl and 3-Butynyl 4,6-Di-O-Acetyl-2,3-Dideoxy-Alpha-D-Erythro-Hex-2-Enopyranosides" *Carbohyd. Res.* 1994, **254**, 311–316.

[9] J. Marcocontelles, "The Pauson-Khand Reaction on Carbohydrate Templates .1. Synthesis of Bis-Heteroannulated Pyranosides" *Tetrahedron Lett.* 1994, **35** (28), 5059–5062.

[10] J. MarcoContelles, "Asymmetric Pauson-Khand reaction. Cobalt-mediated cycloisomerization of 1,6-enynes in carbohydrate templates: Synthesis of bis-heteroannulated pyranosides" *J. Org. Chem.* 1996, **61** (22), 7666–7670.

[11] (a) N. Naz, T. H. Altel, Y. Alabed, W. Voelter, "Synthesis of Polyfunctionalized Bis-Annulated Pyranosides - Useful Intermediates for Triquinane Synthesis" *Tetrahedron Lett.* 1994, **35** (46), 8581–8582; (b) N. Naz, T. H. AlTel, Y. Alabed, W. Voelter, R. Ficker, W. Hiller, "Palladium-cobalt-mediated double annulation process: A new strategy to chiral and polysubstituted bis-cyclopentanoids on carbohydrate precursors" *J. Org. Chem.* 1996, **61** (10), 3250–3255.

[12] C. Mukai, M. Uchiyama, S. Sakamoto, M. Hanaoka, "A highly diastereoselective construction of optically active bicyclo[3.3.0]octenone derivatives from L-ascorbic acid via Pauson-Khand reaction" *Tetrahedron Lett.* 1995, **36** (32), 5761–4.

[13] B. Alcaide, C. Polanco, M. A. Sierra, "Synthesis of fused tricyclic beta-lactams by the Pauson-Khand cyclization of enyne-2-azetidinones" *Tetrahedron Lett.* 1996, **37** (38), 6901–6904.

[14] G. L. Bolton, "Solid phase synthesis of azabicyclo[4.3.0]nonen-8-one amino acid derivatives via intramolecular Pauson-Khand cyclization" *Tetrahedron Lett.* 1996, **37** (20), 3433–3436.

[15] A. R. Kennedy, W. J. Kerr, D. M. Lindsay, J. S. Scott, S. P. Watson, "Stereochemical and mechanistic features of asymmetric Pauson-Khand processes" *Perkin 1* 2000, (24), 4366–4372.

[16] P. Areces, M. A. Duran, J. Plumet, M. B. Hursthouse, M. E. Light, "Synthesis of Enantiopure 3-Azabicyclo[3.3.0]octen-7-one Derivatives via Intramolecular Pauson-Khand Cycloaddition Reaction Using Tri-O-acetyl-D-glucal as Starting Material" *J. Org. Chem.* 2002, **67** (10), 3506–3509.

[17] P. Areces, E. Carrasco, J. Plumet, "From D-galactose to enantiopure 3-azabicyclo[3.3.0]octen-7-one derivatives via Pauson-Khand reaction" *ARKIVOC* (Gainesville, FL, United States) 2005, (9), 165–174.

[18] B. Jiang, M. Xu, "Catalytic Diastereoselective Pauson-Khand Reaction: an Efficient Route to Enantiopure Cyclopenta[c]proline Derivatives" *Org. Lett.* 2002, **4** (23), 4077–4080.

[19] T. Ishikawa, H. Ishii, K. Shimizu, H. Nakao, J. Urano, T. Kudo, S. Saito, "Diastereoselective Total Synthesis of Isocarbacyclin from L-Ascorbic Acid" *J. Org. Chem.* 2004, **69** (23), 8133–8135.

[20] N. Jeong, D. H. Kim, J. H. Choi, "Desymmetrization of meso-dienyne by asymmetric Pauson-Khand type reaction catalysts" *Chem. Commun.* (Cambridge, United Kingdom) 2004, (9), 1134–1135.

[21] R. M. Moriarty, N. Rani, L. A. Enache, M. S. Rao, H. Batra, L. Guo, R. A. Penmasta, J. P. Staszewski, S. M. Tuladhar, O. Prakash, D. Crich, A. Hirtopeanu, R. Gilardi, "The Intramolecular Asymmetric Pauson-Khand Cyclization as a Novel and General Stereoselective Route to Benzindene Prostacyclins: Synthesis of UT-15 (Treprostinil)" *J. Org. Chem.* 2004, **69** (6), 1890–1902.

[22] A. Lanver, H.-G. Schmalz, "A Pauson-Khand approach to new carbocyclic nucleoside analogs" *Eur. J. Org. Chem.* 2005, (7), 1444–1458.

[23] T. Honda, K. Kaneda, "Diastereoselective Formal Synthesis of a Monoterpene Alkaloid, (-)-Incarvilline" *J. Org. Chem.* 2007, **72** (17), 6541–6547.

[24] M. K. Pallerla, J. M. Fox, "Enantioselective Synthesis of (-)-Pentalenene" *Org. Lett.* 2007, **9** (26), 5625–5628.

[25] P. Bremond, E. Girardeau, Y. Coquerel, J. Rodriguez, "Stereoselective synthesis of the core of naturally occurring anti-HIV isolitseane B" *Lett. Org. Chem.* 2007, **4** (4), 232–233.

[26] C. E. Madu, C. J. Lovely, "Steric promotion of the intramolecular Pauson-Khand reaction of aryl enynes" *Synlett* 2007, (13), 2011–2016.

[27] G. R. Dake, E. E. Fenster, B. O. Patrick, "A Synthetic Approach to the Fusicoccane A-B Ring Fragment Based on a Pauson-Khand Cycloaddition/Norrish Type 1 Fragmentation" *J. Org. Chem.* 2008, **73** (17), 6711–6715.

[28] X. Jia, R. M. Williams, "A diastereoselective intramolecular Pauson-Khand approach to the construction of the BC-ring system in tuberostemoninol" *Tetrahedron: Asymmetry* 2008, **19** (24), 2901–2906.

[29] K. Kaneda, T. Honda, "Stereocontrolled synthesis of (+)-α-skytanthine by means of an intramolecular Pauson-Khand reaction" *Tetrahedron* 2008, **64** (51), 11589–11593.

[30] S. Sezer, Y. Guemruekcue, E. Sahin, C. Tanyeli, "Stereoselective synthesis of spirocyclic cyclopentapyrans by the Pauson-Khand reaction on camphor tethered enynes" *Tetrahedron: Asymmetry* 2008, **19** (23), 2705–2710.

[31] A. Vazquez-Romero, J. Rodriguez, A. Lledo, X. Verdaguer, A. Riera, "Enantioselective Syntheses of Carbanucleosides from the Pauson-Khand Adduct of Trimethylsilylacetylene and Norbornadiene" *Org. Lett.* 2008, **10** (20), 4509–4512.

[32] J. Sola, A. Riera, X. Verdaguer, M. A. Maestro, "Phosphine-Substrate Recognition through the C-H•••O Hydrogen Bond: Application to the Asymmetric Pauson-Khand Reaction" *J. Am. Chem. Soc.* 2005, **127** (39), 13629–13633.

[33] K. D. Closser, M. M. Quintal, K. M. Shea, "The Scope and Limitations of Intramolecular Nicholas and Pauson-Khand Reactions for the Synthesis of Tricyclic Oxygen- and Nitrogen-Containing Heterocycles" *J. Org. Chem.* 2009, **74** (10), 3680–3688.

[34] S. Sezer, E. Sahin, C. Tanyeli, "Stereoselective synthesis of optically active cyclopenta[c]pyrans and cyclopenta[c]furans by the intramolecular Pauson-Khand reaction" *Tetrahedron: Asymmetry* 2010, **21** (4), 476–485.

[35] L. Shen, R. P. Hsung, "Pauson-Khand cycloaddition reactions of chiral ynamides. Observation of an unusual endo-addition with norbornadiene" *Tetrahedron Lett.* 2003, **44** (52), 9353–9358.

[36] M. K. Pallerla, J. M. Fox, "Diastereoselective Intermolecular Pauson-Khand Reactions of Chiral Cyclopropenes" *Org. Lett.* 2005, **7** (16), 3593–3595.

[37] P. Areces, E. Carrasco, J. Plumet, "The intermolecular Pauson-Khand reaction of sugar-derived aza-enynes: a case of total diastereoselectivity" *J. Carboh. Chem.* 2006, **25** (2–3), 197–202.

[38] F. Antras, S. Laurent, M. Ahmar, H. Chermette, B. Cazes, "Pauson-Khand Reaction of Allenic Hydrocarbons: Synthesis of 4-Alkylidenecyclopentenones" *Eur. J. Org. Chem.* 2010, (17), 3312–3336.

[39] M. Turlington, Y. Yue, X.-Q. Yu, L. Pu, "Catalytic Asymmetric Synthesis of Chiral Propargylic Alcohols for the Intramolecular Pauson-Khand Cycloaddition" *J. Org. Chem.* 2010, **75** (20), 6941–6952.

5

Asymmetric Intra- and Intermolecular Pauson-Khand Reactions: The Chiral Auxiliary Approach

Albert Moyano

The Pauson-Khand Reaction: Scope, Variations and Applications, First Edition. Edited by Ramon Rios Torres.

For what the hours have given is given, and this
Changeless; howbeit these change, and in good time
Devise new things and good, not one thing still.

—Swinburne, 'Atalanta in Calydon'

5.1 Introduction

The application of the intramolecular Pauson-Khand reaction (PKR) to a concise and highly diastereoselective (but racemic) synthesis of the antitumor sesquiterpene coriolin, reported by Exon and Magnus in 1983,[1] demonstrated the potential of this cobalt-mediated synthesis of 2-cyclopentenones for the stereocontrolled construction of chiral carbocycles.

In effect, both in the intermolecular (Scheme 5.1) and in the intramolecular (Scheme 5.2) version of the PKR, first disclosed by Pauson *et al.*[2] and by Schore and Croudace,[3] respectively, three carbon–carbon bonds and up to two stereogenic centers are created in a single step from achiral precursors (an alkene, an alkyne, and carbon monoxide), in the presence of dicobalt octacarbonyl.

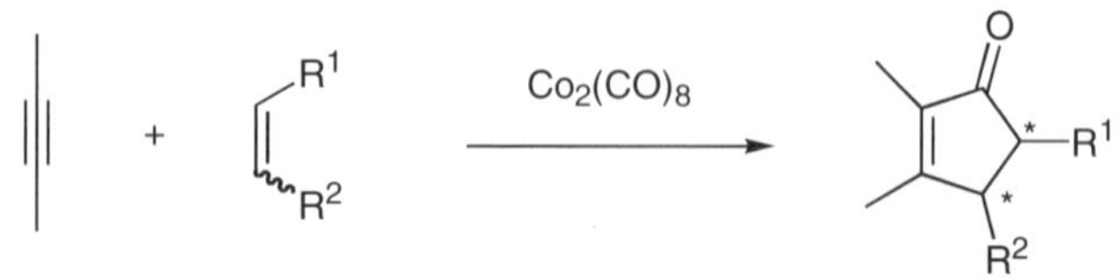

Scheme 5.1

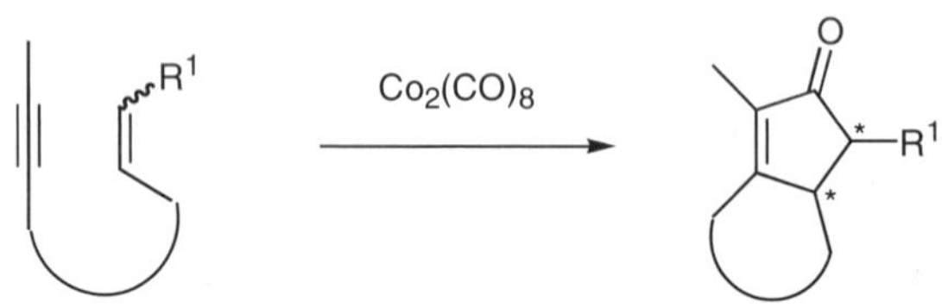

Scheme 5.2

Moreover, it is worth noting that in the case of intermolecular PKRs of symmetrical bicyclic alkenes, two additional stereogenic centers are generated as a consequence of the desymmetrization of the alkene (Scheme 5.3).[4]

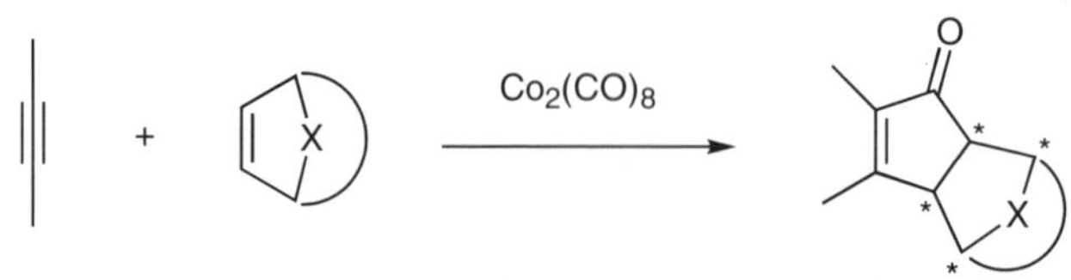

Scheme 5.3

The mechanism of the PKR forwarded by Magnus in 1985[5] and depicted in Scheme 5.4, has been supported both by theoretical studies and by some experimental evidence for the proposed intermediates.[6,7] The first step is the reaction of the alkyne unit with dicobalt octacarbonyl to form the isolable hexacarbonyl complex **I**.[8] Thermal, chemical, or photochemical activation leads to the loss of a carbon monoxide ligand affording a coordinatively unsaturated pentacarbonyl complex (η^2-alkyne)[μ-$Co_2(CO)_5$] (**II**), and the resulting vacant position is occupied by an alkene (**III**). The alkene undergoes 1,2-insertion at a carbon–cobalt bond of the tetrahedral C_2Co_2 cluster, followed by (or concomitantly with) coordination of carbon monoxide, to form the cobaltacycle **IV**, which experiences a carbonyl insertion reaction (**V**) and reductive elimination of a cobalt atom to give **VI**. Finally, reductive elimination of dicobalt hexacarbonyl affords the final cyclopentenone **VII**. As we will see later in more detail, the transformation of **III** into **IV** determines the stereochemical outcome of the reaction. Under certain conditions, the unstable dicobalt hexacarbonyl fragment can take up two molecules of carbon monoxide (or lead directly to the formation of **I**), rendering the whole process catalytic in cobalt.[9] Otherwise, decomposition into inert metallic cobalt leads to the classical experimental procedure of the PKR, that calls for the use of stoichiometric amounts of dicobalt octacarbonyl.[10]

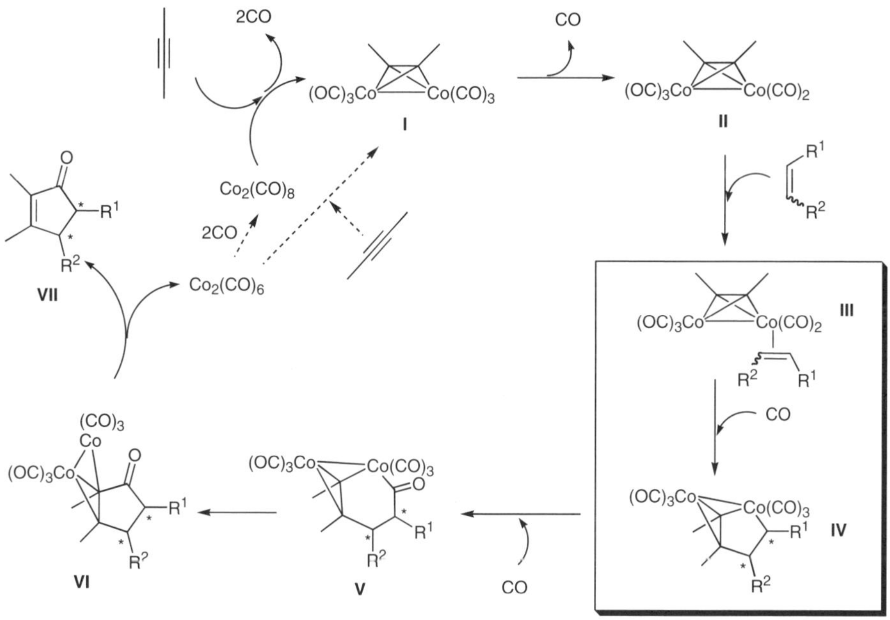

Scheme 5.4

The chiral auxiliary strategy has not only been of paramount importance for the development of asymmetric synthesis, but still provides in many instances the most practical and reliable procedures for the efficient preparation of chiral compounds in highly enantiopure fashion, both in academy and in industry.[11] This is especially true for the PKR, since the use of chiral auxiliaries, that enabled for the first time its application to the enantioselective synthesis of complex natural products, and of chiral ligands continue to be the methods of choice for asymmetric intermolecular PKRs.

Chiral auxiliary-mediated asymmetric PKRs can be classified into three major types, depending on the position of the auxiliary Xc (the first two of these types, in turn, can be performed either in an intermolecular or in an intramolecular fashion):

Type **A**: Chiral auxiliary attached to the alkyne (Scheme 5.5)

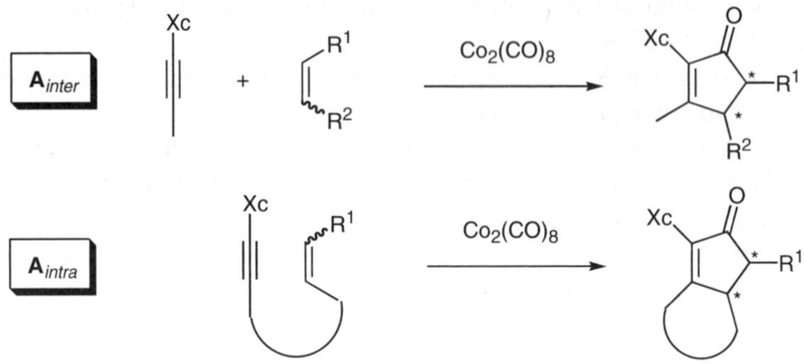

Scheme 5.5

Type **B**: Chiral auxiliary attached to the alkene (Scheme 5.6):

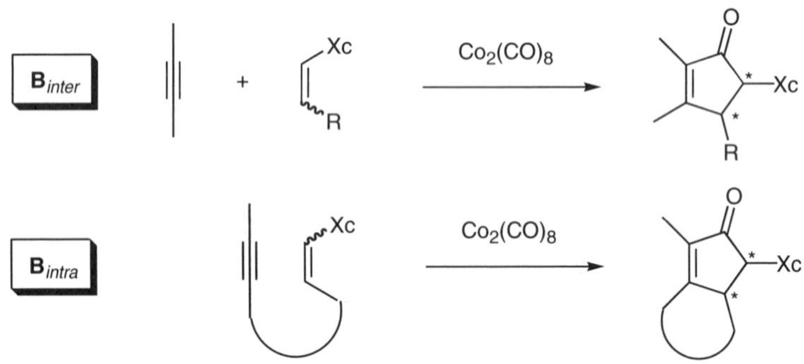

Scheme 5.6

Type **C**: Chiral auxiliary located in the enyne chain (Scheme 5.7):

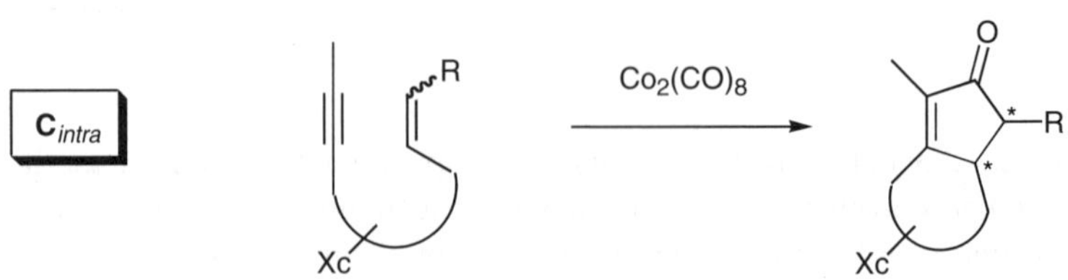

Scheme 5.7

This chapter deals mainly with asymmetric PKRs based on these three approaches. A fourth possibility for rendering the PKR enantioselective consists o using the chiral auxiliary as a ligand in the metal complex either in a stoichiometric or in a catalytic fashion (Scheme 5.8),[12, 13] and is discussed *in extenso* in Chapter 6. Finally, we will comment on the use of chiral reagents for obtaining nonracemic PK adducts.

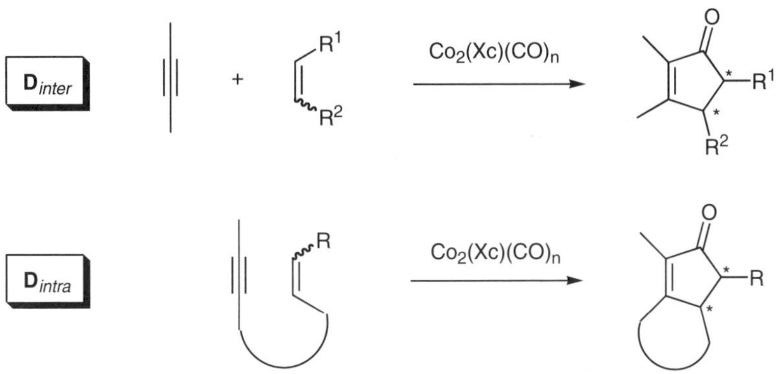

Scheme 5.8

5.2 Asymmetric Intramolecular PKRs with the Aid of Chiral Auxiliaries

5.2.1 Chiral Alkoxyacetylenes

The first ideas concerning the use of chiral alcohols as stereochemical controllers in the PKR crystallized during the fall of 1986 in the laboratory of Prof. Andrew E. Greene in Grenoble, where, in the context of a research program on chiral auxiliary-mediated asymmetric [2+2] cycloadditions, the author of this chapter had developed a convenient synthesis of chiral alkoxyacetylenes (Scheme 5.9).[14, 15]

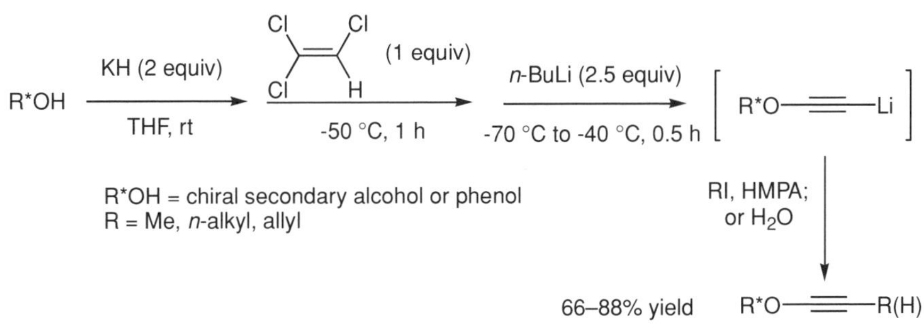

Scheme 5.9

This research set the stage for the first chiral auxiliary-mediated asymmetric intramolecular PKR, disclosed in 1990 (Table 5.1).[16] Best results (52% de) were obtained for compound

Table 5.1

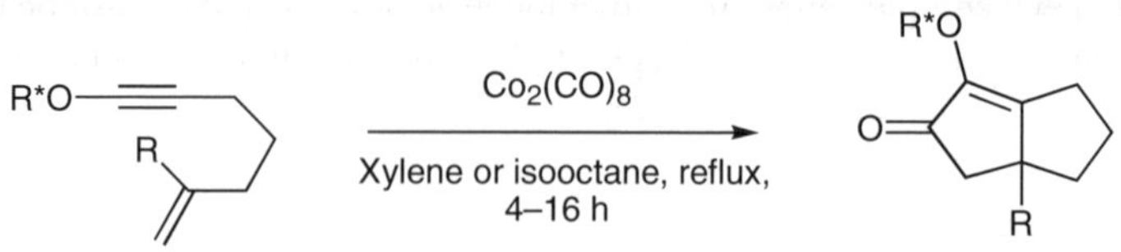

1a–c

Product	R*O-	R	Yield (%)	Dr
1a	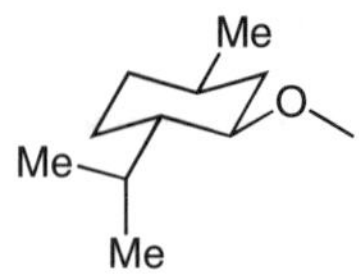	H	30	1.1:1
1b		H	38	3.2:1
1c		Me	28	1.2:1

1b, whose enyne precursor was obtained in 90% yield from (1*S*,2*R*)-2-phenylcyclohexanol and 5-iodopentene by the procedure depicted in Scheme 5.9.[14] The extent of the diastereoselection, however, was observed to be a function of the olefin substitution, and the methyl-substituted bicyclooctenone adduct **1c** was produced with very low diastereoselectivity.

The configuration of the newly created stereogenic center in the major diastereomer of **1b** could be established as shown in Scheme 5.10. The major isomer of **1b**, obtained in diastereomerically pure form after chromatographic purification, was exposed to Yamamoto's methylcopper-boron trifluoride reagent to give the conjugate addition product **2b** (as a mixture of epimers at C2). Reductive cleavage of the auxiliary in **2b** with samarium diiodide then cleanly produced the known (+)-*cis*-1-methylbicyclo[3.3.0]octan-3-one **3** (>95% ee),

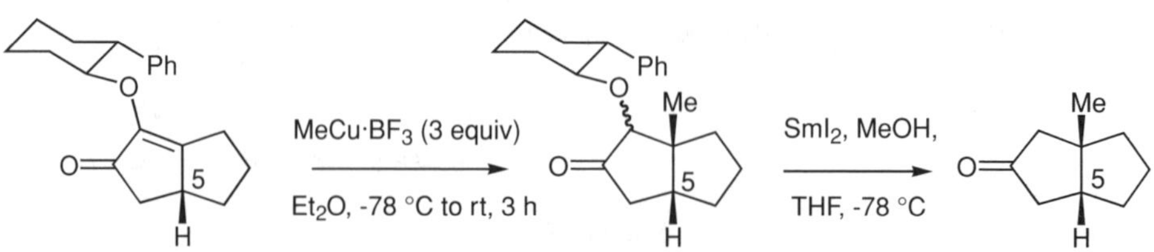

1b (major) **2b** **3**

Scheme 5.10

leading to the recovery of the chiral auxiliary too. Since the absolute configuration of **3** at C5 was known to be (*R*), that at C5 of the major diastereomer of **1b** could also be assigned as (*R*).

This stereochemical outcome could be rationalized in the framework of the currently accepted mechanism for the PKR (see Scheme 5.4 above). Molecular modeling studies showed that the most stable conformation of dicobalt hexacarbonyl complexes of alkoxyethynes derived from *trans*-2-phenycyclohexanol is that in which the bulky [μ-$Co_2(CO)_6$] group is disposed as far as possible from the cyclohexane ring. The analogous conformation in the case of the enynes leading to **1b** and **1c** (Scheme 5.11) places the pro-(*S*) cobalt atom so as to be more accessible to the alkene than the diastereotopic pro-(*R*) one due to shielding by the phenyl group. Assuming that the metallocycle formation is the step that controls the diastereoselectivity of the process, the major diastereomer of **1b** would then arise from a *cis*-cobaltacycle, that is the product of coordination of the pro-(*S*) cobalt with the *Si* face of the olefin. Coordination of the pro-(*S*) cobalt with the *Re*-face of the olefin would lead to a less stable *trans*-cobaltacycle. In the case of **1c**, the energy difference between the *cis*- and the *trans*-cobaltacycles would be diminished due to the steric interaction between the methyl and the exocyclic $Co(CO)_3$ moiety, leading to a reduced diastereoselectivity in the cyclization. It is interesting to note that molecular mechanics calculations on the proposed cobaltacycle intermediates give support to this interpretation.[17]

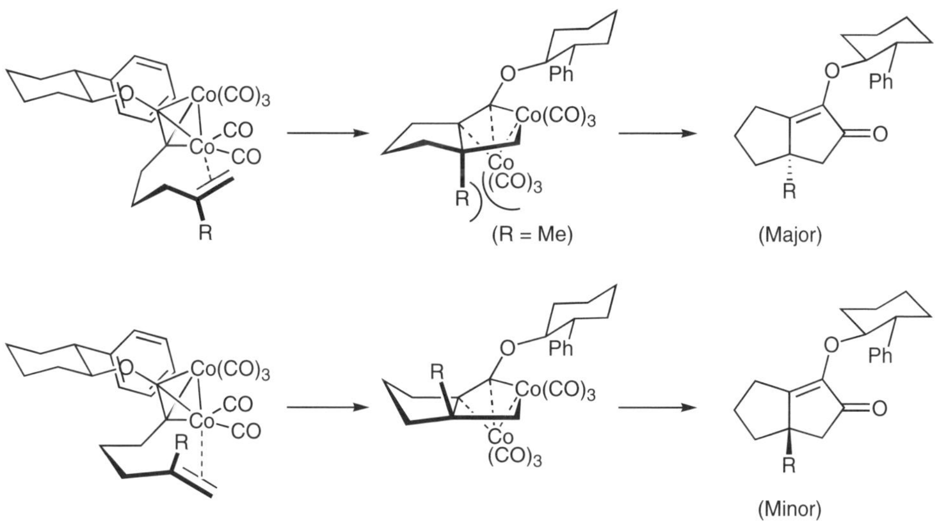

Scheme 5.11

Subsequently, it was found that the intramolecular PKR of ynol ethers of camphor-derived neopentyloxy alcohols **4a** and **4b** took place in milder conditions, leading to higher yields and diastereoselectivities.[18] As depicted in Scheme 5.12, both alcohols were converted to the propargyl alcohols **5a** and **5b**, respectively, by trapping of thc *in situ* generated alkoxyethynyl lithium derivatives[14] with para-formaldehyde. Allylation of these compounds provided the corresponding enynes **6a** and **6b** in good overall yields.

The cobalt-mediated bicyclization of these alkoxyenynes readily took place (Table 5.2). When **6a** was subjected to the thermal PKR conditions at 50 °C, **7a** was obtained as a

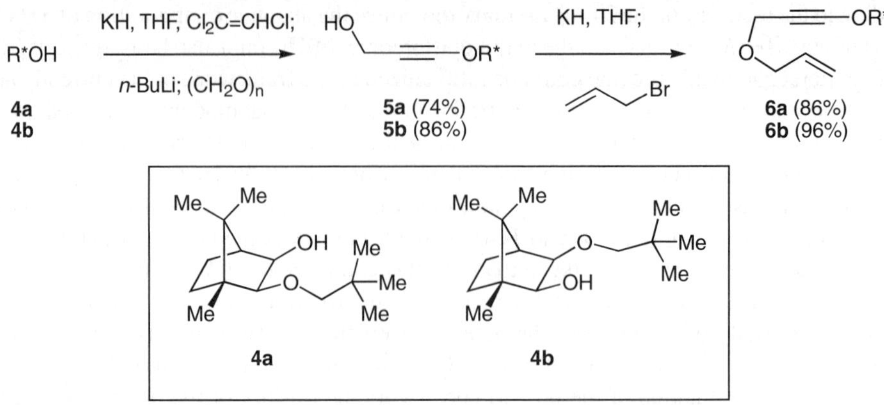

Scheme 5.12

7.3:1 mixture of diastereomers in a 65% yield. Remarkably, with **6b** the bicyclization could be performed at room temperature under nitrogen to give **7b** in a 54% yield and with a diastereomeric ratio of 15.7:1. Moreover, in both cases the diastereomers could be readily separated by simple column chromatography. For **7b**, a single crystal of the minor diastereomer was used to establish the absolute configuration at C5 by X-ray diffraction analysis, and this was found to be (*S*). Accordingly, that at C5 of the major diastereomer of **7b** was found to be (*R*).

Table 5.2

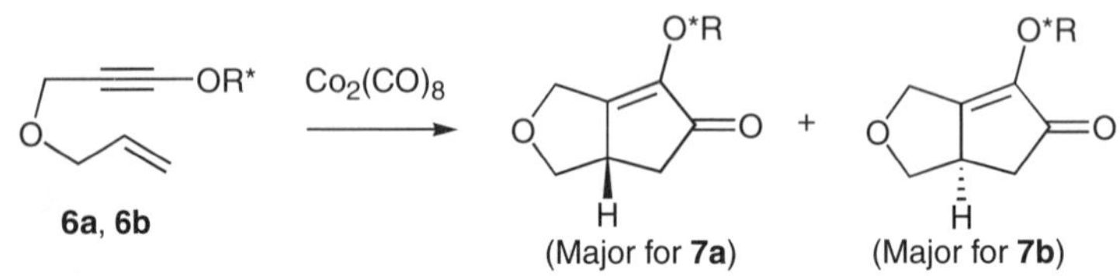

Product	R*O-	Conditions	Yield (%)	Dr
7a	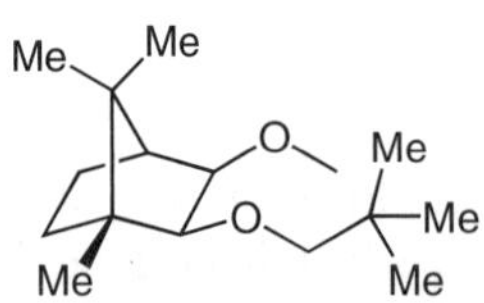	Isooctane, 50 °C, 17 h	65	7.3:1
7b	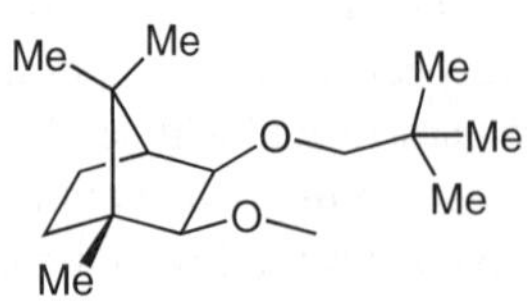	Hexane, rt, 4 h	54	15.7:1

Treatment of enones **7a** and **7b** (major diastereomers) with a catalytic amount of hydrochloric acid in methanol led, with complete regioselectivity and in excellent yield, to both enantiomers of the α-methoxyenone **8**, with a nearly quantitative recovery of the alcohols **4a** and **4b**, respectively (Schemes 5.13 and 5.14). This allowed an unambiguous assignment to be made of the absolute configuration (5*S*) of the major diastereomer of **7a**.

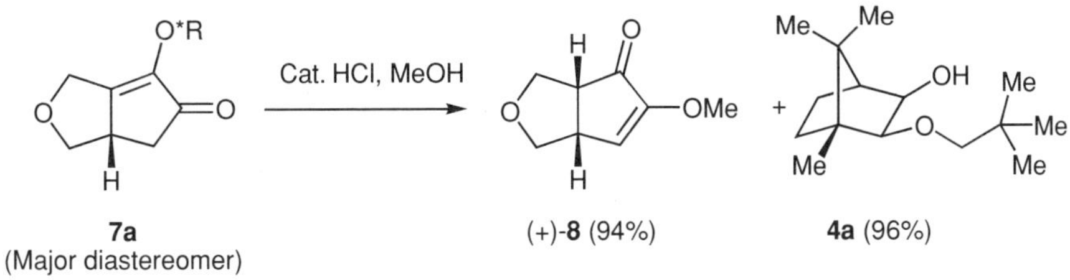

Scheme 5.13

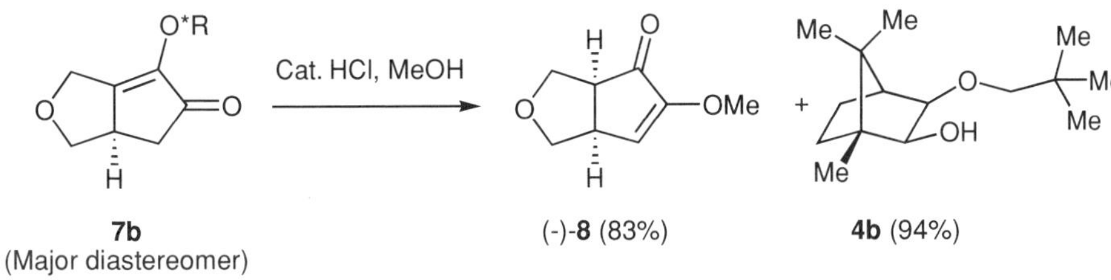

Scheme 5.14

As happened with *trans*-2-phenylcyclohexanol,[16,17] molecular mechanics calculations offered a ready explanation for the stereochemical outcome of the reaction. Thus, the cyclization of **6b** was proposed to occur from a conformation such as that shown in Scheme 5.15, in which the pro-(*R*) cobalt tricarbonyl group is shielded by the neopentyloxy substituent of the chiral auxiliary. Then, interaction of the *Si* face of the olefin with the pro-(*S*) cobalt leads to a (5*R*) configuration in **7b**.

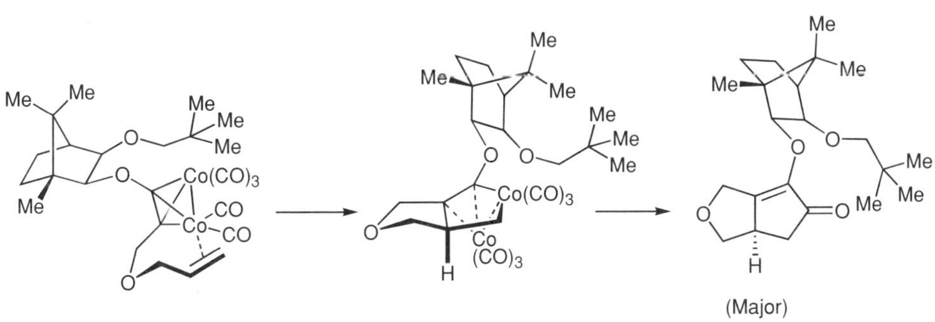

Scheme 5.15

A most useful feature of the intramolecular PKR is its ability to generate, in a single synthetic operation, tricyclic[19] or even tetracyclic[20] frameworks. Camphor-derived alcohol **4b** proved to be a practical chiral auxiliary for the asymmetric synthesis of angular triquinanes (Scheme 5.16).[21]

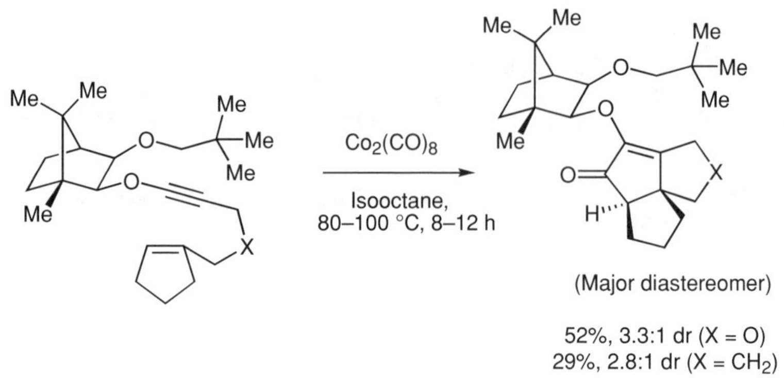

Scheme 5.16

Although both the yields and diastereoselectivities were only moderate, the two diastereomers of the resulting angular triquinane were readily separated by column chromatography. The absolute configuration of the major diastereomer showed that a mechanistic working model similar to that depicted in Scheme 5.15 above was also operative in this case. Other alcohols such as *trans*-2-phenylcyclohexanol or **4a** gave lower diastereoselectivities, without the added benefit of easy separability of the diastereomeric products.

The synthesis of (+)-15-nor-pentalenene, outlined in Scheme 5.17, constituted the first application of chiral 1-alkoxyenynes to the enantioselective preparation of natural products by intramolecular PKR.[22] It is worth noting that the key intramolecular PKR afforded the stereochemically pure major cycloadduct in 69% after chromatographic purification.

10-Methylthioisoborneol **9a**, that as we shall see later on gives outstanding results in intermolecular PKRs, was also tested in intramolecular PK cyclizations.[21,22] As can be seen in Scheme 5.18, very good diastereoselectivities (up to 12:1 dr) were obtained in the cyclization of alkoxyenynes derived from **9a** and leading to angular triquinanes, although the fact that the resulting diastereomeric products were not separable by column chromatography somewhat limited its applicability.

5.2.2 Chiral Acetylene Thioethers

The use of thiols as chiral auxiliaries in intramolecular (as well as in intermolecular) PKRs was briefly examined by Pericàs, Riera and co-workers in 1997.[23] Unfortunately only very low diastereoselectivities could be achieved (Scheme 5.19). As we will see later, somewhat more promising results were obtained in intermolecular reactions.

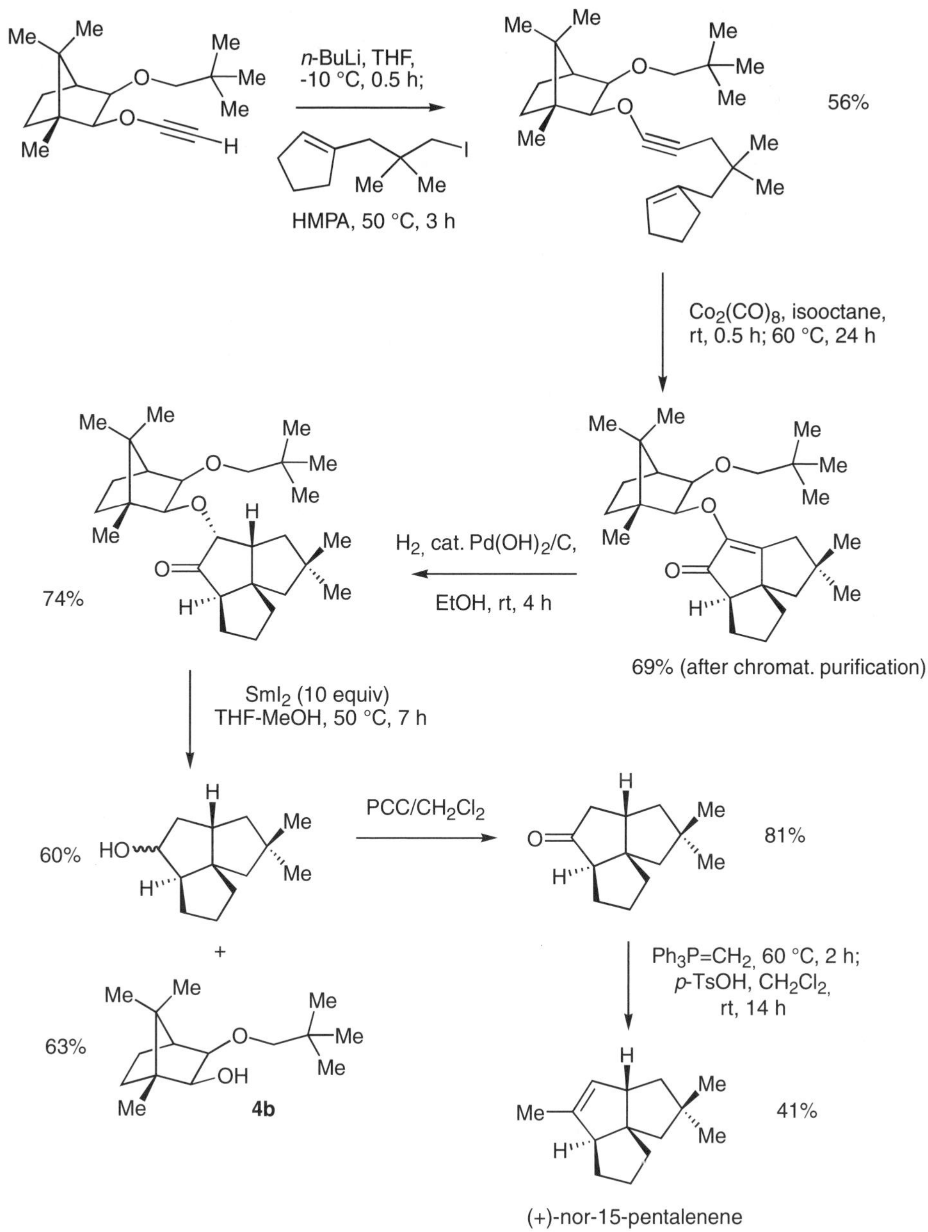

n-BuLi, THF,
-10 °C, 0.5 h;
HMPA, 50 °C, 3 h
56%
Co2(CO)8, isooctane,
rt, 0.5 h; 60 °C, 24 h
69% (after chromat. purification)
H2, cat. Pd(OH)2/C,
EtOH, rt, 4 h
74%
SmI2 (10 equiv)
THF-MeOH, 50 °C, 7 h
60%
+
63%
4b
PCC/CH2Cl2
81%
Ph3P=CH2, 60 °C, 2 h;
p-TsOH, CH2Cl2,
rt, 14 h
41%
(+)-nor-15-pentalenene

Scheme 5.17

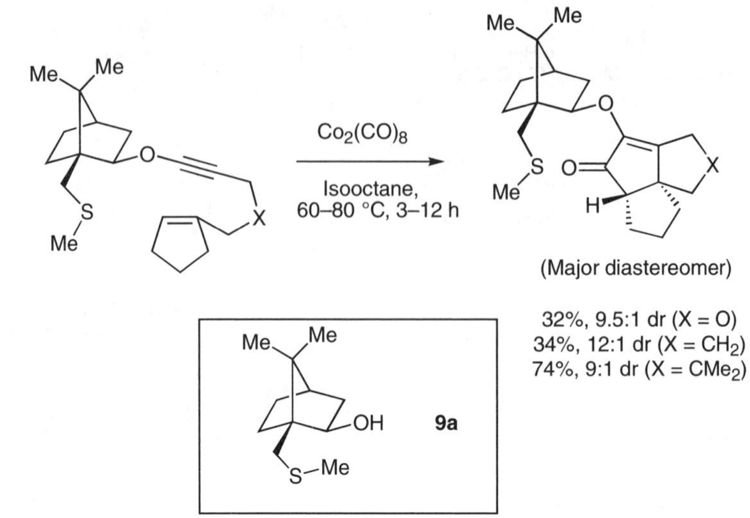

Scheme 5.18

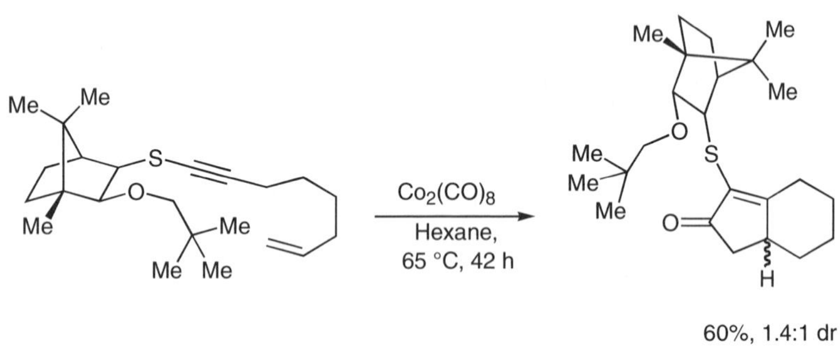

Scheme 5.19

5.2.3 Chiral 2-Alkynoate Derivatives

The use of chiral 2-alkynoate derivatives in intramolecular PKRs was first reported in 1999.[24] Under suitable conditions, 8-nonen-2-ynoate derivatives of chiral oxazolidinones and amines underwent cobalt-mediated bicyclization with acceptable yields, albeit with poor diastereoselectivities (Table 5.3).[24,25] Products **11e** and **11f** were converted into chirally substituted cyclopentadienes.[25]

Another example of a cyclization of a chiral 2-alkynoyl amide was described by Pericàs and co-workers in 1999.[26] The dicobalt hexacarbonyl complex **12**, generated *in situ* by an aminocarbonylation reaction of the dicobalt hexacarbonyl complex of 1-chloro-2-phenylethyne, provided the 3-azabicyclo[3.3.0]oct-1(8)-ene-2,7-dione system **13** with moderate yield and marginal diastereoselectivity by an amine *N*-oxide-promoted[27] intramolecular PKR (Scheme 5.20).

Table 5.3

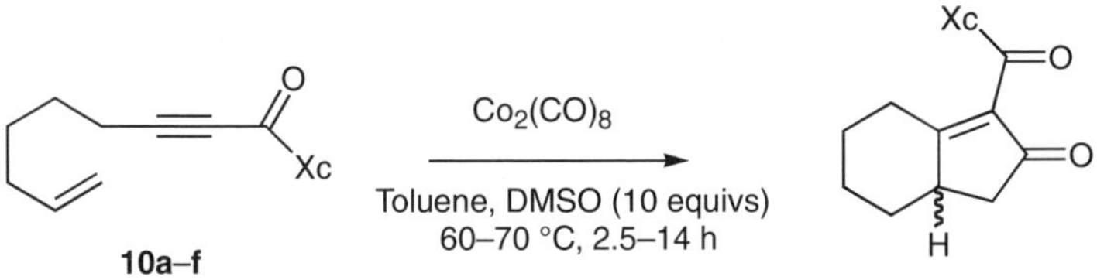

10a–f → **11a–f**

Enyne	Xc–	Product	Yield (%)	Dr
10a	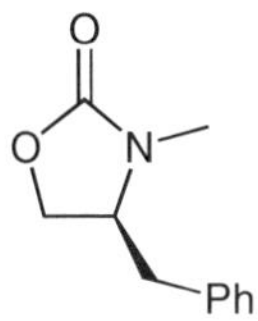	**11a**	92	1:1
10b	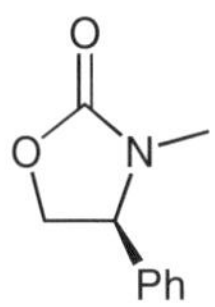	**11b**	80	1.1:1
10c	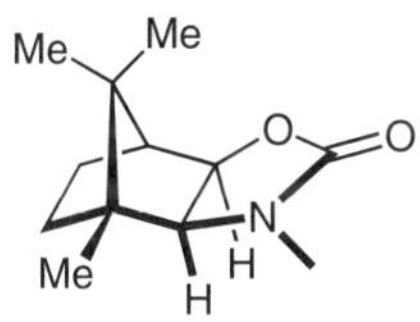	**11c**	80	1.9:1
10d	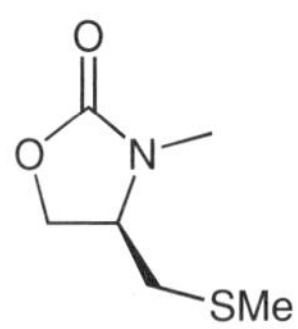	**11d**	59	1.1:1
10e	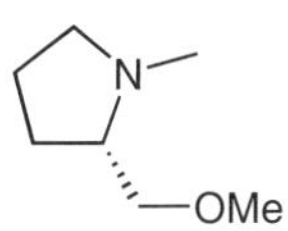	**11e**	71	1:1
10f	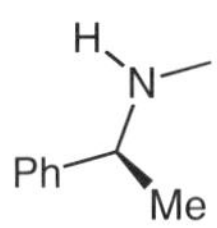	**11f**	85	1:1

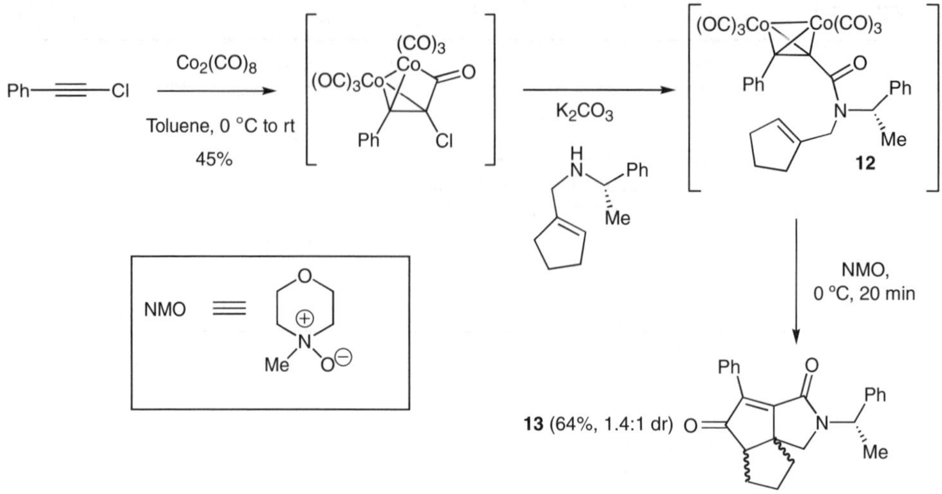

Scheme 5.20

5.2.4 Chiral *O*-Alkyl Enol Ethers

The first enantioselective synthesis of a natural product by means of a chiral auxiliary-mediated intramolecular PKR was reported in 1990 by Moyano, Pericàs, Greene, and co-workers.[28] The chemo- and stereoselective reduction of 1-alkoxyacetylenes into (*E*)- or (*Z*)-*O*-alkyl enol ethers[14,29] enabled the easy preparation of enynes **14a–14d**, derived from (1*S*,2*R*)-2-phenylcyclohexanol. These compounds were stirred with 1.1 equivalents of dicobalt hexacarbonyl in isooctane at room temperature, and the resultant dicobalt hexacarbonyl complexes, without isolation, were then heated at 95 °C for 1–5 h. The (*E*) enol ethers **14a**, **14c**, and **14d** were cyclized in acceptable yields and with moderate to good diastereoselectivity (up to 7:1 dr; Scheme 5.21). It is clear that the fact that in this approximation, contrary to that relying in the cyclization of 1-alkoxyenynes (see Table 5.1 above),[16] the newly created stereogenic center is immediately adjacent to the carbon bearing the chiral auxiliary, leads to a higher degree of stereocontrol.

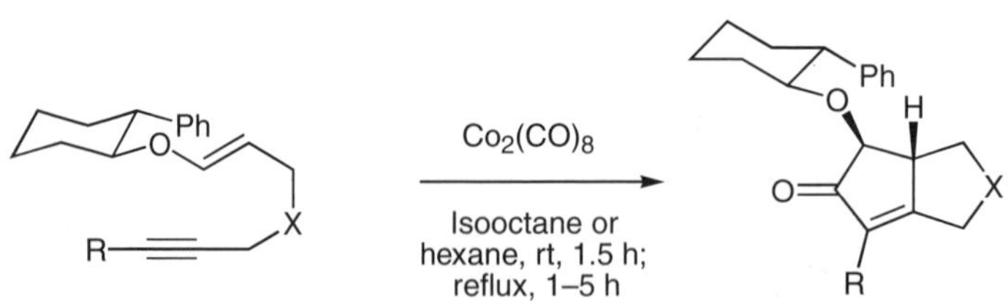

14a (X = CH_2, R = H)
14c (X = O, R = H)
14d (X = O, R = Me)

15a (X = CH_2, R = H): 55%,7:1 dr
15c (X = O, R = H): 41%, 4.5:1 dr
15d (X = O, R = Me): 42%, 4:1 dr

Scheme 5.21

The cyclization of the (*Z*)-configured enol ether **14b** took place in poor yield but still with good diastereoselectivity; this can be attributed to the fact that the bicyclization of this compound requires that the bulky alkoxy group is accommodated in the more sterically confining *endo* position (Scheme 5.22). Interestingly enough, in all instances the resulting diastereomers of the products **15a–d** were easily separable by simple column chromatography.

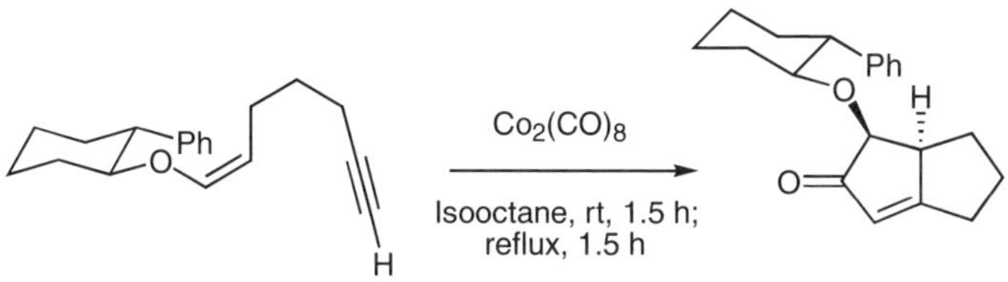

14b

(Major isomer)

15b (20%, 5:1 dr)

Scheme 5.22

The stereochemical identity of the major diastereomer of **15a** was established both by X-ray diffraction analysis of a monocrystal and, as shown in Scheme 5.23, by conversion to optically pure (>99% ee) (-)-(1*R*,5*S*)-1-methylbicyclo[3.3.0]octan-3-one **16**, in a sequence that allowed the recovery of the chiral auxiliary in a 92% yield. When the major diastereomer of **15b** was submitted to the same reaction sequence, the enantiomeric ketone (+)-(1*S*,5*R*)-**16** was obtained.

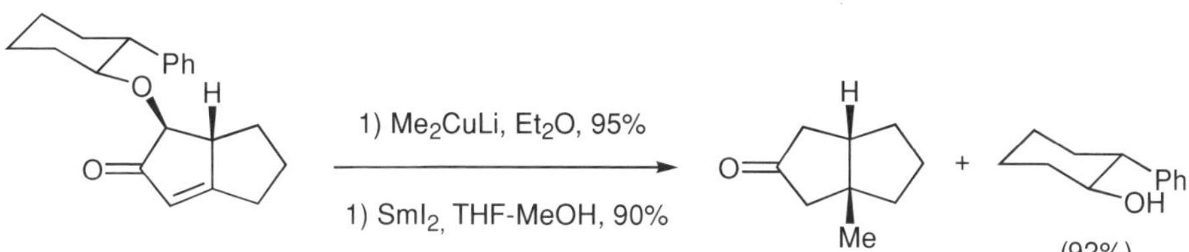

15a (Major isomer)

(-)-**16**

(92%)

Scheme 5.23

The stereochemical outcome of the reaction was rationalized with the aid of molecular mechanics calculations, that showed that the most stable conformation of the enyne **14a** is that (extended *s-trans*) depicted in Scheme 5.24, in which the Cα-*Si*, Cβ-*Re* face of the olefin is shielded by the phenyl group of the auxiliary. The other face of the enol ether is then preferentially coordinated to the cobalt, and evolves into a cobaltacycle that leads to the major diastereomer observed.[17]

The potential of this approach was illustrated by an enantioselective formal total synthesis of the natural product (+)-hirsutene (Scheme 5.25).[28] The acetylenic ether **17**,[14] derived from (1*R*,2*S*)-(-)-2-phenylcyclohexanol, was converted into the iodoacetylene **18**, that was reacted with the zinc reagent derived from the bromide **19** in the presence of

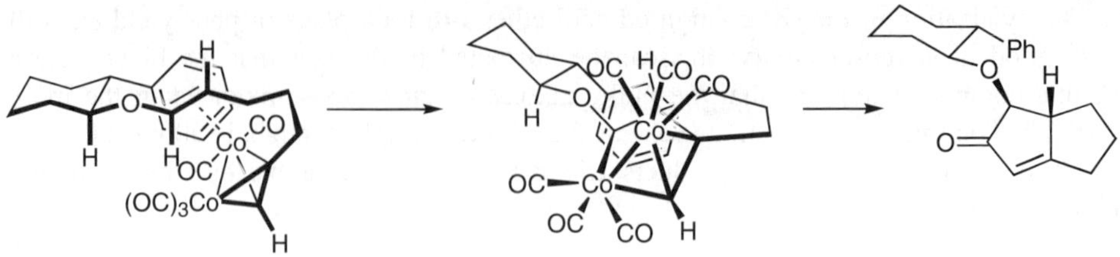

Scheme 5.24

copper(I), providing the diyne **20** in good yield. The chemo- and stereoselective reduction of the alkoxyacetylene moiety in this compound by lithium aluminum hydride afforded the alkoxyenyne precursor **21**. The intramolecular PKR took place under very mild conditions (warming the dicobalt hexacarbonyl complex of **21** at 42 °C for 12 h in hexane) to furnish a 6:1 mixture of two diastereomers, from which stereochemically pure **22** was isolated in 55% yield upon chromatographic purification. Birch reduction of **22** delivered the dihydro derivative **23**, from which the chiral auxiliary could be readily cleaved (91% recovery) to give the dextrorotatory ketone **24**, that in racemic form had been previously converted to hirsutene.

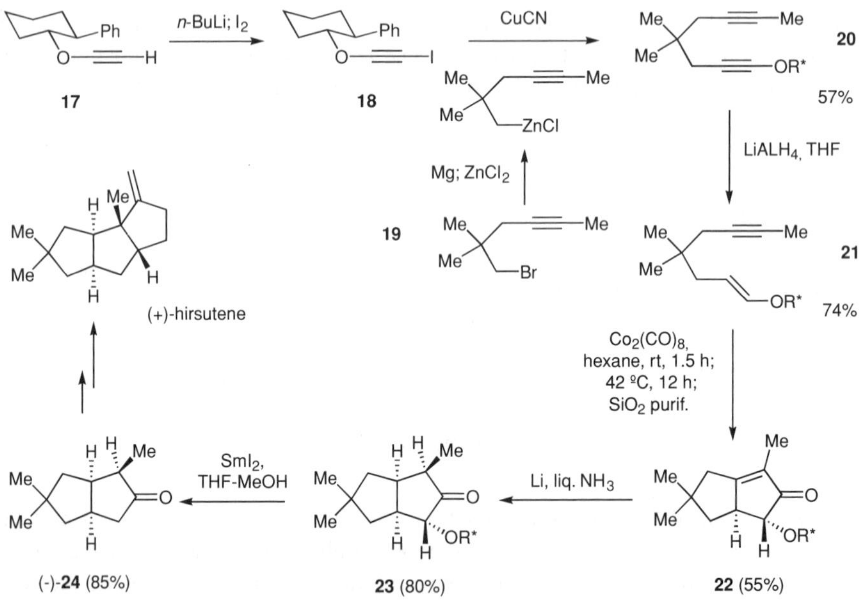

Scheme 5.25

Another application of this strategy in the field of asymmetric natural products synthesis was the enantioselective preparation of (+)-β-cuparenone, reported by Moyano, Pericàs, and co-workers in 1996.[30] This synthesis relied on the intramolecular PKR of the 7-alkoxy-4-thia-6-hepten-1-yne **27**, that was prepared from (1*R*,2*S*)-(-)-2-phenylcyclohexanol by a reaction sequence involving base-promoted isomerization of the propargyl ether **25** and regio- and diastereoselective (9:1 (*E*)/(*Z*) mixture) addition of propargyl thiol to the resulting homochiral allene **26** (Scheme 5.26).

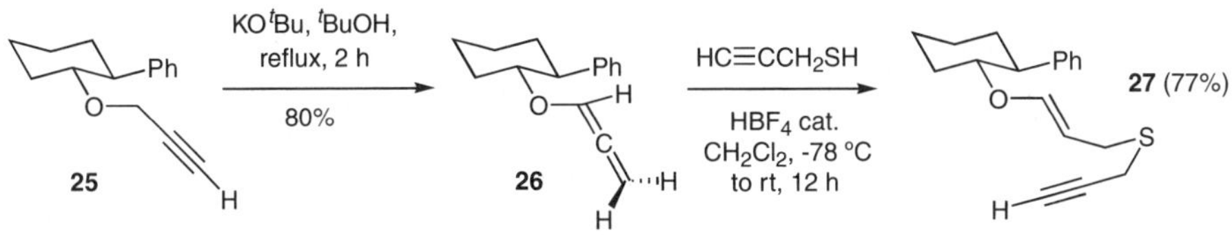

Scheme 5.26

Exposure of **27** to dicobalt octacarbonyl in isooctane, followed by heating at 90 °C for 2.5 h, produced a *ca.* 12:1 mixture of diastereomeric adducts **28** from which the major one could be isolated in 33% yield by column chromatography (Scheme 5.27). No products

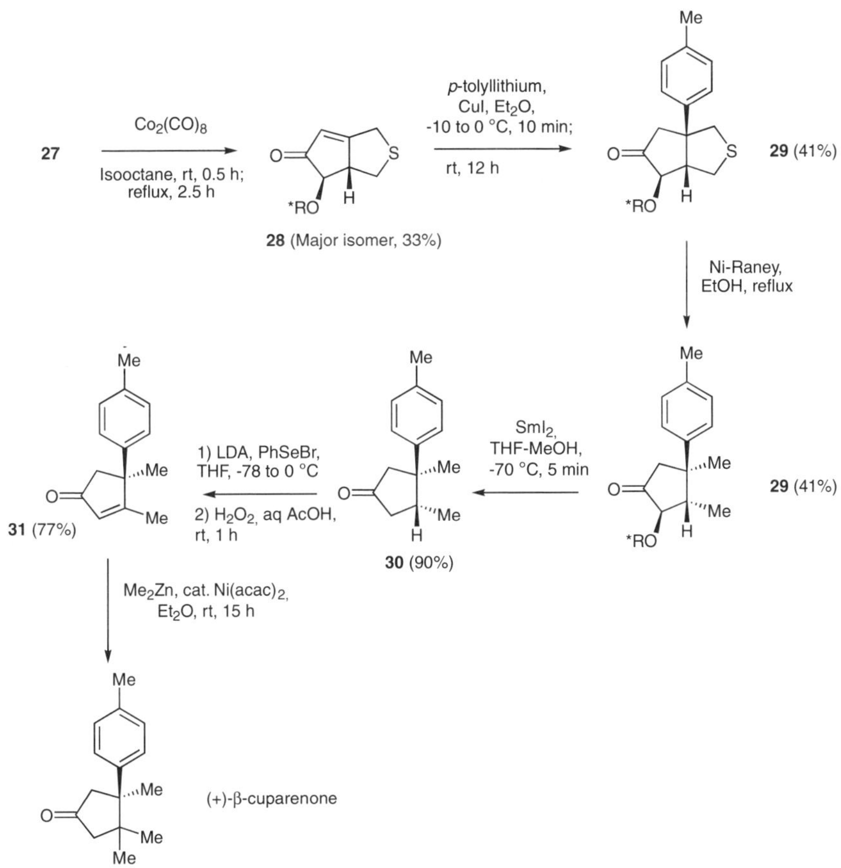

Scheme 5.27

arising from the intramolecular PKR of the minor (*Z*) isomer of **27** were detected, in accordance with the low PK reactivity exhibited by the *cis*-alkoxyenynes **14b**.[28] The absolute configuration of the major isomer of **28** was secured by X-ray diffraction analysis, and was fully coincident to that predicted by the working model depicted in Scheme 5.24 above. The final steps of the synthesis consisted in the conjugate addition of a *p*-tolyl group to **28**, Raney-nickel-promoted transformation of the five-membered ring in **29** into a *vic*-dimethyl moiety, reductive removal (with 95% recovery) of the auxiliary alcohol, introduction of a double bond in **30** via selenium chemistry, and a methyl conjugate addition to the enone **31**.

7-Alkoxy-4-oxy-6-hepten-1-ynes arising from the camphor-derived alcohols **4a** and **4b** undergo cobalt-promoted bicyclization with moderate yields and with good diastereoselectivity (up to 9:1 dr).[18] The asymmetric synthesis of bicyclo[4.3.0]nonan-8-ones by intramolecular PKR was achieved starting from 1-alkoxy-1-octen-7-ynes derived from *trans*-2-phenylcyclohexanol (Scheme 5.28).[28,31]

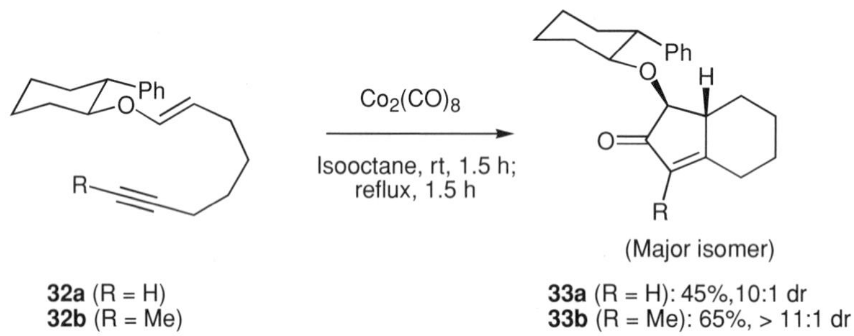

Scheme 5.28

The resulting diastereomers were easily separable by column chromatography, allowing the preparation of enantiopure bicyclononanones. Thus (Scheme 5.29), the major isomer of **33b** was submitted to Birch reduction and to samarium diiodide reductive cleavage to afford the ketone **34** (along with a 98% recovery of the chiral auxiliary). The absolute configuration of **34**, established by circular dichroism, confirmed that the stereochemical outcome of the bicyclization could be correctly predicted by the now familiar mechanistic working model of Scheme 5.24.[31]

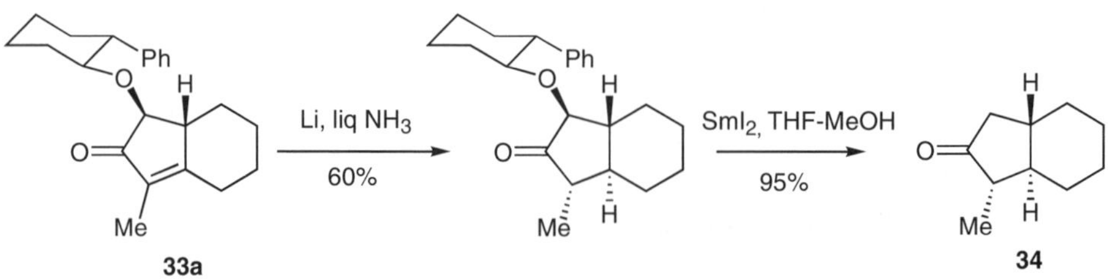

Scheme 5.29

Table 5.4

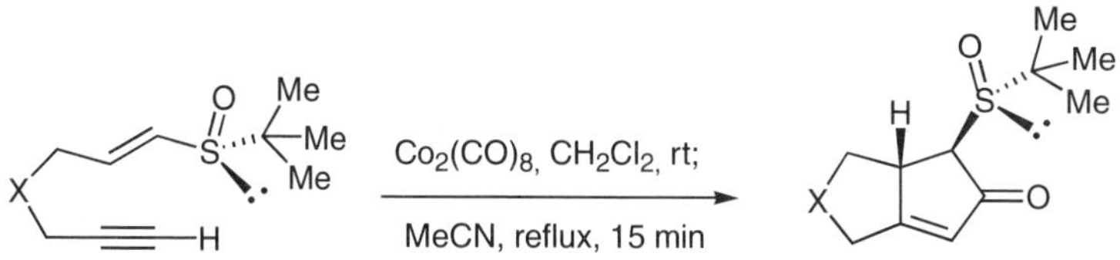

35a–d **36a–d**

Enyne	X	Product	Yield (%)	Dr
35a	CH_2	**36a**	50	> 44:1
35b	CMe_2	**36b**	65	> 44:1
35c	$C(CO_2Et)_2$	**36c**	60	> 44:1
35d	$N(CO_2{}^tBu)$	**36d**	60	> 44:1

5.2.5 Alkenyl Sulfoxides

The use of sulfoxides as chiral controllers in intramolecular PKRs was developed by Adrio and Carretero in 1999.[32] After testing several sulfinyl groups, these authors found that the intramolecular PKR of (*S*)-*trans*-*tert*-butylsulfinyl-1,6-enynes **35a–d** took place with acceptable yields (50–65%) and with excellent diastereoselectivities (> 44:1 dr), when the intermediate dicobalt hexacarbonyl complexes were heated in acetonitrile (Table 5.4). It is worth noting that only terminal acetylenes could be used in the reaction, and that a single example of the cyclization of a *tert*-butylsulfinyl-1,7-enyne took place with practically total diastereoselectivity towards the expected bicyclononenone but with a much reduced yield (30%).[32]

Subsequently, the authors found that the cyclization transcurred without racemization at the sulfur atom, and that the thermal PKRs of the dicobalt hexacarbonyl complexes of the *tert*-butylsulfoxides *cis*-**35a** and *cis*-**35c** gave the same bicyclooctenones **36a** (41% yield) and **36c** (56% yield) obtained from the diastereomeric enynes *trans*-**35a** and *trans*-**35c**. This is particularly gratifying from a practical point of view, since the sulfinylenynes **35** are initially obtained as *trans*/*cis* mixtures in a Wadsworth–Emmons olefination step and this means that these mixtures can be directly used for the PK bicyclization (Scheme 5.30). Most probably, the bicyclooctenone **37a**, initially formed from the *cis*-enyne **35a**, isomerizes in the reaction medium towards the thermodynamically more stable enone **36a**, that places the bulky *tert*-butylsulfinyl substituent in the *exo* face.

The stereochemical outcome of the reaction can be accounted for in the following way (Scheme 5.31): Both computational studies and experimental evidence has established that in *trans* α,β-unsaturated sulfoxides the most stable conformation has a dihedral angle C=C–S–O of *ca.* 0°. Thus, complexation by cobalt takes place exclusively in the

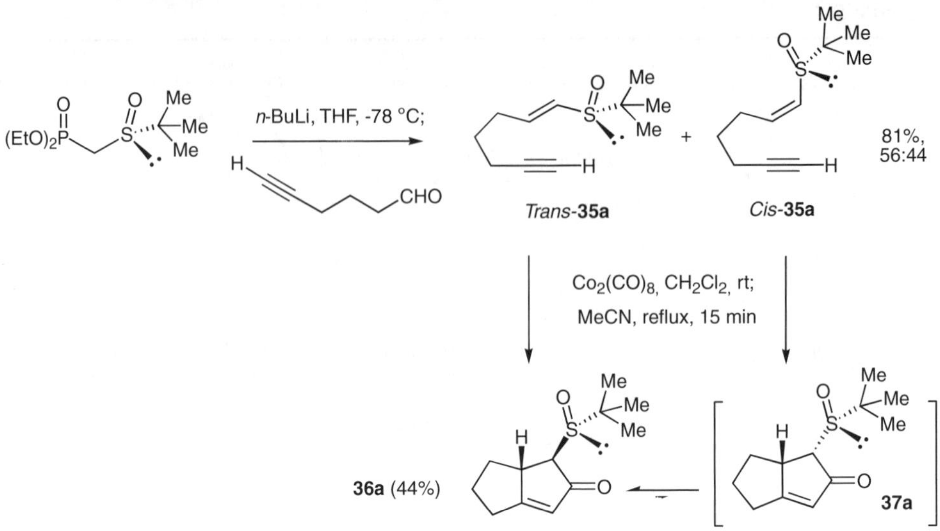

Scheme 5.30

C_β-*Re* face, since the opposite C_β-*Si* face is blocked by the *tert*-butyl group. The resulting cobaltabicyclooctane intermediate then evolves to the observed adduct.[32b]

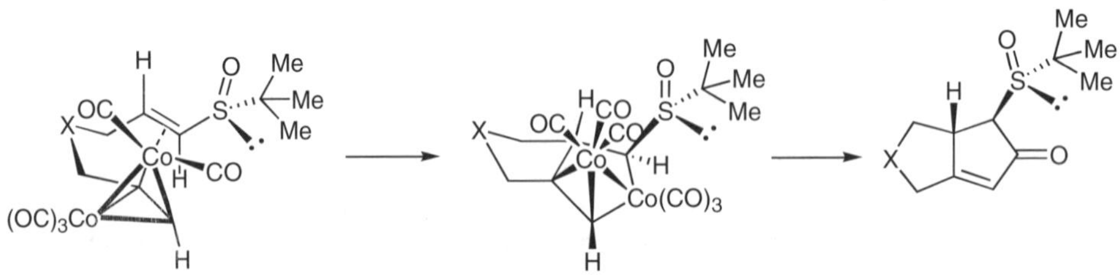

Scheme 5.31

As the last step of these asymmetric PKRs stereocontrolled by sulfoxides, the reductive cleavage of the *tert*-butyl sulfinyl group can be efficiently carried out by treatment with activated zinc, giving the highly enantiopure (> 96% ee) adducts **38** (Scheme 5.32). Thus, the *tert*-butylsulfinyl group acts as a removable, efficient stereochemical controller in the intramolecular PKRs of 1-sulfinylenynes, although *strictu sensu* it can not be considered as a chiral auxiliary, since it cannot be recovered after its removal.

5.2.6 Asymmetric Intramolecular PKRs Mediated by Chiral Auxiliaries Located in the Enyne Tether

This approach to asymmetric PKRs has remained largely unexplored. In 1994, Salaün, de Meijere, and co-workers reported[33] that the cobalt-mediated bicyclizations of 1,6-enynes **39--c**, with a methylenecyclopropane terminator and a chiral acetal moiety adjacent to

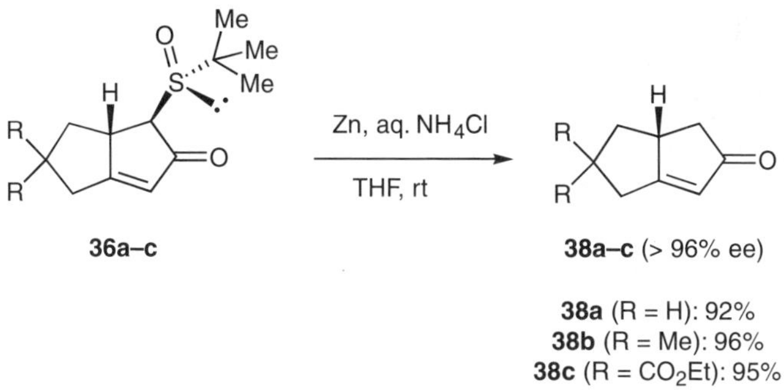

Scheme 5.32

the triple bond, gave the tetracyclic enones **40a–c** in good yields and with poor to good diastereoselectivities (Scheme 5.33)

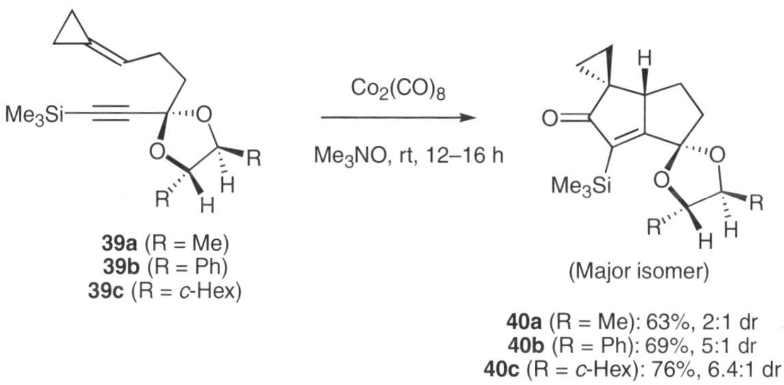

Scheme 5.33

The diastereomers of **40b** and **40c** could be separated by column chromatography. The major isomer of **40b** was transformed into the tricyclic diketone **41**, whose circular dichroism established an (*R*) configuration for the newly created stereogenic center in **40b** (Scheme 5.34). No attempts were apparently made to quantify the recovery of the (*S*,*S*)-1,2-diphenylethane-1,2-diol auxiliary in the acetal cleavage step.

Several years later, Undheim and co-workers described the enantioselective synthesis of the bicyclic α-amino esters **43** and **44** by the intramolecular PKR of an enyne derived from the Schöllkopf bislactim auxiliary **42** (Scheme 5.35).[34] It is worth noting that the stereogenic centers in the bislactim ring did not exert any significant degree of stereocontrol in the bicyclization, but simply allowed the chromatographic separation of the 1:1 diastereomer mixture obtained. The absolute stereochemistry of **43** and **44** was ascertained by single-crystal X-ray diffraction analysis of their immediate precursors.

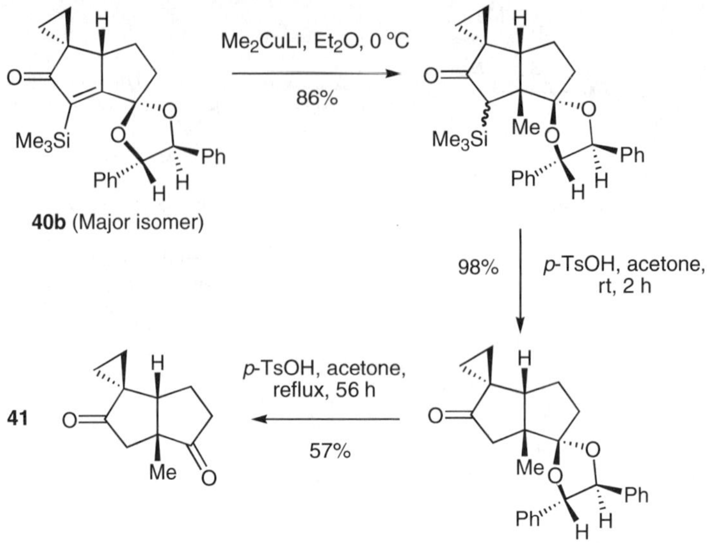

Scheme 5.34

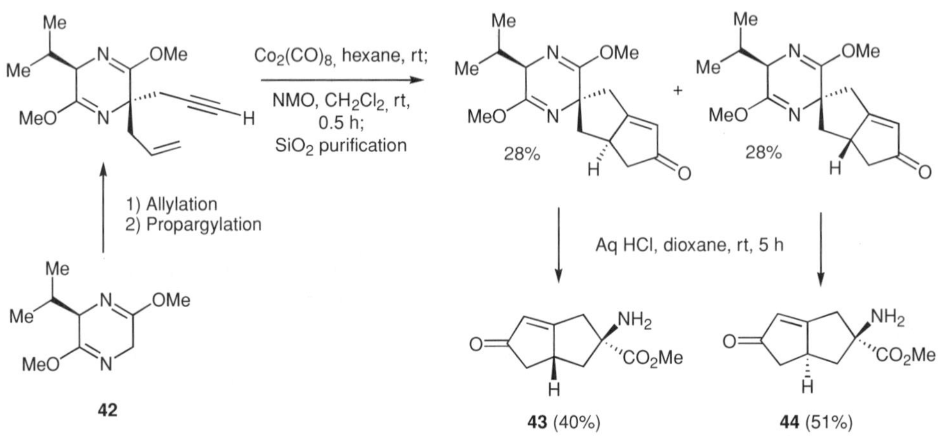

Scheme 5.35

5.3 Asymmetric Intermolecular PKRs with the Aid of Chiral Auxiliaries

5.3.1 Chiral Alkoxyacetylenes

Stereochemical control in intermolecular PKRs, clearly more challenging than in the intramolecular mode, was first addressed with encouraging results through the use of sterically biased chiral alcohols as auxiliaries.[35] It could be rapidly established that the dicobalt hexacarbonyl complexes of chiral alkoxyethynes[14] underwent a totally regioselective intermolecular PKR with norbornene (Table 5.5) and with cyclopentene (Table 5.6), under

Table 5.5

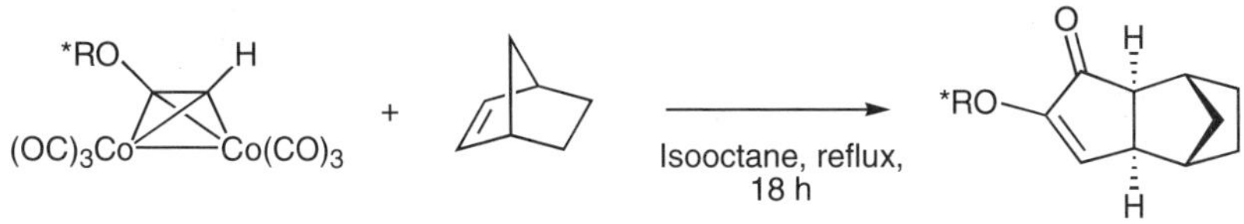

45a–f

Product	R*O-	Yield (%)	Dr
45a		62	3:1
45b	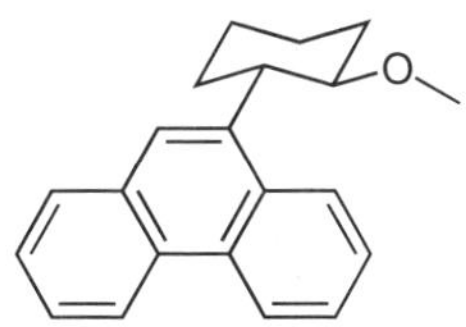	51	> 10:1
45c	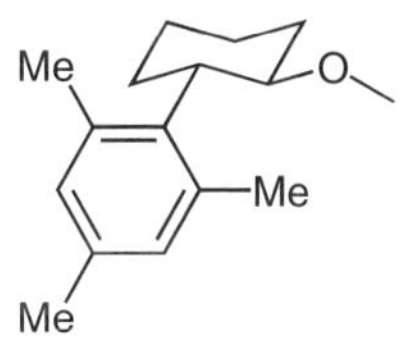	40	3:1
45d	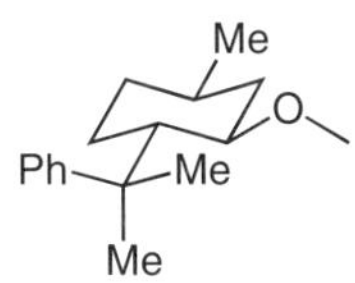	50	2:1
45e	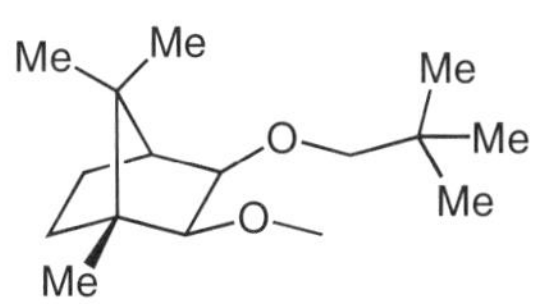	93	1:1
45f	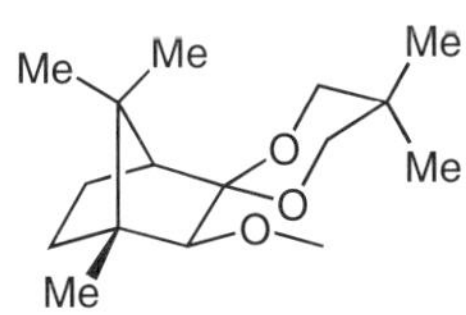	88	2.5:1

Table 5.6

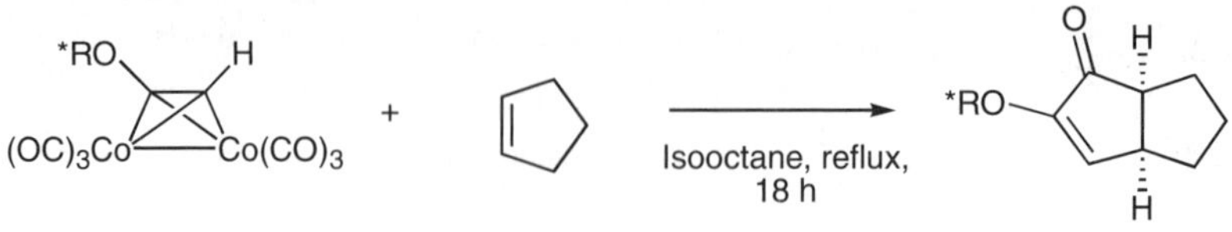

46a–e

Product	R*O-	Yield (%)	Dr
46a		52	2:1
46b	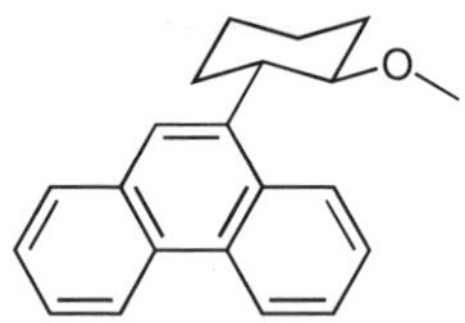	25	> 10:1
46c	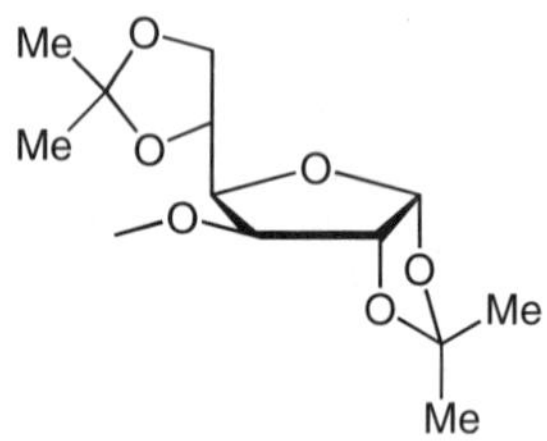	46	2:1
46d	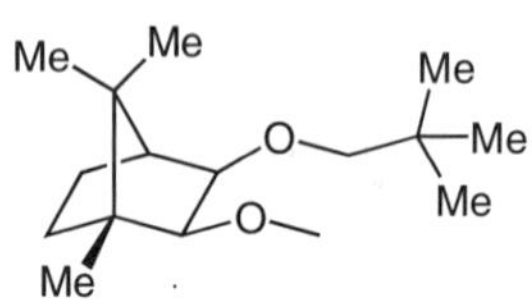	30	3:1
46e	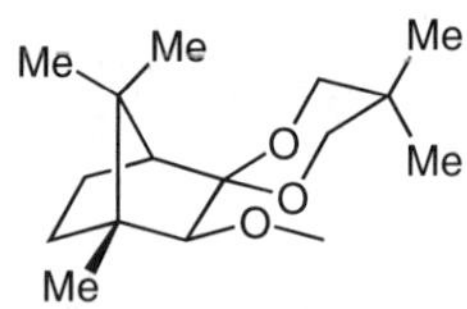	46	1.6:1

classical thermal conditions, to afford the 2-alkoxy-substituted cyclopentenones **45a–f** and **46a–e**, respectively. In accordance with the general behavior of intermolecular PKRs of terminal acetylenes,[10,36] not only was the regioisomer in which the alkyne substituent is vicinal to the carbonyl group predominantly formed, but the *exo*-annulated diastereomers of **45a–f** were the only ones found.

While the use of *trans*-2-(9-phenanthryl)cyclohexanol led to the formation of adducts **45b** and **46b** with complete diastereoselectivity, the difficult preparation of this alcohol in enantiomerically pure form led to the choice of *trans*-2-phenylcyclohexanol, easily available in multigram scale in both enantiomeric forms, as the more practical chiral auxiliary. Moreover, as ihappened in the case of intramolecular PKRs (see Sections **5.2.1** and **5.2.4** above), the diastereomeric PK adducts could invariably be separated by simple column chromatography on silica gel, and they were generally crystalline. The authors then examined the annulation of the alkoxyethyne derived from (1*R*,2*S*)-2-phenylcyclohexanol with norbornadiene, which offered the possibility of effecting a PK-retro-Diels–Alder domino sequence in order to obtain chiral cyclopentenones.[37] The reaction between the dicobalt hexacarbonyl complex of this alkoxyacetylene and norbornadiene (Scheme 5.36) afforded a 2.5:1 mixture of two diastereomers (93% global yield), from which the preponderant one (**47**) was isolated in a 58% yield after chromatographic purification. Boron trifluoride etherate-promoted conjugated addition reactions of methyl, *n*-butyl, *n*-heptyl, and vinyl copper reagents took place exclusively on the *exo*-face of the enone, providing the adducts **48a–d** in 52–80% yields. Samarium (II) iodide in methanol cleanly effected the reductive removal of the chiral alkoxy group, providing the enantiopure ketones **49a–d** (together with the intact alcohol auxiliary for reuse) in 80–89% yield. The application of a Lewis acid-catalyzed retro Diels–Alder reaction to **49c** gave the known levorotatory enone (*S*)-**50** and allowed the assignment of an absolute (1*S*,2*R*,6*R*,7*R*) configuration to **47**.

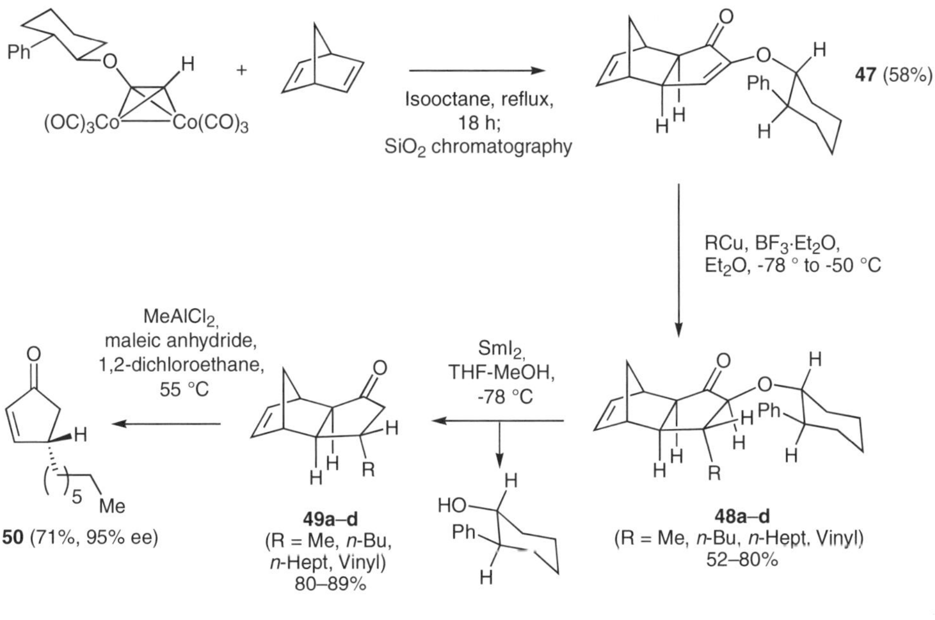

Scheme 5.36

This stereochemical outcome can be explained by assuming a conformation of the complex[17] such as that shown in Scheme 5.37, in which it is seen that the pro-(*S*) cobalt is shielded by the phenyl group. Coordination of norbornadiene, from the *exo*-face, to the pro-(*R*) cobalt and in an orientation such that the methylene bridge avoids steric interactions with the CO ligands (Scheme 5.37) leads to the predominant diastereomer observed.

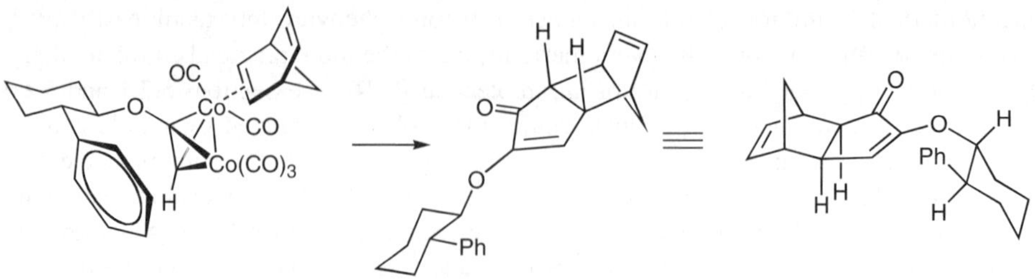

Scheme 5.37

Seeking to enhance stereocontrol in this intermolecular process, and taking into account the ability of amino or of alkylthio groups, tethered either to the alkene[38] or to the alkyne[39] partner, to act as a coordinating ligands in alkyne-dicobalt carbonyl complexes, Pericàs, Riera, and co-workers examined several chiral alcohols bearing an appropriately positioned group able to interact with the metal center along the reaction pathway.[40] They found out that 10-methylthioisoborneol **9a**, a previously known chiral alcohol easily prepared from camphor,[41] appears to function within this conceptual design. The requisite alkoxyacetylene dicobalt hexacarbonyl complex **52a** was obtained in a 62% overall yield from **9a** by the reaction sequence depicted in Scheme 5.38, involving conversion into the 1-alkoxy-1,2-dichloroethene derivative **51a**[14] and one-pot transformation of this compound into **52a** through the trimethylsilylated alkoxy acetylene derivative.[42] The same procedure was applied later on to the preparation of the related complexes **52b** and **52c**.[43]

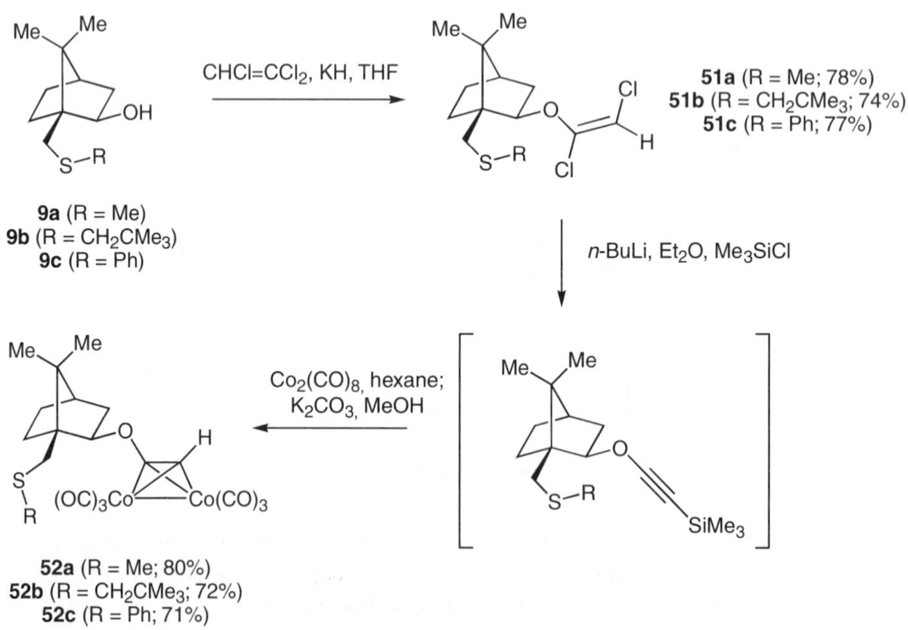

Scheme 5.38

Concordant with the author's expectations, when a solution of **52a** in hexane was allowed to stand at room temperature under a current of dry nitrogen for CO evacuation, it underwent partial transformation to the more polar, chelated dicobalt pentacarbonyl complex **53a** (Scheme 5.39).[40] A maximum 85:15 ratio of **53a** and **52a** was achieved by heating the solution of the parent hexacarbonyl complex **52a** at 50 °C for 1 h with CO removal. When this mixture of complexes was cooled back to room temperature under CO atmosphere, the hexacarbonyl complex **52a** was regenerated. Finally, treating **52a** with an excess of *N*-methylmorpholine-*N*-oxide (NMO) in dichloromethane under nitrogen led to the total formation of the chelated species **53a**. Both by ^{1}H NMR and by TLC analysis, **53a** appeared to be a single diastereomer, probably as a result of thermodynamic control. DFT calculations suggest that the methylthio group coordinates at the axial site of the pro-(*R*) cobalt atom.[43] A similar behavior was observed for complexes **52b** and **52c**.[43]

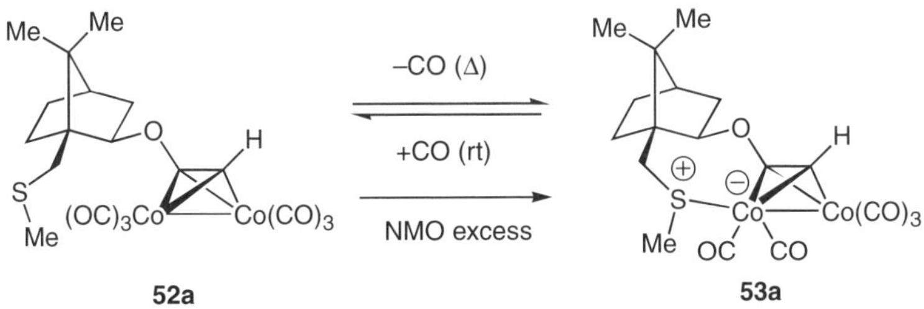

Scheme 5.39

When the preformed complex **53a** was treated at −20 °C with an excess of a strained olefin (norbornene, norbornadiene, or bicyclo[3.2.0]hept-6-ene), the corresponding PK adducts **54**, **55** and **56** were obtained in good yields and with remarkable diastereoselectivities (up to 24:1 dr in the case of **55**; Table 5.7). On the other hand, reaction with a less reactive olefin such as cyclopentene, that required heating of **52a** at 60 °C for 20 h in a sealed tube, afforded the cycloadduct in low yield (14%) and with no selectivity (1:1 dr).[43] In the case of **55**, the major diastereomer could be separated after careful column chromatography.

The absolute configuration of the major isomer of **55** was unambiguously ascertained by X-ray diffraction analysis of a dihydrogenated derivative, as well as by chemical correlation with the known cyclopentenone (-)-**57** (Scheme 5.40), obtained by a sequence of reactions similar to that depicted in Scheme 5.36 above, involving stereoselective cuprate addition from the *exo* face, reductive cleavage of the auxiliary with samarium diiodide and retro-Diels–Alder cycloaddition.[35]

This stereochemical outcome (coincident to that afforded by (1*R*,2*S*)-2-phenylcyclohexanol)[35] could be rationalized with the aid of DFT calculations, that suggested a reaction path involving coordination of norbornadiene (from the *exo* face) to the site occupied by the methylthio group in the pro-(*R*) cobalt (Scheme 5.41).[43]

The application of asymmetric intermolecular PKRs of chiral alkoxyacetylenes to the enantioselective synthesis of natural products was subsequently exemplified by Greene and co-workers in a formal total synthesis of the fungal metabolite (+)-brefeldin A (Scheme 5.42).[44]

Table 5.7

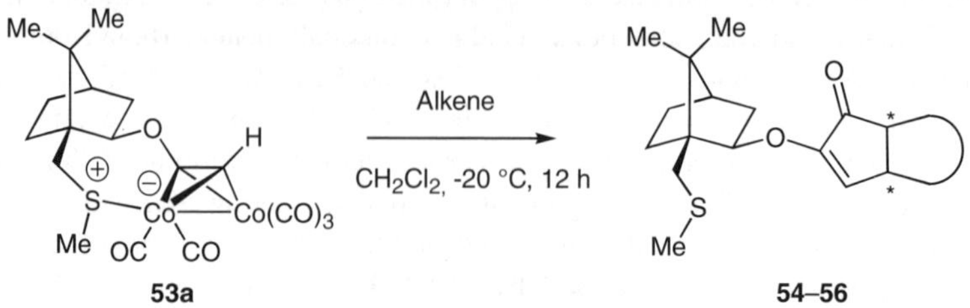

Alkene	Product	Yield (%)	Dr
	**54**	77	11.5:1
	**55**	82	24:1
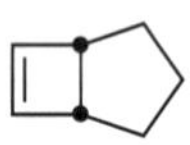	**56**	91	13.3:1

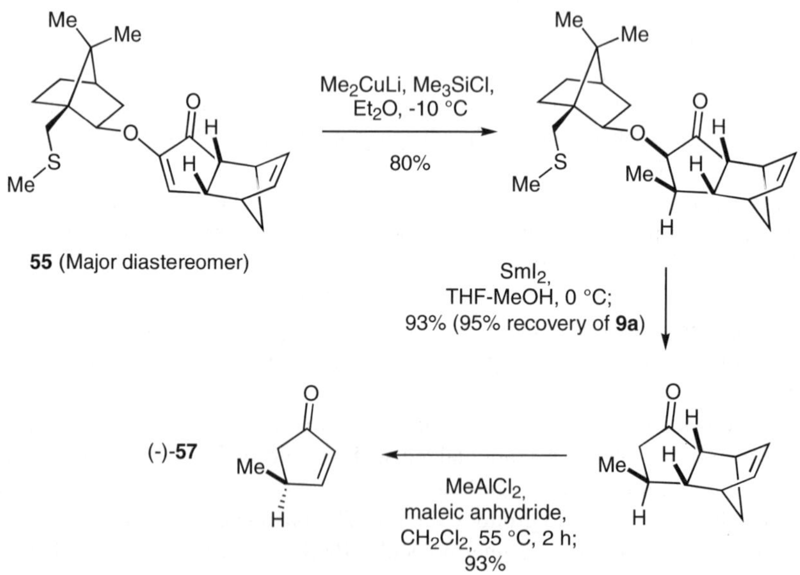

Scheme 5.40

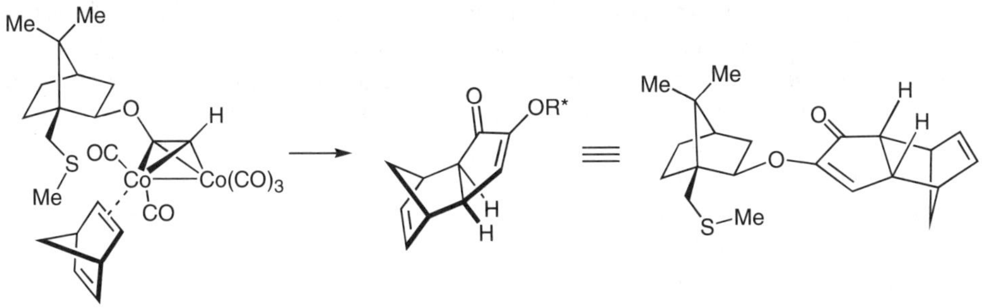

Scheme 5.41

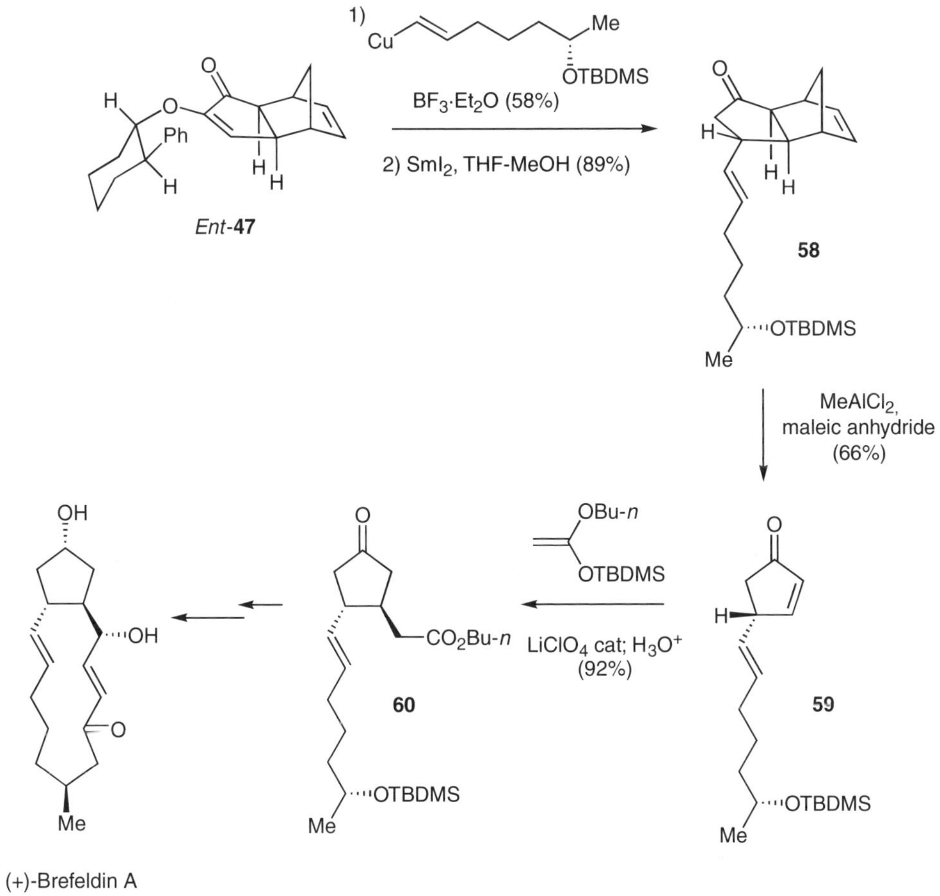

Scheme 5.42

Stereochemically pure adduct *ent*-**47** (obtained in a 58% yield from the intermolecular PKR between norbornadiene and the dicobalt hexacarbonyl complex of (1*S*,2*R*)-(2-phenylcyclohexyloxy)ethyne)[35] was converted into the tricyclic ketone **58** by a two-step sequence involving cuprate addition and samarium-mediated reductive cleavage (52% overall yield). It is worth noting that the same compound was obtained in 92% ee and in 47% overall yield from adduct *ent*-**55**. The Lewis-acid mediated retro Diels–Alder reaction readily furnished the cyclopentenone **59**, which under lithium perchlorate catalysis experienced the stereoselective Michael addition of a ketene acetal to yield on acidic workup the known (+)-brefeldin A precursor **60**.

5.3.2 Chiral Acetylene Thioethers

The first studies on the use of chiral acetylene thioethers in the intermolecular PKR, reported by Pericàs, Riera, and co-workers in 1997,[23] showed that the applicability of these compounds was rather limited. Interestingly enough, however, cycloaddition with relatively unreactive olefins such as cyclopentene took place with promising diastereoselectivities (Scheme 5.43). The reaction of this acetylene thioether with norbornadiene in the same reaction conditions took place with a good yield (85%), but with very poor diastereoselectivities (1.7:1 dr). The diastereomers were not easily separable by column chromatography, and no efforts were made to ascertain the stereochemical sense of induction of the process.

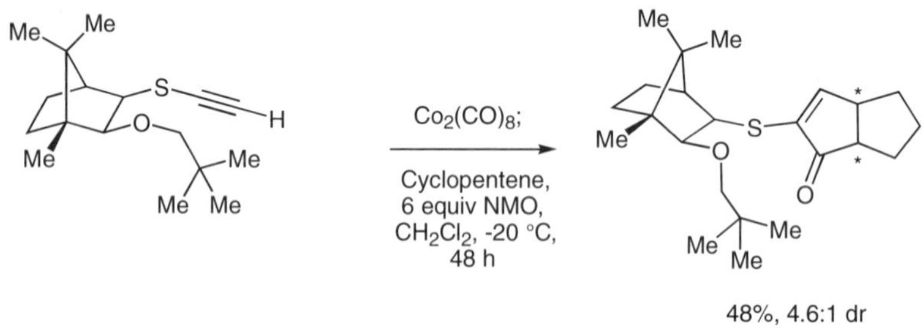

Scheme 5.43

Somewhat better diastereoselectivities (but much lower yields) were achieved through chiral acetylene dithioethers (Scheme 5.44).[45]

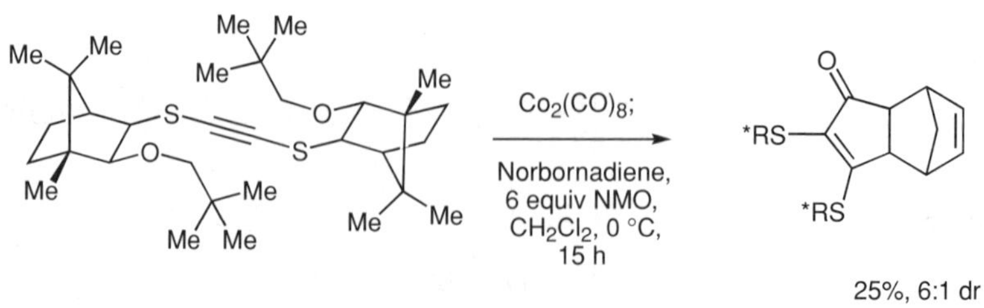

Scheme 5.44

Finally, the thioether analog of the chelated complex **53a** was shown to undergo intermolecular PKRs with the same stereochemical sense of induction but with somewhat lower yields and diastereoselectivities (Scheme 5.45).[46]

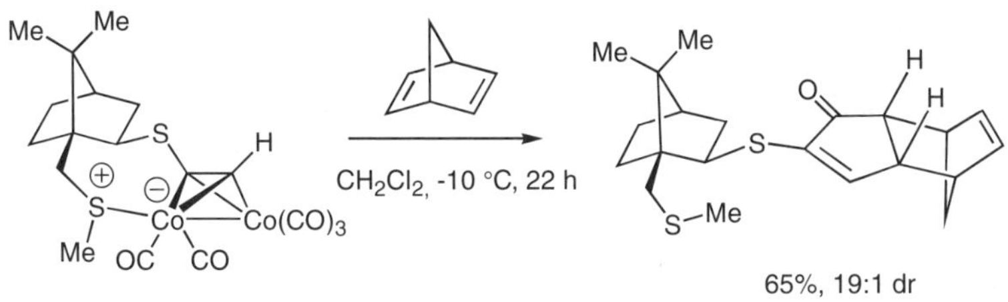

Scheme 5.45

5.3.3 Alkynyl Sulfoxides

The PKR of the dicobalt hexacarbonyl complexes of acetylenic sulfoxides with strained olefins was reported by Pericàs, Riera, and co-workers in 1999.[47] Best results were afforded by the complex **61**, that gave a mixture of the two regioisomeric adducts **62** and **63** (Scheme 5.46). While **62** was formed with no diastereoselectivity, **63** was obtained as a 4.6:1 diastereomeric mixture, and the relative configuration of the major isomer (isolated by column chromatography) could be determined. However, when the major diastereomer of **63** was transformed into the dextrorotatory tricyclic enone **64**, it was found that its enantiomeric excess was only 75%. A polarimetric study of the enantiomeric purity of **61** showed that this compound is configurationally stable and racemizes at an appreciable rate even at room temperature ($k = 2.8\ 10^{-6}\ s^{-1}$ at 25 °C). This unexpected low temperature racemization, completely unprecedented for a non-allylic sulfoxide, seriously limits the application of acetylenic sulfoxides in asymmetric PKRs.

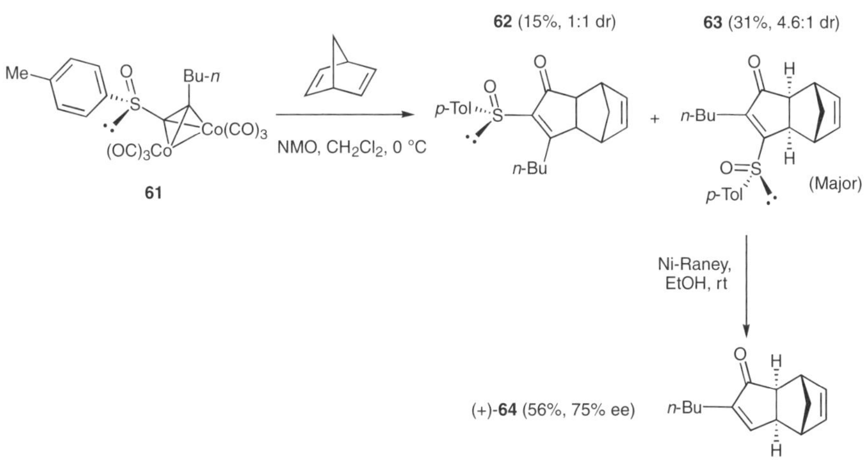

Scheme 5.46

5.3.4 Chiral Ynamines and Ynamides

The synthesis of dicobalt hexacarbonyl complexes of chiral ynamines was reported by Pericàs and co-workers in 2000.[48] The silylated complexes **65a–e** were prepared from a set of chiral secondary amines by a reaction sequence involving *in situ* generation and trapping of dichloroacetylene from trichloroethylene, followed by sequential treatment of the intermediate dichloroenamine with *n*-butyl lithium and trimethylsilyl chloride.[14] Without further purification, the resulting silylated enamines were converted into the complexes **65** by reaction with dicobalt octacarbonyl. The corresponding desilylated complexes **66a–e** could be accessed by exposure of **65** to potassium carbonate and methanol.[42] The green-colored complexes **65** and **66** readily experience loss of carbon monoxide but are stable at room temperature under a positive pressure of this gas (Scheme 5.47).

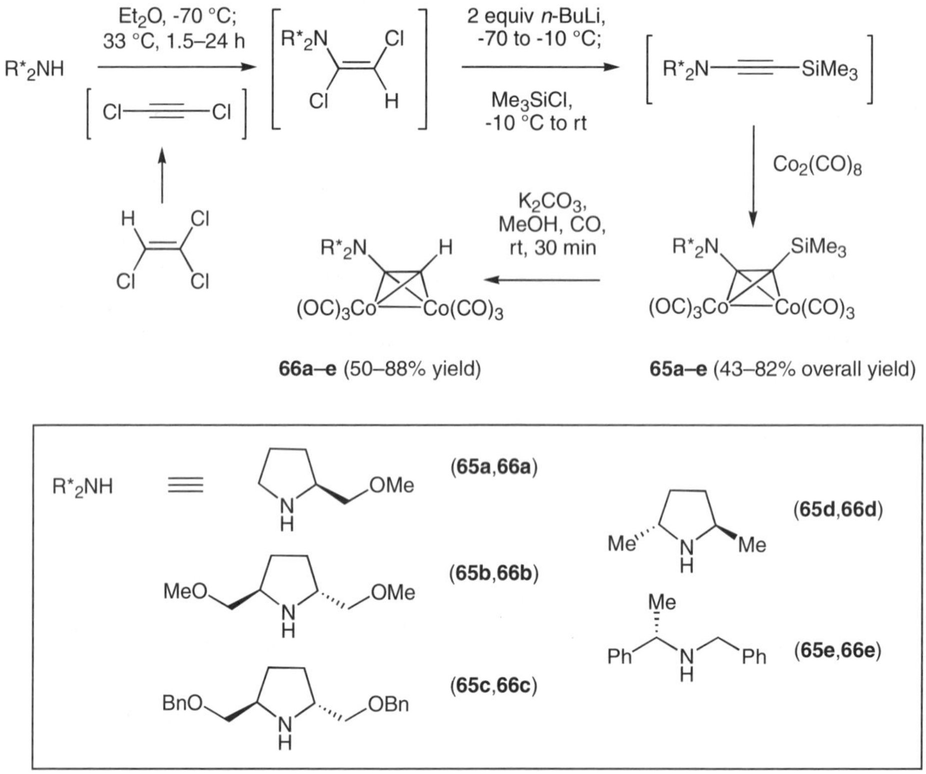

Scheme 5.47

While complexes **65** were too hindered and gave very low yields, intermolecular PKRs of complexes **66** with strained olefins (norbornene, norbornadiene) took place at unprecedented low temperatures (rt, 0 °C, −21 °C, or even −35 °C) to give the expected adducts in poor to moderate yields and with good diastereoselectivities (see Scheme 5.48 for a representative example).

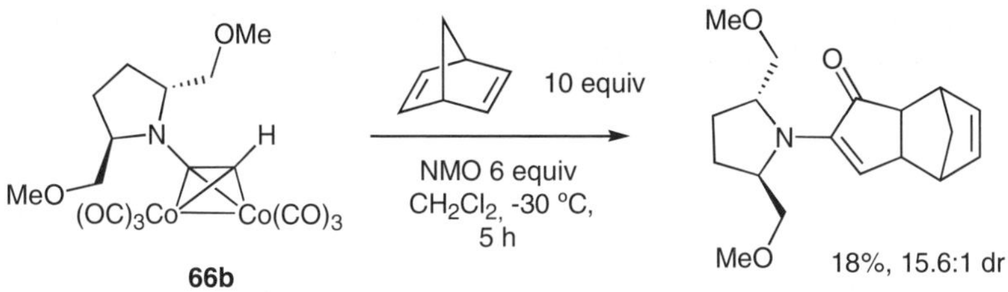

Scheme 5.48

The diastereomeric adducts were not separable, so that from a practical point of view chiral ynamines are much less convenient than chiral alkoxy acetylenes for asymmetric intermolecular PKRs. The high reactivity of complexes **66** is associated to the unusually facile dissociative loss of carbon monoxide that they undergo, a phenomenon that was studied by means of DFT calculations and was found to be triggered by the delocalization of the amine lone pair in the cobalt–carbon cluster.[48]

On the other hand, the dicobalt hexacarbonyl complexes of alkynyl amides appear to be much more stable, due to the reduced availability of the nitrogen lone pair, and they have been shown to undergo efficient intermolecular PKRs.[49] An asymmetric variant of this process was reported by Shen and Hsung in 2003.[50] These authors found that the dicobalt hexacarbonyl complexes of 2-substituted ynamides derived from 4-phenyl-1,3-oxazolidin-2-ones, formed in moderate yields (40–50%) by treatment of the acetylenes with dicobalt octacarbonyl, reacted with norbornadiene under NMO-promoted conditions to give mixtures of a standard *exo*-adduct with an unusual *endo*-adduct (Scheme 5.49). Interestingly enough, in all instances only the 2-amido regioisomers were obtained. The *exo*-adducts were obtained with no diastereoselectivity (1:1 dr), while the *endo*-adducts were isolated as single diastereomers. The absolute configuration of the *endo*-adduct was ascertained in one case (R = Ph) by X-ray diffraction analysis.

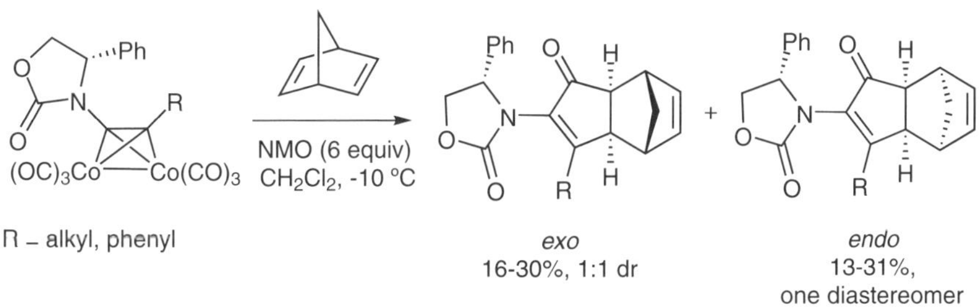

Scheme 5.49

When an unsubstituted *N*-alkynyloxazolidinone was used (R = H), its reaction with norbornadiene afforded the *exo*-2-amido regioisomer in 16% yield and with a 2:1 dr. These results can be interpreted by assuming that the oxazolidinone carbonyl acts as a ligand to the metal, displacing a CO and coordinating preferentially to the pro-(*R*) cobalt.

Norbornadiene would then bind to an equatorial position, probably *trans* to the oxygen, and in the less-hindered *endo*-mode (Scheme 5.50).

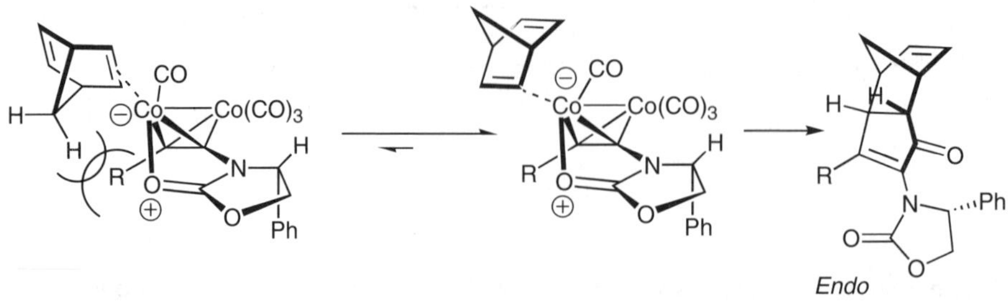

Scheme 5.50

5.3.5 Chiral 2-Alkynoates

In an extensive study of the intermolecular PKRs of 2-alkynoic esters derived from chiral alcohols (including *trans*-2-phenylcyclohexanol, menthol, 8-phenylmenthol and the camphor-derived alcohols **4a**, **4b** and **9a**, among others), Moyano, Pericàs, and co-workers found that under standard thermal conditions (toluene, 40–70°C), the expected *exo*-cycloadducts were obtained with good yields and with excellent regioselectivity, but with low diastereoselectivity (Schemes 5.51 and 5.52).[51]

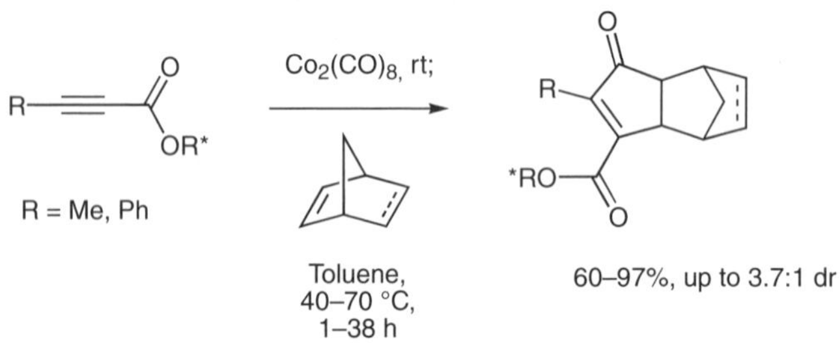

Scheme 5.51

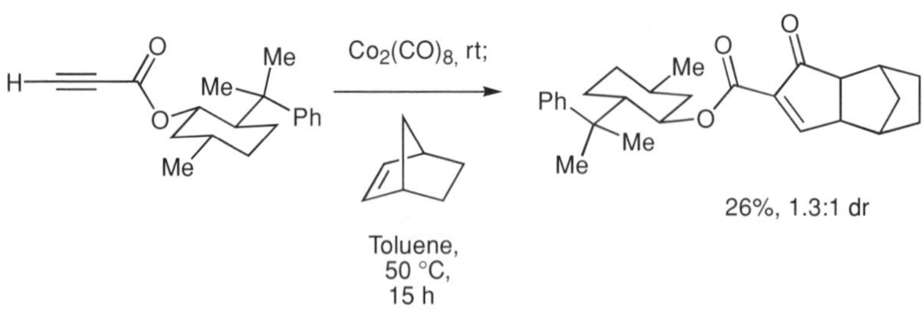

Scheme 5.52

Much better results were subsequently obtained with chiral *N*-(2-alkynoyl) oxazolidinones.[24,52] These compounds, that are readily prepared in a one-pot procedure from 2-alkynoic acids and chiral 1,3-oxazolidin-2-ones,[24,53] efficiently undergo intermolecular PKRs with strained olefins (norbornene, norbornadiene) in good yields and, in several instances, with excellent diastereoselectivities (up to 17.5:1 dr; see Table 5.8 for selected results with norbornadiene). The regioselectivity of the process is the following: with the phenylpropiolate (**67**) or trimethylsilylpropiolate (**68**) derivatives, the

Table 5.8

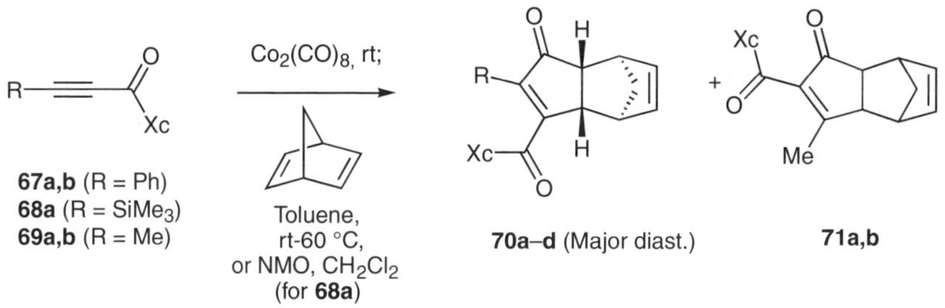

Alkyne	R	Xc–	Product	Yield (%)	Dr
67a	Ph	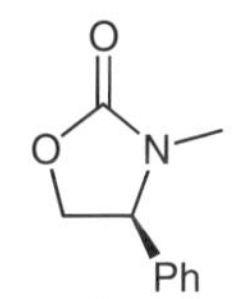	**70a**	96	5.2:1
68a	$SiMe_3$	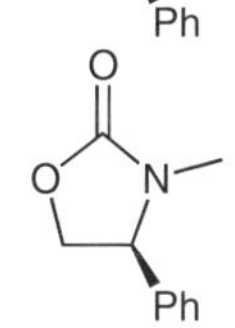	**70b**	88	3.6:1
69a	Me	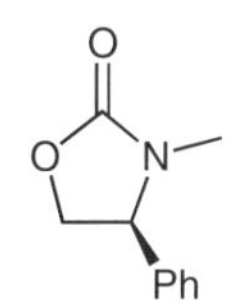	**70c** **71a**	53 23	7.6:1 1.9:1
67b	Ph	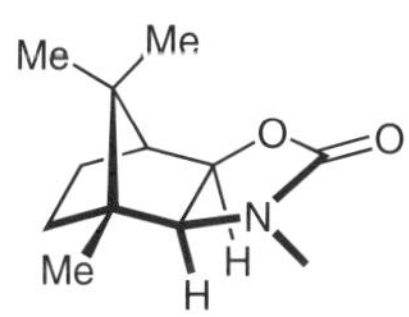	**70d**	97	14:1
69b	Me	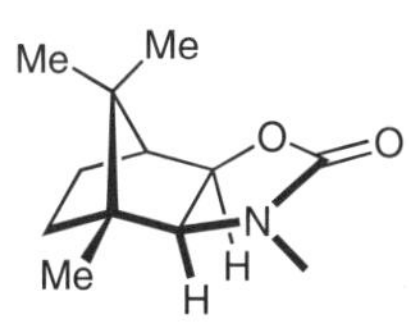	**70e** **71b**	34 38	17.5:1 5.1:1

3-carboxamido regioisomer **70** is exclusively formed; the *N*-(tetrolyl)oxazolidinones **69** give mixtures of 3-carboxamido (**70**) and 2-carboxamido (**71**) regioisomers. These regioisomers are readily separated by column chromatography. In general, the 3-carboxamido regioisomers **70** are obtained with good diastereoselectivity, remarkably those derived from commercially available 4-phenyloxazolidinone. These diastereomers are also readily separated by column chromatography, and in general the major diastereomers of **70** can be isolated in stereochemically pure form in reasonable yields. On the other hand, 2-carboxamido regioisomers **71** are formed with much less diastereoselectivity, and the diastereomers are not separable by column chromatography.

The oxazolidinone chiral auxiliary can be easily cleaved from the PK adducts (Scheme 5.53). Thus, the major diastereomer of **70a** (obtained in 66% yield) was dissolved in anhydrous allyl alcohol and heated at 150 °C in the presence of titanium tetraisopropoxide, to afford the levorotatory allyl ester **72** (43% yield); on the other hand, the major diastereomer of **70b** (obtained in 69% yield) was submitted to catalytic hydrogenation and reacted with basic aqueous hydrogen peroxide, to furnish the ketoacid **73** in a 51% overall yield. The circular dichroism of this compound established its absolute configuration.[24]

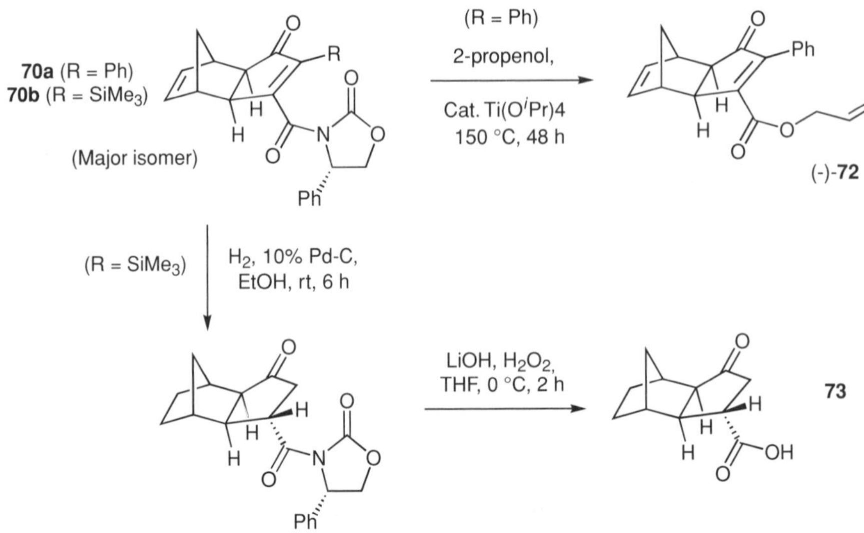

Scheme 5.53

With the aid of semi-empirical PM3(tm) calculations, the authors rationalized the stereochemical outcome of the reaction by means of the norbornadiene–dicobalt pentacarbonyl complex depicted in Figure 5.1.[24] The olefin binds from its *exo*-face to an equatorial site of the less-hindered pro-(*S*) cobalt, in an orientation that minimizes the steric interactions of the methylene bridge with the alkyne substituents.

Hiroi and Watanabe investigated the use of (*S*)-methionine-derived amides as chiral auxiliaries in the intermolecular PKR of phenylpropiolic amides with norbornene.[54] The yields of the adducts were very variable, depending on the amide substituents. Reasonable yields and diastereoselectivities were obtained in some instances (Scheme 5.54). The

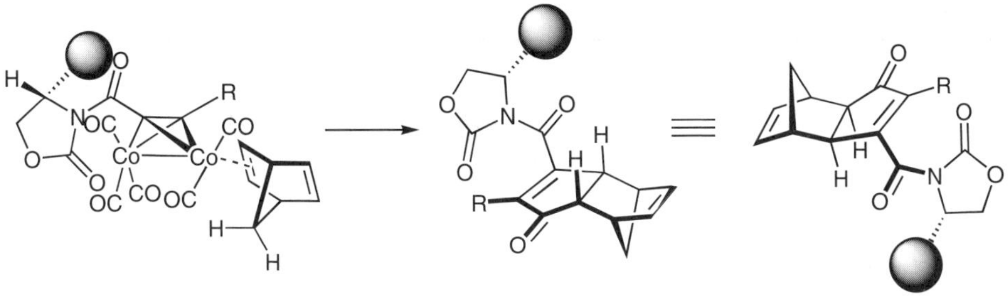

Figure 5.1

absolute configuration of the major diastereomer was ascertained by X-ray diffraction analysis, and the authors, as previously reported by Verdaguer *et al.*,[40,43] suggested that the methionine methylthio group plays a major role in the asymmetric induction, presumably by coordination of the sulfur atom to the cobalt catalyst in the alkyne complex. However, no proofs for the intermediacy of chelated pentacarbonyl complexes similar to **53a** were provided.

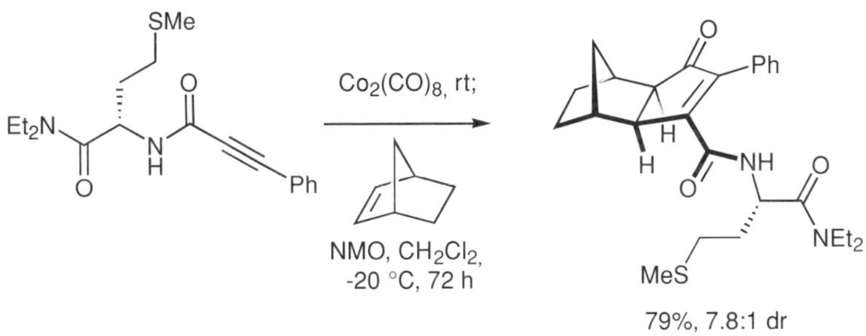

Scheme 5.54

Moyano, Pericàs, and co-workers found that exceptionally high levels of regio- and stereoselectivity can be easily achieved in the intermolecular PKRs of *N*-(2-alkynoyl) derivatives of Oppolzer's 10,2-camphorsultam (Table 5.9).[52,55]

Compounds **74a–d** were readily prepared from bornane-10,2-sultam (commercially available in both enantiomeric forms) and the corresponding 2-alkynoic acids.[53] The reactions of both the phenylpropiolic acid (**74a**) and (trimethylsilyl)propiolic acid derivatives (**74b**) turned out to be completely regioselective, leading exclusively to the 3-carbamoyl cyclopentenones **75a** and **75b**, respectively. The reactions of **74c** and **74d** produced substantial amounts (23–24%) of the 2-carbamoyl regioisomers **76a** and **76b**, that were readily separated from the predominant (52–55%) 3-carbamoyl regioisomers **76c** and **76d** by column chromatography. This behavior closely matches that of the related 2-(alkynoyl)oxazolidinones **67–69**.[24,52] Adducts **75** were formed in all instances with essentially complete stereochemical control (at least 318:1 dr, rigorously established by HPLC). Adducts **76** were obtained with marginal stereoselectivity, and the diastereomers

Table 5.9

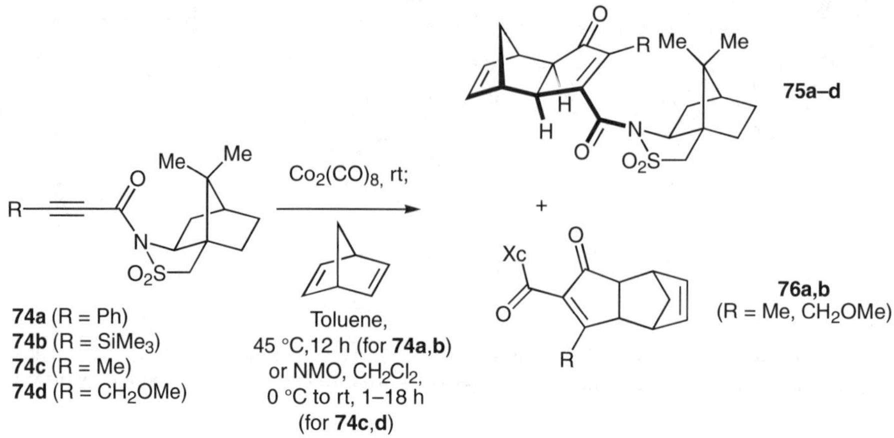

Alkyne	R	Product	Yield (%)	Dr
74a	Ph	**75a**	93	> 800:1
74b	$SiMe_3$	**75b**	78	> 800:1
74c	Me	**75c**	55	318:1
		76a	24	1.7:1
74d	CH_2OMe	**75d**	52	> 800:1
		76b	23	1.1:1

were not separable by chromatography. The reaction of **74a** with norbornene (NMO) took place with somewhat lower yield (54%), but still with excellent diastereoselectivity (125:1 dr). The absolute configuration of adducts **75a** and **75b** were ascertained by a combination of CD spectra, chemical correlation, and X-ray diffraction analysis.

As an illustration of the synthetic applications of these readily obtained stereopure PK adducts, Scheme 5.55 depicts the preparation of the 4-substituted cyclopentenone (*R*)-**80**.[52] Chemoselective hydrogenation of the cyclopentenone double bond and reductive cleavage of the trimethylsilyl group in **75b** were simultaneously achieved by treatment with zinc dust in acetic acid, to give a mixture of epimers at C3 from which the major component

(**77**) could be isolated in 62% yield by column chromatography. Subsequent reaction of this compound with lithium aluminum hydride produced the diol **78** (plus bornane-2,10-sultam), both in essentially quantitative yields. Selective protection of the primary alcohol and oxidation (pyridinium dichromate, PDC) of the secondary gave access to the tricyclic ketone **79** (62% overall yield), which by heating in *o*-dichlorobenzene underwent the expected retro-Diels–Alder reaction. The 4-substituted cyclopentenone **80** thus obtained showed an enantiomeric excess greater than 96%.

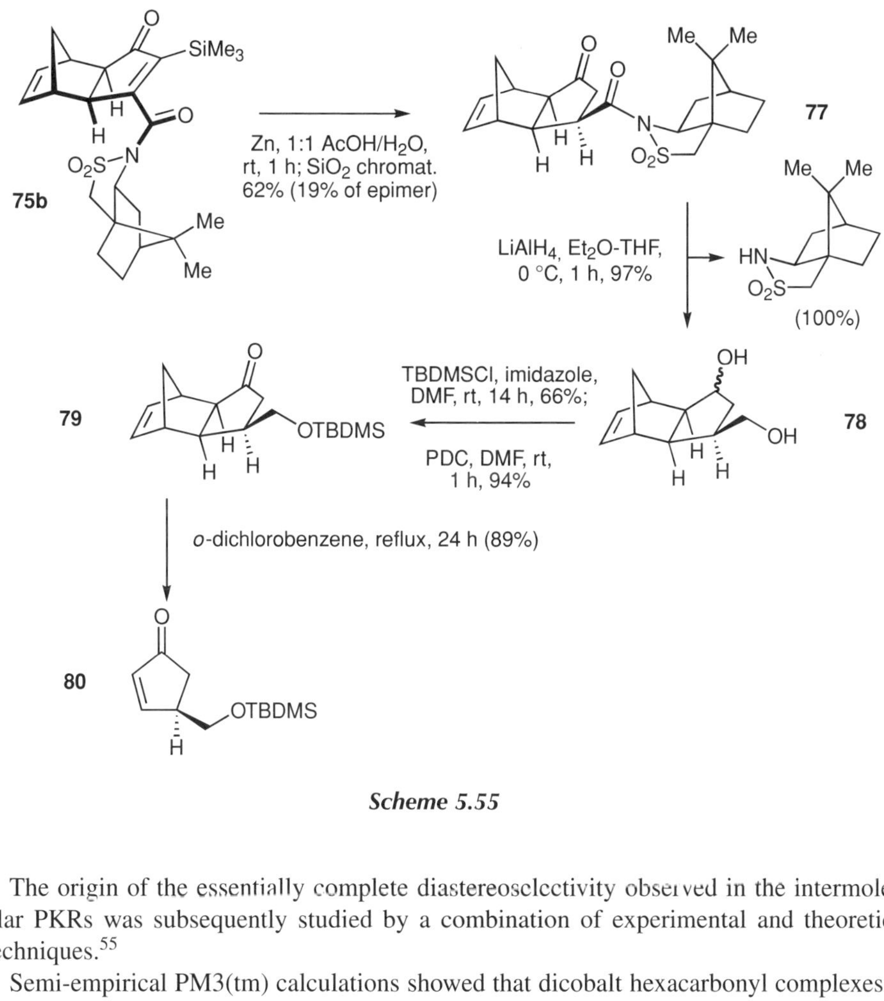

Scheme 5.55

The origin of the essentially complete diastereoselectivity observed in the intermolecular PKRs was subsequently studied by a combination of experimental and theoretical techniques.[55]

Semi-empirical PM3(tm) calculations showed that dicobalt hexacarbonyl complexes of *N*-(2-alkynoyl)sultams **74** had only two geometrically very similar stable conformations; the circular dichroism spectra of these complexes gave support to these calculations. A DFT study of the intermediates in the PKR of the complex derived from **74b** with norbornadiene showed that an oxygen atom of the sultam auxiliary can selectively chelate one of the cobalt atoms (although the intermediate pentacarbonyl complex is too reactive to be isolated), and that coordination of the olefin to the same cobalt from the *exo*-face and with a *syn* orientation is the energetically preferred option. This chelation effect leads to an extremely efficient chirality transfer (Scheme 5.56).

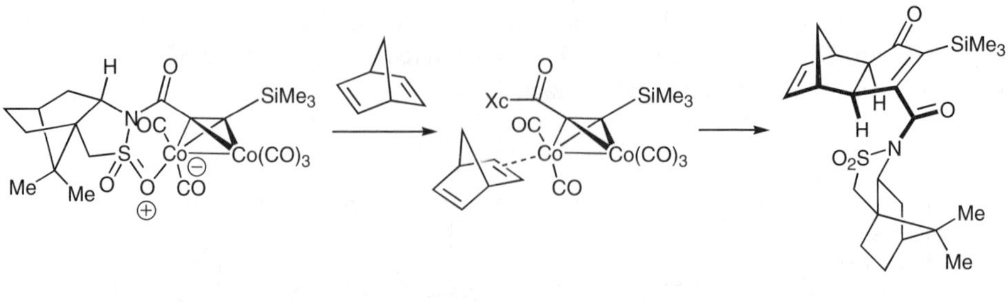

Scheme 5.56

In 2002, Moyano and co-workers reported that the heterobimetallic (Mo–Co) complexes of *N*-(2-alkynoyl) derivatives of chiral oxazolidinones (*Cf.* **69a**) or sultams (*Cf.* **74c**) showed an unprecedented reversal of stereoselectivity in their reaction with norbornadiene, predominantly giving rise to *endo*-fused cyclopentenone adducts in a totally regioselective fashion.[56] These heterobimetallic complexes were easily prepared by *in situ* generation of the dicobalt hexacarbonyl complex in tetrahydrofuran and treatment with 1.5 molar equivs of Na[MoCp(CO)$_3$] (Table 5.10).

Table 5.10

69a
74c

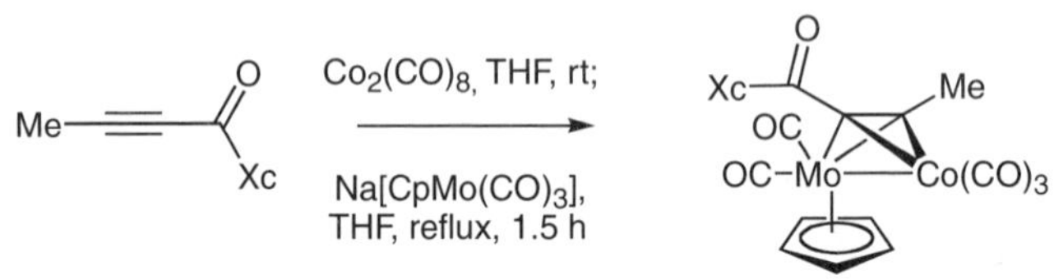

81a,b
82a,b

Alkyne	Xc–	Product	Yield (%)	Dr
Ent-**69a**	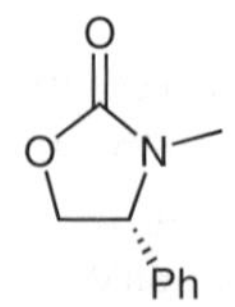	**81a,b**	43	2.3:1
74c	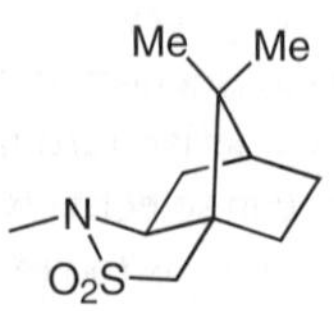	**82a,b**	64	1.2:1

Compound **69a** gave rise to two diastereomeric complexes **81a** and **81b**, differing in the chirality of the heterobimetallic cluster, that were separated by column chromatography. The sultam derivative **74c** also afforded two diastereomeric complexes that could not be separated. When each of the diastereomers of **81** was reacted with norbornadiene under thermal conditions, a mixture of the *endo*-adduct **83a** (only the 2-carbamoyl regioisomer) with the two regioisomers **70c** and **71a** (both as diastereomer mixtures) was isolated. On the other hand, the diastereomeric mixture of **82** gave rise to the *endo*-adduct **83b** and to the *exo*-2-carbamoyl regioisomer **76a** (Table 5.11). It is worth noting that the two diastereomers of **81** display opposite diastereoselectivity, so that the chirality of the C_2CoMo core appears to control the diastereoselectivity of the formation of the *endo*-fused adducts. In the case of **82**, both adducts *endo*-**83b** and *exo*-**76a** were obtained as single diastereomers (by ^{13}C-NMR), which tends to indicate that only one isomer of the heterobimetallic complex **82** had reacted. These results are similar to those discussed above (Scheme 5.49, Section **5.3.4**), reported shortly afterwards (2003) by Shen and Hsung.[50]

It is worth noting that the dimolybdenum complexes of *N*-(2-alkynoyl)oxazolidinones did not react with norbornadiene, suggesting that the cobalt atom is the active one in the intermolecular PKR of the heterobimetallic complexes. Examination of molecular models of these heterobimetallic complexes showed that the presence of an oxazolidinone or sultam moiety could force the bulky cyclopentadienyl ligand to occupy an eaquatorial (i.e., cis to the Mo–Co bond) coordination site. This would then distort the geometry of the ligands around the cobalt and destabilize the coordination of the olefin from the *exo* face (Scheme 5.57).

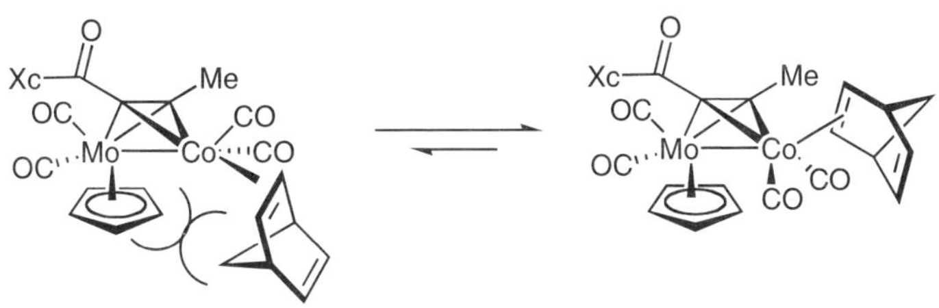

Scheme 5.57

5.3.6 Alkenyl Sulfoxides

In contrast to their successful application in intramolecular PKRs,[28,30,31] chiral *O*-alkyl enol ethers (and chiral alkoxyallenes) give poor results in the intermolecular version of the process.[57] This, together with the fact that in general only strained olefins give good yields in the intermolecular PKR, delayed the development of asymmetric approaches based on chiral auxiliaries bonded to alkenes. It was not until 2003 that Carretero and co-workers demonstrated that vinyl sulfoxides, in particular the potentially cobalt-coordinating 2-(*N*,*N*-dimethylamino)phenyl vinyl sulfoxide (*R*)-**84** (easily available in >99% ee from commercial reagents) reacts with a variety of alkyne–dicobalt hexacarbonyl complexes with high levels of regio- and stereoselectivity (Table 5.12).[58]

Table 5.11

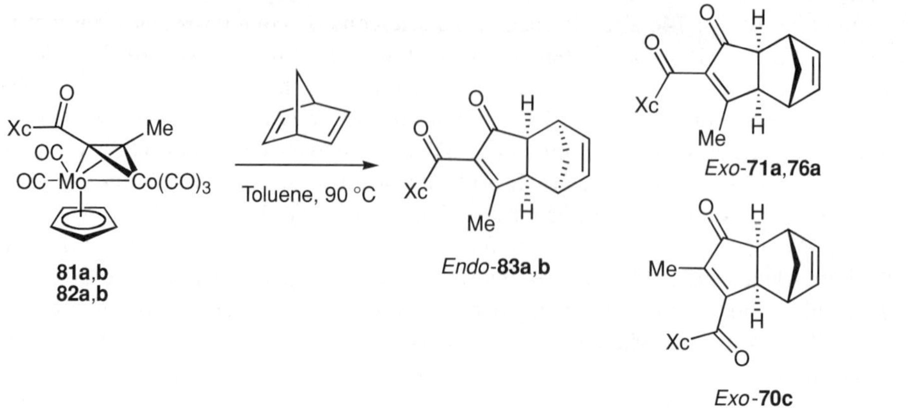

Complex	Xc–	Product	Yield (%)	Dr
81a	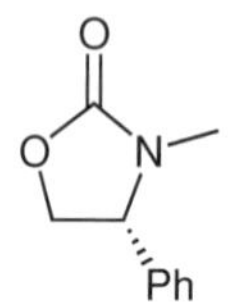	**83a** **71a** **70c**	24 12 62	4:1 > 20:1 5:1
81b	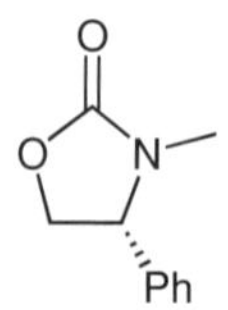	**83a** **71a** **70c**	92 4 3	1:9 n.d. n.d.
82a,b	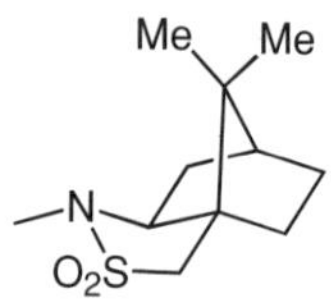	**83b** **76a**	36 18	> 20:1 > 20:1

The reaction takes place in moderate to good yields, with total regioselectivity with respect to the alkyne component, and with excellent diastereoselectivity. With internal alkynes (*Cf.* 2-butyne) the sulfoxide **84** did not react under standard conditions, and high pressures (10 Kbar) were necessary to achieve a 33% yield of the adducts mixture (11.5:1 dr), at room temperature. The authors showed that the reaction took place with retention of the stereochemistry at the sulfur, and that the pairs of diastereomers **85** and **86** differed only in the configuration at the carbon atom. The major diastere-

Table 5.12

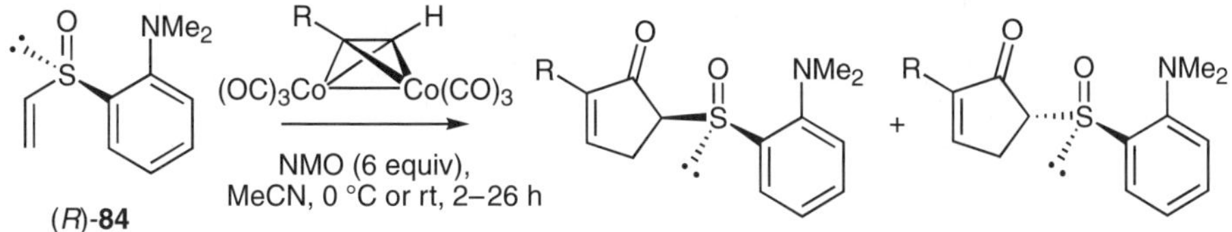

R	Products	Yield (%)	**85**:**86** Ratio
n-Bu	**85a**,**86a**	74	13.3:1
t-Bu	**85b**,**86b**	55	> 49:1
$PhCH_2$	**85c**,**86c**	58	13.3:1
p-Tol	**85d**,**86d**	49	13.3:1
$SiMe_3$	**85e**,**86e**	59	> 49:1
$CH_2OSi^iPr_3$	**85f**,**86f**	62	13.3:1
$CH_2CH_2OSi^iPr_3$	**85g**,**86g**	66	> 49:1:1
$(CH_2)_3Br$	**85h**,**86h**	68	> 49:1:1

omers **85** were the thermodynamically more stable ones. In order to highlight the applicability of the method in natural product synthesis, the authors reported the efficient conversion, in three steps, of the cyclopentenone **85f** (separated from its epimer **86f** by trituration of the mixture with cold hexane) into the natural antibiotic (-)-pentenomycin I (Scheme 5.58).[58a]

The dihydroxylation of **85f** with the stoichiometric pair osmium tetroxide/*N,N,N′,N′*-tetramethylethylenediamine was highly diastereoselective, and took place exclusively on the less hindered face of the double bond. The resulting stable osmate ester **87** underwent a clean sulfoxide pyrolysis in refluxing toluene to afford the enone-osmate diester **88**. Acid hydrolysis of the ester and of the silyl protecting group provided the natural (-)-pentenomycin I in very high enantiomeric purity. Subsequently, **85f** was also converted into the (-)-aminocyclopentitol moiety of bacteriohopanetetrol ether, a triterpenoid found in the membrane of several bacteria, by means of an eight-step stereocontrolled sequence.[58b] As it happened in the intramolecular PKR of 1-sulfinyl enynes,[32] the chiral controller cannot be recovered after the elimination step.

The intermediacy of an internally chelated alkyne-vinyl sulfoxide complex in the reaction mechanism (Figure 5.2) was supported by DFT calculations.[58b]

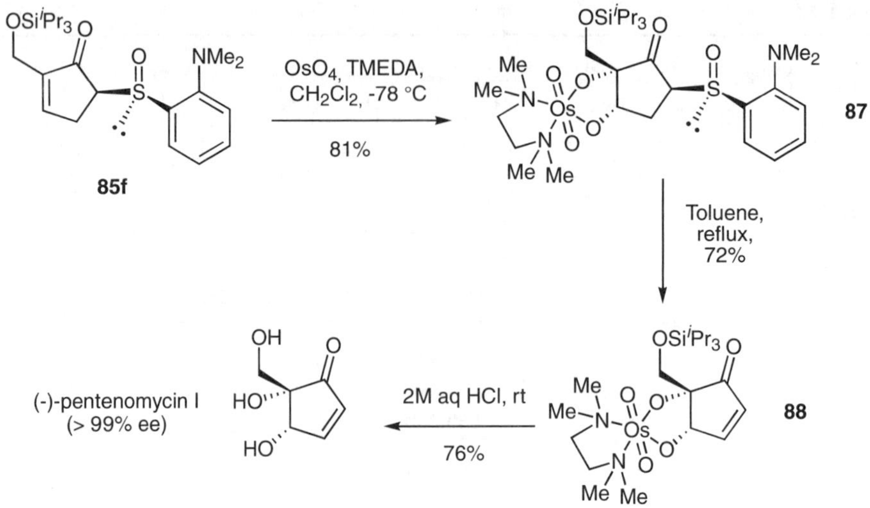

Scheme 5.58

5.4 Chiral Reagents for the Kinetic Resolution of PK Cycloadducts

If the PKR cannot be performed in an enantioselective fashion (either by using a chiral auxiliary or a chiral metal ligand), a kinetic resolution of the racemic PK cycloadduct can also be envisaged. This strategy was pioneered by Hua, 25 years ago, in his remarkable asymmetric total synthesis of (+)-pentalenene (Scheme 5.59).[59] The racemic intramolecular PK adduct *rac*-**89**, obtained in 58% yield from the corresponding enyne precursor, was treated with 0.5 molar equivs of the lithium derivative of (*S*)-allyl *p*-tolyl sulfoxide in THF at −78 °C gave the adduct **90**, derived from (*R*)-**89** (80% yield), and the unreacted (*S*) enantiomer of **89** (45% yield, 82% ee). This compound was subsequently converted into (+)-pentalenene in an eight-step sequence.

Several years later, Schmalz and co-workers applied the kinetic resolution of racemic PK cycloadducts in a synthesis of novel carbocyclic nucleoside analogs.[60] These authors investigated the utility of the asymmetric Corey–Bakshi–Shibata reduction of ketones[61] for this purpose (Table 5.13). When the racemic *exo*-allyloxyenones **91a–c**, obtained in good yields by an intramolecular PKR, were treated with catecholborane (0.80 mol equivs) in the presence of the (*R*)-Me-CBS oxazaborolidine catalyst, the unreacted ketones **91** were recovered in very high enantiomeric purity. The fast-reacting enantiomers of **91** gave rise

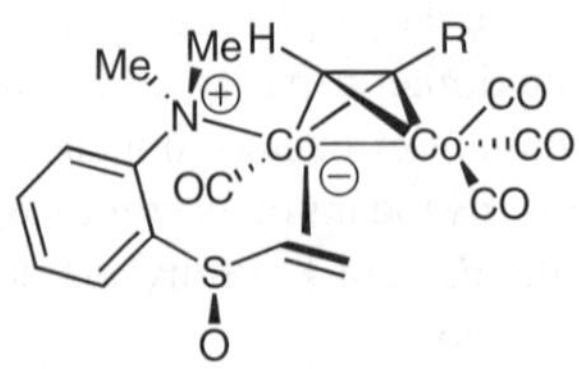

Figure 5.2

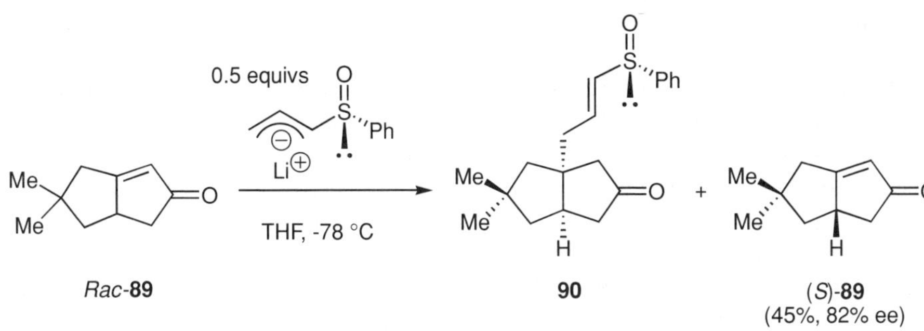

Scheme 5.59

to the alcohols **92** (mainly *exo*), that upon oxidation can give the ketones *ent*-**91** with good enantiomeric purity. An advantage of this method is that the absolute configuration of the slow-reacting enantiomer can be reliably predicted by the transition state working-model of the CBS reduction.[60b]

Table 5.13

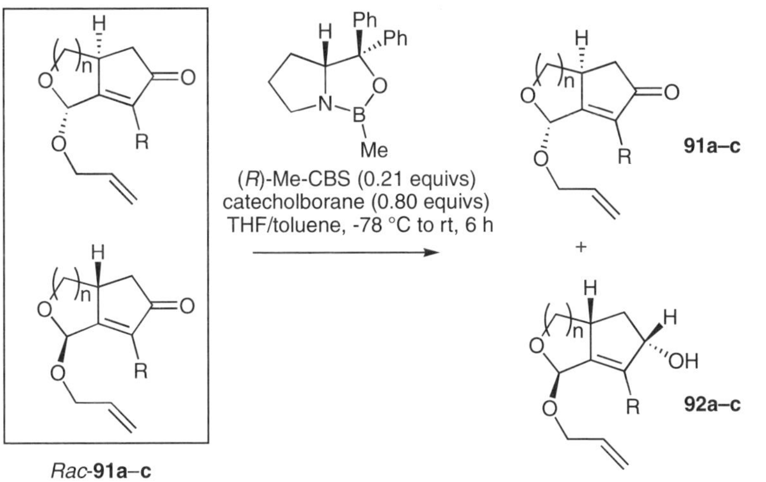

R	n	Unreacted ketone	Yield (%)	Ee (%)	Alcohol	Yield (%)	Ee (%)
$SiMe_3$	1	**91a**	34	> 99	**92a**	58	83
$SiMe_3$	2	**91b**	30	98	**92b**	51	78
Ph	1	**91c**	29	> 99	**92c**	n.d.	n.d.

5.5 Conclusion

This chapter has dealt with the use of chiral auxiliaries, chiral controllers, and chiral reagents in asymmetric versions of the PKR. The chiral auxiliary strategy has played a key role in the applicability of this powerful transition-metal mediated cycloaddition, largely overlooked in the decade following its discovery, to the enantioselective total synthesis of natural products; in this aspect, it is only surpassed by approaches based on the use of "chiral-pool" derived compounds (now including both natural and purely synthetic molecules) as starting materials.[62] Many types of chiral auxiliaries have been explored both for intramolecular and intramolecular PKRs in the past three decades. The Barcelona–Grenoble approach, making use of chiral alcohols attached either to the alkyne or to the alkene components, has been shown to be applicable to both versions of the reaction, and has given rise to several total or formal syntheses of relatively complex natural products. Chiral vinyl sulfoxides, applied as removable chiral controllers in the PKR by Carretero's group, have also given rise to highly useful protocols. Although it is true that recent approaches to the development of PK or of PK-type reactions have been mainly focused on the chemically more efficient use of chiral metal complexes, ideally in catalytic or at least substoichiometric fashion, it must be borne in mind that chiral auxiliaries can still provide very useful approaches for the preparation of certain types of enantiomerically pure PK cycloadducts, especially in the intermolecular version. Thus, chiral *N*-(2-alkynoyl)sultams have given rise to the highest levels of stereocontrol ever achieved in intermolecular PKRs. From a practical point of view, moreover, chiral auxiliaries, both in the PKR as well as in other widely used asymmetric carbon–carbon bond-forming reactions, exhibit a certain number of advantages: a) several of them (*trans*-2-phenylcyclohexanol, 4-phenyl-1,3-oxazolidin-2-one, bornane-10,2-sultam) are commercially available in both enantiomeric forms; b) diastereomeric mixtures of products are in many instances easily separable by crystallization or by chromatography, affording completely enantiopure compounds (getting enantiopure compounds from enantiomerically enriched mixtures is not always feasible); c) they can be recovered from the adduct and reused, while with scarce exceptions chiral phosphine-based ligands are not recoverable from PK reaction mixtures; and d) the stereochemical outcome of the reaction can often be reliably predicted from simple transition state working models.

Acknowledgements

Financial support from MICINN (Project AYA2009-13920-C02-02) is gratefully acknowledged.

References

[1] C. Exon and P. Magnus, Stereoselectivity of intramolecular dicobalt octacarbonyl alkene-alkyne cyclizations: short synthesis of *dl*-coriolin, *J. Am. Chem. Soc.* **105**, 2477 (1983).

[2] I. U. Khand, G. R. Knox, P. L. Pauson, W. E. Watts, and M. I. Foreman, Organocobalt complexes. Part II. Reaction of acetylenehexacarbonyl dicobalt complexes

$(R^1C_2R^2)Co_2(CO)_6$, with norbornene and its derivatives, *J. Chem. Soc. Perkin Trans.* **1**, 977 (1973).

[3] (a) M. C. Croudace and N. E. Schore, General synthesis of ω-acetylenic vinyl esters and ethers, *J. Org. Chem.* **46**, 5357 (1981); (b) N. E. Schore and M. C. Croudace, Preparation of bicyclo[3.3.0]oct-1-ene-3-one and bicyclo[4.3.0]non-1(9)-en-8-one via intramolecular cyclization of α,ω-enynes, *J. Org. Chem.* **46**, 5436 (1981).

[4] For an early example, see: B. E. La Belle, M. J. Knudsen, M. M. Olmstead, H. Hope, M. D. Yanuck, and N. E. Schore, Synthesis of 11-oxatricyclo[5.3.1.$0^{2,6}$]undecane derivatives via organometallic cyclizations, *J. Org. Chem.* **50**, 5215 (1985).

[5] P. Magnus, C. Exon, and P. Albaugh-Robertson, Dicobaltoctacarbonyl-alkyne complexes as intermediates in the synthesis of of bicyclo[3.3.0]octenones for the synthesis of coriolin and hirsutic acid, *Tetrahedron* **41**, 5861 (1985).

[6] C. Ferrer, J. Benet-Buchholz, A. Riera, and X. Verdaguer, Phosphine-alkene ligands as mechanistic probes in the Pauson-Khand reaction, *Chem. Eur. J.* **16**, 8340 (2010); and references cited therein.

[7] See the Chapter by M. A. Pericás *et al.* in this book.

[8] H. Greenfield, H. W. Sternberg, R. A. Friedel, J. H. Wotiz, R. Marby, and I. Wender, Acetylenic dicobalt hexacarbonyls. Organometallic compounds derived from alkynes and dicobalt octacarbonyl, *J. Am. Chem. Soc.* **78**, 120 (1955).

[9] (a) V. Rautenstrauch, P. Megard, J. Conesa, and W. Kuster, 2-Pentylcyclopent-2-en-1-one by catalytic Pauson-Khand reaction, *Angew. Chem. Int. Ed. Engl.* **29**, 1413 (1990); (b) N. Jeong, S. H. Hwang, Y. S. Lee, and Y. K. Chung, Catalytic conversion of the intramolecular Pauson-Khand reaction, *J. Am. Chem. Soc.* **116**, 3159 (1994); (c) N. Jeong, H. H. Hwang, Y. W. Lee, and J. S. Lim, Catalytic Pauson-Khand reaction in super critical fluids, *J. Am. Chem. Soc.* **119**, 10549 (1997).

[10] P. L. Pauson, The Khand reaction. A convenient and general route to a wide range of cyclopentenone derivatives, *Tetrahedron* **41**, 5855 (1985).

[11] See: M. Christmann and S. Bräse (Eds.), *Asymmetric Synthesis – The Essentials, 2nd ed.*, Wiley-VCH, Weinheim (2008); Part I: Chiral auxiliaries in asymmetric synthesis.

[12] Seminal references: (a) P. Bladon, P. L. Pauson, H. Brunner, and R. Eder, Optisch aktive übergangsmetall-komplexe. XCV. Optisch reines 3a,4,5,6,7,7a-hexahydro-2-phenyl-4,7-methano-1H-inden-1-on durch asymmetrische Khand-reaktion, *J. Organomet. Chem.* **355**, 449 (1988); (b) A. M. Hay, W. J. Kerr, G. G. Kirk, and D. Middlemiss, Highly efficient enantioselective Pauson-Khand reactions, *Organometallics* **14**, 4986 (1995); (c) J. Castro, A. Moyano, M. A. Pericàs, A. Riera, A. Alvarez-Larena, and J. F. Piniella, The first alkyne-dicobaltcarbonyl complex with a bidentate chiral ligand with Co–P and Co–N coordination, *J. Organomet. Chem.* **585**, 53 (1999); (d) J. Castro, A. Moyano, M. A. Pericàs, A. Riera, A. Alvarez-Larena, and J. F. Piniella, Acetylene–dicobaltcarbonyl complexes with chiral phosphinooxazoline ligands: structural characterization, and application to enantioselective intermolecular Pauson-Khand reactions, *J. Am. Chem. Soc.* **122**, 7944 (2000); (e) K. Hiroi, T. Watanabe, R. Kawagishi, and I. Abe, Catalytic use of chiral phosphine ligands in asymmetric Pauson-Khand reactions, *Tetrahedron: Asymmetry* **11**, 797 (2000); (f) X. Verdaguer, A. Moyano, M. A. Pericàs, A. Riera, M. A. Maestro, and J. Mahía, A new chiral bidentate (P,S) ligand for the asymmetric intermolecular Pauson-Khand reaction, *J. Am. Chem. Soc.* **122**, 10242 (2000).

[13] For the use of chiral ligands attached to metals other than cobalt, see: (a) F. A. Hicks and S. L. Buchwald, Highly enantioselective catalytic Pauson-Khand type formation of bicyclic cyclopentenones, *J. Am. Chem. Soc.* **118**, 11688 (1996); (b) S. J. Sturla and S. L. Buchwald, Catalytic asymmetric cyclocarbonylation of nitrogen-containing enynes, *J. Org. Chem.* **64**, 5547 (1999); (c) F. A. Hicks and S. L. Buchwald, An intramolecular titanium-catalyzed asymmetric Pauson-Khand type reaction, *J. Am. Chem. Soc.* **121**, 7026 (1999); (d) N. Jeong, B. K. Sung, and Y. K. Choi, Rhodium(I)-catalyzed asymmetric intramolecular Pauson-Khand-type reaction, *J. Am. Chem. Soc.* **122**, 6771 (2000); (e) T. Shibata and K. Takagi, Iridium–chiral diphosphine complex catalyzed highly enantioselective Pauson-Khand-type reaction, *J. Am. Chem. Soc.* **122**, 9852 (2000).

[14] A. Moyano, F. Charbonnier, and A. E. Greene, A simple preparation of chiral acetylenic ethers, *J. Org. Chem.* **52**, 2919 (1987).

[15] For another approach to alkoxyacetylenes, see: M. A. Pericàs, F. Serratosa, and E. Valentí, An efficient synthesis of *tert*-alkoxyethynes, *Tetrahedron* **43**, 2311 (1987).

[16] M. Poch, E. Valentí, A. Moyano, M. A. Pericàs, J. Castro, A. DeNicola, and A. E. Greene, Asymmetric induction studies in the intramolecular Pauson-Khand cyclization of 1-alkoxy-1-hepten-6-ynes, *Tetrahedron Lett.* **31**, 7505 (1990).

[17] J. Castro, A. Moyano, M. A. Pericàs, and A. Riera, A qualitative molecular mechanics approach to the stereoselectivity in intramolecular Pauson-Khand reactions, *Tetrahedron* **51**, 6541 (1995).

[18] X. Verdaguer, A. Moyano, M. A. Pericàs, A. Riera, A. E. Greene, J. F. Piniella, and A. Alvarez-Larena, Camphor-derived alcohols as chiral auxiliaries for asymmetric Pauson-Khand bicyclizations. Enantioselective synthesis of α-methoxyenones, *J. Organomet. Chem.* **433**, 305 (1992).

[19] First examples: (a) M. J. Knudsen and N. E. Schore, Synthesis of the angularly fused triquinane skeleton via intramolecular organometallic cyclization, *J. Org. Chem.* **49**, 5025 (1984); (b) E. Carceller, V. Centellas, A. Moyano, M. A. Pericàs, and F. Serratosa, Direct entry to the all-cis tricyclo[5.2.1.0^{4,10}]decane (perhydrotriquinacene) skeleton by a cobalt mediated intramolecular cyclization, *Tetrahedron Lett.* **26**, 2475 (1985); (c) W. A. Smit, A. S. Gybin, A.S. Shashkov, Y. T. Strychkov, L. G. Kyz'mina, G. S. Mikaelian, R. Caple, and E. D. Swanson, New route to the synthesis of polycyclic compounds based on a stepwise Ad_E-reaction of dicobalt hexacarbonyl complexes of conjugated enynes with a subsequent intramolecular Khand-Pauson reaction, *Tetrahedron Lett.* **27**, 1241 (1986); (d) E. Carceller, M. L. García, A. Moyano, M. A. Pericàs, and F. Serratosa, Synthesis of triquinacene derivatives. A new approach towards the synthesis of dodecahedrane, *Tetrahedron* **42**, 1831 (1986); (e) S. L. Schreiber, T. Sammakia, and W. E. Crowe, Lewis acid mediated version of the Nicholas reaction: synthesis of *syn*-alkylated products and cobalt-complexed cycloalkynes, *J. Am. Chem. Soc.* **108**, 3128 (1986); (f) N. E. Schore and M. J. Knudsen, Two approaches to angularly fused triquinanes via intramolecular Pauson-Khand reaction, *J. Org. Chem.* **52**, 569 (1987); (g) C. Almansa, A. Moyano, and F. Serratosa, Synthesis and chiroptical properties of C_3-perhydrotriquinacene derivatives, *Tetrahedron* **44**, 2657 (1988); (h) N. E. Schore and E. G. Rowley, Diastereofacial selectivity in intramolecular Pauson-Khand cycloadditions: highly stereoselective synthesis of pentalenene, *J. Am. Chem. Soc.* **110**, 5224 (1988).

[20] (a) M. Thommen, P. Gerber, and R. Keese, Synthesis of a [5.5.5.5]fenestrane *via* Pauson-Khand reaction, *Chimia* **45**, 21 (1991); (b) A. van der Waals and R. Keese, Synthesis of a [5.5.5.5]fenestranedione *via* tandem Pauson-Khand tetracyclisation, *J. Chem. Soc., Chem. Commun.* 570 (1992); (c) M. Thommen, A. L. Veretenov, R. Guidetti-Grept, and R. Keese, The tandem *Pauson-Khand* reaction, *Helv. Chim. Acta* **79**, 461 (1996); (d) S. G. Van Ornum and J. M. Cook, Generation of four five-membered rings in a one-pot process. Studies directed toward the synthesis of dicyclo[a,d]pentalene via the tandem Pauson-Khand reaction, *Tetrahedron Lett.* **37**, 7185 (1996); (e) S. G. Van Ornum, M. M. Bruendl, and J. M. Cook, The Livinghouse catalytic approach to the tandem Pauson-Khand reaction. Entry into the parent ring systems of dicyclopenta[*a,f*]pentalene and dicyclopenta[*a,e*]pentalene, *Tetrahedron Lett.* **39**, 6649 (1998); (f) M. M. Bruendl, S. G. Van Ornum, T.-M. Chan, and J. M. Cook, Extension of the tandem Pauson-Khand reaction to [5.6.6.5]tetracycles: a unique entry into *cis*-fused decalins, *Tetrahedron Lett.* **40**, 1113 (1999); (g) S. H. Hong, J. W. Kim, D. S. Choi, Y. K. Chung, and S.-G. Lee, Construction of tetracyclic ring systems by $Co_2(CO)_8$–catalyzed tandem [2+2+1]/[2+2+2] cycloaddition reaction of diynes: a simple one-pot reaction, *Chem. Commun.* 2099 (1999); (h) S. G. Van Ornum, M. M. Bruendl, H. Cao, M. Reddy, D. S. Grubisha, D. W. Bennett, and J. M. Cook, Utility of the tandem Pauson-Khand reaction in the construction of tetracycles, *J. Org. Chem.* **65**, 1957 (2000); (i) S. Tanimori, K. Fukubayashi, and M. Kirihata, A new pathway to chiral tetracyclic indolidines via Pauson-Khand reaction, *Tetrahedron Lett.* **42**, 4013 (2001); D. H. Kim, S. U. Son, Y. K. Chung, and S.-G. Lee, Catalytic one-pot synthesis of [5.5.5.6]fenestrane systems via a dicobalt octacarbonyl-catalyzed tandem cycloaddition of dienediynes, *Chem. Commun.* 56 (2002).

[21] J. Tormo, X. Verdaguer, A. Moyano, M. A. Pericàs, and A. Riera, Chiral auxiliary-induced stereocontrol in intramolecular Pauson-Khand reactions leading to angular triquinanes, *Tetrahedron* **52**, 14021 (1996).

[22] J. Tormo, A. Moyano, M. A. Pericàs, and A. Riera, Enantioselective construction of angular triquinanes through an asymmetric intramolecular Pauson-Khand reaction. Synthesis of (+)-15-nor-pentalenene, *J. Org. Chem.* **62**, 4851 (1997).

[23] E. Montenegro, M. Poch, A. Moyano, M. A. Pericàs, and A. Riera, Chiral acetylene thioethers: synthesis and Pauson-Khand reactions, *Tetrahedron* **53**, 8651 (1997).

[24] S. Fonquerna, R. Rios, A. Moyano, M. A. Pericàs, and A. Riera, Stereoselective inter- and intramolecular Pauson-Khand reactions of *N*-(2-alkynoyl) derivatives of chiral oxazolidin-2-ones, *Eur. J. Org. Chem.* **12**, 3459 (1999).

[25] R. Rios, A. Moyano, and M. A. Pericàs, An intramolecular Pauson-Khand approach to the synthesis of chiral cyclopentadienes, *Tetrahedron Lett.* **43**, 1023 (2002).

[26] J. Balsells, A. Moyano, A. Riera, and M. A. Pericàs, Tandem aminocarbonylation/Pauson-Khand reaction of haloacetylenes, *Org. Lett.* **1**, 1981 (1999).

[27] (a) S. Shambayati, W. E. Crowe, and S. L. Schreiber, *N*-Oxide promoted Pauson-Khand cyclizations at room temperature, *Tetrahedron Lett.* **31**, 5289 (1990); (b) N. Jeong, Y. K. Chung, B. Y. Lee, S. H. Lee, and S.-E. Yoo, A dramatic acceleration of the Pauson-Khand reaction by trimethylamine *N*-oxide, *Synlett* **3**, 204 (1991)

[28] J. Castro, H. Sörensen, A. Riera, C. Morin, A. Moyano, M. A. Pericàs, and A. E. Greene, Asymmetric approach to Pauson-Khand bicyclization. Enantioselective formal synthesis of hirsutene, *J. Am. Chem. Soc.* **112**, 9388 (1990).

[29] L. Solà, J. Castro, A. Moyano, M. A. Pericàs, and A. Riera, Divergent stereoselective synthesis of (*E*) and (*Z*) *O*-alkyl enol ethers, *Tetrahedron Lett.* **33**, 2863 (1992).

[30] J. Castro, A. Moyano, M. A. Pericàs, A. Riera, A. E. Greene, A. Alvarez-Larena, and J. F. Piniella, Asymmetric approach to (+)-β-cuparenone by intramolecular Pauson-Khand reaction, *J. Org. Chem.* **61**, 9016 (1996).

[31] J. Castro, A. Moyano, M. A. Pericàs, A. Riera, and A. E. Greene, Asymmetric synthesis of bicyclo[4.3.0]nonan-8-ones by intramolecular Pauson-Khand reaction, *Tetrahedron: Asymmetry* **5**, 307 (1994).

[32] (a) J. Adrio and J. C. Carretero, The *tert*-butylsulfinyl group as a highly efficient chiral auxiliary in asymmetric Pauson-Khand reactions, *J. Am. Chem. Soc.* **121**, 7411 (1999); (b) J. C. Carretero and J. Adrio, Asymmetric Pauson-Khand cyclizations of 1-sulfinylenynes, *Synthesis* 1888 (2001).

[33] A. Stolle, H. Becker, J. Salaün, and A. de Meijere, Enantioselective construction of spiro(cyclopropane-1,4′-bicyclo[3.3.0]oct-1-en-3-ones), *Tetrahedron Lett.* **35**, 3521 (1994).

[34] J. Wang, M. L. Falck-Pedersen, C. Rømming, and K. Undheim, The Pauson-Khand reaction in synthesis of bicyclic rigid α-amino acids, *Synth. Commun.* **31**, 1141 (2001).

[35] V. Bernardes, X. Verdaguer, N. Kardos, A. Riera, A. Moyano, M. A. Pericàs, and A. E. Greene, Practical asymmetric version of the intermolecular Pauson-Khand reaction, *Tetrahedron Lett.* **35**, 575 (1994).

[36] N. E. Schore, The Pauson-Khand cycloaddition reaction for the synthesis of cyclopentenones, *Org. React.* **40**, 1 (1991).

[37] (a) N. E. Schore, A short convenient synthesis of cyclopentenones *via* derivatives of *exo*-dicyclopentadiene, *Synth. Commun.* **9**, 41 (1979); (b) L. Daalman, R. F. Newton, P. L. Pauson, and A. Wadsworth, Alkyne–cobalt complexes as prostaglandin synthons. Part 4. A possible approach *via* 2-phenylthiocyclopent-2-enone, *J. Chem. Research* (S), 346 (1984); (c) D. C. Billington, P. Bladon, I. M. Helps, P. L. Pauson, W. Thomson, and D. Willison, Syntheses of jasmone, jasmonic acid, and some analogues from alkyne–cobalt complexes *via* the Khand reaction, *J. Chem. Research* (S), 326 (1988).

[38] (a) M. E. Krafft, Regiocontrol in the intermolecular cobalt-catalyzed olefin–acetylene cycloaddition, *J. Am. Chem. Soc.* **110**, 968 (1988); (b) M. E. Krafft, C. A. Juliano, I. L. Scott, C. Wright, and M. D. McEachin, The directed Pauson-Khand reaction, *J. Am. Chem. Soc.* **113**, 1693 (1991).

[39] M. E. Krafft, I. L. Scott, and R. H. Romero, Acceleration of the thermal Pauson-Khand reaction by coordinating ligands, *Tetrahedron Lett.* **33**, 3829 (1992).

[40] X. Verdaguer, A. Moyano, M. A. Pericàs, A. Riera, V. Bernardes, A. E. Greene, A. Alvarez-Larena, and J. F. Piniella, A dual–function, highly efficient chiral controller for stereoselective, intermolecular Pauson-Khand reactions, *J. Am. Chem. Soc.* **116**, 2153 (1994).

[41] N. Furukawa, Y. Sugihara, and H. Fujihara, Camphoryl sulfide as a chiral auxiliary and a mediator for one-step synthesis of optically active 1,2-diaryloxiranes, *J. Org. Chem.* **54**, 4222 (1989).

[42] V. Bernardes, X. Verdaguer, A. Moyano, M. A. Pericàs, A. Riera, and A. E. Greene, A convenient synthesis of hexacarbonyldicobalt complexes of chiral (non-racemic) terminal alkoxyacetylenes, *J. Organomet. Chem.* **470**, C12 (1994).

[43] X. Verdaguer, J. Vázquez, G. Fuster, V. Bernardes-Génisson, A. E. Greene, A. Moyano, M. A. Pericàs, and A. Riera, Camphor-derived, chelating auxiliaries for the highly diastereoselective intermolecular Pauson-Khand reaction: experimental and computational studies, *J. Org. Chem.* **63**, 7037 (1998).

[44] V. Bernardes, N. Kann, A. Riera, A. Moyano, M. A. Pericàs, and A. E. Greene, Asymmetric Pauson-Khand cyclization: a formal total synthesis of natural brefeldin A, *J. Org. Chem.* **60**, 6670 (1995).

[45] J. Balsells, A. Moyano, M. A. Pericàs, and A. Riera, Efficient synthesis of chiral acetylene dithioethers in enantiomerically pure form, *Tetrahedron: Asymmetry* **8**, 1575 (1997).

[46] E. Montenegro, M. Poch, A. Moyano, M. A. Pericàs, and A. Riera, Highly diastereoselective Pauson-Khand reactions of a stable, internally chelated, dicobalt pentacarbonyl complex of a chiral acetylene thioether, *Tetrahedron Lett.* **39**, 335 (1998).

[47] E. Montenegro, A. Moyano, M. A. Pericàs, A. Riera, A. Alvarez-Larena, and J.-F. Piniella, Studies on the Pauson-Khand reaction of alkynyl sulfoxides. Unexpectedly easy racemization of their dicobalt hexacarbonyl complexes, *Tetrahedron: Asymmetry* **10**, 457 (1999).

[48] J. Balsells, J. Vázquez, A. Moyano, M. A. Pericàs, and A. Riera, Low-energy pathway for Pauson-Khand reactions: synthesis and reactivity of dicobalt hexacarbonyl complexes of chiral ynamines, *J. Org. Chem.* **65**, 7291 (2000).

[49] B. Witulski and M. Göβmann, Intermolecular Pauson-Khand reactions with *N*-alkynylamides – electronically tuned variants of ynamines, *Synlett* 1793 (2000).

[50] L. Shen and R. P. Hsung, Pauson-Khand reactions of chiral ynamides. Observation of an unusual *endo*-addition with norbornadiene, *Tetrahedron Lett.* **44**, 9353 (2003).

[51] S. Fonquerna, A. Moyano, M. A. Pericàs, and A. Riera, Diastereoselectivity in the intermolecular Pauson-Khand reaction of chiral 2-alkynoates, *Tetrahedron* **51**, 4239 (1995).

[52] S. Fonquerna, A. Moyano, M. A. Pericàs, and A. Riera, Totally stereocontrolled intermolecular Pauson-Khand reactions of *N*-(2-alkynoyl) sultams, *J. Am. Chem. Soc.* **119**, 10225 (1997)

[53] S. Fonquerna, A. Moyano, M. A. Pericàs, and A. Riera, A convenient preparation of *N*-(2-alkynoyl) derivatives of chiral oxazolidin-2-ones and bornane-10,2-sultam, *Tetrahedron: Asymmetry* **8**, 1685 (1997).

[54] K. Hiroi and T. Watanabe, Participation of an organosulfur functionality in asymmetric Pauson-Khand reactions using (*S*)-methionine-derived amides as enantiocontrollable substrates, *Tetrahedron Lett.* **41**, 3935 (2000).

[55] J. Vázquez, S. Fonquerna, A. Moyano, M. A. Pericàs, and A. Riera, Bornane-2,10-sultam: a highly efficient chiral controller and mechanistic probe for the intermolecular Pauson-Khand reaction, *Tetrahedron: Asymmetry* **12**, 1837 (2001).

[56] R. Rios, M. A. Pericàs, A. Moyano, M. A. Maestro, and J. Mahía, Reversing the stereoselectivity of the intermolecular Pauson-Khand reaction: formation of *endo*-fused norbornadiene adducts, *Org. Lett.* **4**, 1205 (2002).

[57] A. Moyano and J. Castro, unpublished.

[58] (a) M. Rodríguez Rivero, J. C. de la Rosa, and J. C. Carretero, Asymmetric intermolecular Pauson-Khand reactions of unstrained olefins: the

(*o*-dimethylamino)phenylsulfinyl group as an efficient chiral auxiliary, *J. Am. Chem. Soc.* **125**, 14992 (2003); (b) M. Rodríguez Rivero, I. Alonso, and J. C. Carretero, Vinyl sulfoxides as stereochemical controllers in intermolecular Pauson-Khand reactions: applications to the enantioselective synthesis of natural products, *Chem. Eur. J.* **10**, 5443 (2004).

[59] D. H. Hua, Asymmetric total synthesis of (+)-pentalenene via chiral sulfinylallyl anions. Hydrolitic ring closure of enol thioether ketones, *J. Am. Chem. Soc.* **108**, 3835 (1986).

[60] (a) J. Velcicky, A. Lanver, J. Lex, A. Prokop, T. Wieder, and H.-G. Schmalz, Transition-metal-mediated synthesis of novel carbocyclic nucleoside analogs with antitumoral activity, *Chem. Eur. J.* **10**, 5087 (2004); (b) A. Lanver and H.-G. Schmalz, A Pauson-Khand approach to new carbocyclic nucleoside analogs, *Eur. J. Org. Chem.* **7**, 1444 (2005).

[61] For a review, see: D. Laurent and M. Srebnik, Asymmetric boron-catalyzed reactions, *Chem. Rev.* **93**, 741 (1993).

[62] Selected recent examples: (a) D. Pizzarani, T. Kaya, P. A. Clemons, and S. L. Schreiber, Stereochemical and skeletal diversity arising from amino propargylic alcohols, *Org. Lett.* **12**, 2822 (2010); (b) A. Farwick and G. Helmchen, Enantioselective total synthesis of (-)-α-kainic acid, *Org. Lett.* **12**, 1108 (2010); (c) A. Nakayama, N. Kogure, M. Kitajima, and H. Takayama, First asymmetric total syntheses of fawcettimine-type *Lycopodium* alkaloids, lycoposerramine-C and phlegmariurine-A, *Org. Lett.* **11**, 5554 (2009); (d) Y. Kavanagh, M. O'Brien, and P. Evans, Stereocontrolled preparation of bicyclic alkaloid analogs: an approach towards the kinabalurine skeleton, *Tetrahedron* **65**, 8259 (2009); (e) K. Kaneda and T. Honda, Stereocontrolled synthesis of (+)-α-skytantine by means of an intramolecular Pauson-Khand reaction, *Tetrahedron* **64**, 11589 (2008); (f) G. R. Drake, E. E. Fenster, and B. O. Patrick, A synthetic approach to the fusicoccane A–B ring fragment based on a Pauson-Khand cycloaddition/Norrish type I fragmentation, *J. Org. Chem.* **73**, 7611 (2008); (g) K. A. Miller, C. S. Shanahan, and S. F. Martin, The Pauson-Khand reaction as a new entry to the synthesis of bridged bicyclic heterocycles: application to the enantioselective total synthesis of (–)-alstonerine, *Tetrahedron* **64**, 6884 (2008); (h) Y. Hayashi, N. Miyakoshi, S. Kitagaki, and C. Mukai, Stereoselective total synthesis of uncommon sesquiterpenoids isolated from *Jathropa neopauciflora*, *Org. Lett.* **10**, 2385 (2008).

6

The Enantioselective Pauson-Khand Reaction

Agustí Lledó, Xavier Verdaguer and Antoni Riera

The Pauson-Khand Reaction: Scope, Variations and Applications, First Edition. Edited by Ramon Rios Torres.

6.1 Introduction

In the Pauson-Khand reaction (PKR), one or two new chiral sp^3 carbons (originating from the alkene component) arise at the newly formed cyclopentenone core. Soon after the first PK adducts were fully characterized by Pauson and co-workers in 1973 (Scheme 6.1),[1] research and development on stereoselective versions of the reaction began. The bulk of this work has comprised diastereoselective reactions, including syntheses based on chiral auxiliaries (covered in the previous chapter), and stereospecific syntheses, which require optical resolution or asymmetric transformations prior to the cyclization. Despite the numerous examples of the latter type in the literature (chiefly, as intramolecular reactions), they will not be discussed here.[2]

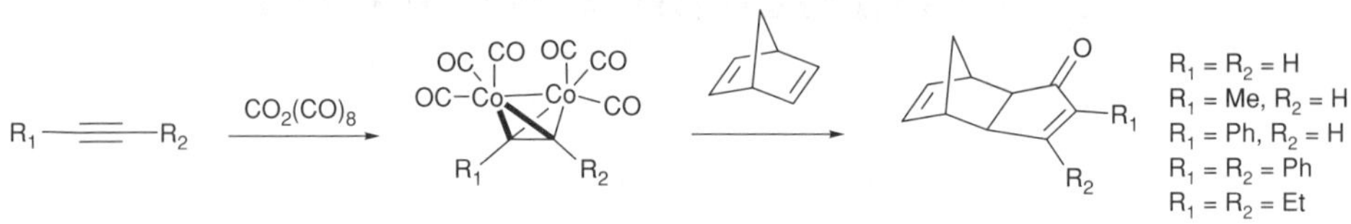

Scheme 6.1 *The first-ever reported carbonylative cycloadditions (Pauson* et al.*).*

In this chapter we aim to cover those transformations in which prochiral or racemic substrates are transformed into homochiral (or enantioenriched) products through cobalt-promoted PK reactions. In the absence of covalently attached chiral auxiliaries, this chemistry obviously requires a chiral promoter or ligand. Research on dicobalt clusters with chiral ligands dates back to 1988, with the pioneering work of Brunner.[3] Since then, numerous related strategies have been devised that deliver high levels of asymmetric induction. However, it is fair to say that the rapid progression of asymmetric catalysis has left the PKR behind. Synthetically useful induction levels are only attained in a few intramolecular cyclizations of enynes. Nevertheless, the current state-of-the-art in ligand design for asymmetric PKRs holds promise that these limitations will be overcome in the near future.

6.2 Mechanistic Considerations. Topology of Alkyne-Dicobalt Clusters

Based on the general mechanism initially proposed by Magnus[4] and subsequently confirmed by molecular orbital calculations,[5] we can outline the different pathways in the reaction of a terminal alkyne with norbornadiene to give each enantiomer (Scheme 6.2). As observed, the dicobalt scaffold in the complex **I** clamps the alkyne substrate by orthogonally fixing it to the Co–Co axis, thereby giving it less conformational freedom than that of monometallic alkyne complexes. The presence of two cobalt centers raises the degrees of freedom in the system, since the alkene can substitute carbon monoxide at either center, leading to the intermediates **IIA** and **IIB**. The stereochemistry of the newly formed chiral centers is determined during the next step, in which the olefin is inserted into the less substituted Co–C bond. Depending on the orientation of the norbornadiene, **IIA** and **IIB** can each form two cobaltacycles (**anti-IIIA** and **syn-IIIA**; or **anti-IIIB** and **syn-IIIB**, respectively). Interestingly, **syn IIIA** and **anti-IIIB** would provide one enantiomer, whereas **anti-IIIA** and **syn-IIIB** would provide the opposite one. However, molecular orbital calculations have shown that *anti* cobaltacycles are much more stable than *syn* cobaltacycles[6] (for a detailed mechanistic discussion, see Chapter 2). In summary, enantioselective PKRs have three main requirements: stereoselective coordination of the olefin to one of the two cobalt atoms; regioselective insertion of the olefin into one of the two Co–C bonds; and diastereoselective insertion through an *anti* or *syn* mode.

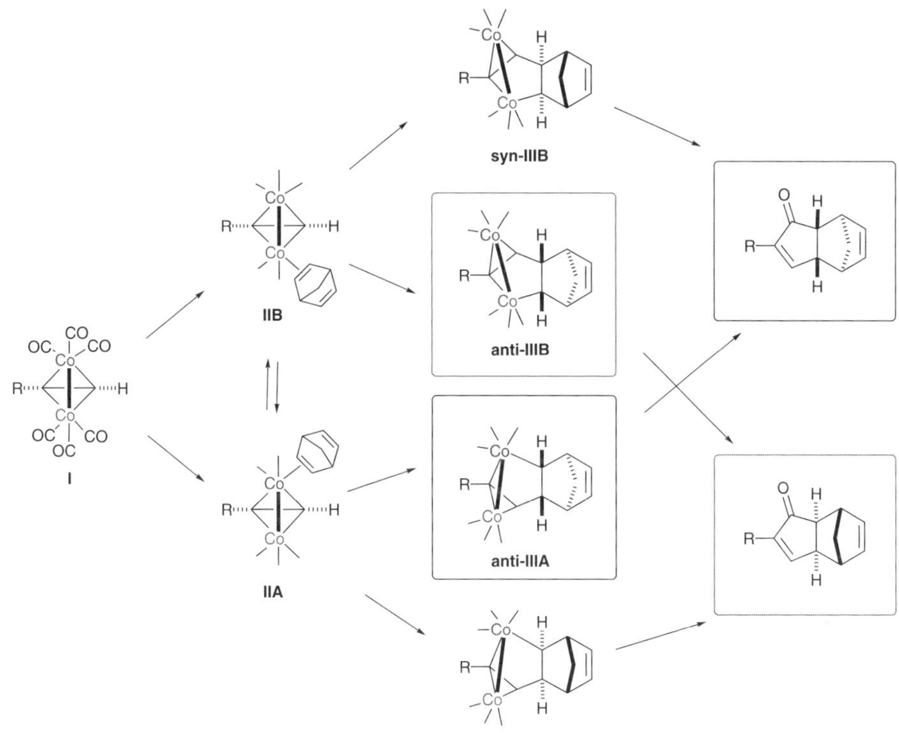

Scheme 6.2 *Stereochemical mechanistic pathways leading to each enantiomer of the cobaltacycle **III**. Carbonyl ligands on intermediates have been omitted for clarity.*

Since coordination of a Lewis base into the dicobalt cluster nearly always drives olefin coordination to (and subsequent insertion at) the distal cobalt center, diastereoselective coordination with a ligand is extremely important in a ligand-based strategy, especially for catalytic applications. Although turnstile rotation can enable ligand exchange between equatorial and axial positions within a cobalt vertex, this is irrelevant to the stereochemistry of the complexes. In contrast, exchange between cobalt sites may occur at elevated temperatures, leading to isomerization processes that are often deleterious for efficient asymmetric induction into the final PK adduct. Interestingly, when a CO is replaced with a different ligand (**L**), dicobalt complexes of terminal or asymmetrically disubstituted acetylenes yield a pair of enantiomers, indicating that the cobalt vertices are prochiral (enantiotopic) (Scheme 6.3). However, if the ligand is chiral (**L***), then two diastereomeric complexes arise. Using symmetric alkynes initially simplifies the stereochemical challenge: coordination of an achiral ligand yields a single achiral complex whereas the use of a chiral ligand provides a single chiral complex (Scheme 6.3). In this case, once the olefin has coordinated to the unsubstituted cobalt, diastereoselective insertion of the olefin into one of the two (almost-equally hindered) Co–C bonds becomes much more difficult, as we will discuss later on.

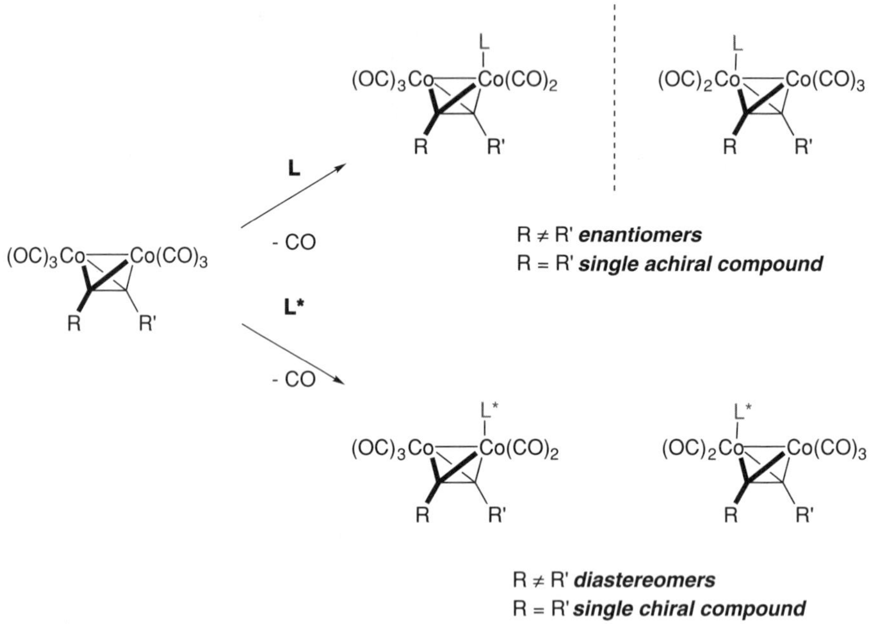

Scheme 6.3 *Formation of stereoisomeric dicobalt-alkyne complexes with an arbitrary ligand L.*

In summary, diastereoselective coordination of the olefin to one of the two enantiotopic cobalt atoms is one of various requirements for any enantioselective PKR. In the following sections, we cover different versions of the asymmetric PKR. Rather than following a strict chronological order, we explore them according to the approach used for desymmetrization of the alkyne-dicobalt cluster.

6.3 Intrinsically Chiral Dicobalt Clusters

6.3.1 Resolution of Monophosphine Complexes

The groups of Chung[7] and Kerr[8] have described preparation of intrinsically chiral dicobalt complexes bearing *achiral* phosphines using a chiral auxiliary approach. Reaction of the hexacarbonyl dicobalt complex of a chiral acetylene (menthyl propargyl ether [**1a**] or phenylprop-2-yn-1-ol [**1b**]) with simple phosphines or phosphites affords a mixture of diastereomeric complexes that can be separated by column chromatography (Scheme 6.4). PK reaction of the individual complexes with norbornene (NBN) or norbornadiene (NBD) yields the diastereomeric adducts **2/2'** or **3/3'**, respectively, which have opposite stereochemistry around the cyclopentenone core (Scheme 6.4). Although these are formally diastereoselective reactions, the results indicate that the chirality of the product is dictated by the chirality around the cobalt cluster. This was corroborated when the hexacarbonyldicobalt complexes of the same chiral alkynes were subjected to the PKR: very little or no diastereoselectivity was observed. Importantly, the *N*-oxide protocol is required, as thermal activation would most likely lead to equilibration of the diastereomeric complexes through ligand exchange, which would in turn lead to an enantiomeric excess as low as the thermodynamic ratio of the diastereomeric pair in question.

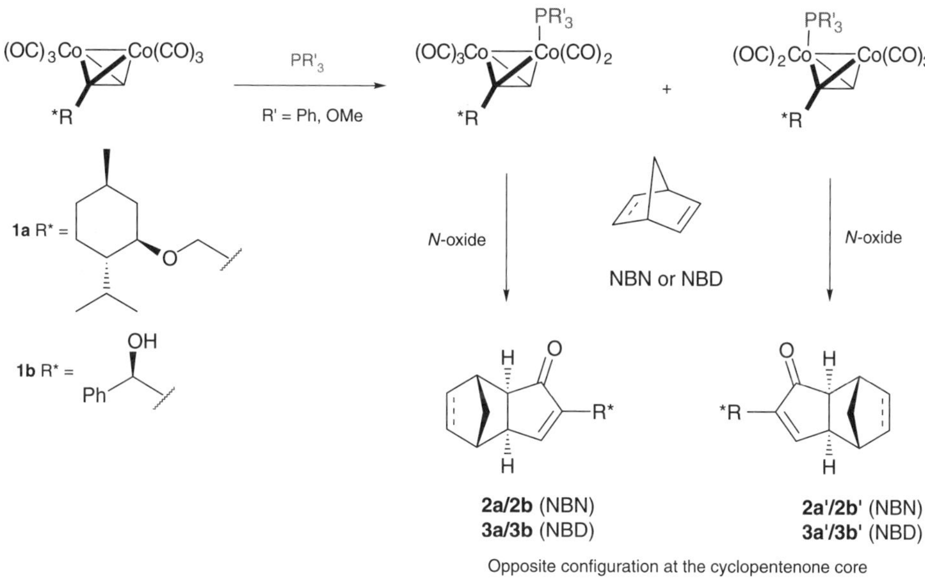

Scheme 6.4 *Desymmetrization of dicobalt-alkyne clusters driven by a chiral auxiliary.*

6.3.2 Synthesis of Chiral Monophosphine Complexes Using Chiral *N*-Oxides

Chiral resolution can also be accomplished by using a chiral *N*-oxide without chiral auxiliaries, as demonstrated by Nicholas in the work on his namesake reaction.[9] Kerr exploited this approach for the PKR[10] (Scheme 6.5). His results demonstrate that the chirality of the final product is dictated solely by olefin coordination to, and subsequent insertion at, either of the enantiotopic cobalt vertices, away from the phosphine ligand.

6.3.3 Heterobimetallic Complexes

A third approach to the preparation of intrinsically chiral complexes entails a heterobimetallic PKR.[11] Desymmetrization of dicobalt-alkyne through iso-lobular replacement

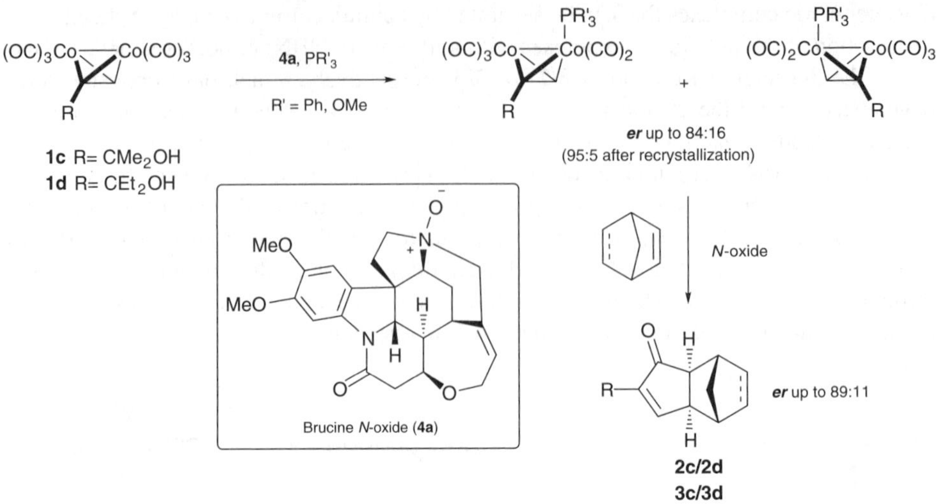

Scheme 6.5 *Desymmetrization of dicobalt-alkyne clusters driven by a chiral* N*-oxide.*

with molybdenum allowed Christie *et al.* to obtain the diastereomeric Co-Mo clusters **5a** and **5a'** from the cobalt hexacarbonyl complex of menthyl propargyl ether (**1a**). Separation of these clusters also yielded diastereomeric products (the cyclopentenones **3a** and **3a'**, respectively) with opposite configuration at the newly created stereocenters (Scheme 6.6).

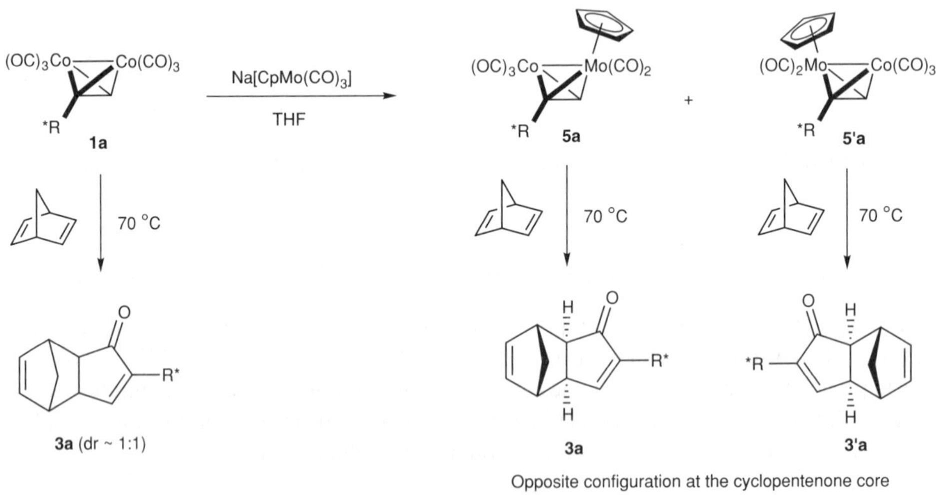

Scheme 6.6 *Desymmetrization of dicobalt-alkyne through iso-lobular replacement with molybdenum.*

6.4 Chiral Promoters

6.4.1 Sulfoxides

Sulfoxides can be used as PK promoters. For example, Jeong and Pauson attempted to use the chiral sulfoxide **7** to promote intramolecular PKR of enyne **6a**, which resulted in no induction.[12] The hemilabile coordinating behavior of sulfoxides on dicobalt clusters has recently been documented and anticipates the failure of this approach (Scheme 6.7).

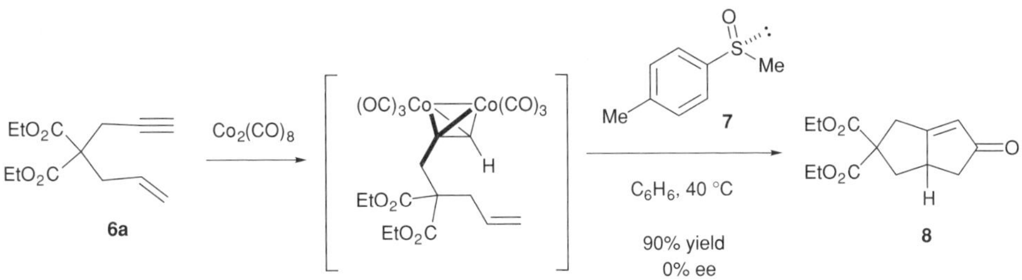

Scheme 6.7 Non enantioselective PKR promoted by a chiral sulfoxide.

6.4.2 Chiral *N*-Oxides

Tertiary amine *N*-oxides are the most effective promoters reported to date for stoichiometric PKRs. These reactions can be run at temperatures as low as −70 °C, depending on the substrate. In 1998, Kerr reported the first ever use of a chiral *N*-oxide (brucine *N*-oxide [**4a**]) as a promoter in the asymmetric PKR.[13] The Laschat group later extended this approach to a series of structurally diverse alkaloid *N*-oxides.[14] However, they did not achieve better enantioselectivity: their best result, obtained with brucine *N*-oxide, was 78% enantiomeric excess (Scheme 6.8).

Moreover, this methodology seems to be restricted to propargylic substrates. Researchers have postulated that the preferential attack to the CO ligand at one of the enantiotopic cobalt vertices by the *N*-oxide is facilitated by tethering of the reagent to the cobalt complex through hydrogen bonding between the *N*-oxide and the hydroxyl group of the latter. The reaction must be run at low temperature to prevent isomerization (through CO migration) of the highly unstable (chiral) unsaturated complex that results from CO loss. Olefin insertion onto the opposite cobalt atom would lead to inversion in the absolute configuration of the product (Scheme 6.9).

Whether the resulting chiral tertiary amine byproduct contributes to the asymmetric induction has been the subject of debate. However, recent attempts to use chiral primary amines to promote asymmetric PKRs have failed. For example, cyclization of the cobalt complex **10** promoted by the chiral amines **11a–c** afforded the tricyclic enone **12** in racemic form (Scheme 6.10).[15]

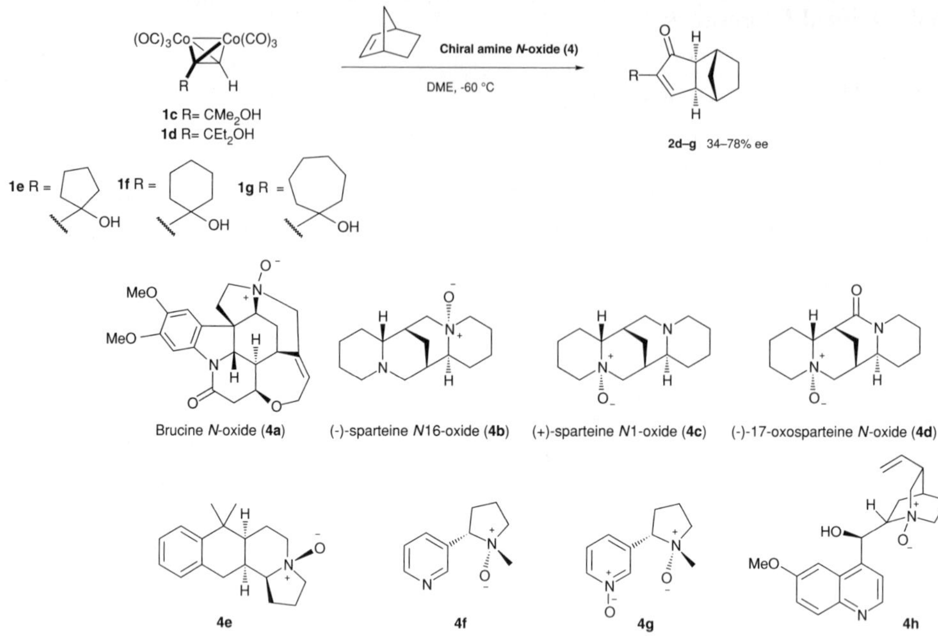

Scheme 6.8 Asymmetric PKRs promoted by chiral N-oxides. The N-oxides depicted are those reported in the literature.

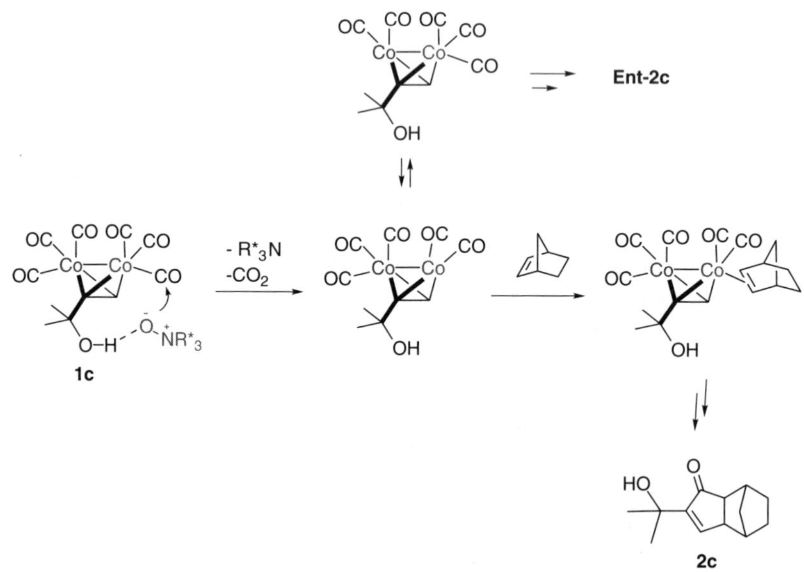

Scheme 6.9 PKR promoted by a chiral N-oxide and anchimerically assisted by propargylic substrates.

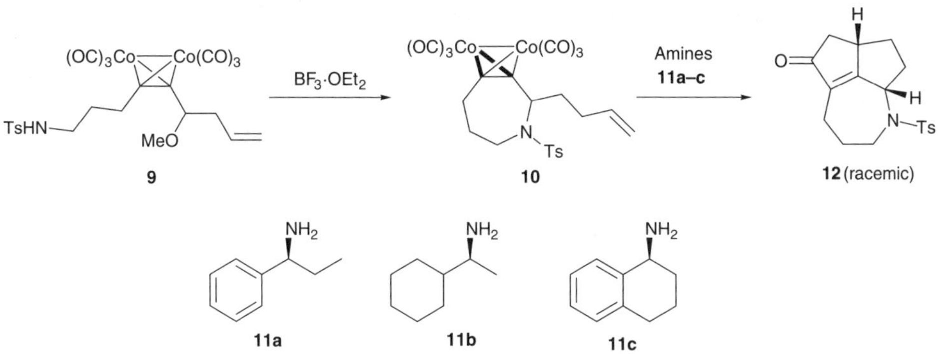

Scheme 6.10 *Non enantioselective PKRs promoted by chiral amines.*

6.5 Chiral Ligands

Here we describe chiral molecules that have been specifically designed to coordinate to dicobalt clusters, as well as common Lewis bases for which evidence of stable coordinated cobalt clusters exists.

6.5.1 Sulfides

The promising results obtained with thioether-containing chelating auxiliaries[16] prompted the group of Pericàs and Riera to explore sulfides as ligands. However, the intrinsically weak coordination properties of thioethers preclude the isolation of stable complexes, except for the case of alkynes substituted with strong electron-withdrawing groups. A stable complex **15** from bis(*tert*-butyl sulfonyl)acetylene dicobalt hexacarbonyl complex (**13**) and the camphor-derived chiral thioether **14** could be prepared[17] (Scheme 6.11). Unfortunately, when this complex was subjected to PKR with norbornadiene, no PK adducts could be isolated—presumably, due to the severe steric congestion imposed by the two *tert*-butyl sulfonyl groups.

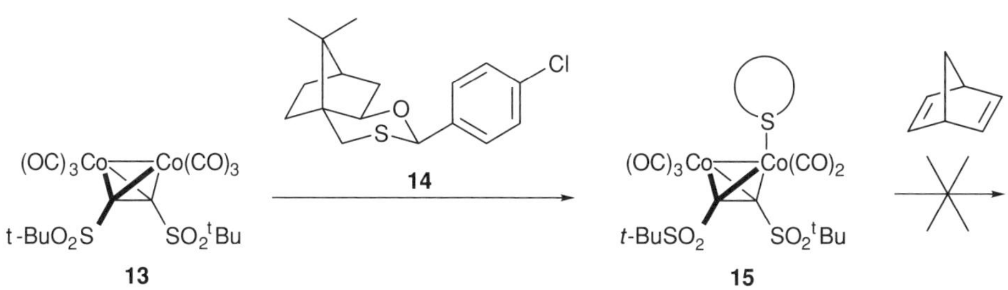

Scheme 6.11 *Failed attempt at using a chiral sulfide as a PKR ligand.*

6.5.2 Cyclopentadienes (Heterobimetallic)

Here we include an example based on the heterobimetallic approach.[18] The isolobal substitution of a $Co(CO)_3$ vertex of the alkyne **16** by a molybdenum fragment bearing a chiral cyclopentadiene ligand (Cp^c) affords a mixture of diastereomeric complexes with moderate selectivity (Scheme 6.12). If the diastereomers can be separated, then each one will yield opposite enantiomers of the corresponding PK adduct in excellent ee. This reaction is particular in that it furnishes *endo* fused tricyclic structures, an unusual feature for intermolecular PKRs; indeed, only certain substrates such as ynamides give significant amounts of the *endo* isomer.[19]

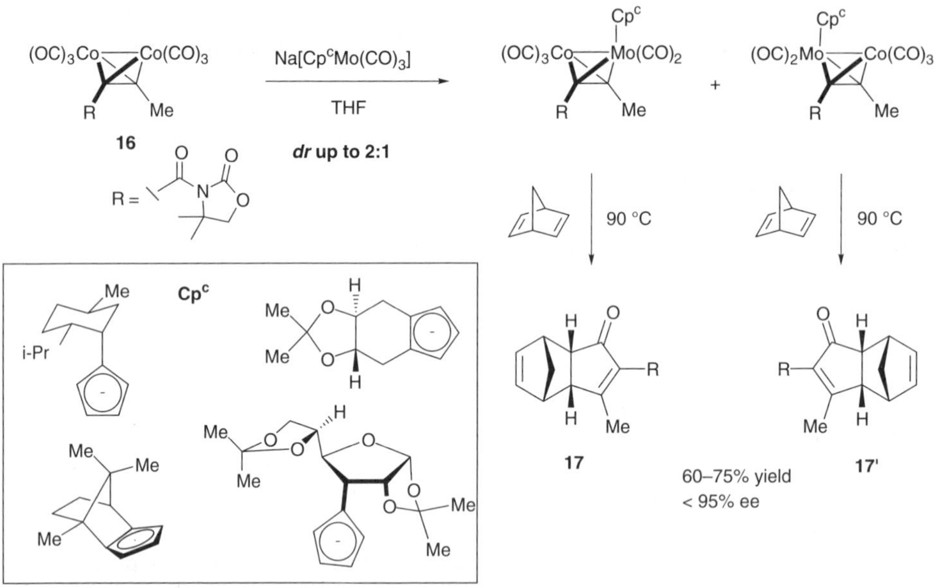

Scheme 6.12 Desymmetrization through isolobal replacement of a $Co(CO)_3$ fragment by a molybdenum bearing a chiral cyclopentadiene ligand (Cp^c).

6.5.3 C-Chiral and Axially Chiral Phosphines

6.5.3.1 Intramolecular PKR

In 2000 the Hiroi group pioneered studies on asymmetric cobalt-catalyzed PKRs through a chiral ligand approach.[20] After screening a structurally diverse set of readily available chiral phosphines, they found that BINAP offered unparalleled results in the cyclization of 1,6-enynes (Scheme 6.13, Table 6.1). Other atropoisomeric binaphthyl derivatives that are weaker ligands (e.g. BINOL and binaphthyl diamine) gave poor enantioselectivities. Extending this work, Consiglio achieved similar enantiomeric induction levels by exploiting the closely related atropoisomeric bis(phosphine) MeO-BIPHEMP.[21] A later study by Gibson[22] completed a thorough compendium of atropoisomeric chiral bis(phosphines), which have proven to be valuable in this transformation (Scheme 6.13, Table 6.1). Since then, the BINAP-Co combination has become a workbench system for studying catalytic PKRs (*vide infra*), although proficient applications have been restricted to the intramolecular version. Importantly, these examples all require high catalyst loadings (20 mol%). A related methodology that employs a C_2-symmetric bidentate phosphite was described in 2002 by Buchwald.[23] The combination of the BINOL-derived bis(phosphite) **18** and catalytic amounts of dicobalt octacarbonyl effected intramolecular PKR of diverse enyne substrates with disparate degrees of enantiocontrol. The most significant improvement offered by this methodology is that it entails lower catalyst loadings (6 mol%). Interestingly, Buchwald discovered a case of matched/mis-matched chirality upon incorporating BINOL fragments of different configurations into diastereomers of **18** (Scheme 6.13).

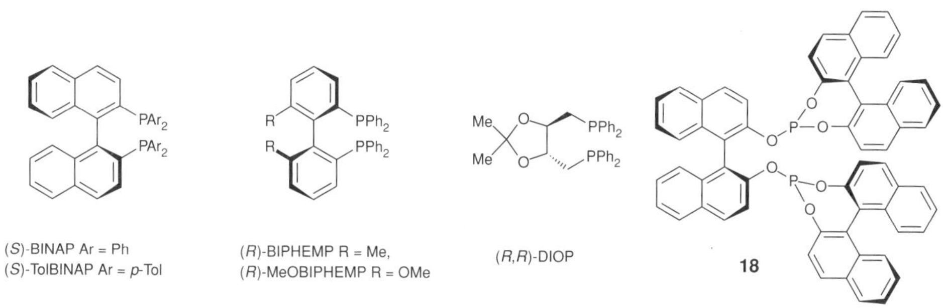

Scheme 6.13 *Ligands used in cobalt-catalyzed intramolecular asymmetric PKRs (see Table 6.1).*

At first glance (Table 6.1), the relative success of C_2-symmetric ligands in these transformations could be rationalized in terms of the general ability of highly symmetric ligands to reduce the number of diastereomeric transition states through symmetry elements. However, numerous kinetic and theoretical studies on this system do not corroborate this premise. In 1999 Laschat demonstrated that although ligand exchange reactions of dicobalt-alkyne clusters could produce bridged complexes with BINAP and other achiral bis(phosphines),[24] these clusters did not engage in any PK chemistry under standard conditions. Conversely, in an extended series of kinetic studies,[25] Gibson showed that chelated coordination of BINAP (which would yield C_1-symmetric dicobalt-alkyne clusters downstream to PK adducts) is responsible for the catalytic activity of this catalyst-ligand combination in the cyclization of enynes. This thorough investigation included identification (by ^{31}P NMR and X-ray crystallography) of the chelated BINAP-$Co_2(CO)_6$ cluster as a PK precatalyst; this precatalyst gave the bicyclic enone in 88% ee[22] (Scheme 6.14). A recent theoretical study by Maseras, Pericàs et al. rationalized the outcome of asymmetric intramolecular PKRs using the cobalt-BINAP system.[26]

In terms of cobalt-catalyzed reactions, the BINAP system is the only one that delivers useful levels of asymmetric induction. However—and very much like every other asymmetric intramolecular PKR—chemistry of the BINAP system is strongly limited in substrate scope: it only works for some 1,6-enynes. Further substitution at either the alkene or the

Table 6.1 *Cobalt-catalyzed intramolecular asymmetric PKRs with bis(phosphine) ligands.*

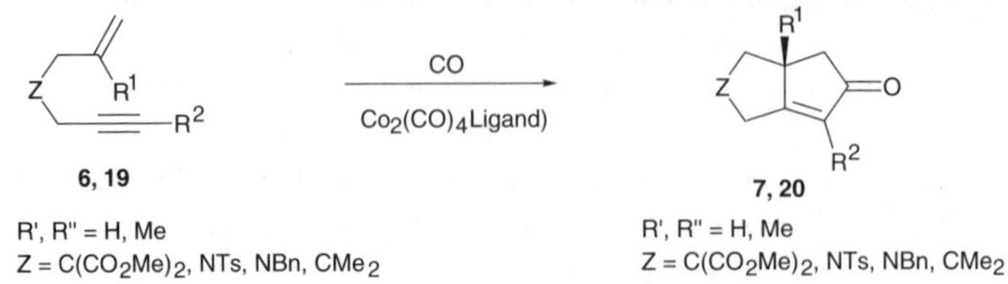

Substrate	Z	R^1	R^2	Ligand	Product	yield (%)	ee (%)
6a	$C(COOEt)_2$	H	H	**BINAP**	**7a**	55	90
6a	$C(COOEt)_2$	H	H	**DIOP**	**7a**	76	5
6b	$C(COOEt)_2$	Me	H	**BINAP**	**7b**	31	63
6b	$C(COOEt)_2$	Me	H	**DIOP**	**7b**	54	40
6c	$C(COOEt)_2$	Me	Me	**BINAP**	**7c**	90	0
6c	$C(COOEt)_2$	Me	Me	**DIOP**	**7c**	47	31
19a	N-Ts	H	H	**BINAP**	**20a**	60	93
19b	N-Ts	Me	H	**BINAP**	**20b**	13	62
19c	N-Bn	H	H	**BINAP**	**20a**	36	82
6a	$C(COOEt)_2$	H	H	**BIPHEMP**	**7a**	90	61
6a	$C(COOEt)_2$	H	H	**MeO-BIPHEMP**	**7a**	86	91
6a	$C(COOEt)_2$	H	H	**18**	**7a**	87	22
6d	$C(COOEt)_2$	H	Ph	**18**	**7d**	75	75
6e	$C(COOEt)_2$	H	Me	**18**	**7e**	97	7

alkyne terminus often results in a substantial decrease in enantioselectivity as well as unpredictable changes in relative reactivity (see Table 6.1).

6.5.3.2 Intermolecular PKR

The first attempts to develop a chiral ligand based approach of the PKR date back to the late 1980s and were focused on the intermolecular version. However, despite three decades of continued research in this direction, practical methodologies have only just been devised over the past few years. Indeed, a catalytic protocol with reasonable scope and induction level remains elusive.

The first example stemmed from a collaboration between the groups of Brunner and Pauson himself, who harnessed the glycerol-derived monophosphine GLYPHOS as desymmetrizing element for the dicobalt-alkyne cluster.[3,27] Ligand exchange and thermodynamic equilibration of the phenylacetylene-derived complex **1h** yielded the diastereomers **21** and **21'** in a 6:4 ratio (Scheme 6.15). After fractional crystallization and/or chromatographic separation, followed by reaction with norbornene, each diastereomer furnished opposite enantiomers of the corresponding cyclopentenone. High induction levels (albeit with low yields) were obtained by thermal activation at 45 °C. However, the use of temperatures above the 60 °C required for the equilibration of the diasteromeric mixture of carbonylic complexes ($t_{1/2}$ ~ 170 min in toluene), resulted in some loss in enantiomeric excess. A more

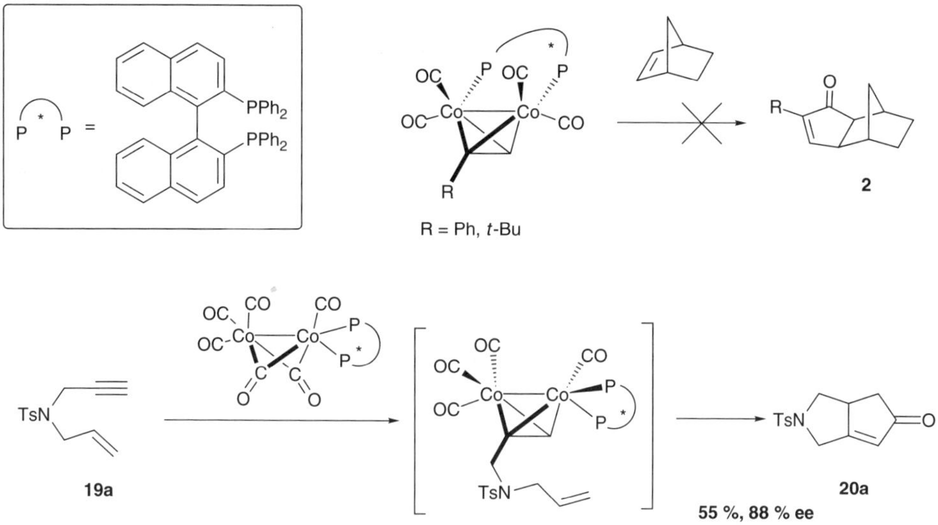

Scheme 6.14 Bridged *vs.* chelated *bis(phosphine) cobalt-alkyne complexes in the PKR.*

challenging alkene substrate, 2,5-dihydrofuran, was employed in a subsequent study, offering moderate success.[28] Kerr and co-workers later combined this approach with *N*-oxide activation,[29] observing higher yields and good enantiomeric excess for several cyclopentenones.

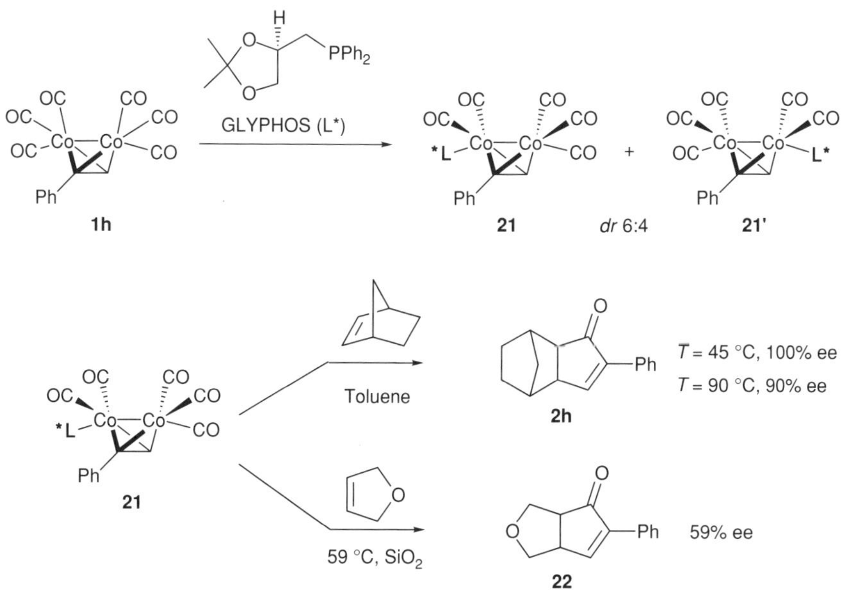

Scheme 6.15 *Ligand exchange and PKRs with GLYPHOS.*

In 1999 and 2000 the group of Moyano reported that bidentate 2-(2-diphenylphosphinophenyl)oxazolines are superior chiral ligands for this transformation.[30] Their chelating coordination mode is the basis for the high bias achieved in the equilibration of the diastereomeric complexes (85:15 using the phenyl derivative ligand **23a** and the complex **1h**; Scheme 6.16). Reactions of the *tert*-butyl derivative **23b** showcase the advantage of chelation control: its higher steric demand provided the monocoordinated complexes **25** in nearly 1:1 diastereomeric mixture. The chelated complexes are much less prone to isomerization but, paradoxically, their PKRs provide only moderate enantiomeric excess (51 to 57% ee). More striking is the complete loss of enantiocontrol under *N*-oxide activation. Despite the low diastereomeric excesses obtained in the ligand exchange reaction, the monocoordinated complexes **25** provide high levels of asymmetric induction, especially under *N*-oxide activation. One could speculate that the free imino nitrogen provides some form of anchimeric assistance in these reactions.

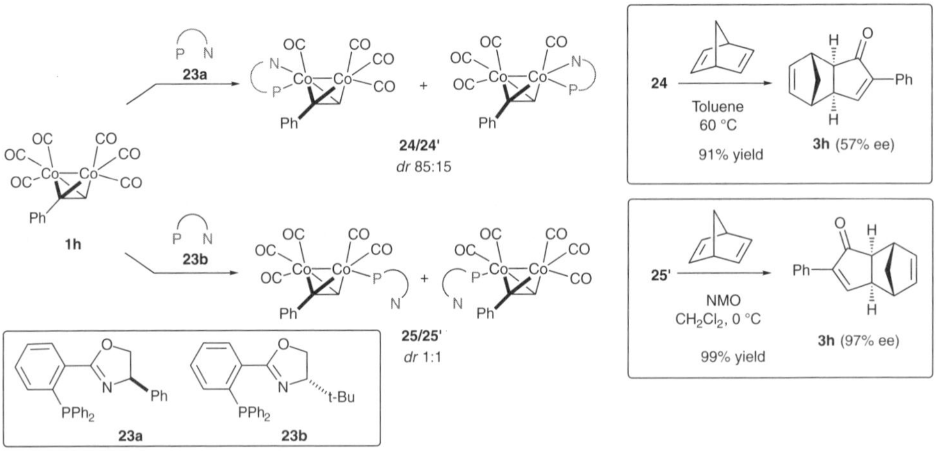

Scheme 6.16 (P,N) *bidentate ligands that can function as either monodentate or chelating ligands, thereby offering distinct diastereo- and enantio-selectivities.*

Phosphoramidites are another type of chiral ligand used in intermolecular PKRs. BINOL-derived phosphoramidites are highly modular compounds that have been used in numerous asymmetric transformations. In a strategy devised by Gimbert et al., two of these ligands are incorporated into the dicobalt-alkyne cluster, thereby obviating diastereomer separation[31] (Scheme 6.17). Although this strategy provides only moderate to low induction in stoichiometric PKRs with norbornene, it remains conceptually interesting. Firstly, it demonstrates that in terms of reactivity, phosphoramidites are promising PKR ligands, as indicated by the fact that coordination of related phosphines at both cobalt centers leads to a drastic reduction in reactivity. Secondly, this strategy showcases a feasible approach for future catalytic versions of the PKR, via intermediacy of a single diastereomer at the dicobalt-alkyne complex stage within the catalytic cycle.

Attempts to extend the well-established BINAP chemistry to the intermolecular asymmetric PKR have met little success. In a seminal contribution, Hiroi screened the same

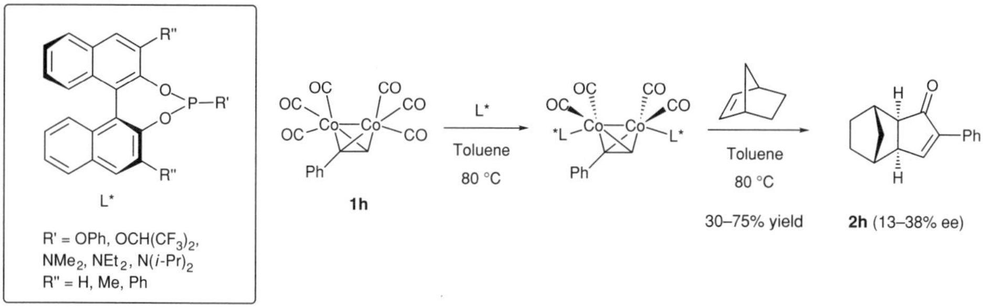

Scheme 6.17 *Asymmetric PKR based on double substitution of a dicobalt hexacarbonyl complex with a monodentate chiral phosphoramidite ligand.*

library of chiral phosphines used for the intermolecular PKR against the archetypical phenylacetylene-norbornene diad, obtaining less than 10% ee with each phosphine[19a] (Scheme 6.18). In a more recent report, reaction of high nuclearity cobalt clusters with various chiral phosphines gave little or no induction.[32]

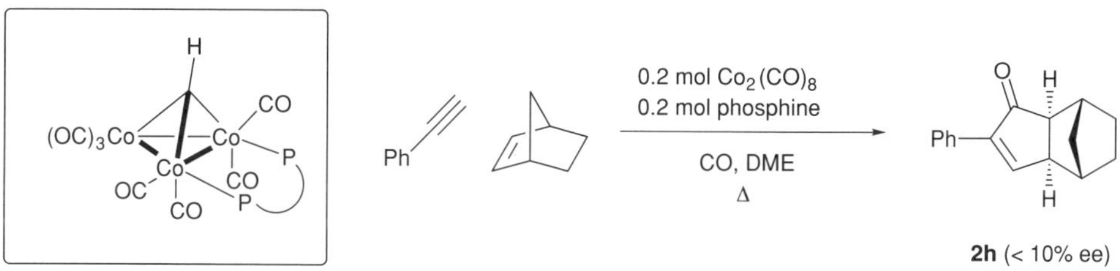

Scheme 6.18 *Catalytic intermolecular PKR with bis(phosphines) as ligands: poor enantioselectivities are observed, even when other chiral cobalt clusters (left) are employed.*

6.5.4 C_2-Symmetric Bridging Bis-phosphines

The advantage of using C_2-symmetric dicobalt-alkyne intermediates was first exploited by Meijere in the context of diastereoselective reactions using C_2-chiral auxiliaries.[33] Prior to their studies on monodentate phosphoramidites, Greene, Gimbert et al. described a ligand-based route to such precursors that avoids formation of diastereomeric complexes. They reasoned that successful implementation of C_2-symmetric ligands would require a single atom spacer between the Lewis basic sites in order to enforce a bridged coordination with the cobalt cluster (as bridging would result in a five-membered cobaltacycle, whereas chelation would lead to a four-membered ring). To overcome the (reported) lack of reactivity of bis(phosphine) bridged dicobalt clusters, they focused on less electron-donating aminophosphines and phosphoramidites. The resulting complexes showed reasonable reactivities, providing PK adducts in good yields. Preliminary results with a ligand derived from α-methylbenzylamine displayed some asymmetric induction (Scheme 6.19). However, neither tuning the stereoelectronic properties of the ligand, nor incorporating chirality in the phosphine moiety (in the form of a BINOL-derived phosphoramidite), offered any significant improvement in the results.[31]

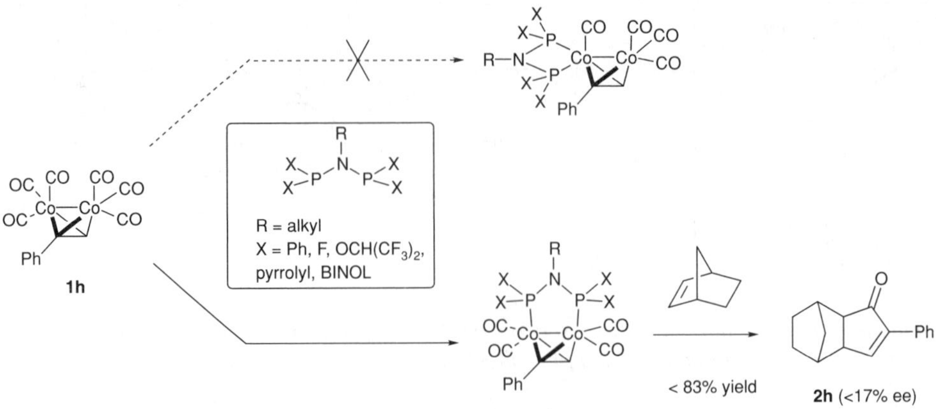

Scheme 6.19 *Preparation of C_2-symmetric dicobalt-alkyne clusters using bis(phosphoramidite) ligands that enforce bridged coordination, and their subsequent PKR.*

6.5.5 C-Chiral Bidentate (P,S) Ligands

Since 2000 Riera and Verdaguer's group has carved out a niche in the area of asymmetric intermolecular PKRs, through the introduction of a series of bidentate ligands that combine phosphorus and sulfur coordination. These ligands incorporate two key features that we introduced earlier in this chapter: a bidentate scaffold containing a single-atom linker, to enforce bridged coordination; and a sulfide moiety for hemilabile coordination to cobalt, which is beneficial for reactivity (as previously demonstrated in the context of chiral auxiliaries).[34]

A relevant precedent of this work was found in a paper from Mays describing the coordination of the alkylthiomethyl diphenylphosphines **27** with electron-deficient acetylenes such as **26**[35] (Scheme 6.20). The ligands **27** can be considered as achiral analogs of the ligands subsequently developed by the Riera and Verdaguer group. An X-ray diffraction study of the resulting dicobalt-alkyne complex **28a** confirmed the bridged coordination of these ligands. Although these complexes have not been evaluated for the PKR, the reversibility of the sulfide coordination in solution inspired development of the chiral P,S ligands and could foreshadow success of these scaffolds as PKR ligands.

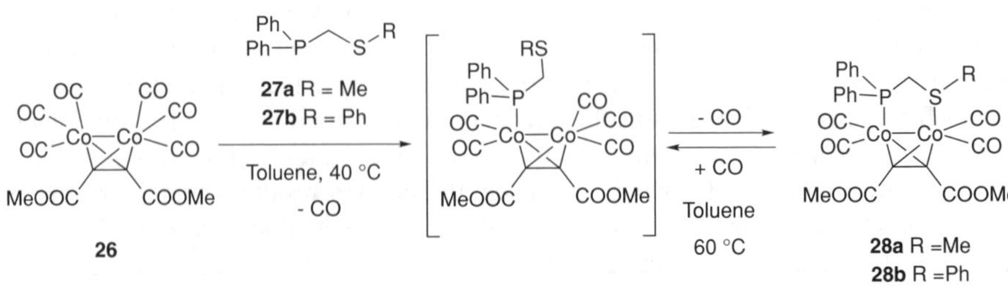

Scheme 6.20 *Bridged coordination of achiral (P,S) bidentate ligands.*

Bidentate (P,S) ligands with a chiral carbon skeleton were prepared in a few synthetic steps from commercially available pulegone or camphorsulfonic acid. Importantly, total diastereoselectivity was achieved in the key phosphinylation step to form the oxathianes **29** and **30**. The resulting ligands are best isolated as their borane adducts[36] (Scheme 6.21).

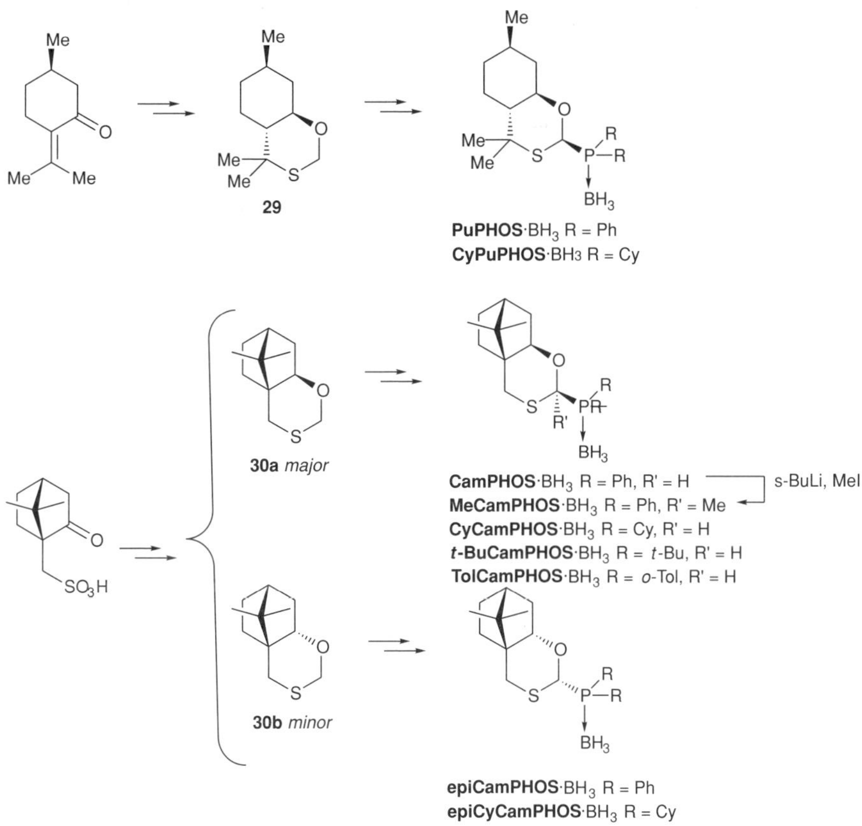

Scheme 6.21 *Bidentate (P,S) ligands derived from pulegone and camphorsulfonic acid.*

The corresponding P,S bridged dicobalt-alkyne complexes **31** were conveniently prepared by *in-situ* deprotection of the phosphine-borane ligands showed in Scheme 6.21 with 1,4-diazabicyclooctane (DABCO). Cobalt complexes **31** derived from terminal acetylenes bearing bulky substituents (**1i**, **1j** and **1c**) afforded, after thermal equilibration, good diastereoselectivities (Table 6.2). Most importantly, the resulting diastereomers were easily separated by chromatography or crystallization, and were bench stable. The methyl-substituted derivative MeCamPHOS, which imposes a greater steric demand around the tetrahedral cobalt-alkyne cluster, provided excellent diastereoselectivities (up to 20:1) for all three complexes.

The diastereoselectivity of these ligands has been dramatically improved through use of terminal propynamides as alkyne partners.[37] With these substrates, a non-classical hydrogen

Table 6.2 *Ligand exchange reactions and diastereoselectivites with hemilabile bidentate ligands.*

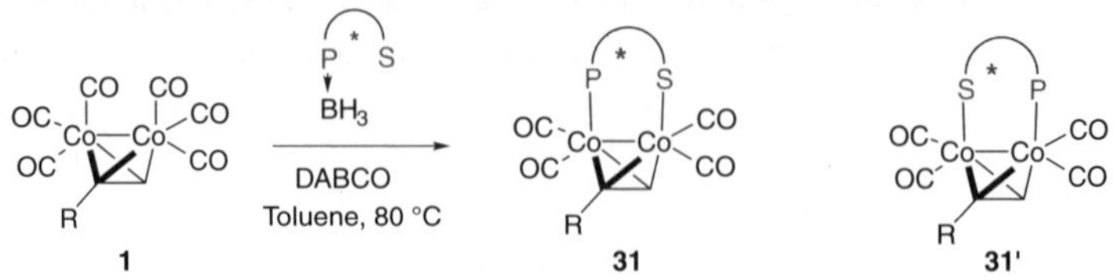

SM*	R	Ligand	Product	dr
1i	TMS	PuPHOS	**31ia**	3:1
1i	TMS	CyPuPHOS	**31ib**	1.6:1
1i	TMS	CamPHOS	**31ic**	5:1
1i	TMS	MeCamPHOS	**31id**	**12:1**
1i	TMS	CyCamPHOS	**31ie**	1.5:1
1i	TMS	t-BuCamPHOS	**31if**	1.3:1
1i	TMS	TolCamPHOS	**31ig**	1:1
1i	TMS	epiCamPHOS	**31ih**	2:1
1i	TMS	epiCyCamPHOS	**31ii**	3.4:1
1j	*tert*-Bu	PuPHOS	**31ja**	3:1
1j	*tert*-Bu	CamPHOS	**31jc**	2:1
1j	*tert*-Bu	MeCamPHOS	**31jd**	**20:1**
1c	CMe_2OH	PuPHOS	**31ca**	4.5:1
1c	CMe_2OH	MeCamPHOS	**31cd**	13:1

*starting material

bonding interaction between the amide carbonyl group and the highly polarized methine group in the ligand (connected to three heteroatoms) is established in only one of the diastereomers. This weak stabilizing interaction leads to much higher diastereomeric ratios, and in some cases, near total diastereoselection (see Table 6.3). Other hydrogen bond acceptors (e.g. sulfonyl groups) are also suitable directing groups.[38]

The resulting diastereomerically pure complexes **31** smoothly react with norbornadiene to yield the enantioenriched cyclopentenones **3** (see Table 6.4 for selected results). For the complexes derived from unfunctionalized alkynes (**31ia** and **31id**), *N*-oxide activation yields the corresponding adducts in excellent yields and moderate to excellent enantioselectivities; however, thermal activation gives slightly lower enantiomeric excesses, probably due to the relatively rapid isomerization of the complex. Conversely, when propynamides are used as starting alkyne (**31ka** and **31kb**), thermal activation provides comparable or even better results than *N*-oxide activation, although the reaction is not completely stereospecific (90 to 92% ee).

The unusually high diastereoselectivities attained in the formation of the intermediate chiral dicobalt-alkyne complex when a non-classical H-bond forms between the ligand and the alkyne-substrate prompted Riera, Verdaguer et al. to run catalytic asymmetric PKRs with alkynylamides (Scheme 6.22).[39] When 5 mol% of complex **31lc**, which gave the highest diastereoselectivity in the complexation (200:1) as a catalyst, was used, the

Table 6.3 *Enhancement of the diastereoselectivity of hemilabile ligands through hydrogen bonding interactions.*

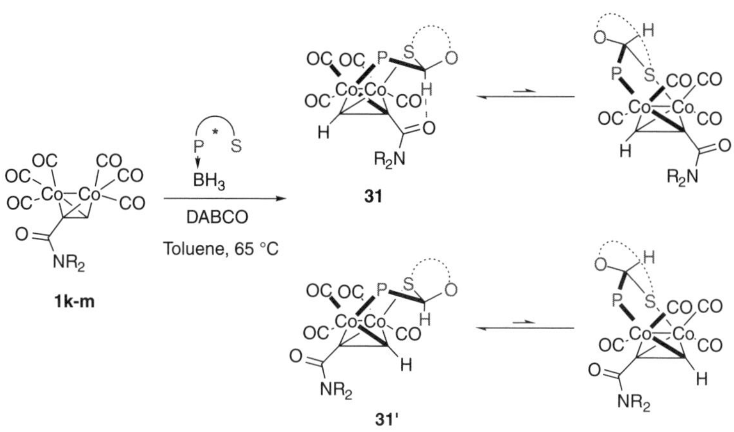

SM*	R	Ligand	Product	dr (31/31′)
1k	Et	PuPHOS	**31ka**	32:1
1k	Et	CamPHOS	**31kc**	19:1
1k	Et	TolCamPHOS	**31kg**	**200:1**
1k	Et	CyCamPHOS	**31ke**	3.1:1
1k	Et	t-BuCamPHOS	**31kf**	3.1:1
1l	*i*-Pr	PuPHOS	**31la**	19:1
1l	*i*-Pr	CamPHOS	**31lc**	**200:1**
1m	$-(CH_2)_5-$	PuPHOS	**31ma**	7.3:1
1m	$-(CH_2)_5-$	CamPHOS	**31mc**	**200:1**

*starting material

intermolecular PK adduct **3l** was obtained in high yield (Scheme 6.22). Although the induction level attained (23%) was still far from what would be considered synthetically useful, this method offers the best overall results (i.e. yield plus ee) achieved to date.

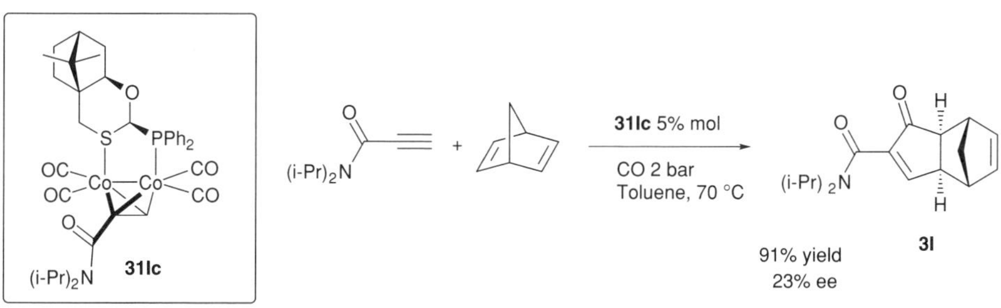

Scheme 6.22 *Catalytic asymmetric PKR of diisopropylpropynamide and norbornadiene.*

Table 6.4 Asymmetric PKRs of norbornadiene and cobalt tetracarbonyl complexes of (P,S) bidentate ligands.

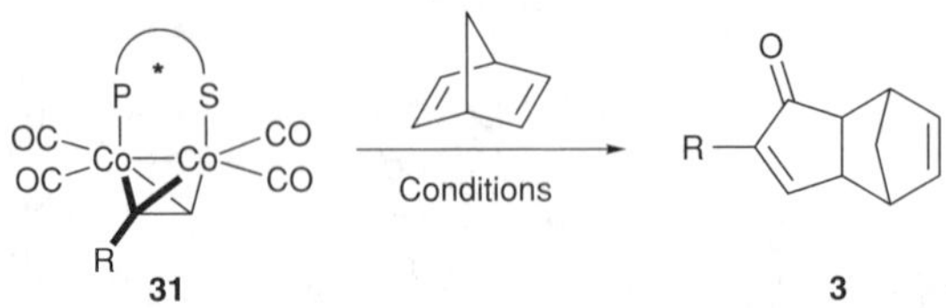

31 3

SM*	R	Ligand	de**	Conditions	Product	Yield	ee
31ia	TMS	PuPHOS	100%	toluene/50 °C	**3i**	92%	57%
31ia	TMS	PuPHOS	100%	CH_2Cl_2/NMO/rt	**3i**	93%	97%
31id	TMS	MeCamPHOS	85%	C_6D_6/65 °C	**3i**	68%	56%
31id	TMS	MeCamPHOS	85%	CH_2Cl_2/NMO/rt	**3i**	72%	79%
31ca	CMe_2OH	PuPHOS	100%	CH_2Cl_2/NMO/rt	**3c**	98%	70%
31cd	CMe_2OH	MeCamPHOS	86%	CH_2Cl_2/NMO/rt	**3c**	90%	50%
31ka	$CONEt_2$	PuPHOS	100%	toluene/65 °C	**3k**	33%	90%
31ka	$CONEt_2$	PuPHOS	100%	CH_2Cl_2/NMO/rt	**3k**	81%	73%
31kc	$CONEt_2$	CamPHOS	100%	toluene/DMSO/70 °C	**3k**	68%	92%
31kc	$CONEt_2$	CamPHOS	100%	CH_2Cl_2/NMO/rt	**3k**	81%	80%

* Starting material; ** Diastereomeric excess of the starting complex

The P,S ligand system was further evolved with the stereoselective oxidation of the sulfur in CamPHOS to afford the novel sulfoxide-phosphine ligand **32**[40] (Scheme 6.23). During tests on its coordination capacity, ligand **32** was found to react with the alkyne dicobalt complex **13** (already used for coordination with chiral sulfides; see Scheme 6.11). The X-ray structure of the resulting complex (**33**) showed a bridged P,S coordination to cobalt by the lone pair at the sulfur and the phosphorous atom. Coordination of the sulfoxide to cobalt set the stage for yet another generation of chiral P,S ligands, in which the chiral information resides at the sulfur atom.

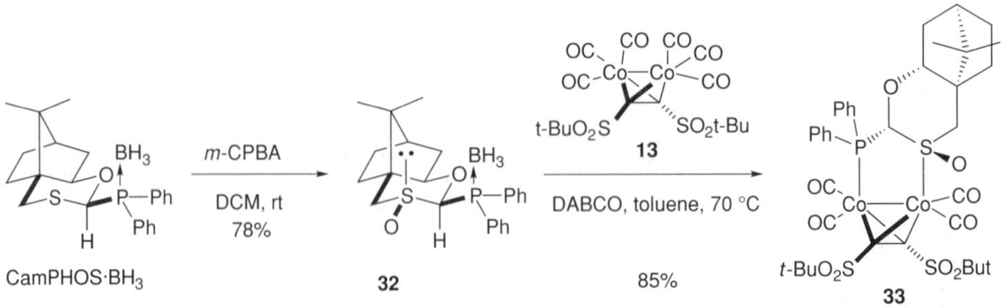

Scheme 6.23 Oxidation of CamPHOS, and subsequent coordination of the product to hexacarbonyl dicobalt bis(tert-butylsulfonyl)acetylene (**13**).

6.5.6 S-Chiral Bidentate (P,S) Ligands

The discovery that the phosphine-sulfoxide ligand **32**, which has a one atom tether separation, worked as P,S ligand with alkyne dicobalt-carbonyl complexes prompted Riera, Verdaguer et al. to develop *N*-phosphinosulfinamide ligands (PNSO), which they designed as more versatile ligands for the enantioselective PKR. Sulfinamides are among the most accessible and useful chirogenic compounds. In fact, chiral *tert*-butylsulfinamide (**34a**) and *p*-tolylsulfinamide (**34b**) have emerged as efficient chiral controllers in numerous processes[41] (Scheme 6.24). Reductive amination of commercially available sulfinamides, followed by phosphinylation at the nitrogen, and finally, protection with borane, yielded a series of PNSO ligands in which the phosphorus is protected.[42] Whereas the *t*-Bu-PNSO ligands remain stable upon deprotection (removal of borane with DABCO), the corresponding deprotected *p*-Tol-PNSO analogs are prone to hydrolysis during purification on silica gel. Interestingly, the PNSO ligand system is highly modular; the groups attached to the phosphorous, the nitrogen and the sulfur can be varied independently.

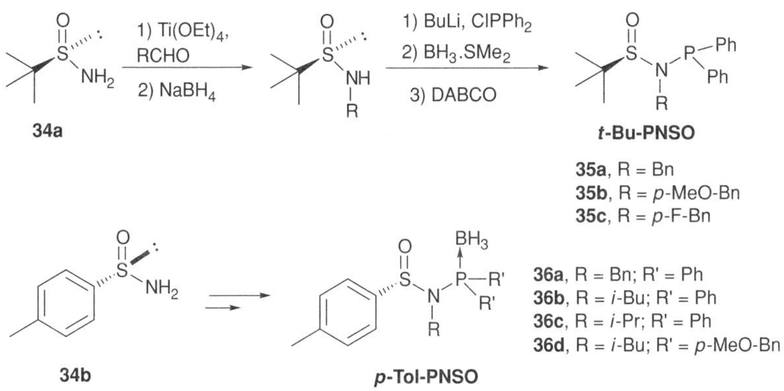

Scheme 6.24 *Synthesis of PNSO ligands.*

Ligand exchange of PNSO ligands with cobalt complexes derived from terminal alkynes (**1i,c,h,n**) provided the corresponding bridged diastereomers **37** or **38** in excellent yield. With the *t*-Bu-PNSO ligands **35**, the reaction gave good to excellent yields, affording the bridged complexes **37** with selectivities between 7:1 and 12:1 (Table 6.5). Conveniently, the major diastereomer can be isolated (usually, in 50% overall yield) by crystallization of the crude product after filtration on silica. In contrast, the *p*-Tol-PNSO ligands **36** were not selective upon coordination to terminal alkyne complexes, often affording 1:1 diastereomeric mixtures (see Table 6.6). The discrepancy in reactivity between the *t*-Bu-PNSO ligands **35** and the *p*-Tol-PNSO analogs **36** was attributed to the greater steric bias and more hemilabile character of the former.

PK reaction of the diastereomerically pure tetracarbonyl complexes **37** with norbornadiene was highly stereoselective, affording cyclopentenones in up to 99% ee (Table 6.7). The cycloaddition can be run using either thermal activation, or *N*-oxide activation with *N*-methylmorpholine *N*-oxide (NMO) at room temperature. The parent *N*-benzyl

Table 6.5 *Ligand exchange reaction of* t*-Bu-PNSO ligands.*

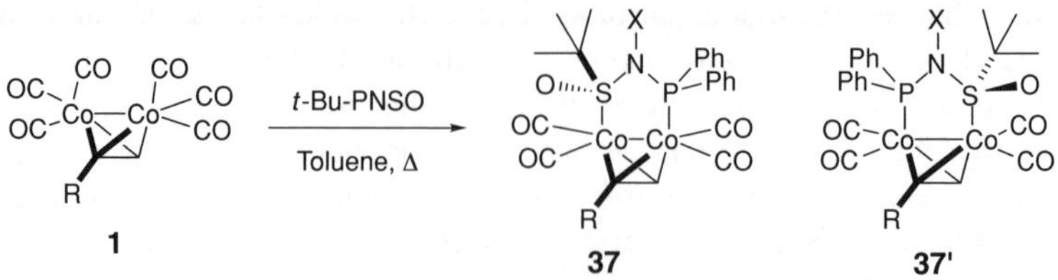

SM*	R	Ligand	X	Conditions	Product	Yield	d.r.
1i	TMS	**35a**	Bn	70 °C, 16h	**37ia**	78%	12:1
1i	TMS	**35b**	*p*-MeO-Bn	65 °C, 24h	**37ib**	75%	12:1
1i	TMS	**35c**	*p*-F-Bn	65 °C, 40h	**37ic**	99%	10:1
1c	CMe_2OH	**35a**	Bn	65 °C, 16h	**37ca**	95%	7:1
1h	Ph	**35a**	Bn	65 °C, 5h	**37ha**	81%	8:1
1n	$CH_2OTBDPS$	**35a**	Bn	65 °C, 6h	**37na**	85%	10:1

*Starting material

ligand (**35a**) provided excellent results for most substrates; however, in some cases, the *p*-methyoxybenzyl analog (**35b**) showed higher selectivity.

Although the *p*-Tol-PNSO ligands are not as selective or efficient as the *t*-Bu ligands in PKRs of terminal alkynes, they have proven useful in the cyclization of symmetrically substituted alkynes.[43] Internally symmetric alkynes are challenging substrates: firstly, they are intrinsically less reactive than terminal alkynes; and secondly to achieve high selectivity with them, coordination of the olefin to a single cobalt center is not sufficient—the

Table 6.6 *Ligand exchange reaction of* p*-Tol-PNSO ligands.*

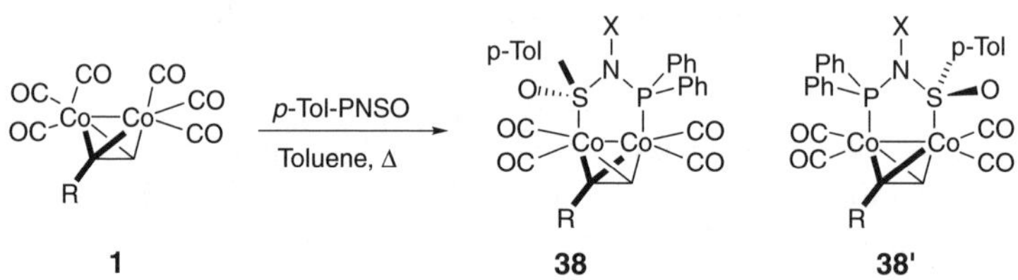

SM*	R	Ligand	X	Conditions	Product	Yield (%)	d.r.
1i	TMS	**36a**	Bn	65 °C, 2 h	**38ia**	88	1:1
1i	TMS	**36b**	*i*-Bu	65 °C, 2 h	**38ib**	62	2:1
1i	TMS	**36c**	*i*-Pr	65 °C, 7 h	**38ic**	80	3:1
1c	CMe_2OH	**36b**	*i*-Bu	65 °C, 2 h	**38cb**	86	1:1
1h	Ph	**36b**	*i*-Bu	65 °C, 2 h	**38hb**	93	1:1
1c	CMe_2OH	**36c**	*i*-Pr	65 °C, 5.5 h	**38cc**	81	1.3:1
1h	Ph	**36c**	*i*-Pr	65 °C, 2 h	**38hc**	80	1:1

*Starting material

Table 6.7 *PKR of t-Bu-PNSO-dicobalt complexes with norbornadiene.*

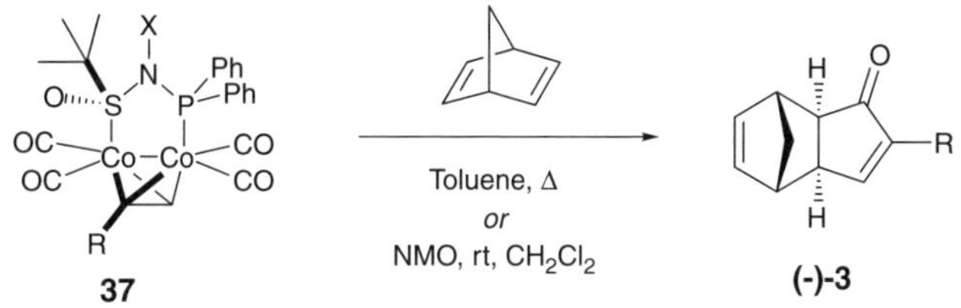

S.M.	R	X	Conditions	Product	Yield (%)	ee (%)
37ia	TMS	Bn	NMO, 16 h	**3i**	99	97
37ib	TMS	*p*-MeO-Bn	70 °C, 40 h	**3i**	99	> 99
37ic	TMS	*p*-F-Bn	70 °C, 48 h	**3i**	99	78
37ca	CMe_2OH	Bn	70 °C, 16 h	**3c**	95	92
37ca	CMe_2OH	Bn	NMO, 16 h	**3c**	87	93
37ha	Ph	Bn	rt, 4 h	**3h**	99	99
47na	$CH_2OTBDPS$	Bn	rt, 5 h	**3n**	99	92

olefin must also insert into a defined Co–C bond. As mentioned above, due to its higher symmetry the ligand exchange reaction of PNSO ligands **35a** or **36b** with hexacarbonyl(μ-diphenylacetylene)dicobalt (**39**) produced single complexes **40** or **42** (see Scheme 6.25). Thermal cyclization of these complexes with norbornadiene provided the corresponding adduct **41** in low enantiomeric excess. However, somewhat surprisingly, the *N*-oxide-promoted reactions gave excellent enantioselectivity (up to 96% ee). In this case, ligand **36b** gave a better overall yield but slightly lower selectivity than did **35a**. The modular nature of these ligands enabled fine tuning to develop a bis-(p-methoxybenzylphosphinyl) ligand (**36d**), which, in the best case, gave the desired adduct (**41**) in 77% yield and 94% ee.

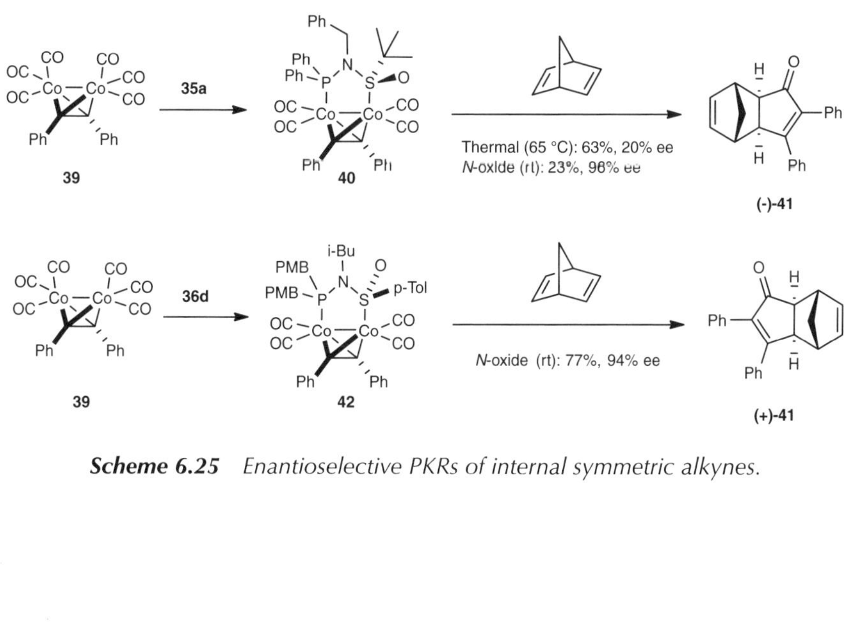

Scheme 6.25 *Enantioselective PKRs of internal symmetric alkynes.*

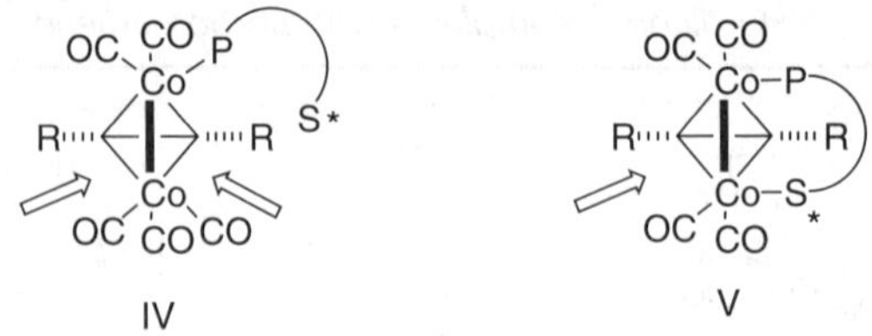

Figure 6.1 *Possible bonds for olefin insertion in PNSO complexes derived from symmetrically substituted alkynes: as hemilabile (left) and solid bridging (right).*

Although PNSO ligands are usually hemilabile, the high enantiomeric excesses, better results obtained under N-oxide activation, and long reaction times observed in the reactions described above led Riera, Verdaguer et al. to propose an alternative mechanism. Another factor suggesting that the mechanism was not hemilabile was that, in contrast with terminal alkynes, hemilabile ligands could not direct olefin insertion (i.e. it would not be able to discriminate between the two Co–C bonds), because the sulfur chirality would be too far from the reaction center, as shown in **IV** (Figure 6.1). Thus, they suggested that the PNSO ligand works as a solid bridging ligand, directing olefin coordination to the same cobalt center to which the sulfinyl moiety is bound, as in **V** (Figure 6.1). Once the olefin has bound to the cobalt atom, the sulfur chirality directs insertion of the olefin into a single Co–C bond.

X-ray analysis of the PNSO cobalt complexes **37ia**, **38ib** and **43ia** furnished interesting clues to the mode of operation of the PNSO ligands.[44] The Co–S bond distance reflects the bond strength between the sulfur and cobalt atoms and can be used to estimate the hemilabile character of the ligand. Electron-deficient groups on the sulfur lead to a shorter and stronger Co–S bond, due to the relatively higher π-acidity of the sulfinyl group (Figure 6.2). Within the series of PNSO ligand studied, the *t*-Bu-PNSO leads to the weakest sulfur–cobalt bond, and therefore, is the most hemilabile. Conversely, the CF_3-PNSO ligand behaves more like a solid bridged ligand (in other words, less hemilabile). Reactivity studies and CO stretching frequency measurements corroborate this hypothesis.

The success of PNSO ligands in asymmetric PKRs prompted the Riera and Verdaguer group to prepare the corresponding analogs with a carbon tether.[45] The sulfinylmethyl

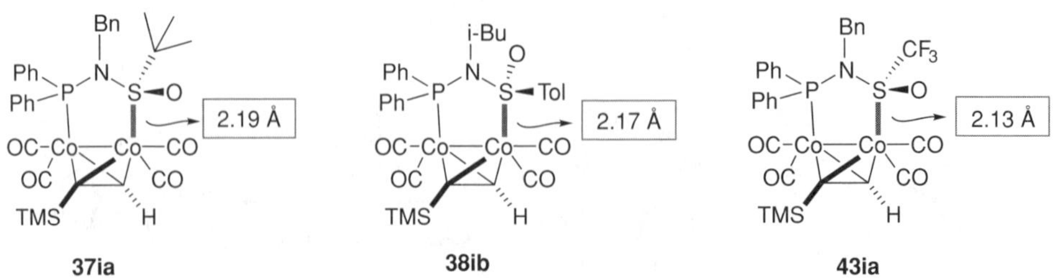

Figure 6.2 *Co–S bond lengths in different PNSO-alkyne cobalt complexes (based on X-ray analysis).*

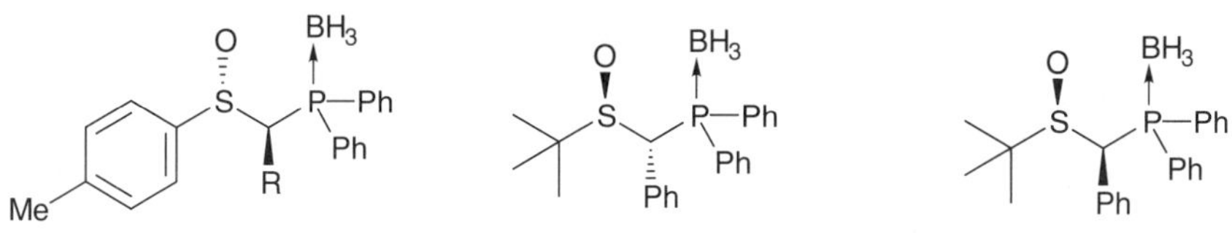

44a, R = H
44b, R = Ph
44c, R = Bn

45a

45b

Figure 6.3 *Sulfinylmethyl phosphine (PCSO) ligands studied in intermolecular PKRs.*

phosphine (PCSO) ligands enabled introduction of an additional chiral center between the phosphorous and sulfur atoms (Figure 6.3). Synthesis of the ligands **44** and **45**, and studies on their ligand exchange and PK reactions, revealed that the PCSO ligands are neither as easy to prepare nor as selective as the PNSO ligands. Nevertheless, this work underscored the importance of the substitution at the central carbon atom. Compound **44a** did not provide the bridged P,S complex, whereas ligands **44b** and **44c** did. In ligands **45a** and **45b**, each of which bears a *tert*-butylsulfoxide group, the chirality of the central carbon atom is crucial to formation of the bridged species: the desired complex can only be formed if the adjacent phenyl and *tert*-butyl groups adopt the *trans* conformation due to steric repulsion between the *tert*-butyl and the phenyl. Thus, whereas **45a** does not provide P,S-bridged complexes, its corresponding epimer **45b** does.

6.6 Synthetic Applications

Given that only a few practical asymmetric PK methodologies have been developed, enantioselective PKRs of complex molecules are scarce. The PK cycloadduct **3i**, formed from trimethylsilylacetylene and norbornadiene, has proven to be the most synthetically useful to date. It can be prepared in high optical purity and on a multi-gram scale using either PuPHOS or PNSO ligands. Using PuPHOS, the diastereoselectivity in the ligand exchange reaction is 3:1, but gram scale amounts of diastereomerically pure compound can be obtained through a dynamic resolution process in which crystallization and re-equilibration steps are alternatively employed (Scheme 6.26);[46] in fact, the major diastereomer can be obtained in 66% overall yield after two cycles. The crystalline complex gives the PKR with norbornadiene in 96% yield and 96% ee, which can easily be increased by recrystallization. Working with the ligand prepared from natural pulegone, the Riera and Verdaguer group obtained the dextrorotatory enantiomer, whose absolute configuration is shown in Scheme 6.26.

The PNSO ligands offer three main advantages over the PuPHOS ligands: they are much easier to prepare, offer higher diastereoselectivity and both enantiomers are equally available. When cobalt complex **1i** (prepared from trimethylsilylacetylene) was reacted with ligand **35b**, the bridging complex **37ib** was formed in 67% yield and with a diastereoselectivity of 14:1 (Scheme 6.27). Therefore, this process does not require a second

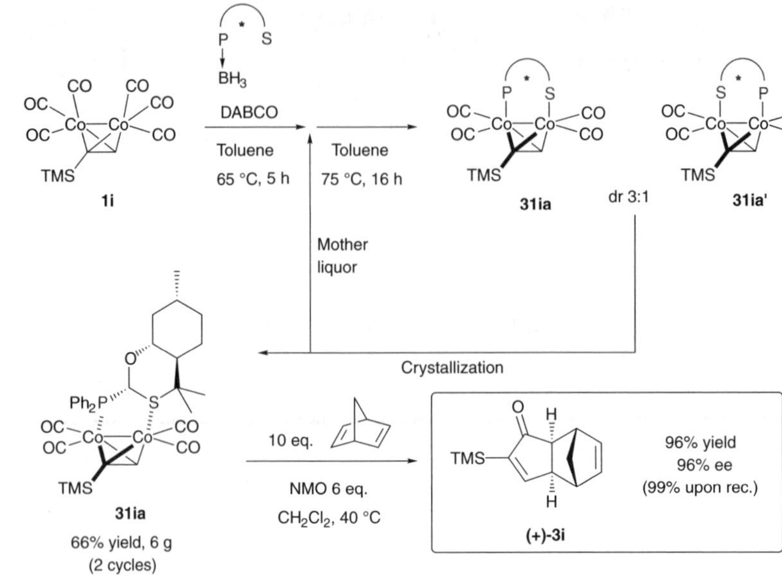

Scheme 6.26 *Formation of the Pauson-Khand adduct **3i** using PuPHOS. Preparation of diastereomerically pure complex **3ia** via dynamic resolution.*

equilibration/crystallization. Starting from (*R*)-*tert*-butylsulfinamide for the preparation of ligand **35b**, the levorotatory cycloadduct **3i** was obtained in 98% ee (again, the ee can be further increased by recrystallization).

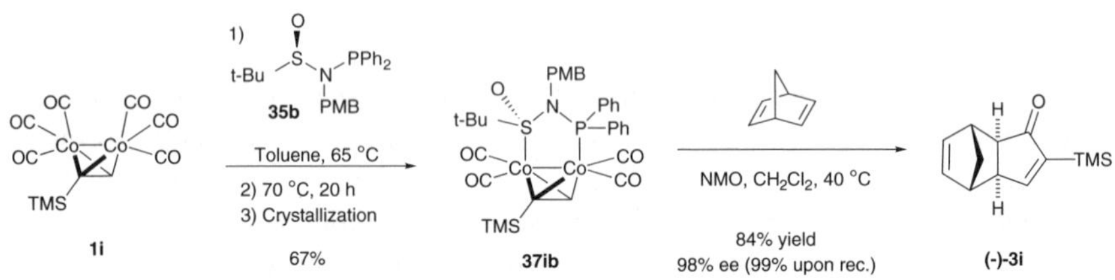

Scheme 6.27 *Multi-gram enantioselective synthesis of the PK cycloadduct **3i** using the PNSO ligand **35b**.*

Cycloadduct **3i** can be considered a chiral synthon of cyclopentadienone. The enone functionality enables a wide range of chemistries; in particular, it is an excellent substrate for conjugate additions. Additionally, the trimethylsilyl group can be easily removed and the second enone can be unmasked (via retro-Diels–Alder reaction) (Scheme 6.28). Thus, organocuprate addition, followed by simple treatment with TBAF, gave bicyclic ketones that, after retro-Diels–Alder reaction, provided enantioenriched 4-substituted cyclopentenones.[46] However, access to α,β-disubstituted cyclopentenones has proved to be problematic, due to lack of reactivity and to formation of polyalkylation side-products during introduction of the α-side chain. To date, the best reported methodology for overcoming this problem is a one-pot combination of conjugate addition and Peterson olefination, devised by P. Evans.[47] In this approach, the intermediate enolate (formed by treatment of the adduct with organocuprate reagents) is trapped with an aldehyde, thereby facilitating access to cross-conjugated enones.

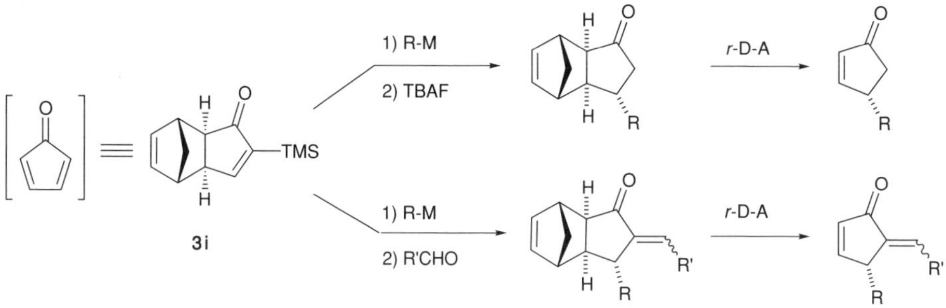

Scheme 6.28 *Formation of homochiral substituted cyclopentenones through a sequence comprising conjugate addition, Peterson olefination and retro-Diels–Alder reaction. The PK cycloadduct* ***3i*** *can be regarded as a chiral cyclopentadienone synthon.*

Natural products containing a five-membered carbon ring are ubiquitous. These include prostaglandins and phytoprostanes, which constitute an important family of biologically active compounds. Some of these contain a cyclopentenone ring substituting two side chains at positions 4 and 5. Evans, in collaboration with the Riera and Verdaguer group envisioned that the conjugate addition/Peterson olefination methodology would be ideal for preparing unsaturated prostaglandins and phytoprostanes. They exploited this method to synthesize two of these compounds, the dehydrophytoprostanes dPPJ$_1$-I and dPPJ$_1$-II, in enantiomerically pure form from the PK cycloadduct (+)-**3i**.[48] As shown in Scheme 6.29, treatment of the starting material with lithium diethyl cuprate, followed by addition of the

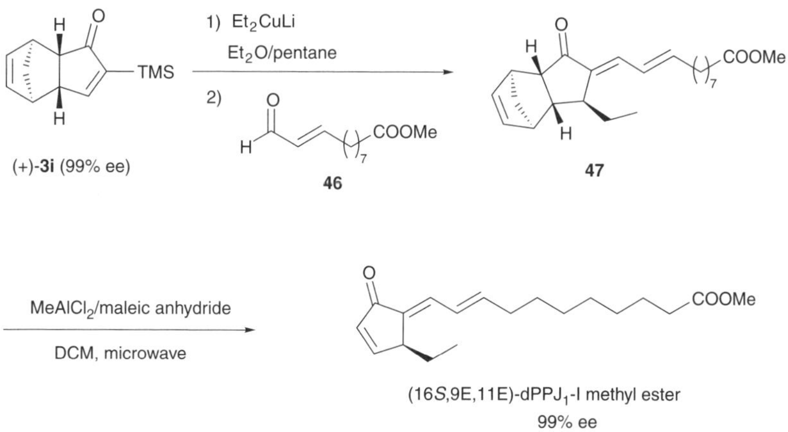

Scheme 6.29 *Synthesis of dPPJ$_1$-I methyl ester from enantiopure (+)-**3i**.*

unsaturated aldehyde **46**, afforded, in a one-pot reaction, the exocyclic enone with complete stereoselectivity. The retro-Diels–Alder reaction, using microwave-assisted chemistry and a modified version of the conditions developed by Grieco, provided the desired phytoprostane in good yield and excellent enantiomeric purity.

Phytoprostane dPPJ$_1$ type II was synthesized with the same approach. In this case, since the β side chain holds the ester functionality, the conjugate addition had to be performed using a silyl protected alcohol as a precursor of the ester. After the cuprate addition, 2-pentenal was added to the reaction mixture to afford the exocyclic enone **48** in excellent yield (Scheme 6.30). The silyl ether was converted into the methyl ester via high-yielding standard protocols: acid deprotection of the silyl ether; oxidation to the aldehyde with Dess-Martin periodinane (DMP); oxidation to the acid with sodium chlorite; and finally, esterification with trimethylsilyl diazomethane. The retro-Diels–Alder reaction (again, using the modified version of Grieco's conditions), afforded phytoprostane dPPJ$_1$-II methyl ester.

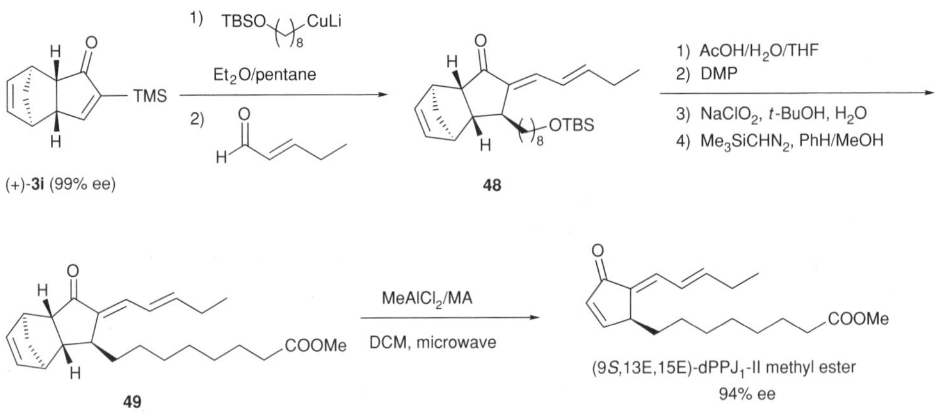

Scheme 6.30 *Synthesis of dPPJ$_1$-II methyl ester from enantiopure (+)-**3i**.*

Carbanucleosides are an important class of nucleoside analogs, some of which, whether natural (e.g. Aristeromycin and Neplanocin) or synthetic (e.g. Carbovir and Abacavir), exhibit anti-tumor and anti-viral activities. In fact, Abacavir has been approved and launched for the treatment of HIV. The Riera and Verdaguer group reasoned that the highly functionalized five-membered rings featured in these compounds made them excellent targets for synthesis via the PKR.[49] They envisaged that for reaction of the PK cycloadduct **3i** as chiral starting material, the key reactions would comprise introduction of the nucleobase and addition of the hydroxymethyl side chain. Although many d^1-synthons can be added to **3i** with high yields and complete stereoselectivity, the simplest transformation is photochemical conjugate addition of methanol, a procedure originally developed by Fraser–Reid.[50] Thus, photochemical conjugate addition of methanol to **3i**, using one equivalent of benzophenone as a triplet sensitizer,[51] proceeded cleanly. Subsequent cleavage of the silyl group by TBAF afforded the bicyclic compound hydroxymethylcyclopentenone **50** in good yield. Protection of the alcohol as the silyl ether, followed by retro-Diels–Alder, gave the key intermediate **52**. This protected cyclopentenone was diastereoselectively reduced with DIBALH to give the allyl alcohol **53** (Scheme 6.31).

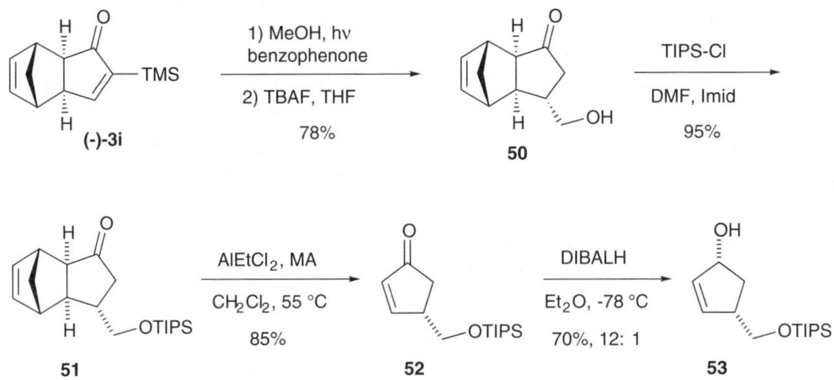

Scheme 6.31 *Preparation of the homochiral allyl alcohol **53** from the PK cycloadduct (-)-**3i**.*

The nucleobase was then introduced via palladium-catalyzed allylic substitution onto the methylcarbonate of alcohol **53** (which gave better yields than the corresponding acetate). When 6-chloro-2-purinamine was used as nucleophile, and sodium hydride as base, the nucleoside **54** was obtained in good yield and in excellent optical purity (> 99% ee, as determined by HPLC). Compound **54** was ultimately transformed into the final products Abacavir and Carbovir by standard chemistries (Scheme 6.32).

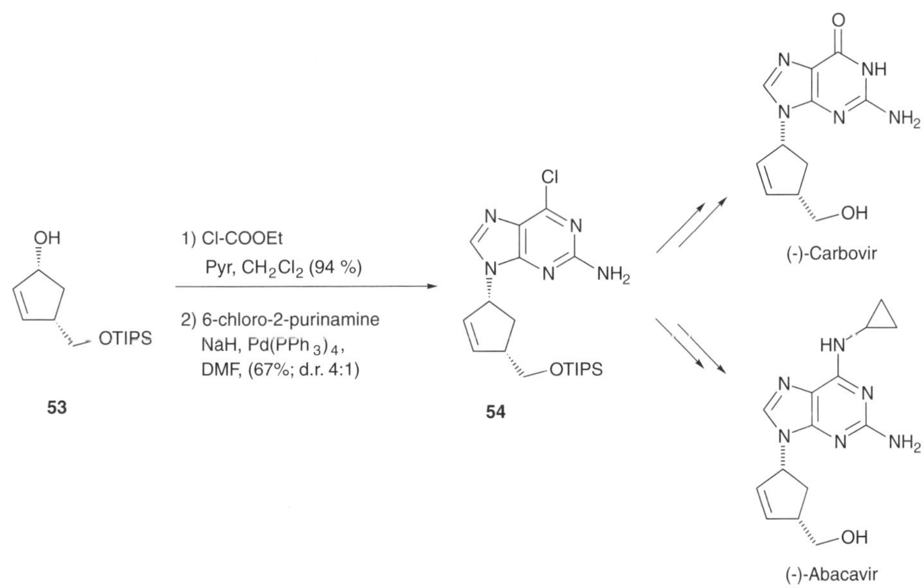

Scheme 6.32 *Synthesis of Carbovir and Abacavir from the homochiral allyl alcohol **53**.*

6.7 Conclusion

A general catalytic methodology for the enantioselective PKR has not yet been developed. However, research efforts in this area have proved invaluable, not only for revealing the difficulty of this endeavor, but has also for providing enlightening mechanistic insight that should help pave the way for asymmetric PKRs in the future. Nonetheless, high enantioselectivities have been achieved in a few intramolecular PKRs using catalytic amounts of BINAP as ligand. Moreover, stoichiometric reaction of bidentate (P,S)-ligands has proven to be a reliable procedure, providing ready access to useful synthetic intermediates. The PK cycloadduct **3i** (formed by reaction of trimethylsilylacetylene and norbornadiene) has recently been used to synthesize biologically active compounds, including the phytoprostanes dPPJ$_1$–I and dPPJ$_1$–II and the carbanucleosides Carbovir and Abacavir, underscoring its value as a chiral starting material.

References

[1] I. U. Khand, G. R. Knox, P. L. Pauson, and W. E. Watts, Organocobalt complexes. Part I. Arene complexes derived from dodecacarbonyltetracobalt, *J. Chem. Soc., Perkin Trans.* **1**, 975–977 (1973).

[2] (a) H.-W. Lee and F.-Y. Kwong, A Decade of Advancements in Pauson-Khand-Type Reactions, *Eur. J. Org. Chem.*, **5**, 789–811 (2010). (b) S. Laschat, A. Becheanu, T. Bell, and A. Baro, Regioselectivity, stereoselectivity and catalysis in intermolecular Pauson-Khand reactions: Teaching an old dog new tricks, *Synlett*, **17**, 2547–2570 (2005). (c) J. Blanco-Urgoiti, L. Anorbe, L. Perez-Serrano, G. Dominguez and J. Perez-Castells, The Pauson-Khand reaction, a powerful synthetic tool for the synthesis of complex molecules, *Chem. Soc. Rev.*, **33**, 32–42 (2004).

[3] P. Bladon, P. L. Pauson, H. Brunner, and R. Eder, Optically active transition metal complexes. XCV. Optically pure 3a,4,5,6,7,7a-hexahydro-2-phenyl-4,7-methano-1H-inden-1-one from an asymmetric Khand reaction, *J. Organomet. Chem.*, **355**, 449–454 (1988).

[4] P. Magnus, C. Exon, and P. Albaugh-Robertson, Dicobaltoctacarbonyl-alkyne complexes as intermediates in the synthesis of bicyclo[3.3.0]octenones for the synthesis of coriolin and hirsutic acid, *Tetrahedron*, **41**, 5861–5869 (1985).

[5] (a) B. T. J. M. de, C. Michel, K. Vekey, A. E. Greene, Y. Gimbert, and A. Milet, First C-C bond formation in the Pauson-Khand reaction: Influence of carbon-carbon triple bond polarization on regiochemistry, *J. Organomet. Chem.*, **691**, 4281–4288 (2006). (b) B. T. J. M. de, A. Milet, A. E. Greene, and Y. Gimbert, Insight into the Reactivity of Olefins in the Pauson-Khand Reaction, *J. Org. Chem.*, **69**, 1075–1080 (2004). (c) Y. Gimbert, D. Lesage, A. Milet, F. Fournier, A. E. Greene, and J.-C. Tabet, On Early Events in the Pauson-Khand Reaction, *Org. Lett.*, **5**, 4073–4075 (2003). (d) M. A. Pericàs, J. Balsells, J. Castro, I. Marchueta, A. Moyano, A. Riera, J. Vazquez, and X. Verdaguer, Toward the understanding of the mechanism and enantioselectivity of the Pauson-Khand reaction. Theoretical and experimental studies, *Pure Appl. Chem.*, **74**, 167–174 (2002). (e) F. Robert, A. Milet, Y. Gimbert, D. Konya, and A. E. Greene, Regiochemistry in the Pauson-Khand Reaction: Has a Trans Effect

Been Overlooked?, *J. Am. Chem. Soc.*, **123**, 5396–5400 (2001). (f) M. Yamanaka and E. Nakamura, Density Functional Studies on the Pauson-Khand Reaction, *J. Am. Chem. Soc.*, **123**, 1703–1708 (2001).

[6] Jordi Vázquez, *Theoretical Study on the Mechanism of the Pauson-Khand Reaction*, Doctoral Thesis, University of Barcelona (Spain), 1999. (b) Torstein Fjermestad, *Computational Studies on the Mechanism of the Pauson-Khand Reaction*, Doctoral Thesis, Rovira i Virgili University (Spain), 2010.

[7] H.-J. Park, B. Y. Lee, Y. K. Kang, and Y. K. Chung, Stereoselective Pauson-Khand Reactions via the Chiral [(alkyne)$Co_2(CO)_5$\{$P(OMe)_3$\}], *Organometallics*, **14**, 3104–3107 (1995).

[8] A. R. Kennedy, W. J. Kerr, D. M. Lindsay, J. S. Scott, and S. P. Watson, Stereochemical and mechanistic features of asymmetric Pauson-Khand processes, *J. Chem. Soc., Perkin Trans.* **1**, 4366–4372 (2000).

[9] N. E. Carpenter and K. M. Nicholas, Some studies involving the use of chiral amine oxides for the enantioselective preparation of (propargyl alcohol)$Co_2(CO)_5(PR_3)$ complexes, *Polyhedron*, **18**, 2027–2034 (1999).

[10] D. R. Carbery, W. J. Kerr, D. M. Lindsay, J. S. Scott, and S. P. Watson, Preparation and reaction of desymmetrised cobalt alkyne complexes, *Tetrahedron Lett.*, **41**, 3235–3239 (2000).

[11] D. T. Rutherford and S. D. R. Christie, Heterobimetallic alkyne complexes in organic synthesis: an asymmetric variant of the Pauson-Khand reaction, *Tetrahedron Lett.*, **39**, 9805–9808 (1998).

[12] Y. K. Chung, B. Y. Lee, N. Jeong, M. Hudecek, and P. L. Pauson, Promoters for the (alkyne) hexacarbonyldicobalt-based cyclopentenone synthesis, *Organometallics*, **12**, 220–223 (1993).

[13] (a) D. R. Carbery, W. J. Kerr, D. M. Lindsay, J. S. Scott, and S. P. Watson, Preparation and reaction of desymmetrised cobalt alkyne complexes, *Tetrahedron Lett.*, **41**, 3235–3239 (2000). (b) W. J. Kerr, G. G. Kirk, and D. Middlemiss, Enantioselective Pauson-Khand reactions mediated by a chiral amine N-oxide, *Synlett*, **10**, 1085–1086 (1995).

[14] (a) V. Derdau and S. Laschat, Alkaloid N-oxide promoted asymmetric cobalt-mediated Pauson-Khand reaction, *J. Organomet. Chem.*, **642**, 131–136 (2002). (b) V. Derdau, S. Laschat, and P. G. Jones, Amine N-oxides derived from alkaloids as chiral promoters in enantioselective Pauson-Khand reactions, *Heterocycles*, **48**, 1445–1453 (1998).

[15] K. D. Closser, M. M. Quintal, and K. M. Shea, The Scope and Limitations of Intramolecular Nicholas and Pauson-Khand Reactions for the Synthesis of Tricyclic Oxygen- and Nitrogen-Containing Heterocycles, *J. Org. Chem.*, **74**, 3680–3688 (2009).

[16] (a) X. Verdaguer, A. Moyano M. A. Pericàs, A. Riera, V. Bernardes, A. E. Greene, A. Alvarez-Larena, and J. F. Piniella, A Dual-Function, Highly Efficient Chiral Controller for Stereoselective Intermolecular Pauson-Khand Reactions. *J. Am. Chem. Soc.*, **116**, 2153–2154 (1994). (b) I. Marchueta, E. Montenegro, D. Panov, M. Poch, X. Verdaguer, A. Moyano, M. A. Pericàs, and A. Riera, Asymmetric Pauson-Khand reactions using camphor-derived chelating thiols as chiral controllers. *J. Org. Chem.*, **66**, 6400–6409 (2001).

[17] X. Verdaguer, A. Moyano, M. A. Pericàs, A. Riera, A. Alvarez-Larena, and J.-F. Piniella, Alkyne Dicobalt Carbonyl Complexes with Sulfide Ligands. Synthesis, Crystal Structure, and Dynamic Behavior, *Organometallics*, **18**, 4275–4285 (1999).

[18] R. Rios, S. Paredes, M. A. Pericàs, and A. Moyano, Chiral cyclopentadiene-mediated approach to enantioselective heterobimetallic Pauson-Khand reactions, *J. Organomet. Chem.*, **690**, 358–362 (2005).

[19] J. Solà, A. Riera, M. A. Pericàs, X. Verdaguer, and M. A. Maestro, Stereoselectivity in the intermolecular Pauson-Khand reaction of electron-deficient terminal alkynes, *Tetrahedron Lett.*, **45**, 5387–5390 (2004).

[20] (a) K. Hiroi, T. Watanabe, R. Kawagishi, and I. Abe, Catalytic use of chiral phosphine ligands in asymmetric Pauson-Khand reactions, *Tetrahedron: Asymm.*, **11**, 797–808 (2000). (b) K. Hiroi, T. Watanabe, R. Kawagishi, and I. Abe, Asymmetric catalytic Pauson-Khand reactions with chiral phosphine ligands: dramatic effects of substituents in 1,6-enyne systems, *Tetrahedron Lett.*, **41**, 891–895 (2000).

[21] T. M. Schmid and G. Consiglio, Asymmetric cyclocarbonylation of 1,6-enynes with cobalt catalysts, *Tetrahedron: Asymm.*, **15**, 2205–2208 (2004).

[22] S. E. Gibson, S. E. Lewis, J. A. Loch, J. W. Steed, and M. J. Tozer, Identification of an Asymmetric Pauson-Khand Precatalyst, *Organometallics*, **22**, 5382–5384 (2003).

[23] S. J. Sturla and S. L. Buchwald, Cobalt-Phosphite-Catalyzed Asymmetric Pauson-Khand Reaction, *J. Org. Chem.*, **67**, 3398–3403 (2002).

[24] V. Derdau, S. Laschat, I. Dix, and P. G. Jones, Cobalt-Alkyne Complexes with Diphosphine Ligands as Mechanistic Probes for the Pauson-Khand Reaction, *Organometallics*, **18**, 3859–3864 (1999).

[25] (a) S. E. Gibson, K. A. C. Kaufmann, J. A. Loch, and A. Miyazaki, Metal carbonyl complexes and chirality, *P. App. Chem.*, **80**, 903–911 (2008). (b) S. E. Gibson, D. J. Hardick, P. R. Haycock, K. A. C. Kaufmann, A. Miyazaki, M. J. Tozer, and A. J. P. White, A study of (binap)(enyne)tetracarbonyldicobalt(0) complexes, *Chem. Eur. J.*, **13**, 7099–7109 (2007). (c) S. E. Gibson, K. A. C. Kaufmann, J. A. Loch, J. W. Steed, and A. J. P. White, A study of [Co_2(alkyne)(binap)$(CO)_4$] complexes (BINAP = (1,1′-binaphthalene)-2,2′-diylbis(diphenylphosphine)), *Chem. Eur. J.*, **11**, 2566–2576 (2005).

[26] T. Fjermestad, M. A. Pericàs and F. Maseras, Origin of enantioselectivity in asymmetric Pauson-Khand reactions catalyzed by [(BINAP)$Co_2(CO)_6$], *J. Mol. Catal. A: Chem.*, **324**, 127–132 (2010).

[27] H. Brunner, Right or left - this is the question (enantioselective catalysis with transition metal compounds), *Pure Appl.Chem.*, **62**, 589–594 (1990).

[28] H. Brunner and Niedernhuber, Optically active transition-metal complexes. 102. Enantioselective synthesis of 3-oxabicyclo[3.3.0]oct-7-en-7-phenyl-6-one via Khand-Pauson reaction, *Tetrahedron: Asymm.*, **1**, 711–714 (1990).

[29] A. M. Hay, W. J. Kerr, G. G. Kirk, and D. Middlemiss, Highly Efficient Enantioselective Pauson-Khand Reactions, *Organometallics*, **14**, 4986–4988 (1995).

[30] (a) J. Castro, A. Moyano, M. A. Pericàs, A. Riera, A. Alvarez-Larena, and J. F. Piniella, Acetylene−Dicobaltcarbonyl Complexes with Chiral Phosphinooxazoline Ligands: Synthesis, Structural Characterization, and Application to

Enantioselective Intermolecular Pauson-Khand Reactions, *J. Am. Chem. Soc.*, **122**, 7944–7952 (2000). (b) J. Castro, A. Moyano, M. A. Pericàs, A. Riera, A. Alvarez-Larena, and J. F. Piniella, The first alkyne-dicobaltcarbonyl complex with a bidentate chiral ligand with Co-P and Co-N coordination, *J. Organomet. Chem.*, **585**, 53–58 (1999).

[31] D. Konya, F. Robert, Y. Gimbert, and A. E. Greene, Asymmetric Pauson-Khand reaction with chiral, electron-deficient mono- and bis-phosphine ligands, *Tetrahedron Lett.*, **45**, 6975–6978 (2004).

[32] V. Moberg, M. A. Mottalib, D. Sauer, Y. Poplavskaya, D. C. Craig, S. B. Colbran, A. J. Deeming, and E. Nordlander, Chiral and achiral phosphine derivatives of alkylidyne tricobalt carbonyl clusters as catalyst precursors for (asymmetric) inter- and intramolecular Pauson-Khand reactions, *Dalton Trans.*, 2442–2453 (2008).

[33] A. Stolle, H. Becker, J. Salaün, and A. de Meijere, Enantioselective construction of spiro{cyclopropane-14′-bicyclo[3.3.0]oct-1-en-3-ones}, *Tetrahedron Lett.*, **35**, 3521–3524 (1994).

[34] (a) I. Marchueta, E. Montenegro, D. Panov, M. Poch, X. Verdaguer, A. Moyano, M. A. Pericàs, and A. Riera, Asymmetric Pauson-Khand reactions using camphor-derived chelating thiols as chiral controllers, *J. Org. Chem.*, **66**, 6400–6409 (2001). (b) X. Verdaguer, J. Vázquez, G. Fuster, V. Bernardes-Genisson, A. E. Greene, A. Moyano, M. A. Pericàs, and A. Riera, Camphor-Derived, Chelating Auxiliaries for the Highly Diastereoselective Intermolecular Pauson-Khand Reaction: Experimental and Computational Studies, *J. Org. Chem.*, **63**, 7037–7052 (1998). (c) E. Montenegro, M. Poch, A. Moyano, M. A. Pericàs, and A. Riera, Highly diastereoselective Pauson-Khand reactions of a stable, internally chelated, dicobalt pentacarbonyl complex of a chiral acetylene thioether, *Tetrahedron Lett.*, **39**, 335–338 (1998). (d) X. Verdaguer, A. Moyano, M. A. Pericàs, A. Riera, V. Bernardes, A. E. Greene, A. Alvarez-Larena, and J. F. Piniella, A Dual-Function, Highly Efficient Chiral Controller for Stereoselective Intermolecular Pauson-Khand Reactions, *J. Am. Chem. Soc.*, **116**, 2153–2154 (1994).

[35] A. J. Edwards, S. R. Mack, M. J. Mays, C.-Y. Mo, P. R. Raithby, and M.-A. Rennie, A comparison of the reactions of PPh_2CH_2SR (R = Me, Ph) with alkyne-bridged dicobalt carbonyl complexes and with an iron dicobalt [mu]3-thioxo carbonyl complex, *J. Organomet. Chem.*, **519**, 243–252 (1996).

[36] (a) X. Verdaguer, M. A. Pericàs, A. Riera, M. A. Maestro, and J. Mahía, Design of New Hemilabile (P,S) Ligands for the Highly Diastereoselective Coordination to Alkyne Dicobalt Complexes: Application to the Asymmetric Intermolecular Pauson-Khand Reaction, *Organometallics*, **22**, 1868–1877 (2003). (b) X. Verdaguer, A. Moyano, M. A. Pericàs, A. Riera, M. A. Maestro and J. Mahía, A New Chiral Bidentate (P,S) Ligand for the Asymmetric Intermolecular Pauson-Khand Reaction, *J. Am. Chem. Soc.*, **122**, 10242–10243 (2000).

[37] J. Solà, A. Riera, X. Verdaguer and M. A. Maestro, Phosphine−Substrate Recognition through the C−H···O Hydrogen Bond: Application to the Asymmetric Pauson-Khand Reaction, *J. Am. Chem. Soc.*, **127**, 13629–13633 (2005)

[38] J. Solà, A. Riera, X. Verdaguer, and M. A. Maestro, C−H···O Hydrogen Bond-Directed Ligand Exchange Reaction: Diastereoselective Synthesis of P,S-Bridged (μ-alkyne)$Co_2(CO)_4$ Complexes, *Organometallics*, **25**, 5795–5799 (2006).

[39] A. Lledó, J. Solà, X. Verdaguer, A. Riera, and M. A. Maestro, PuPHOS and CamPHOS Ligands in the Intermolecular Catalytic Pauson-Khand Reaction, *Adv. Synth. Catal.*, **349**, 2121–2128 (2007).

[40] J. Solà, A. Riera, and X. Verdaguer, Synthesis of a new camphor derived P,S(O) ligand. The importance of C-H. . .O bonding in the ligand exchange reactions with $Co_2(\mu\text{-alkyne})(CO)_6$ complexes, *J. Organomet. Chem.*, **695**, 23772380 (2010).

[41] M. T. Robak, M. A. Herbage, and J. A. Ellman, Synthesis and applications of tert-butanesulfinamide, *Chem. Rev.*, **110**, 3600–3740 (2010)

[42] (a) M. Revés, T. Achard, J. Solà, A. Riera, and X. Verdaguer, N-phosphino-p-tolylsulfinamide ligands: synthesis, stability, and application to the intermolecular Pauson-Khand reaction, *J. Org. Chem.*, **73**, 7080–7087 (2008). (b) J. Solà, M. Revés, A. Riera and X. Verdaguer, N-phosphino sulfinamide ligands: an efficient manner to combine sulfur chirality and phosphorus coordination behavior, *Angew. Chem. Int. Ed.*, **46**, 5020–5023 (2007).

[43] Y. Ji, A. Riera, and X. Verdaguer, Asymmetric Intermolecular Pauson-Khand Reaction of Symmetrically Substituted Alkynes, *Org. Lett.*, **11**, 4346–4349 (2009).

[44] M. Revés, A. Riera, and X. Verdaguer, PNSO ligands as a tool to study metal bonding of electron-deficient sulfinyl groups, *Eur. J. Inorg. Chem.*, **29**, 4446–4453 (2009).

[45] C. Ferrer, A. Riera, and X. Verdaguer, Sulfinylmethyl Phosphines as Chiral Ligands in the Intermolecular Pauson-Khand Reaction, *Organometallics*, **28**, 4571–4576 (2009).

[46] X. Verdaguer, A. Lledó, C. Lopez-Mosquera, M. A. Maestro, M. A. Pericàs, and A. Riera, PuPHOS: a synthetically useful chiral bidentate ligand for the intermolecular Pauson-Khand reaction, *J. Org. Chem.*, **69**, 8053–8061 (2004).

[47] M. Iqbal and P. Evans, Conjugate addition-Peterson olefination reactions: expedient routes to cross conjugated dienones, *Tetrahedron Lett.*, **44**, 5741–5745 (2003).

[48] (a) M. Iqbal, P. Duffy, P. Evans, G. Cloughley, B. Allan, A. Lledó, X. Verdaguer, and A. Riera, The conjugate addition-Peterson olefination reaction for the preparation of cross-conjugated cyclopentenone, PPAR-γ ligands, *Org. Biomol. Chem.*, **6**, 4649–4661 (2008). (b) M. Iqbal, P. Evans, A. Lledó, X. Verdaguer, M. A. Pericàs, A. Riera, C. Loeffler, A. K. Sinha, and M. J. Mueller, Total Synthesis and Biological Activity of 13,14-Dehydro-12-Oxo-Phytodienoic Acids (Deoxy-J_1-Phytoprostanes), *ChemBioChem*, **6**, 276–280 (2005).

[49] A. Vázquez-Romero, J. Rodríguez, A. Lledó, X. Verdaguer, and A. Riera, Enantioselective Syntheses of Carbanucleosides from the Pauson-Khand Adduct of Trimethylsilylacetylene and Norbornadiene, *Org. Lett.*, **10**, 4509–4512 (2008).

[50] Z. Benko, B. Fraser-Reid, P. S. Mariano, and A. L. J. Beckwith, Conjugate addition of methanol to alpha -enones: Photochemistry and stereochemical details. *J. Org. Chem.*, **53**, 2066–2072 (1988).

[51] A. Lledo, J. Benet-Buchholz, A. Sole, S. Olivella, X. Verdaguer, and A. Riera, Photochemical rearrangements of norbornadiene Pauson-Khand cycloadducts *Angew. Chem., Int. Ed.*, **46**, 5943–5946 (2007).

7

Recent Advancement of Catalytic Pauson-Khand-type Reactions

*Fuk Loi Lam, Hang Wai Lee, Jun Wang and Fuk Yee Kwong**

7.1 Introduction

Cycloaddition reaction is one of the most attractive synthetic protocols for accessing a variety of scientifically interesting and pharmaceutically useful ring-structured skeletons. This transformation is an addition reaction in which the waste materials (side products) generated are expected to be minimized and thus is environmentally favourable. Indeed, these significances have been highlighted by Trost: 'an ideal chemical reaction must be not only selective, but also a simple addition in which the other reactants are required only in catalytic amounts.'[1] In fact, transition-metal-mediated catalytic reactions can play a versatile role in addressing the atom economy issue. One of the earliest examples is the

The Pauson-Khand Reaction: Scope, Variations and Applications, First Edition. Edited by Ramon Rios Torres.

[2+2+1] carbonylative cyclization reaction. This type of reaction was pioneered by Pauson and Khand in 1971.[2] This transition metal-mediated reaction is a cycloaddition of three components, an alkyne, alkene and a carbon monoxide moiety, for generating a variety of synthetically useful cyclopentenones (Scheme 7.1).[3]

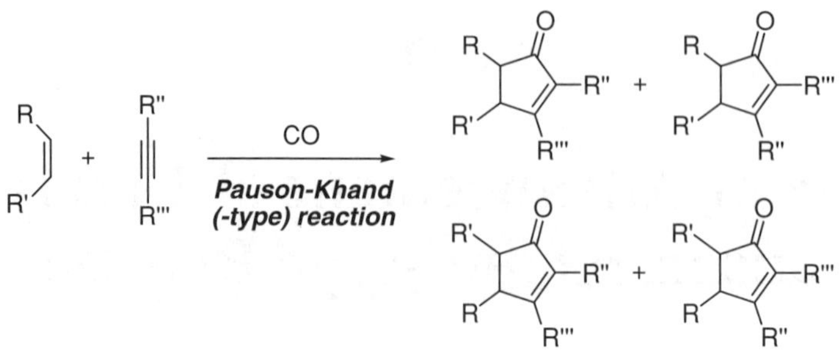

Scheme 7.1 *Pauson-Khand type reaction.*

In this pioneering study of an intermolecular-type reaction, only activated and symmetrical alkenes such as ethylene and norbornene were used, since regioisomers are expected to be formed when unsymmetrical alkyne and alkene are applied. Nevertheless, previous research studies showed that the regiochemistry from the alkyne partner is predictable and does not give a mixture of products in general. In addition to the intermolecular reaction, the intramolecular version of this reaction is expected to avoid the formation of regioisomeric mixture and thus only gives single bicyclic cyclopentenone (Scheme 7.2). In this context, we briefly summarize the early developments on the catalytic Pauson-Khand reaction. Thus we focus on the evolution of the catalytic systems and their potential applications from the late 1990s to early 2011.

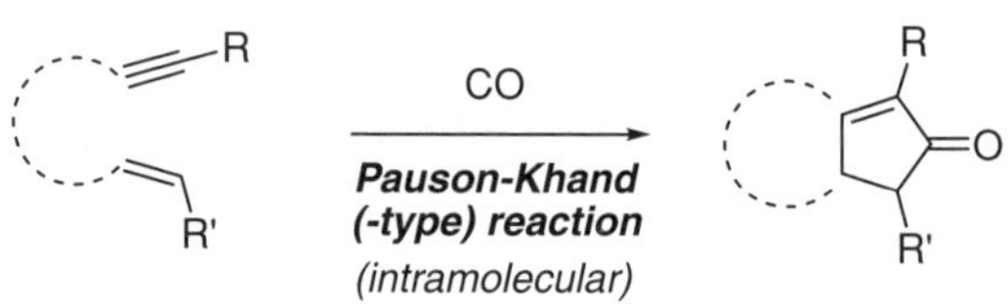

Scheme 7.2 *Intramolecular Pauson-Khand reaction.*

7.2 Rhodium-Catalyzed Pauson-Khand-Type Cyclizations

The first successful Rh-catalyzed asymmetric intramolecular Pauson-Khand–type reaction was reported by Jeong and co-workers in 2000.[4] They disclosed a complex of $[RhCl(CO)_2]_2$/(*S*)-BINAP which was highly effective for the enantioselective transformation of tethered 1,6-enynes substrates to give corresponding cyclopentenones (Scheme 7.3). Excellent yields and enantioselectivities (up to 96% ee) were obtained from an array of *C*-, *O*- and *N*-tethered 1,6-enynes.

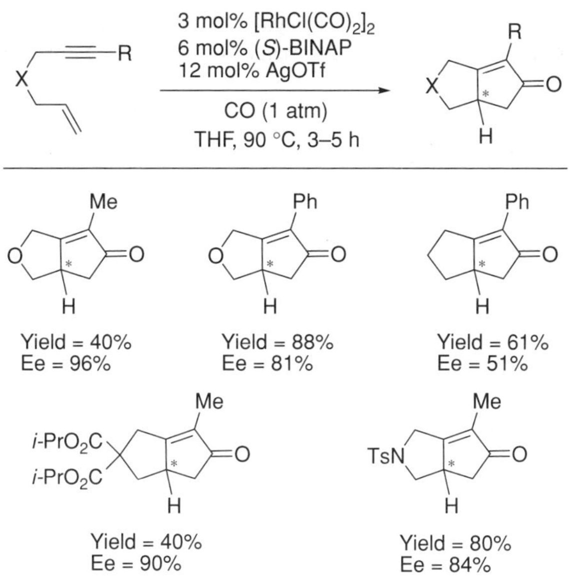

Scheme 7.3 *Rh-catalyzed Pauson-Khand reaction of 1,6-enynes.*

They proposed a cationic pathway for the asymmetric PKR mechanism. 1,6-enyne was suggested to bind with $[Rh(CO)(S)\text{-BINAP}]^+$ complex and transform to octahedral Rh(III)-metallocyclopentene intermediate. Subsequent migratory insertion of CO and reductive elimination furnished the desired carbonylative cycloadduct.[4] Based on the success of the $[RhCl(CO)_2]_2$ complex in intramolecular cycloaddition of 1,6-enynes, Narasaka *et al.* further extended its application to intermolecular Pauson-Khand-type cyclization.[5] They reported that the reaction between norbornene and 1-phenylpropyne gave the desired cyclic enones in 69% product yield as a mixture of regioisomers (Scheme 7.4). This Rh-catalyzed reaction was accelerated by reducing CO partial pressure to less than 1 atm.

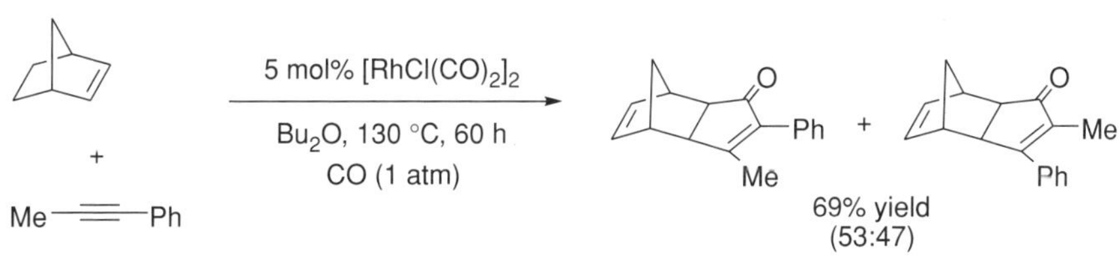

Scheme 7.4 *Rh-catalyzed intermolecular PKR of norbornene and 1-phenylpropyne.*

In 2003, Chung and co-workers established a recyclable Pauson-Khand-type protocol assisted by the entrapped Rh complexes, which were prepared by a sol-gel process.[6] The metal complex could be recovered and reused at least ten times without loss of catalytic activity. However, the stereo-induction offered by this catalytic system was poor. Only 13% ee was obtained in intramolecular Pauson-Khand-type cyclization of aromatic 1,6-enyne.

In the same year, the Chung group reported a relatively more environmentally benign Rh-catalyzed PKR under a water/dioxane reaction medium.[7] Surfactant (e.g. SDS) was necessary for this successful transformation.

Consiglio *et al.* disclosed an interesting system for the Pauson-Khand-type chemistry.[8] They employed CO-saturated molecular sieves as a CO reservoir for the carbonylative cyclization of 1,6-enynes. Particularly noteworthy is that this protocol represented the lowest temperature of Pauson-Khand-type reaction achieved so far (Scheme 7.5). At 0 °C, the desired product was obtained in 97% ee with 99% substrate conversion.

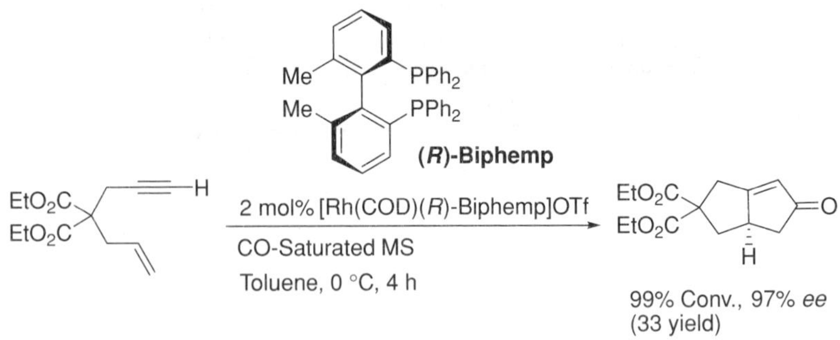

Scheme 7.5 *Rh-catalyzed PKR with a molecular sieve as the CO reservoir.*

In 2008, Ratovelomanan-Vidal, Genêt, and Jeong reported an efficient asymmetric Pauson-Khand-type reaction mediated by a Rh(I) catalyst at ambient temperatures (18–20 °C) under a reduced partial pressure of CO.[9] They employed (*R*)-xylyl-BINAP and $[RhCl(CO)_2]_2$ as the catalyst system with 0.1 atm CO partial pressure to give bicyclic cyclopentenones in up to 99% yield and enantioselectivity (Scheme 7.6). In addition to 1,6-enynes substrates, the asymmetric desymmetrization of dienynes was also accomplished by a Rh-catalyzed enantioselective Pauson-Khand-type reaction at ambient temperatures, in which (*S*)-DifluorPhos was found to be the best ligand to obtain the target adducts in up to 95% *ds* (Scheme 7.7).[9] In 2009, Jeong *et al.* further demonstrated that solvents could significantly influence this asymmetric reaction.[10] The coordinating solvents such as THF provided a faster rate of reaction and better enantioselectivity than non-coordinating solvents.

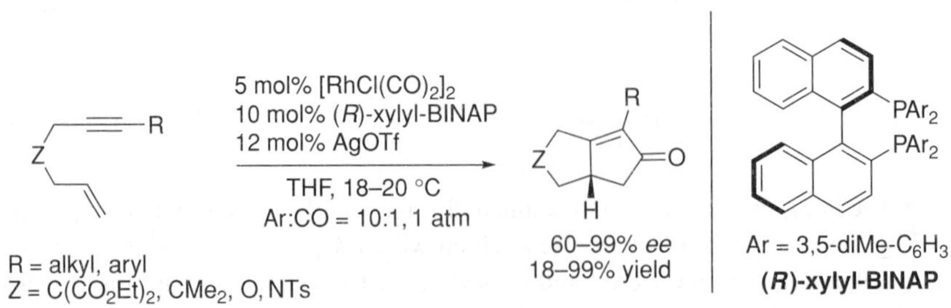

Scheme 7.6 *Rh-catalyzed PKR under a reduced partial CO pressure.*

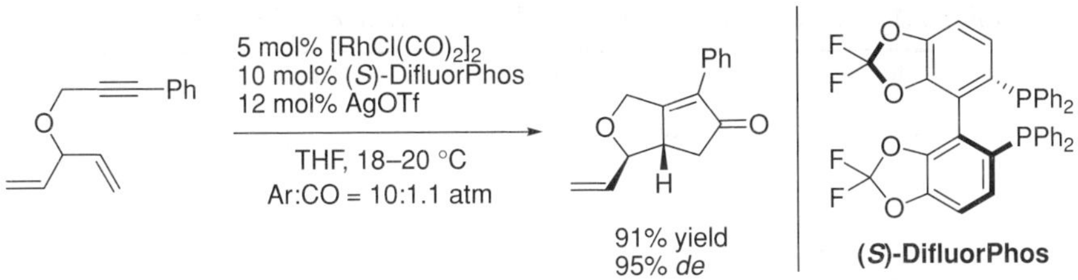

Scheme 7.7 *Asymmetric Rh-catalyzed desymmetrization of dienynes.*

In 2010, Ratovelomanan-Vidal and Jeong *et al.* showed an alternative asymmetric Rh-catalyzed PKR of 1,6-enynes using sterically hindered (*S*)-dtbm-MeO-Biphep as the chiral ligand (Scheme 7.8).[11] This catalytic system was an improvement on the previously reported system (c.f. Scheme 7.6). The product yield and/or enantioselectivity were higher, especially for the substrates containing an aryl substituent on an alkyne terminus, though a higher reaction temperature was required for this system.

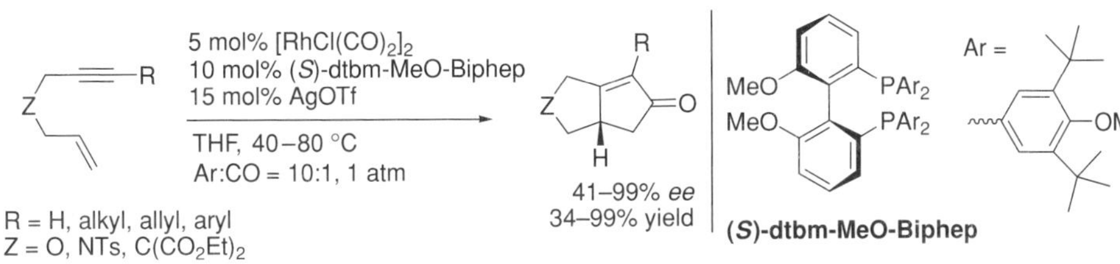

Scheme 7.8 *Asymmetric Rh-catalyzed PKR of enynes.*

To illustrate the applicability of the Pauson-Khand reaction in the synthesis of natural compounds, the same research group reported an efficient asymmetric desymmetrization of the prochiral acetal.[12] The desymmetrization of acetal was sensitive to the reaction conditions and decomposition of the acetal occurred under thermal reaction conditions. Interestingly, cinnamaldehyde can serve as the CO surrogate[18] for the carbonylative cycloaddition to give the target products with satisfactory yields and moderate *ee* value (Table 7.1).[12]

Apart from the chiral bidentate phosphine ligands, Zhou and co-workers developed a chiral monodentate phosphoramidite ligand (**SIPHOS**) which was found to be effective in the Rh-catalyzed asymmetric carbonylative cyclization (Scheme 7.9).[13] The finding represented the first successful example of chiral monodentate ligand in asymmetric Rh-catalyzed intramolecular PKR. Thus the chiral monodentate ligands have a high potential for the application in related enanatioselective reactions.

Transition metal-catalyzed carbonylation has proved successful in the area of organic synthesis, and has been recognized as a powerful protocol for the direct synthesis of a wide variety of carbonyl-containing compounds.[14] However, the use of highly poisonous carbon monoxide constitutes a drawback to these methodologies. In 2002, two independent groups, Morimoto/Kakiuchi[15] and Shibata,[16] respectively, illustrated a conceptual evolution in the

Table 7.1 *Rh-catalyzed asymmetric desymmetrization of acetal.*

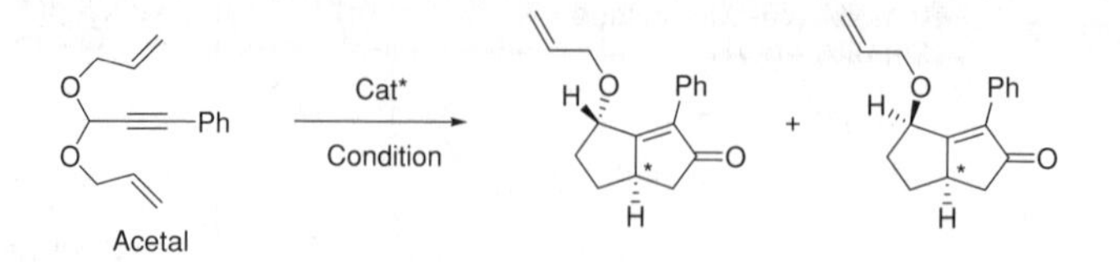

Conditions	Yield/ee[%]	Yield/ee[%]
A	No Rxn, (acetal decomposition)	
B	41/86	25//70

A: $[Rh(CO)_2Cl]_2$(3 mol%), *(R)*-BINAP (9 mol%), Ag0Tf (12 mol%), THF, 90°C,CO (1atm)
B: $[Rh(CO)_2Cl]_2$(5 mol%), *(R)*-BINAP (12 mol%),cinnamaldehyde (20 eq), 120°C, Ar (1atm)

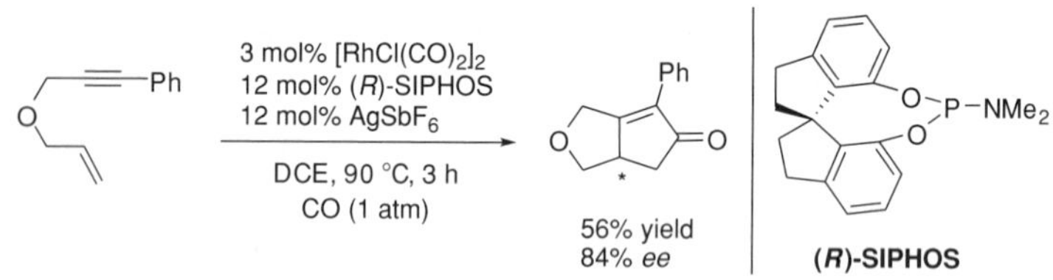

Scheme 7.9 *Asymmetric Rh-catalyzed PKR using (R)-SIPHOS as a chiral ligand.*

use of metal-carbonyl to attain PKR in the absence of gaseous carbon monoxide.[17,18] This hypothesis employed an *in situ* aldehyde decarbonylation process and a subsequent CO transfer reaction (Scheme 7.10).

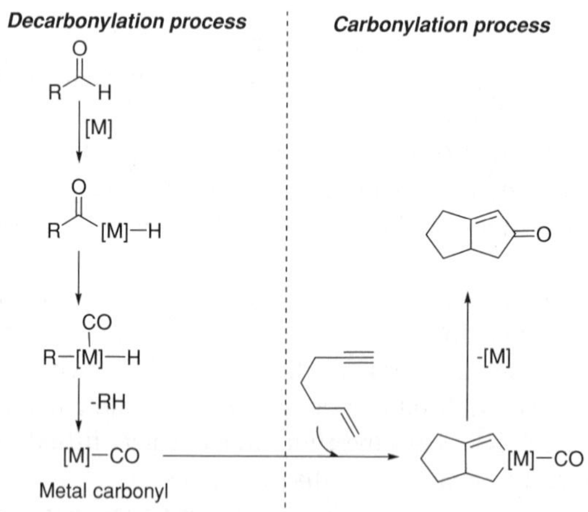

Scheme 7.10 *The proposed mechanism of PKR with an aldehyde as CO source.*

Morimoto/Kakiuchi *et al.* reported pentafluorobenzaldehyde was an efficient CO surrogate for the Pauson-Khand-type cyclization, whereas the Shibata group showed cinnamaldehyde was the best CO donor (Scheme 7.11). Moreover, Shibata *et al.* later developed a protocol using (*S*)-tol-BINAP as chiral ligand for asymmetric Pauson-Khand-type reaction to give the desired product in up to 90% ee.[19]

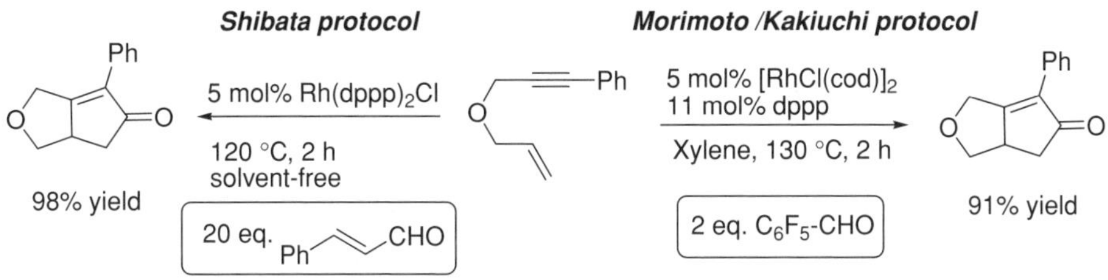

Scheme 7.11 *Sequential decarbonylation of aldehyde and carbonylation of enyne.*

Apart from aromatic aldehyde, the Morimoto/Kakiuchi group further reported that formaldehyde (in water) was an effective CO reservoir for Rh-catalyzed PKR.[20] They used dppp/TPPTS (an organic-soluble and a water-soluble ligand, respectively) as a mixed-phosphine catalyst system with the aid of SDS to obtain the desired cycloadduct in up to 97% chemical yield. They proposed a micelle formation[21] during the course of the reaction, and the tandem decarbonylation and the carbonylation processes took place independently in the aqueous and micellar phase, respectively. In 2004, the same research group further disclosed the enantioselective version of this reaction by employing (*S*)-tol-BINAP as a catalyst to give the desired product good to excellent enantioselectivity (up to 95% ee).[22]

Recently, considerable attention has been given to aqueous transition metal-catalyzed reactions.[23] However, no catalytic asymmetric systems that enable the use of water as the only solvent (without a surfactant) in the PKR have been developed since 2005. Kwong/Chan *et al.* did indeed present a Rh-catalyzed asymmetric PKR in water using chiral dipyridyl diphosphine ligand, P-Phos (Scheme 7.12).[24] Varieties of *O*-, *N*-,and *C*-tethered 1,6-enynes were converted to the corresponding cyclopentenones with up to 95% ee.

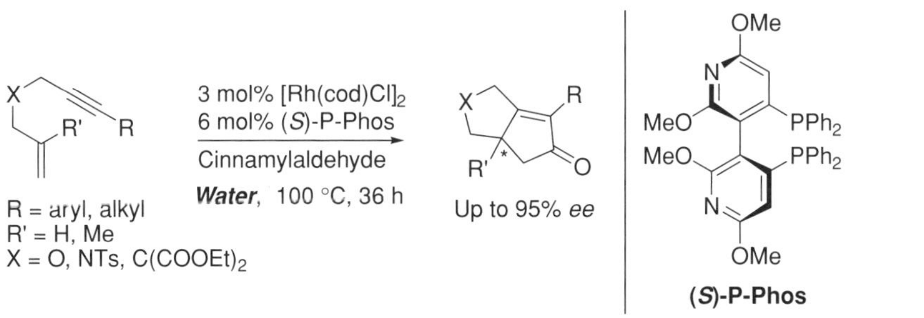

Scheme 7.12 *Asymmetric Rh-catalyzed PKR of 1,6-enynes in water.*

In addition to the aqueous system, Kwong/Chan also demonstrated a homogeneous asymmetric Rh-catalyzed PKR using an environmentally benign alcoholic solvent. The combination of (*S*)-BisbenzodioxanPhos (also named (*S*)-SYNPHOS by Genêt *et al.*) with

Rh precatalyst as catalytic system gave the bicyclic cyclopentenones in up to 96% ee (Scheme 7.13).[25]

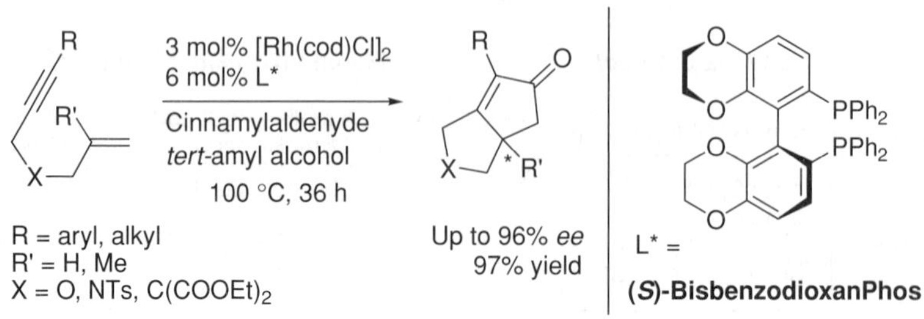

Scheme 7.13 *Asymmetric Rh-catalyzed PKR in an alcoholic medium.*

The electronic effect in asymmetric catalysis is one of the most important parameters to control the stereo-outcome of the desired product.[26] In fact, this effect may be attributed to both substrates and/or catalysts (ligand) which can significantly alter the extent of product enantioselectivity. Recently, Kwong/Chan first disclosed the substrate electronic effect of 1,6-aromatic enynes which could alter the level of enantioselectivity of the desired product, such as bicyclic cyclopentenone (Scheme 7.12).[24] A higher enantioselectivity was obtained when the electron-rich substituent was attached to the aromatic ring. The Hammett study showed that a linear free energy relationship was found (Figure 7.1).[27]

Moreover, they proposed the structure of the metal complex leading to the desired cycloadducts. The electron-rich *O*-tethered 1,6-enynes were suggested to bind the Rh metal center more closely and thus have a better stereoinduction, whereas electron-poor enynes were loosely coordinated (Figure 7.2).[28]

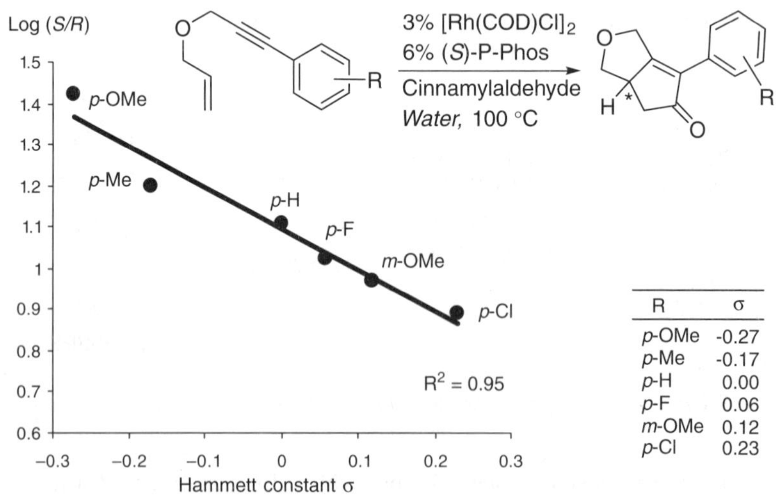

R	σ
p-OMe	-0.27
p-Me	-0.17
p-H	0.00
p-F	0.06
m-OMe	0.12
p-Cl	0.23

Figure 7.1 *Hammett plot of Rh-catalyzed asymmetric PKR in water.*

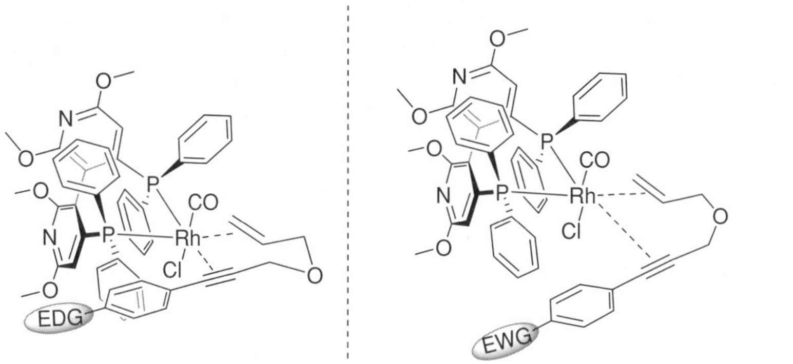

Figure 7.2 *Suggested transition-state models of electronically controlled Rh-catalyzed PKR.*

Apart from the substrate electronic effects, Ratovelomanan-Vidal, Genêt, and Jeong reported the electronic effect of chiral ligands in asymmetric Rh-catalyzed PKR.[29] They demonstrated that both the rate and the enantioselectivity were significantly dependent on the electronic density of the axially chiral BINAP class ligands. With ligands bearing a relatively electron-deficient phosphorus donor (i.e. with *para*-electron-withdrawing substituent(s) on the phosphorus-attached aromatic ring), the reaction rate was slower, but the PKR products were obtained in a higher enantioselectivity. Recently, the same research groups carried out a detailed study on the dihedral angle effect of ligands towards the enantioselectivity of the PKR cycloadducts.[30] They found that the steric properties of the chiral biaryl diphosphine ligands, especially the value of its dihedral angle, played a crucial role in the stereochemical outcome of Rh-mediated PKR. Chiral ligands, such as **SYNPHOS** (also named as BisbenzodioxanPhos by Chan *et al.*) and **DifluoroPhos** with a narrower dihedral angle than BINAP-type ligands, gave improved reactivity and enantioselectivity for most 1,6-enyne substrates (Figure 7.3). Based on the influences of electronic properties of ligands and substrates on the enantioselectivity of cycloadducts, they developed an efficient kinetic resolution system for the racemic 1-arylallyl propargyl ethers by enantioselective Rh- or Ir-catalyzed PKR.[31] The choice of catalyst was highly dependent on the enyne substrates applied. The cationic Rh(I) catalyst was suitable for substrates having aryl substituents on the alkyne terminus, while a neutral Ir(I) catalyst was found effective in handling substrates having alkyl substituents.

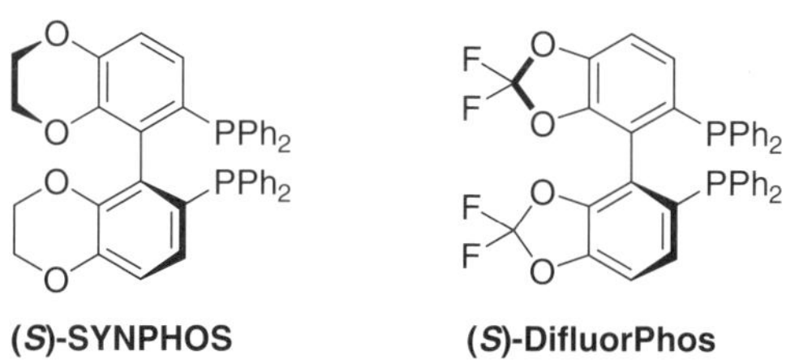

Figure 7.3 *Axially chiral ligands possess a narrower dihedral angle than BINAP.*

In addition to using aldehyde as the CO surrogate, the Kwong group recently disclosed that benzyl formate could serve as a CO source for a cooperative dual catalysis.[32] The decarbonylation of formate by the Rh complex and subsequent carbonylation of the enyne offered the carbonylative cycloadduct in up to 94% *ee* (Scheme 7.14).

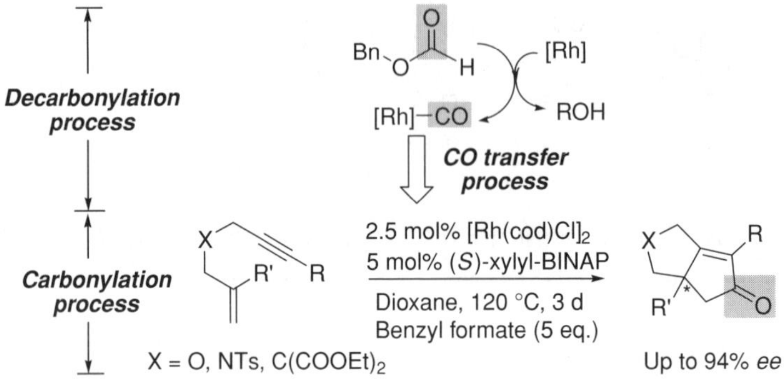

Scheme 7.14 *The Rh-catalyzed PKR with formate as CO source.*

They proposed the cooperative catalytic cycle as shown in Scheme 7.15. The formate coordinates to the Rh center and subsequent C-H oxidative addition gives the Rh(III) complex. The decarbonylation of the Rh-acyl complex yields the Rh-carbonyl species. The Rh-CO complex enters a carbonylation process of the enyne to generate cyclopentenone as the usual PKR mechanism. In addition to the suggested mechanism, they also performed ^{13}C labelled experiments to further confirm the source of CO moiety.[32]

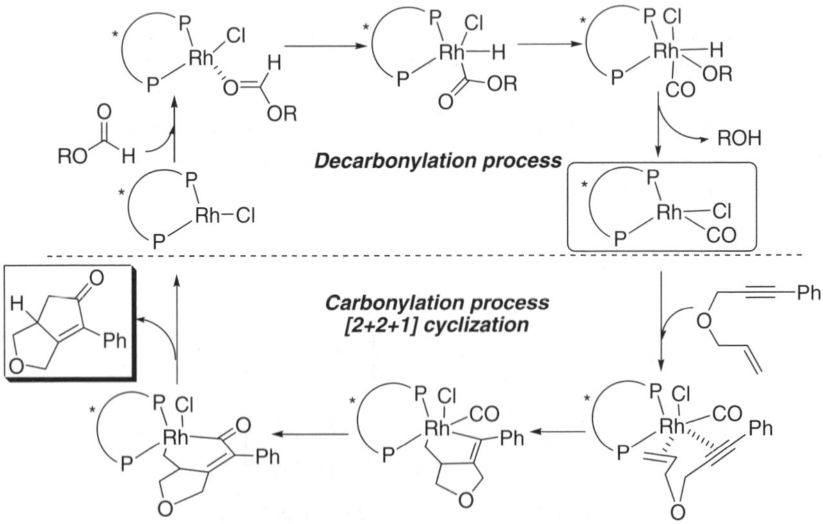

Scheme 7.15 *The proposed mechanism of PKR with formate as CO source.*

Microwave heating has been applied to many different areas of chemical research in the last decade.[33] Since the pioneering reports on microwave-accelerated organic transformation by the groups of Gedye and Giguere/Majetich in 1986,[34] the application of microwave dielectric heating to microwave-assisted organic synthesis (MAOS) has attracted considerable attention in recent years.[35] In 2008, Kwong and co-workers reported a microwave-assisted Rh-diphosphine complex-catalyzed PKR.[36,37] The cooperative process provided [2+2+1] cycloadducts by sequential decarbonylation of aldehyde or formate and carbonylation of enynes within a short period of time. Various *O*-, *N*- and *C*-tethered enynes were transformed to corresponding products in good yields. This catalysis represents the first enantioselective version of a microwave-accelerated cascade cyclization. In the presence of the chiral complex of Rh-L* [L*=(*S*)-BisbenzodioxanPhos], up to 90% ee of bicyclic cyclopentenones were achieved (Scheme 7.16).[37]

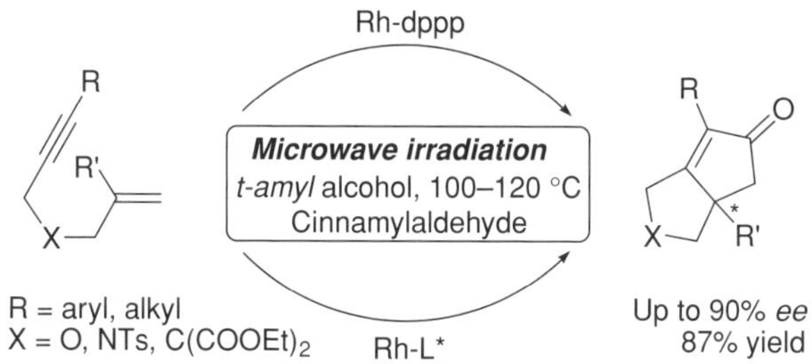

Scheme 7.16 *The microwave-assisted asymmetric PKR.*

In 2008, Evans and Baik reported a mechanistic study on the diastereoselective Rh-catalyzed PKR with the role of a coordination number for governing the stereo-outcomes.[38] They figured out a theoretical analysis of Rh-catalyzed PKR that could provide two mechanistic scenarios for the origin of diastereoselectivity (Table 7.2). The optimum selectivity could be attributed to a five- rather than a four-coordinating organorhodium complex. These findings offer a foundation to determine relevant controlling factors for improving the stereochemical outcome of future Pauson-Khand reactions.

Table 7.2 *The diastereoselectivity of PKR adducts governed by CO/Ar ratio.*

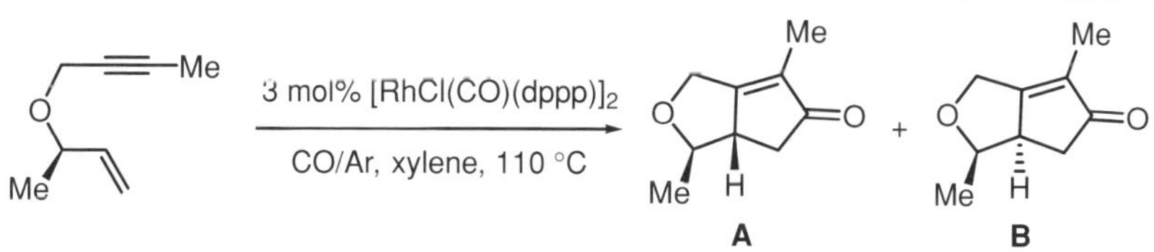

CO pressure	Ar pressure	%yield	dr (A/B)
1.00	0.00	88	99.1
0.10	0.90	51	58.1
0.05	0.95	44	57.1

Recently, Shi *et al.* reported the synthesis of 4,5-disubstituted triazoles (Tria) as organic-anionic ligands, and formed an air- and moisture-stable $[Rh(cod)(Tria)]_2$ complex as a precatalyst for a Pauson-Khand reaction of 1,6-enyne (Scheme 7.17).[39] The best result was obtained with ligand Tria-2 in up to 88% isolated yield.

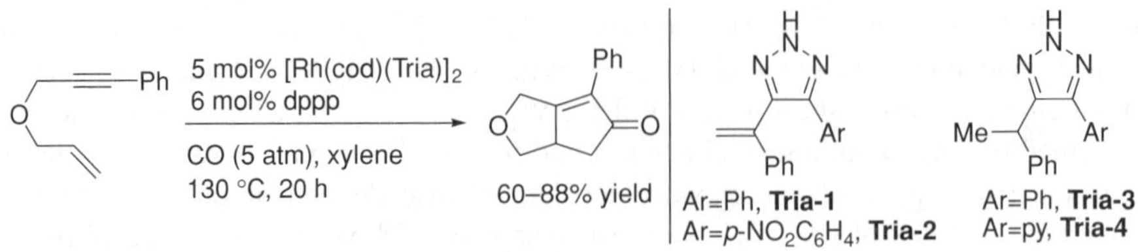

Scheme 7.17 *[Rh(Tria)dppp]-catalyzed PRK of 1,6-enyne.*

In 2002, Mukai[40] and Brummond[41] independently reported Rh-catalyzed Pauson–Khand-type cyclization involving allenyl moiety. By using the appropriate allenynes (allene-alkyne substrates), the construction of seven-membered ring adducts was produced in acceptable yields (Scheme 7.18).[40]

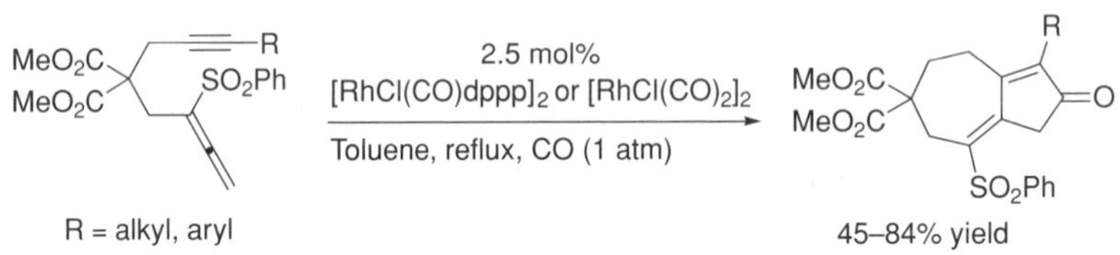

Scheme 7.18 *The Rh-catalyzed PKR of allenynes.*

In 2006, Mukai *et al.* further demonstrated a Rh-catalyzed intramolecular PK cycloaddition of allenenes.[42] The intramolecular coupling of an allene–alkene substrate, which is tethered by three or four atoms, afforded a bicyclic cyclopentenone with a 6-5 or 7-5 fused ring system, respectively, along with the isomerization of carbon–carbon double bond. In 2010, the same research group further demonstrated a $[\{RhCl(CO)dppp\}_2]$-catalyzed intramolecular PKR of bis(phenylsulfonylallene)s and afforded the desired bicyclic adducts in up to >99% isolated yield (Scheme 7.19).[43] They suggested that the two phenylsulfonyl groups on the allenyl functionalities were crucial to perform this carbonylative [2+2+1] cycloaddition smoothly.

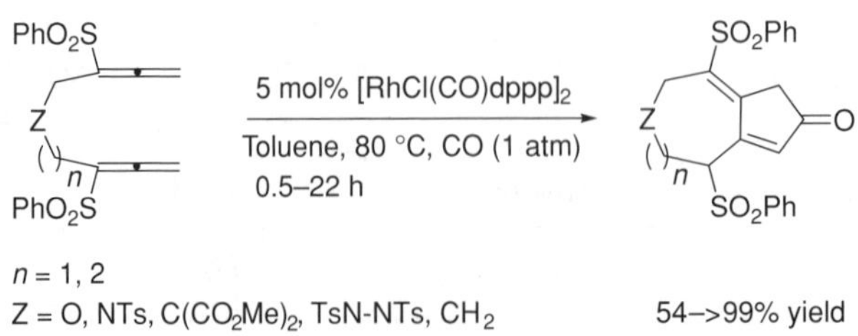

Scheme 7.19 *The Rh-catalyzed PKR of allene–alkene.*

In 2010, Ma *et al.* reported that the Rh-catalyzed PKR of 1,3,4-alkatrien-2-yldihydrofurans was achievable (Scheme 7.20).[44] The polyusubstituted dihydrofuryl cyclopententones were obtained in high yields. The product structures are commonly found in natural products and bioactive molecules.

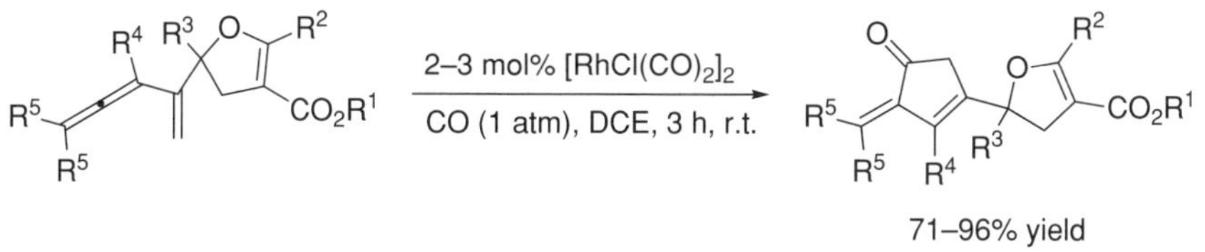

Scheme 7.20 *Rh-catalyzed PKR of allene-alkene.*

In 2010, Wang *et al.* illustrated a Rh-catalyzed [3+2+1] carbonylative cycloaddition of tethered ene-cyclopropene (Scheme 7.21).[45] The cycloadducts were obtained in up to 89% yield.

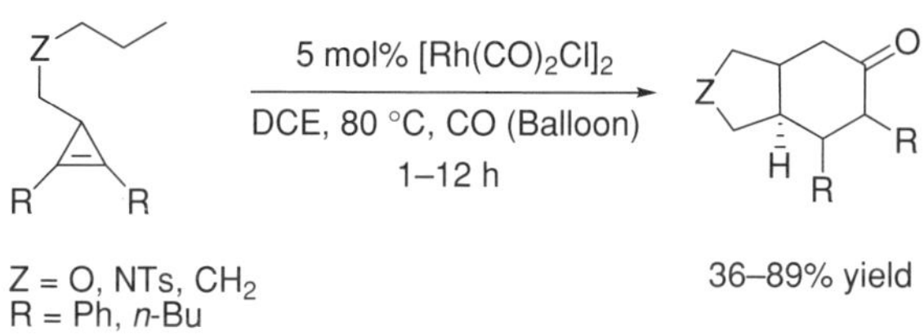

Scheme 7.21 *Rh-catalyzed PKR of tethered ene-cyclopropene.*

Yu and co-workers disclosed a Rh-catalyzed [(3+2)+1] homologous Pauson-Khand reaction of 1-yne/ene-vinylcyclopropanes in 2010 (Scheme 7.22).[46] The desired bicyclic cycloadducts were afforded in up to 92% isolated yield. The product derived from this method was further utilized for the synthesis of furanoid sesquiterpene natural product, α-agarofuran.

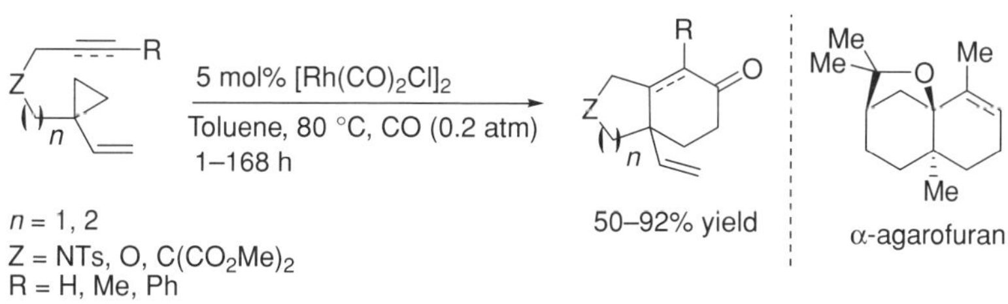

Scheme 7.22 *Rh-catalyzed PKR of 1-yne/ene-vinylcyclopropanes.*

Saito and co-workers first demonstrated the Rh-catalyzed intramolecular PKR of alkynyl-carbodiimide for the synthesis of pyrroloindolones and pyrrolopyrrolinones in up to 85%

isolated yield (Scheme 7.23).[47] In 2010, the same research group further extended this methodology to access potentially biologically active pyrroloquinolines with up to 85% yield, with the application of $[RhCl(CO)_2]_2$ as a precatalyst.[48]

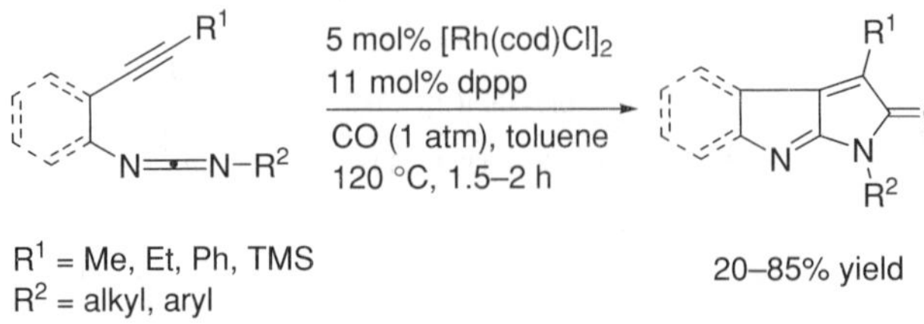

Scheme 7.23 *Rh-catalyzed PKR of alkynyl-carbodiimide.*

In 2010, a pioneering work was developed by Chung and co-workers showing that primary alcohols (such as cinnamyl alcohol) could be used as a CO surrogate for the Rh-catalyzed intramolecular PK-type reaction (Scheme 7.24).[49] The $[\{RhCl(CO)dppp\}_2]$ complex was found to effectively carry out this cascade reaction, in which alcohol sequentially conducted an oxidative dehydrogenation, decarbonylation and Pauson-Khand reaction in one-pot. Various enyne substrates including *C*-, *O*- and *N*-tethered enynes were examined. Up to 99% isolated yields were obtained.

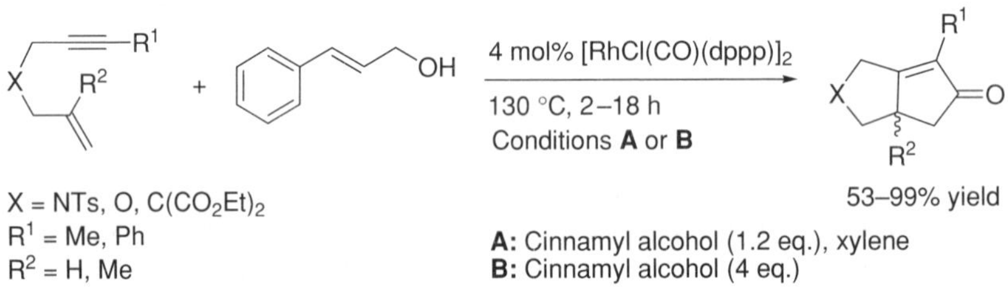

Scheme 7.24 *Rh-catalyzed PKR of 1,6-enynes with cinnamyl alcohol as CO source.*

In order to have a better picture for comparison between the rhodium catalysts, we have summarized the turn-over activity of each catalyst as shown in Table 7.3.

7.3 Iridium-Catalyzed Pauson-Khand-Type Cyclizations

In 2000, Shibata and co-workers described the first Ir-catalyzed enantioselective intra- and intermolecular Pauson-Khand-type reaction.[50] They used $[Ir(cod)Cl]_2$ and (*S*)-tol-BINAP as the catalyst system for Pauson-Khand-type transformation and afforded the corresponding cycloadducts in excellent ee, up to 98% (Scheme 7.25). Notably, the intermolecular product yields and/or enantioselectivities were further improved under reduced partial CO pressure (up to 93% ee at 0.2 CO atm).[51]

Table 7.3 *A summary of rhodium catalysts and their activities.*

Entry	Inter-PKR	Intra-PKR	Catalyst system	p(CO) [atm]	TON[a]	Ref.
1	X	√	$[Rh(dppp)_2Cl]$[b]	0	(11–20)	16
2	X	√	$[Rh(dppp)_2Cl]$[c]	0	(7–10)	15
3	X	√	$[Rh(COD)Cl]_2$ + (*S*)-P-Phos[c]	0	(8–15)	24
4	X	√	$[Rh(COD)Cl]_2$ + (*S*)-BisbenzodioxanPhos[c]	0	(10–16)	25
5	X	√	$[Rh(COD)Cl]_2$ + (*S*)-xyl-BINAP[d]	0	(1–7)	32
6	X	√	$[Rh(COD)Cl]_2$ + dppp[e]	0	(3–15)	37
7	X	√	$[Rh(COD)Cl]_2$ + dppp[f]	0	(3–7)	36
8	X	√	[Rh(COD)Biphemp]X[g] X = BF_4^-, PF_6^-, OTf^-	0	(16–25)	8
9	X	√	$[Rh(COD)Cl]_2$ + dppp + TPPTS + SDS[h]	0	(6–9)	20
10	X	√	$[Rh(COD)Cl]_2$[i]	5	(7–9)	6
11	√	√	$[Rh(CO)_2Cl]_2$	1	(4–6) (7–28)	5
12	X	√	$[Rh(CO)_2Cl]_2$ + (*R*)-SIPHOS	1	(5–12)	13
13	X	√	$[Rh(CO)(dppp)Cl]_2$[j]	1	(7–15)	38
14	X	√	$[Rh(CO)_2Cl]_2$ + (*S*)-BINAP	1	(6–16)	4
15	X	√	$[Rh(CO)_2Cl]_2$[k]	1	(3–10)	9

[a]TONs calculated with respect to the number of moles of Rh_1, TONs in parentheses are turnover numbers of intramolecular PKR
[b]Using pentafluorobenzaldehyde as CO source
[c]Using cinnamaldehyde as CO source under conventional heating
[d]Using benzylformate as CO source under conventional heating
[e]Using cinnamaldehyde as CO source under microwave irradiation
[f]Using p-chlorobenzylformate as CO source under microwave irradiation
[g]Using 4Å M.S. entrapped with CO
[h]Using formaldehyde as CO source in aqueous medium
[i]Rh catalyst entrapped by silica-sol gel
[j]Under an (Ar:CO) atmosphere, demonstrating the effects of CO pressure on product yields
[k]Under 1 atm (Ar:CO = 10:1) atmosphere

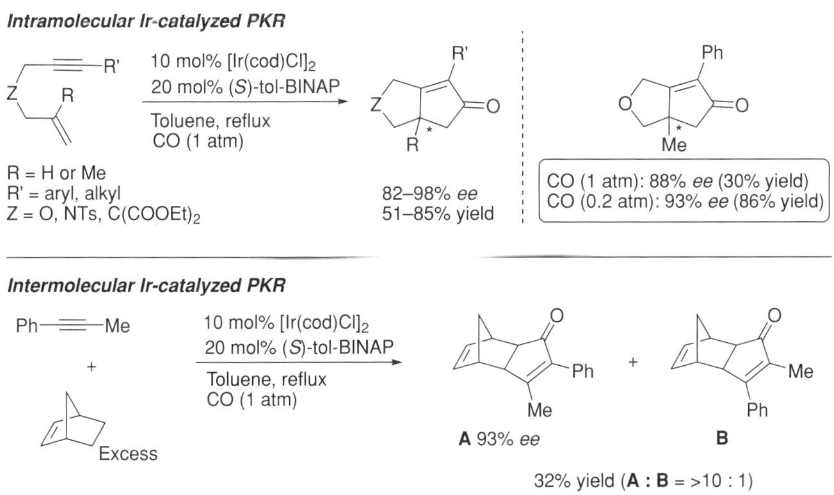

Scheme 7.25 *Asymmetric Ir-catalyzed intra- and intermolecular PKR.*

Jeong applied the iridium system in the desymmetrization of meso-dienynes.[52] A highly enantio- and diastereoselective Pauson-Khand-type reaction proceeded smoothly to afford vinyl-substituted bicyclic cyclopentenones with two stereogenic centers (Scheme 7.26).

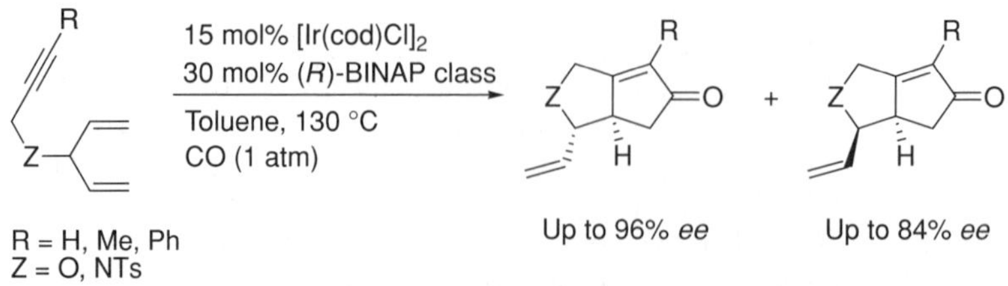

Scheme 7.26 *Ir-catalyzed desymmetrization of meso-dienynes.*

Shibata further revealed that the chiral Ir-complex derived from (*S*)-tol-BINAP ligand proved to be an efficient catalyst for decarbonylation of aldehyde and the corresponding metal carbonyl was utilized for PK cyclization (Scheme 7.27).[51] Though the product yield was generally moderate, the enantioselectivity was high (85–95%) which exceeded that of Rh-catalyzed PKR.[16] Independently, the Kwong group also reported a similar protocol to access a series of optically active bicyclic cyclopentenones by using an (*S*)-BINAP-iridium complex.[53]

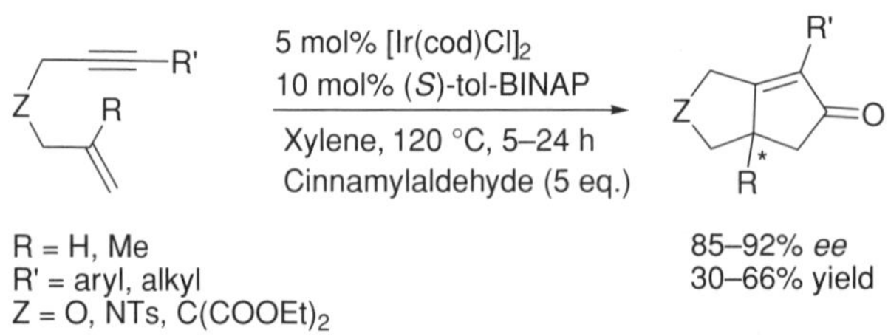

Scheme 7.27 *Asymmetric Ir-catalyzed PKR with aldehyde as CO source.*

Recently, Pfaltz reported a chiral cationic Ir(phox) complex for application in an intramolecular asymmetric Pauson-Khand-type reaction (Scheme 7.28).[54] Under optimized reaction conditions, the desired products were obtained with high yields and enantioselectivities. Lowering of the catalyst loading was possible for *O*-tethered 1,6-enyne substrates (91% ee with 1 mol% of Ir catalyst). The influences of CO pressure and counter anion were also studied. The CO pressure significantly affected the reaction rate rather than enantioselectivity. A higher product yield was obtained under a higher CO pressure. Moreover, the study of counter anion indicated that hexafluoroantimonate was found to be the most effective counter pair for the complex.

Apart from the carbonylative alkene–alkyne coupling, Shibata realized an alkyne–alkyne coupling with carbon monoxide effectively (Scheme 7.29).[55] They applied Vaska's complex $IrCl(CO)(PPh_3)_2$ to perform the intramolecular [2+2+1] cycloaddition of diyne. The

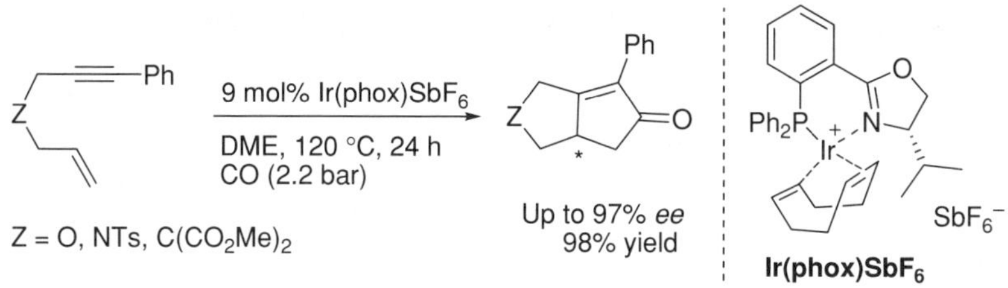

Scheme 7.28 *Asymmetric Ir(phox)SbF_6 catalyzed PKR of 1,6-enynes.*

desired cyclopentadienones were obtained in up to 99% isolated yield. Notably, they also reported that isocyanides could substitute CO and formed the corresponding iminocyclopentadienes by a $[Rh(cod)Cl]_2$ complex.

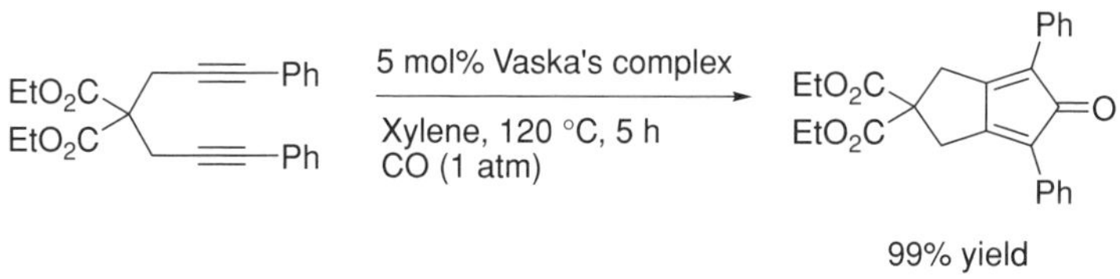

Scheme 7.29 *Vaska's complex-catalyzed coupling alkyne–alkyne with CO.*

To sum up, the Ir- and Rh-catalyzed Pauson-Khand-type reactions interestingly showed different regioselectivitives in allenyne cyclization. Allenyne possessing two terminal substituents on the allene was favoured to be catalyzed by an Ir-complex under a low partial pressure of carbon monoxide. The internal π-bond of the allene moiety was the major reaction site and the bicyclic cyclopentenone with an alkylidene substituent was obtained (Scheme 7.30).[56] In contrast, when $RhCl(CO)(PPh_3)_2$ was applied, instead of $IrCl(CO)(PPh_3)_2$ as the catalyst under the same reaction conditions, the reaction of external π-bond of allene moiety was mainly dominant. Further to these complementary investigations, the Shibata group also demonstrated the enantioselective version of this protocol by using a chiral BINAP ligand instead of triphenylphosphine.[57]

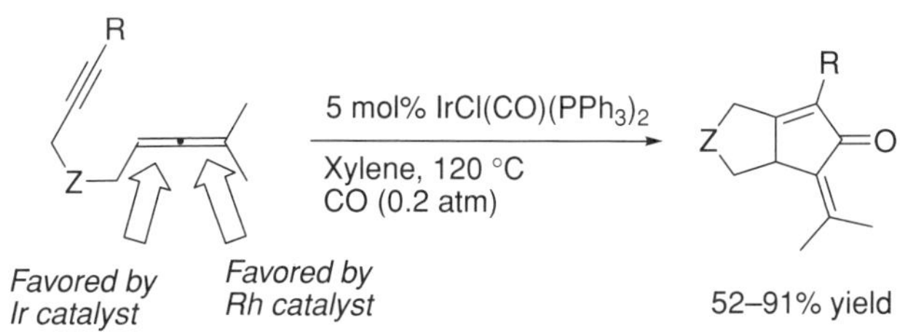

Scheme 7.30 *Regioselectivity of allenyne substrate in PKR governed by Ir-/ Rh-catalyst.*

7.4 Titanium-Catalyzed Pauson-Khand-Type Cyclizations

In 1996, Buchwald and co-workers reported a Ti-catalyzed intramolecular coupling of 1,6-enynes under an atmosphere of carbon monoxide, generating bicyclic cyclopentenones in good to excellent yields, up to 95% (Scheme 7.31).[58,59] In a different approach, Negishi *et al.* showed the stoichiometric Cp_2Zr-mediated bicyclic zirconacyclopentenes could be transformed into the corresponding cyclopentenones in the presence of CO in up to 65% isolated yield.[60]

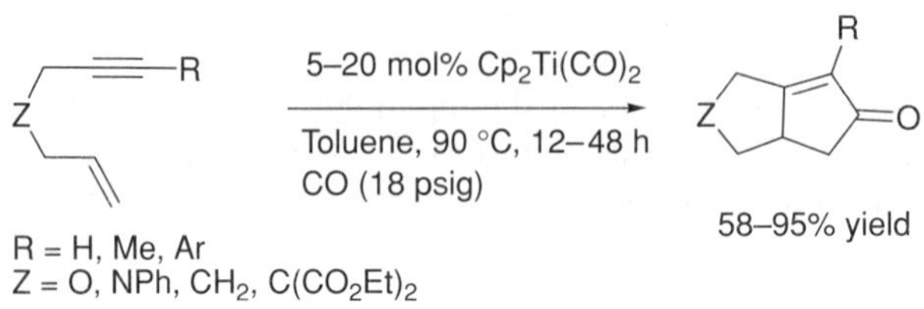

Scheme 7.31 *Ti-catalyzed PKR of 1,6-enynes under CO atmosphere.*

A pioneering establishment of a catalytic enantioselective Pauson-Khand reaction was achieved by Buchwald in 1996.[61,62] In fact, an effective enantioselective reaction under a CO atmosphere using a transition metal catalyst with a chiral ligand was rather problematic at that time, because the chiral ligand could be easily dissociated from the metal center due to the excess amount of strong CO donor, and thus part of the reaction might proceed under an achiral environment. Notably, Buchwald tackled these difficulties by employing a chiral Ti-complex in which the metal center and chiral scaffold were bridged with a σ-bond (Scheme 7.32). The 1,6-enynes were transformed into desired cyclopentenones in good to excellent enantioselectivity. However, this PK reaction must be performed in a glovebox because the low valent Ti-complex containing a Ti-C σ-bond is highly sensitive to air and moisture.

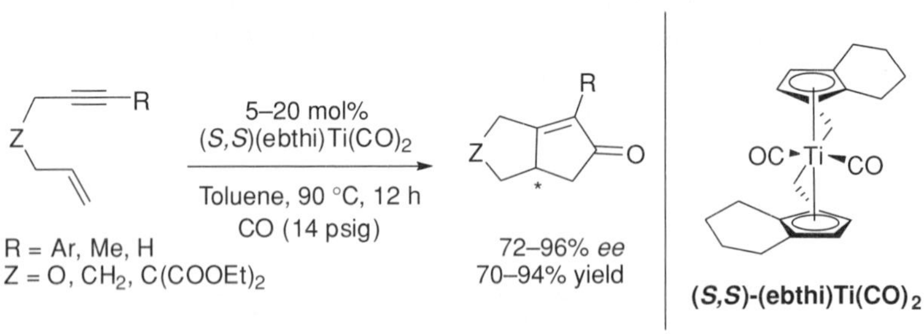

Scheme 7.32 *Enantioselective Ti-catalyzed PKR of 1,6-enynes.*

In addition to *C*- and *O*-tethered enynes, Buchwald further disclosed the use of a modified Ti-catalyst for the Pauson-Khand-type cyclization of *N*-tethered enynes

(Scheme 7.33).[63] The desired nitrogen-containing cyclopentenone was obtained in good yield and high enantioselectivity (up to 95% ee). Although various kinds of 1,6- and 1,7-enynes were successfully converted to their corresponding products, the application of dicyclopentadienyl- titanium dicarbonyl complex was limited in sterically hindered alkynes and/or olefins. Thereafter, Buchwald and co-workers developed new monocyclopentadienyltitanium aryloxide complexes to apply in the intramolecular Pauson-Khand reaction (Scheme 7.34).[64]

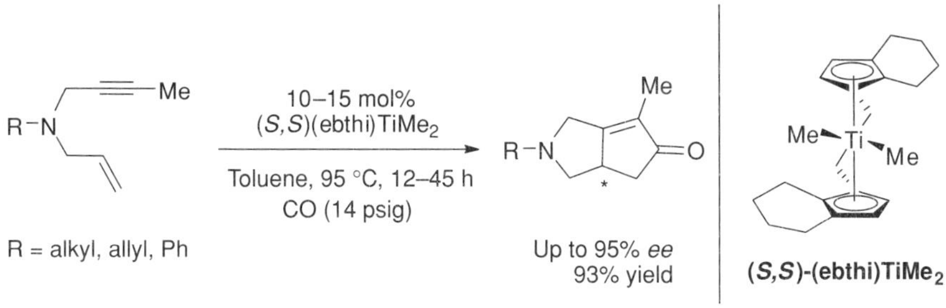

Scheme 7.33 *Enantioselective Ti-catalyzed PKR of* N*-tethered 1,6-enynes.*

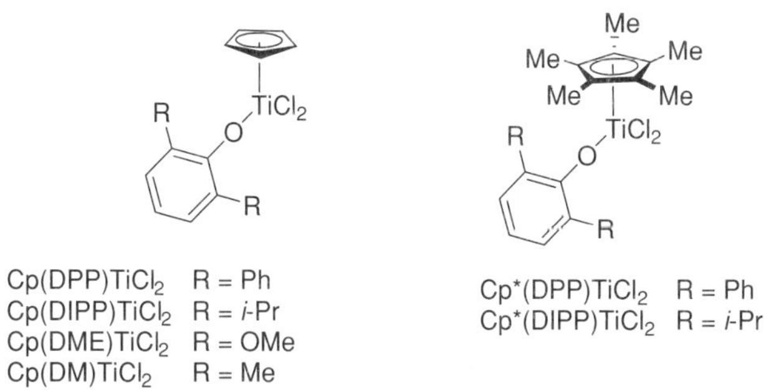

Scheme 7.34 *A series of monocyclopentadienyltitanium aryloxide complexes.*

They found that pentamethylmonocyclopentadienyl complex Cp*(DPP)$TiCl_2$ failed to react with enyne to form metallocycle while the monocyclopentadienyl complex Cp(DPP)$TiCl_2$ was feasible (Scheme 7.34). The Buchwald group examined a series of these complexes for the formation of metallocycles. Interestingly their catalytic activities were specific toward different enynes. Complex Cp(DPP)$TiCl_2$ was found to mediate cyclocarbonylation of trimethylsilyl-substituted enynes effectively. Employing Cp(DIPP)$TiCl_2$ assisted the cyclocarbonylation of trisubsituted olefins while complex Cp(DME)$TiCl_2$ mediated the cyclocarbonylation of phenyl-substituted enynes. Complex Cp(DM)$TiCl_2$ was particularly suitable for mediating the diene cycloisomerization.

7.5 Ruthenium-Catalyzed Pauson-Khand-Type Cyclizations

In 1997, Mitsudo[65] and Murai[66] independently disclosed Ru-complexes that were effective in promoting intramolecular Pauson-Khand-type cyclizations. Under a high atmosphere of carbon monoxide and at a high temperature, the Ru-catalyzed PK reaction was successfully achieved and gave the desired bicyclic products in high yields (Scheme 7.35). In contrast, the complex of [RuCl(cod)Cp*], which was generally an effective catalyst for [2+2] cycloaddition of norbornenes with alkynes,[67] was totally ineffective in this carbonylative [2+2+1] reaction.

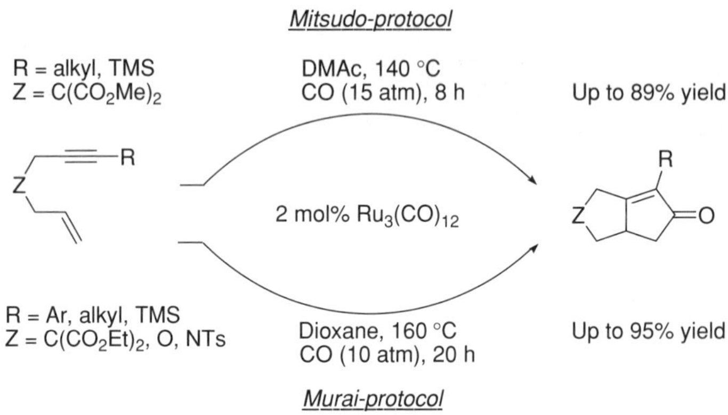

Scheme 7.35 *Ru-catalyzed PKR of 1,6-enynes.*

In 2006, Kondo *e. al.* demonstrated the first Ru-catalyzed intermolecular [2+2+1] co-cyclization of isocyanates, alkyne and CO affording the corresponding maleimides in up to 98% isolated yield.[68] This process allows a rapid and atom-economical protocol to generate a series of unsymmetrical and polysubstituted maleimides in one-pot (Scheme 7.36).

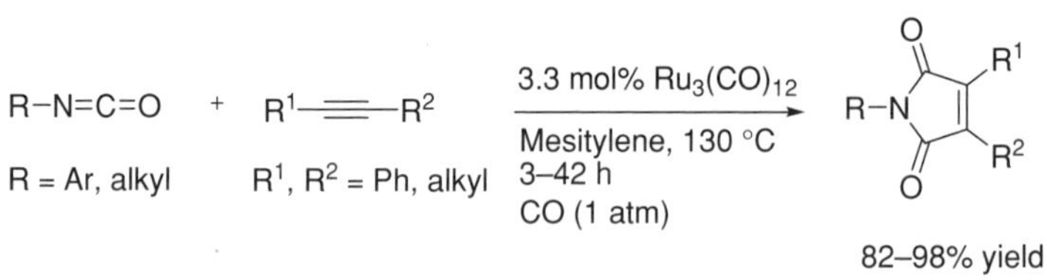

Scheme 7.36 *Ru-catalyzed intermolecular [2+2+1] co-cyclication.*

Kondo and co-workers also postulated a mechanism for this newly developed co-cyclization (Scheme 7.37). With the evidence of^{13}C-labelled experiments to hand, they suggested the formation of azaruthenacyclopentenones was as a mixture of regioisomers. The formation of α-phenylsubstituted intermediate was verified to be more favourable than α-methylsubstituted intermediate, as judged by^{13}C NNE spectra. The insertion of CO into a Ru-C(sp^2) bond rather than Ru-N bond predominantly occurred to give azaruthenacyclohexenediones, followed by a reductive elimination to release the desired maleimides and regeneration of the Ru-catalyst.

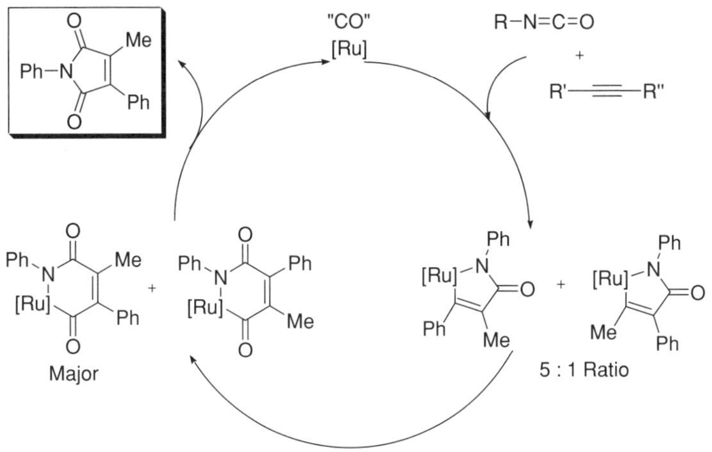

Scheme 7.37 *Proposed mechanism of Ru-catalyzed co-cyclization.*

In 2008, Mastrorilli and Braunstein reported tri- and tetra-nuclear heterometallic Ru clusters with a formula of $NEt_4[RuCo_3(CO)_{12}]$ which was effective as a precatalyst for an intramolecular Pauson-Khand-type reaction of diethyl (allylpropargyl)malonate.[69] The best result was obtained in the presence of 2 mol% precatalyst under 8 bar of CO at 70 °C. The desired cyclopentenones were afforded in up to 93% isolated yield.

7.6 Nickel- and Palladium-Catalyzed Pauson-Khand-Type Cyclizations

The nickel-assisted Pauson-Khand-type [2+2+1] cyclization of enynes was first achieved by Buchwald and co-workers in 1996.[70] They demonstrated that a stoichiometric amount of Ni(0)-complex was required to convert an enyne and isocyanide into the corresponding iminocyclopentene (Scheme 7.38). Turning the supporting ligand from phosphine to highly electron-donating bidentate nitrogen ligands, in which *rac*-diimine **L** was superior in this transformation gave the target product in 80% yield.

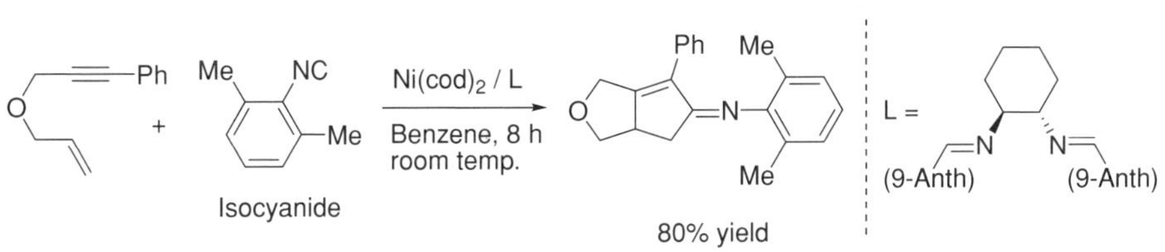

Scheme 7.38 *Ni-assisted cyclization of 1,6-enyne and isocyanide.*

Buchwald mentioned that the Ni-mediated PK reaction was a failure in a catalytic manner, because of the formation of nickel-isocyanide complexes, which prevent further coordination, and hence cyclization of enyne. Upon using other isocyanide sources, such as trialkylsilyl cyanides, the coupling transformation was effectively accomplished catalytically. The bicyclic cyclopentenones (after acid hydrolysis) were obtained in up to 85% isolated yield (Scheme 7.39).[70] They hypothesized that trialkysilyl cyanides generate a sufficient yet relatively low equilibrium concentration of the trialkylsilyl isocyanide isomer ($K \approx 0.01$) to effect the PK coupling reaction.[71]

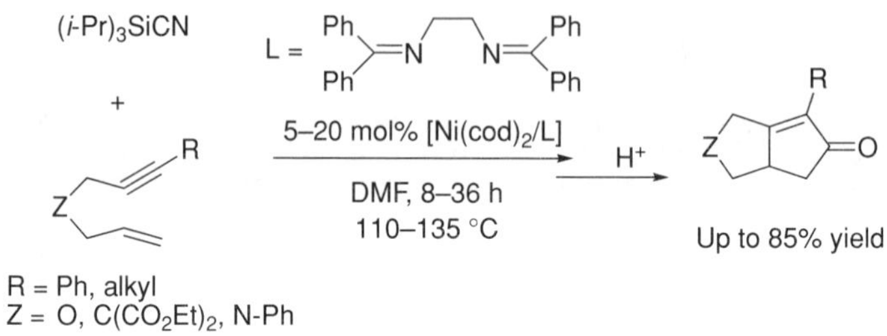

Scheme 7.39 *Ni-catalyzed cyclization of 1,6-enynes.*

In 2003, Grigg *et al.* reported the $Pd(PPh_3)_2$-catalyzed intramolecular cycloaddition of 1,6-heptadiynes with CO (Scheme 7.40).[72] The corresponding bicyclic octadienones were furnished in up to 74% yield

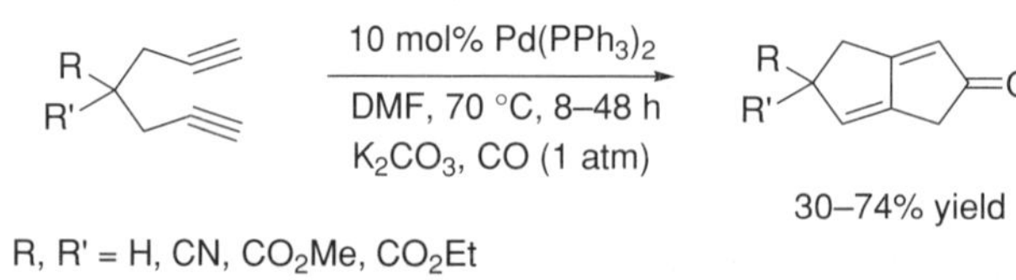

Scheme 7.40 *Catalytic Pd-catalyzed PKR of 1,6-diynes with CO.*

In 2005, the first successful Pd-catalyzed intramolecular Pauson-Khand-type cyclization was achieved by Chen/Yang and co-workers.[73] They showed that the Pd-complexes derived from the *N,N*-tetrasubstituted thiourea ligands with $PdCl_2$ in 1:1 ratio was critical for the success of this Pd-catalyzed PKR. Ligand tmtu was the best ligand of choice to give the desired products in up to 96% yield (Scheme 7.41). They pointed out that *C*- and *O*-tethered 1,6-enynes as substrates were ineffective in this reaction, while the *N*-tethered enynes were efficiently transformed to corresponding bicyclic adducts in good yields. In addition, the *N*-tethered substrates containing an electron-deficient group on an alkyne terminus gave better results than other substrates.

Later the same research group mentioned that the addition of LiCl as an additive into the Pd-catalyzed intramolecular PK reaction could significantly enhance the corresponding product yields (Scheme 7.42).[74] This catalytic system was found to be more general than the previously developed Pd catalyst system.[73]

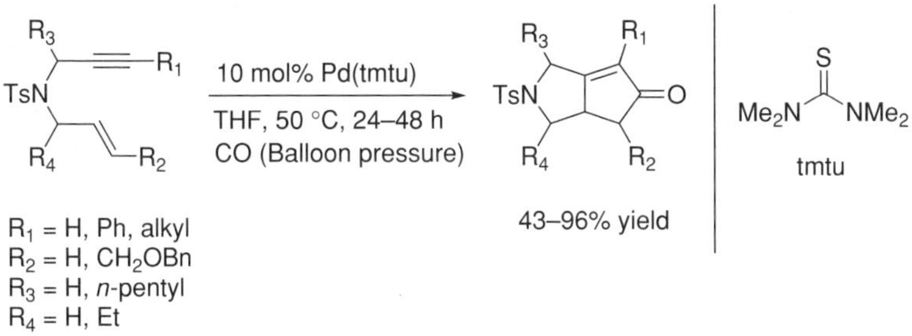

Scheme 7.41 *Pd(tmtu)-catalyzed PKR of* N*-tethered 1,6-enynes.*

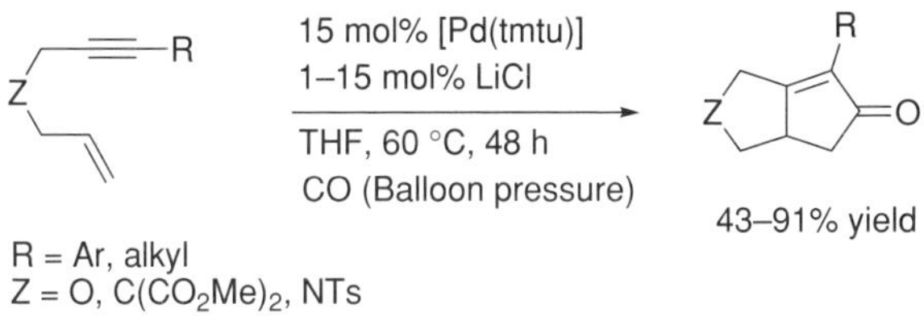

Scheme 7.42 *Pd-catalyzed PKR with LiCl additive.*

In 2009, Wiest/Yang/Wu and co-workers proposed a mechanism for a Pd-catalyzed intramolecular Pauson-Khand-type reaction based on both DFT calculations and experimental studies.[75] The first step is suggested to involve the *cis*-halometallation of the alkyne moiety, and is then followed by sequential alkene and carbonyl insertions. The rate-determining step is an intramolecular C-Cl oxidative addition, in generating Pd(IV) species (Scheme 7.43). Finally, a reductive elimination as usual gives the desired product Table 7.4.

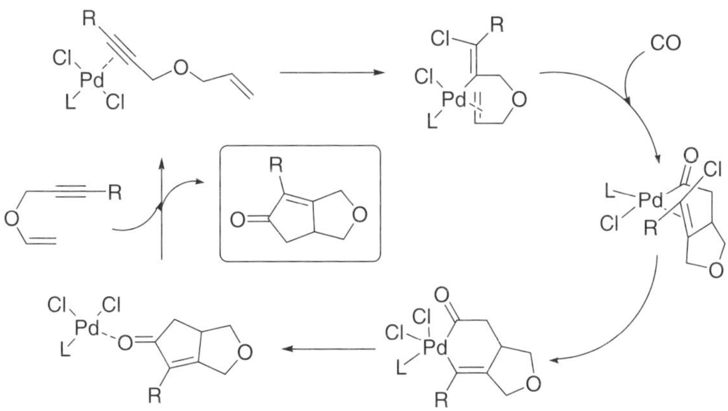

Scheme 7.43 *Proposed mechanism of Pd-catalyzed 1,6-enynes PKR.*

Table 7.4 *A summary of Ti, Ru, Ni, Pd, Ir catalysts and their activities.*

Entry	Inter-PKR	Intra-PKR	Catalyst system	p(CO) [atm]	TON[a]	Ref.
1	X	√	$[Cp_2Ti(CO)_2]$	1.22	(9–19)	58,59
2	X	√	$[Ru_3(CO)_{12}]$	10–15	(16–7)	66
3	X	√	$[RuCo_3(CO)_{12}]X$ X = NEt_4, bmim, H	8	(32–46)	69
4	X	√	$Ni(COD)_2$ + L[b]	0	(8–17)	71
			L = 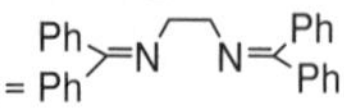			
5	X	√	$PdCl_2$-tmtu	1	(2–4)	72
			tmtu = 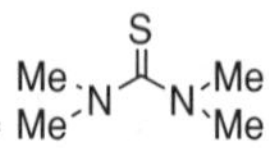			
6	X	√	$PdCl_2$-tmtu + LiCl	1	(3–30)	74
7	X	√	$[Ir(COD)Cl]_2$ + Phox	2.2	(9–11)	54
			Phox = 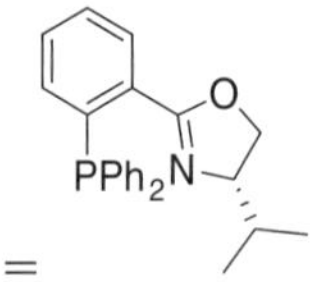			
8	X	√	$[Ir(COD)Cl]_2$ + (*S*)-tol-BIANP[c]	1	(2–9)	50
9	X	√	$[Ir(COD)Cl]_2$ + (*S*)-tol-BIANP[d]	0	(2–5)	51
10	X	√	$[Ir(COD)Cl]_2$ + (*S*)-BINAP[e]	0	(2–7)	53

[a]TONs calculated with respect to the number of moles of metals, TONs in parentheses are turnover numbers of intramolecular PKR
[b]Insertion of 2,6-dimethlphenyl isocyanide followed by hydrolysis to obtain desired cyclopentenones
[c]Under 1 atm (AR:CO = 8:2) atmosphere
[d]Using cinnamaldehyde as CO source
[e]Using nonylaldehyde as CO source

7.7 Tandem Reactions and Miscellaneous (other than Co complex)

The Pauson-Khand-type reactions have been successfully developed as effective methods to construct cyclopentenone frameworks for generating many pharmaceutically useful and scientifically interesting molecules.[76,77] Based on the principle of atom economy, tandem reactions has become one of the hot topics of recent decades. In 2000, Jeong and co-workers showed a bimetallic catalyst comprising of $[Pd_2(dba)_3(CHCl_3)]$ and $[RhCl(CO)(dppp)]_2$. The alkyne malonate and allyl acetate were converted into the desired cyclopentenones in a tandem manner, in up to 92% yield (Scheme 7.44).[78]

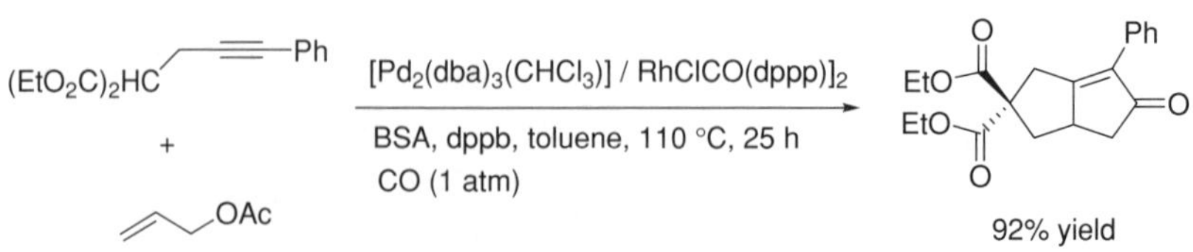

Scheme 7.44 *A bimetallic catalyst system for PKR.*

Evans and co-workers reported a regio- and diastereoselective tandem Rh-catalyzed alkylation/ Pauson-Khand annulation reaction. They employed $[RhCl(CO)dppp]_2$ as a catalyst to reacted allylic carbonate with the sodium salt of α-branched malonate under a carbon monoxide atmosphere to furnish the desired cyclopentenones in good yields and diastereoselectivities (Scheme 7.45).[79]

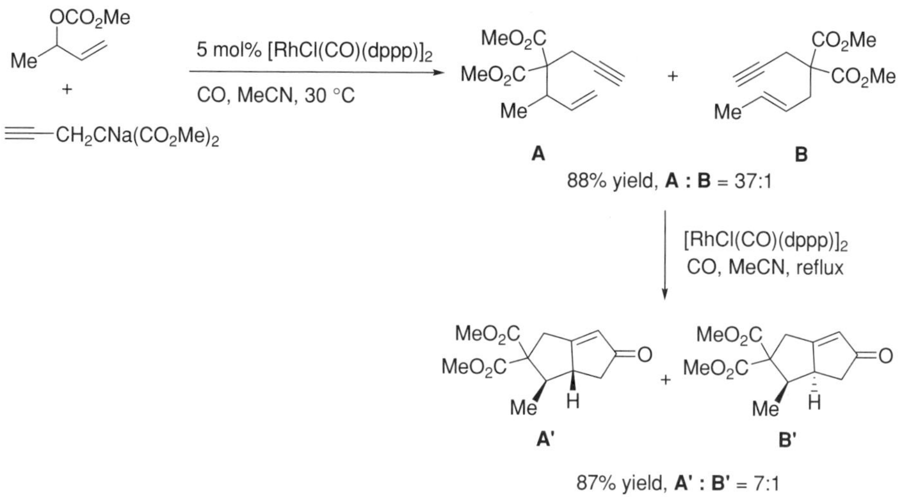

Scheme 7.45 *Rh-catalyzed tandem alkylation/ PK reaction.*

The Chung group attempted to utilize a combined chiral Pd(II) and Co/C catalytic system for the asymmetric allylic alkylation and cascade carbonylative cyclization (Scheme 7.46).[80]

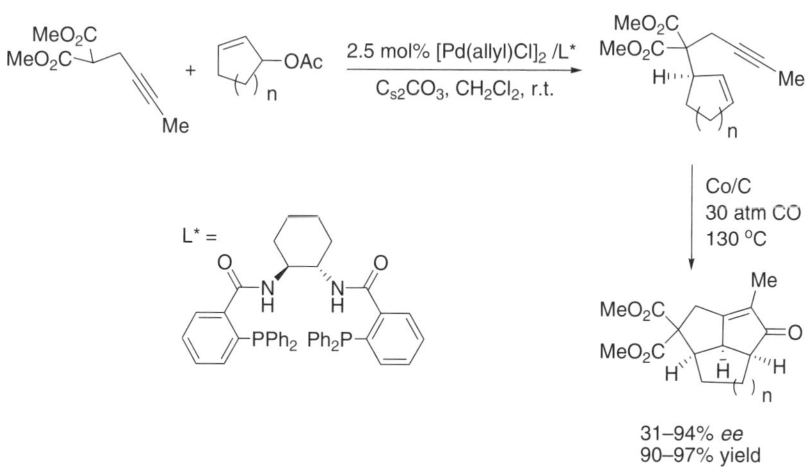

Scheme 7.46 *Tandem alkylation-carbonylative cyclization.*

Surprisingly tricyclic compounds can also be obtained in good yields and moderate ee. Afterwards the same research group reported a Pauson-Khand reaction of α,β-unsaturated aldehydes (such as acrolein) with alkynes in the presence of a bimetallic Co_2Rh_2 catalyst to afford the desired substituted cyclopentenones (Scheme 7.47).[81]

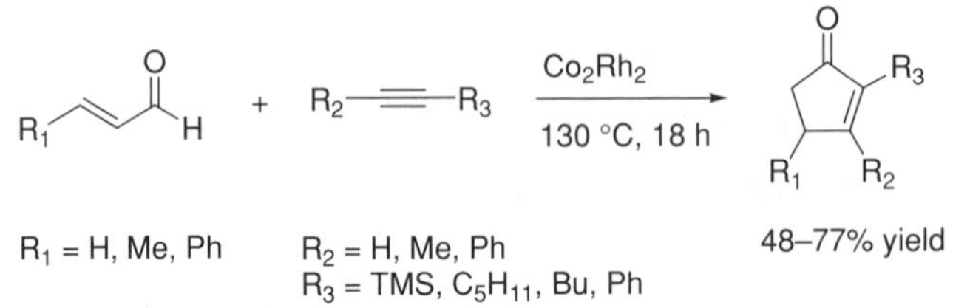

Scheme 7.47 *Co_2Rh_2-catalyzed PK reaction of α,β-unsaturated aldehydes with alkynes.*

7.8 Conclusion

This chapter illustrates the current evolution of the catalytic Pauson-Khand-type reaction and the recent catalyst advances for these reactions. Although some significant improvements have been achieved in intramolecular PK reaction, a path to the perfect intermolecular PK reaction, especially with the elements of environmentally friendly, enantioselective and low catalyst loading, is still a long way away. We believe the summary of the TON tables of each type of catalysts provide useful information for readers to choose an appropriate catalyst system in this reaction type for genuine synthetic applications. Further ligand (chiral) designs are expected to be more practical this decade.

Acknowledgements

We thank the Research Grants Council of Hong Kong (GRF: PolyU 5014/10P), State Key Laboratory of Chirosciences, European Commission (CATAFLU.OR), PolyU internal funding (A-PA8R, G-U243), postdoctoral fellowship (G-YX1L) and the University Grants Committee Areas of Excellence Scheme (AoE/P-10/01) for financial support.

References

[1] B. M. Trost, *Science*, 1991, **254**, 1471.
[2] For initial reports from Pauson and Khand, see: a) I. U. Khand, G. R. Knox, P. L. Pauson, W. E. Watts, *J. Chem. Soc., Chem. Commun.* 1971, 36; b) I. U. Khand, G. R. Knox, P. L. Pauson, W. E. Watts, M. I. Foreman, *J. Chem. Soc., Perkin Trans.* **1** 1973, 977.
[3] For reviews on PKR, see: a) H.-W. Frühauf, *Chem. Rev.* 1997, **97**, 523; b) O. Geis, H.-G. Schmalz, *Angew. Chem.* 1998, **110**, 955; *Angew. Chem. Int. Ed.* 1998, **37**, 911; c) N. Jeong, In *Transition Metals In Organic Synthesis: Building Blocks and Fine Chemicals*, M. Beller, C. Bolm, Eds., Wiley-VCH: Weinheim, 1998, Vol. 1,

p.560; d) Y. K. Chung, *Coord. Chem. Rev.* 1999, **188**, 297; e) S. L. Buchwald, F. A. Hicks, In *Comprehensive Asymmetric Catalysis*, Vol. II, E. N. Jacobsen, A. Pfaltz, H. Yamamoto, Eds., Springer, 1999, p.491; f) K. M. Brummond, J. L. Kent, *Tetrahedron* 2000, **56**, 3263; g) A. J. Fletcher, S. D. R.Christie, *J. Chem. Soc., Perkin Trans.* **1** 2000, 1657; h) M. R. Rivero, J. Adrio, J. C. Carretero, *Eur. J. Org. Chem.* **7**, 2002, 2881; i) B. E. Hanson, *Comments Inorg. Chem.* **2002**, *23*, 289; j) L. V. R. Boñaga, M. E. Krafft, *Tetrahedron* **2004**, *60*, 9795; k) D. Strübing, M. Beller, In *Transition Metals In Organic Synthesis: Building Blocks and Fine Chemicals*, M. Beller, C. Bolm, Eds., 2nd Ed. Wiley-VCH: Weinheim, **2004**, Vol. 1, p.619; l) S. E. Gibson, S. E. Lewis, N. Mainolfi, *J. Organomet. Chem.*, **2004**, *689*, 3873; m) K. H. Park, Y. K. Chung, *Synlett* **2005**, 545; n) S. E. Gibson, N. Mainolfi, *Angew. Chem. Int. Ed.* **2005**, *44*, 3022; o) S. E. Gibson, A. Stevenazzi, *Angew. Chem. Int. Ed.* **2003**, *42*, 1800; p) D. Strübing, M. Beller, in *Topic in Organometallic Chemistry: Catalytic Carbonylation Reactions*, M. Beller, Ed., Springer-Verlag: Berlin Heideberg, **2006**, Vol. 18, pp 165; q) J. Perez-Castells, in *Topics in Organometallic Chemistry: Metal Catalyzed Cascade Reactions*, Springer-Verlag: Berlin Heideberg, 2006, Vol. 19, pp 207; r) T. Shibata, *Adv. Synth. Catal.* 2006, **348**, 2328; s) N. Jeong, in *Modern Rhodium-Catalyzed Organic Reactions*, P. A. Evans, Ed. Wiley-VCH: Weinheim, 2005; pp 215 t) H. W. Lee, F. Y. Kwong, Eur. J. Org. Chem. 2010, 789.

[4] N. Jeong, B. K. Sung, Y. K. Choi, *J. Am. Chem. Soc.* 2000, **122**, 6771.

[5] T. Kobayashi, Y. Koga, K. Narasaka, *J. Organomet. Chem.* 2001, **624**, 73.

[6] K. H. Park, S. U. Son, Y. K. Chung, *Tetrahedron Lett.* 2003, **44**, 2827.

[7] W. H. Suh, M. Choi, S. I. Lee, Y. K. Chung, *Synthesis*, 2003, **14**, 2169.

[8] T. M. Schmid, G. Consiglio, *Chem. Commun.* 2004, **20**, 2318.

[9] D. E. Kim, I. S. Kim, V. Ratovelomanana-Vidal, J.-P. Genet, N. Jeong, *J. Org. Chem.* 2008, **73**, 7985.

[10] Y. H. Choi, J. Kwak, N. Jeong, *Tetrahedron Lett.* 2009, **50**, 6068.

[11] D. E. Kim, V. Ratovelomanan-Vidal, N. Jeong, *Adv. Synth. Catal.* 2010, **352**, 2032.

[12] D. E. Kim, B. H. Lee, M. Rajagopalasarma, J.-P. Genet, V. Ratovelomanana-Vidal, N. Jeong, *Adv. Synth. Catal.* 2008, **350**, 2695.

[13] B.-M. Fan, J.-H. Xie, S. Li, Y.-Q. Tu, Q.-L. Zhou, *Adv. Synth. Catal.* 2005, **347**, 759.

[14] For a general overview, see: H. M. Colquhoun, D. J. Thompson, M. V. Twigg, *Carbonylation*; Plenum Press: New York, 1991.

[15] T. Morimoto, K. Fuji, K. Tsutsumi, K. Kakiuchi, *J. Am. Chem. Soc.* 2002, **124**, 3806.

[16] T. Shibata, N. Toshida, K. Takagi, *Org. Lett.* 2002, **4**, 1619.

[17] J. Tsuji, In *Organic Synthesis via Metal Carbonyls*, P. Wender, P. Pino, Eds.; Wiley: New York, 1977, Vol. 2.

[18] For a recent review, see: T. Morimoto, K. Kakiuchi, *Angew. Chem. Int. Ed.* 2004, **43**, 5580.

[19] T. Shibata, N. Toshida, K. Takagi, *J. Org. Chem.* 2002, **67**, 7446.

[20] K. Fuji, T. Morimoto, K. Tsutsumi, K. Kakiuchi, *Angew. Chem. Int. Ed.* 2003, **42**, 2409.

[21] For a review describing micelles, see: U. M. Lindstrom, *Chem. Rev.* 2002, **102**, 2751.

[22] K. Fuji, T. Morimoto, K. Tsutsumi, K. Kakiuchi, *Tetrahedron Lett.* 2004, **45**, 9163.

[23] a) C. J. Li, T. H. Chan, *Organic Reactions in Aqueous Media*, John Wiley & Sons: New York, 1997; b) B. Cornils, W. A. Herrmann, Eds., *Aqueous-Phase Organometallic Catalysis*, Wiley-VCH: Weinheim, 1998; c) U. M. Lindström, *Chem. Rev.* 2002, **102**, 2751.

[24] F. Y. Kwong, Y. M. Li, W. H. Lam, L. Qiu, H. W. Lee, C. H. Yeung, K. S. Chan, A. S. C. Chan, *Chem. Eur. J.* 2005, **11**, 3872.

[25] F. Y. Kwong, H. W. Lee, L. Qiu, W. H. Lam, Y. M. Li, H. L. Kwong, A. S. C. Chan, *Adv. Synth. Catal.* 2005, **347**, 1750.

[26] For recent selected references of electronic effects in asymmetric catalysis, a) E. N. Jacobsen, W. Zhang, M. L. Güler, *J. Am. Chem. Soc.* 1991, **113**, 6703; b) H. C. Zhang, F. Xue, T. C. W. Mak, K. S. Chan, *J. Org. Chem.* 1996, **61**, 8002; c) M. M. C. Lo, G. C. Fu, *J. Am. Chem. Soc.* 1998, **120**, 10270; d) H. Doucet, E. Fernandez, T. P. Layzell, J. M. Brown, *Chem. Eur. J.* 1999, **5**, 1320; e) F. Y. Kwong, Q. Yang, T. C. W. Mak, A. S. C. Chan, K. S. Chan, *J. Org. Chem.* 2002, **67**, 2769; g) F. L. Lam, T. T. L. Au-Yeung, F. Y. Kwong, K. Y. Wong, A. S. C. Chan, *Angew. Chem. Int. Ed.* 2008, **47**, 1280.

[27] For a review of Hammett study, see: a) C. Hansch, A. Leo, R. W. Taft, *Chem. Rev.* 1991, **91**, 165; b) A. J. Gordon, R. A. Ford, *The Chemist's Companion* John Wiley: New York, **1972.**

[28] For evidences for the role of dπ-pπ(bonding in rhodium-mediated reaction, see: a) K. Burgess, W. A. van der Donk, M. B. Jarstfer, M. J. Ohlmeyer, *J. Am. Chem. Soc.* 1991, **113**, 6139. For models that assume the complexation of the alkene/alkyne (π-interaction) to the metal-center is the rate stereochemical-determining step of the reaction, see: b) W. R. Moser, D. W. Slocum, Eds. *Homogeneous Transition Metal Catalyzed Reactions*, Advances in Chemistry Series 230, American Chemical Society: Washington, D. C., 1992, pp 163.

[29] D. E. Kim, C. Choi, I. S. Kim, S. Jeulin, V. Ratovelomanana-Vidal, J. P. Genet, N. Jeong, *Synthesis* 2006, **23**, 4053.

[30] D. E. Kim, C. Choi, I. S. Kim, S. Jeulin, V. Ratovelomanana-Vidal, J. P. Genet, N. Jeong, *Adv. Synth. Catal.* 2007, **349**, 1999.

[31] D. E. Kim, J. Kwak, I. S. Kim, N. Jeong, *Adv. Synth. Catal.* 2009, **351**, 97.

[32] H. W. Lee, A. S. C. Chan, F. Y. Kwong, *Chem. Commun.* 2007, **25**, 2633.

[33] D. Adam, *Nature* 2003, **421**, 571.

[34] a) R. Gedye, F. Smith, K. Westaway, H. Ali, L. Baldisera, L. Laberge, J. Rousell, *Tetrahedron Lett.* 1986, **27**, 279; b) R. J. Giguere, T. L. Bray, S. M. Ducan, G. Majetich, *Tetrahedron Lett.* 1986, **27**, 4945.

[35] For recent reviews, see: a) C. O. Kappe, *Angew. Chem. Int. Ed.* 2004, **43**, 6250; b) C. O. Kappe, D. Dallinger, *Nature Rev. Drug Discov.* 2006, **5**, 51.

[36] H. W. Lee, F. Y. Kwong, A. S. C. Chan, *Synlett.* 2008, **10**, 1553.

[37] H. W. Lee, L. N. Lee, A. S. C. Chan, F. Y. Kwong, *Eur. J. Org. Chem.* 2008, **19**, 3403.

[38] H. Wang, J. R. Sawyer, P. A. Evans, M.-H. Baik, *Angew. Chem. Int. Ed.* 2008, **47**, 342.

[39] H. Duan, S. Sengupta, J. L. Petersen, X. Shi, *Organometallics* 2009, **28**, 2352.

[40] C. Mukai, I. Nomura, K. Yamanishi, M. Hanaoka, *Org. Lett.* 2002, **4**, 1755.

[41] K. M. Brummond, H. Chen, K. D. Fisher, A. D. Kerekes, B. Rickards, P. C. Sill, S. J. Geib, *Org. Lett.* 2002, **4**, 1931.

[42] F. Inagaki, C. Mukai, *Org. Lett.* 2006, **8**, 1217.
[43] T. Kawamura, F. Inagaki, S. Narita, Y. Takahashi, S. Hirata, S. Kitagaki, C. Mukai, *Chem. Eur. J.* 2010, **16**, 5173.
[44] W. Shu, S. Ma, *Tetrahedron*, 2010, **66**, 2869.
[45] C. Li, H. Zhang, J. Feng, Y. Zhang, J. Wang, *Org. Lett.* 2010, **12**, 3082.
[46] L. Jiao, M. Lin, L.-G. Zhuo, Z.-X. Yu, *Org. Lett.* 2010, **12**, 2528.
[47] T. Saito, K. Sugizaki, T. Otani, T. Suyama, *Org. Lett.* 2007, **9**, 1239.
[48] T. Saito, N. Furukawa, T. Otani, *Org. Biomol. Chem.* 2010, **8**, 1126.
[49] J. H. Park, Y. Cho, Y. K. Chung, *Angew. Chem. Int. Ed.* 2010, **49**, 5138
[50] T. Shibata, K. Takagi, *J. Am. Chem. Soc.* 2000, **122**, 9852.
[51] T. Shibata, N. Toshida, M.Yamasaki, S. Maekawa, K. Takagi, *Tetrahedron* 2005, **61**, 9974.
[52] N. Jeong, D. H. Kim, J. H. Choi, *Chem. Commun.* 2004, **9**, 1134.
[53] F. Y. Kwong, H. W. Lee, W. H. Lam, L. Qiu, A. S. C. Chan, *Tetrahedron: Asymmetry*, 2006, **17**, 1238.
[54] Z.-L. Lu, E. Neumann, A. Pfaltz, *Eur. J. Org. Chem.* 2007, **25**, 4189.
[55] T. Shibata, K. Yamashita, E. Katayama, K. Takagi, *Tetrahedron* 2002, **58**, 8661.
[56] T. Shibata, S. Kadowaki, M. Hirase, K. Takagi, *Synlett* 2003, **4**, 573.
[57] Unpublished results reported by Shibata and co-workers, cited in T. Shibata, *Adv. Synth. Catal.* 2006, **348**, 2328.
[58] F. A. Hicks, S. C. Berk, S. L. Buchwald, *J. Am. Chem. Soc.* 1996, **118**, 9450.
[59] F. A. Hicks, N. M. Kablaoui, S. L. Buchwald, *J. Am. Chem. Soc.* 1999, **121**, 5881.
[60] a) E.-i. Negishi, S. J. Holmes, J. M. Tour, J. A. Miller, *J. Am. Chem. Soc.* 1985, **107**, 2568; b) E.-i. Negishi, S. J. Holmes, J. M. Tour, J. A. Miller, F. E. Cederbaum, D. R. Swanson, T. Takahashi, *J. Am. Chem. Soc.* 1989, **111**, 3336.
[61] F. A. Hicks, S. L. Buchwald, *J. Am. Chem. Soc.* 1996, **118**, 11688.
[62] F. A. Hicks, S. L. Buchwald, *J. Am. Chem. Soc.* 1999, **121**, 7026.
[63] S. J. Sturla, S. L. Buchwald, *J. Org. Chem.* 1999, **64**, 5547.
[64] S. J. Sturla, S. L. Buchwald, *Organometallics* 2002, **21**, 739.
[65] T. Kondo, N. Suzuki, T. Okada, T. Mitsudo, *J. Am. Chem. Soc.* 1997, **119**, 6187.
[66] T. Morimoto, N. Chatani, Y. Fukumoto, S. Murai, *J. Org. Chem.* 1997, **62**, 3762.
[67] T. Mitsudo, H. Naruse, T. Kondo, Y. Ozaki, Y. Watanabe, *Angew. Chem. Int. Ed. Engl.* 1994, 33, 580.
[68] T. Kondo, M. Nomura, Y. Ura, K. Wada, T. Mitsudo, *J. Am. Chem. Soc.* 2006, **128**, 14816.
[69] R. Paolillo, V. Gallo, P. Mastrorilli, C. F. Nobile, J. Rose, P. Braunstein, *Organometallics* 2008, **27**, 741.
[70] M. Zhang, S. L. Buchwald, *J. Org. Chem.* 1996, **61**, 4498.
[71] J. K. Rasmussen, S. M. Heilmann, L. R. Krepski, *Advances in Silicon Chemistry*, L. G. Larson Ed.; JAI: Greewich, 1991, Vol 1, pp. 67.
[72] R. Grigg, L. Zhang, S. Collard, A. Keep, *Chem. Commun.* 2003, **3**, 1902.
[73] Y. Tang, L. Deng, Y. Zhang, G. Dong, J. Chen, Z. Yang, *Org. Lett.* 2005, **7**, 1657.
[74] J. Liu, J.-Q. Huang, Y. Hu, M. Chen, Y. Lan, J.-H. Chen, A. Lei, Z. Yang, *Synthesis*, 2007, **16**, 2565.
[75] Y. Lan, L. Deng, J. Liu, C. Wang, O. Wiest, Z. Yang, Y.-D. Wu, *J. Org. Chem.* 2009, **74**, 5049.

[76] For a pertinent review, see: a) J. Habermann, *Curr. Org. Chem.* 2010, **14**, 1139; b) J. Blanco-Urgoiti, L. Añorbe, L. Pérez-Serrano, G. Domínguez, J. Pérez-Castells, *Chem. Soc. Rev.* 2004, **33**, 32.

[77] For recent selected references, see: a) B. Jiang, M. Xu, *Angew. Chem.* 2004, **116**, 2597; *Angew. Chem. Int. Ed.* 2004, **43**, 2543; b) J. Velcicky, A. Lanver, J. Lex, A. Prokop, T. Wieder, H.-G. Schmalz, *Chem. Eur. J.* 2004, **10**, 5087; c) J. D. Winkler, E. C. Y. Lee, L. I. Nevels, *Org. Lett.* 2005, **7**, 1489; d) D. H. Kim, K. Kim, Y. K. Chung, *J. Org. Chem.* 2006, **71**, 8264; e) E. McNeill, I. Chen, A. Y. Ting, *Org. Lett.* 2006, **8**, 4593.

[78] a) N. Jeong, S. D. Seo, J. Y. Shin, *J. Am. Chem. Soc.* 2000, **122**, 10220; b) N. Jeong, B. K. Sung, J. S. Kim, S. B. Park, S. D. Seo, J. Y. Shin, K. Y. Choi, *Pure Appl. Chem.* 2002, **74**, 85.

[79] P. A. Evans, J. E. Robinson, *J. Am. Chem. Soc.* 2001, **123**, 4609.

[80] S. U. Son, K. H. Park, H. Seo, Y. K. Chung, S.-G. Lee, *Chem. Commun.* 2001, **23**, 2440.

[81] K. H. Park, L. G. Jung, Y. K. Chung, *Org. Lett.* 2004, **6**, 1183.

8

Recent Adventures with the Pauson-Khand Reaction in Total Synthesis

Scott G. Van Ornum,[*] Sarah Hoerner, and James M. Cook

The Pauson-Khand Reaction: Scope, Variations and Applications, First Edition. Edited by Ramon Rios Torres.

8.1 Introduction

The total synthesis of target molecules of natural or of computational interest has been a long standing challenge for the synthetic organic chemist. Starting with simple precursors which culminate in a successful synthesis has been described as both an art and a science.[1] Not only are the chemical reactions to form carbon–carbon bonds a challenge, stereochemical strategies must also be considered when a successful total synthesis is developed. Published results have benefited both academia and the chemical and pharmaceutical industries for they have provided key chemical transformations for the benefit of all, as well as new strategies for molecular construction.

The Pauson-Khand reaction, as described in 1971, became one of the most important methods for the synthesis of cyclopentenones and bicylic enones.[2] In the stoichiometric Pauson-Khand reaction, the alkyne is first reacted with dicobalt octacarbonyl to form a thermally stable alkyne metal complex. When the reaction mixture is heated in the presence of an alkene and carbon monoxide, this generated cyclopentenones or bicyclic enones in yields ranging from 40–60%.[3] Basic examples of both inter- and intramolecular Pauson-Khand reactions are shown in Scheme 8.1. Numerous methods have been developed to promote the cyclization process catalytically and improve stereoselectivity with various metals or cyclization processes.[3]

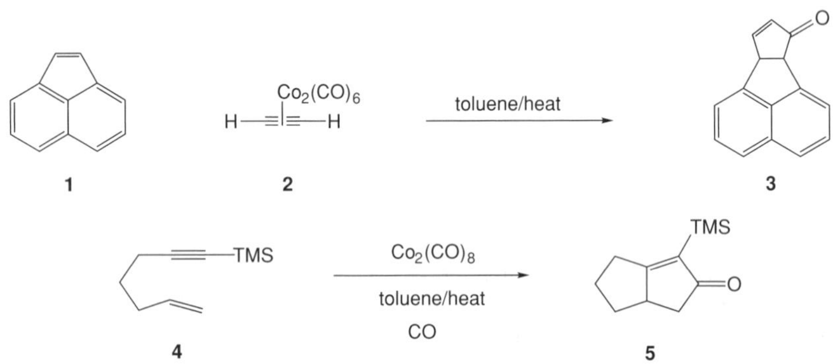

Scheme 8.1 *Inter- and intramolecular Pauson-Khand reaction examples.*

The scope of this chapter has been restricted to the last 15 years highlighting synthetic routes in which the Pauson-Khand reaction has been instrumental in the total synthesis of the chosen target molecule. Noteworthy synthetic examples are showcased in detail, followed by a section which highlights key Pauson-Khand reactions to divert the reader to further examples.

8.2 (+)-Epoxydictymene

The diterpene (+)-epoxydictymene **16**,[4] isolated from the brown alga *Dictyota dichomata*, contains a 5-8-8-5 ring substructure which encases a *trans*-3-oxabicyclo[3.3.0]octane system. A successful synthesis of this moiety was reported by Schreiber using a Lewis acid-promoted Nicholas reaction followed by a cobalt-mediated Pauson-Khand reaction and is shown in Scheme 8.2.[5] (*R*)-Pulegone was chosen as the asymmetric starting material to synthesize key intermediate **6**. The lithium anion of **7** was reacted through displacement of a triflate ester derived from alcohol **6** to furnish alkyne **8** in 74% as a 1:1 mixture of diastereomers at C(10). Treatment of alkyne **8** with dicobalt octacarbonyl provided the corresponding metallocycle and this was followed by a Lewis-acid mediated cyclization to provide a mixture of diastereomeric ethers represented by **9** and by-product **10**. The selectivity of the reaction was 91:9 for **9:10**. Numerous Lewis acids and solvents were screened and the best conditions were obtained by the use of trimethylsilyl triflate in ether at −78 °C to provide good yield and site-selective combination.

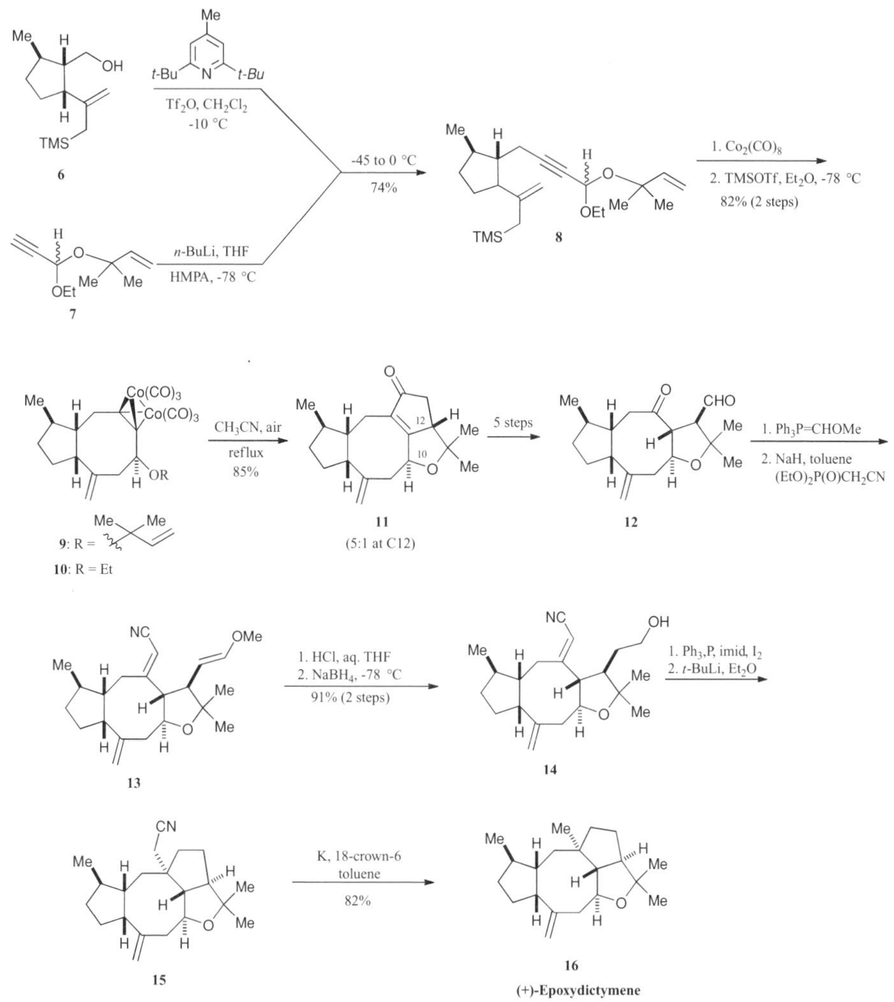

Scheme 8.2 *The total synthesis of (+)-epoxydictymene.*

The intramolecular Pauson-Khand reaction of metallocycle **9** to enone **11** was carried out under thermal,[3] oxidative,[6] and ultrasonic[7] conditions. All methods provided the desired product with differences in yield and stereoselectivity at C(12). The oxidative approach developed by Schreiber[6] employed the use of *N*-methylmorpholine *N*-oxide at room temperature and these conditions afforded a 70% yield of **11** with a mixture of 11:1 diastereomer selectivity at C(12). Ultrasonic conditions in acetonitrile furnished a much lower yield of 45% but with 3:1 diastereomer selectivity at C12. The highest yield, albeit with slightly lower diastereoselectivity, incorporated an atmosphere of air and thermal cyclization in acetonitrile at reflux to furnish an 85% yield and 5:1 selectivity at C(12). Attempts to epimerize the beta hydrogen at C(12) were unsuccessful.

To complete the synthesis of (+)-epoxydictymene, the stereochemistry of the enone system in ring C had to be adjusted to provide the *trans* ring junction. From tetracycle **11**, a five step sequence was executed to provide ketoaldehyde **12** which contained the appropriate functionality to complete this synthesis. Through a series of functional group transformations on [5.8.5] tricyclic **12**, alcohol **14** was received in good yield. Ring closure of alcohol **14** to generate the *trans*-fused 5-5 C ring was achieved by anionic cyclization to give nitrile **15** which was subsequently reduced with potassium in 18-crown-6 ether to complete the total synthesis of (+)-epoxydictymene **16** in 82% yield for this last step.

8.3 (±)-Pentalenene and (-)-Pentalenene

Pentalenene **25**, a triquinane natural product of the pentanolactone family of sesquiterpene antiobiotics, was synthesized by the Weiss reaction[8] and by various other methods[9]. Depicted in Scheme 8.3 is the early synthesis of (±)-pentalenene carried out by Schore and Rowley using the Pauson-Khand strategy.[10] The 5-methylcyclopentyllithium **17** was readily available and was added to the BHT protected methylacrylate **18**. This was followed by methylation of the enolate which provided **19** in 90% yield. Treatment of ester **19** with sodium in liquid ammonia afforded alcohol **20**. Conversion of alcohol **20** to enyne **21** was achieved in 34% overall yield. Enyne **21** was then reacted with dicobalt octacarbonyl followed by heating to 110 °C in heptanes to furnish a mixture of tricyclic enones **22** and **23** in 51% yield. Reduction of the two isomers with Li/NH_3 gave a mixture of saturated ketones represented by **24**. The conversion of tricyclic **24** into (±)-pentalenene **25** was carried out analogously to steps previously reported in the literature.[9]

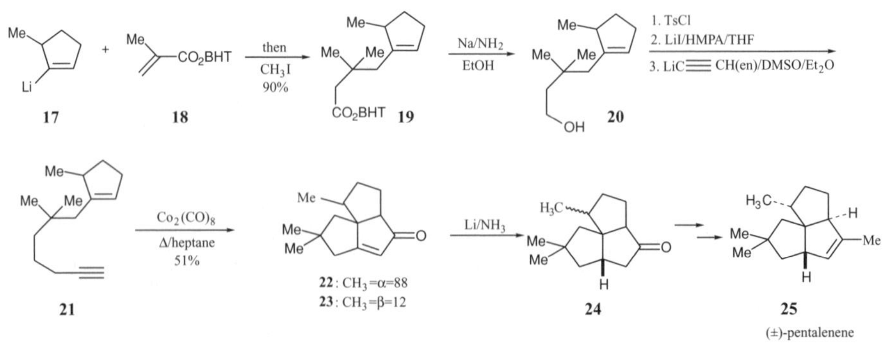

Scheme 8.3 *The Pauson-Khand approach to (±)-pentalenene **25**.*

An enantioselective synthesis of (-)-pentalenene **28** was reported by Fox and Pallerla as shown in Scheme 8.4.[11] Key cyclopropene **26** was synthesized in chiral fashion in order to provide the required stereochemical features at the quaternary carbon centers after an intramolecular Pauson-Khand reaction. Consequently, cyclopropane **26** was treated with 60 mol% of dicobalt octacarbonyl and gently heated in the presence of tetramethylthiourea (TMTU) in toluene to afford a 64% yield of the desired diastereomer **27** (4:1 selectivity dr). Various promoters were screened while NMO and methyl sulfide proved less useful than the highlighted conditions. The subsequent steps provided (-)-pentalenene **28** as reported.[11]

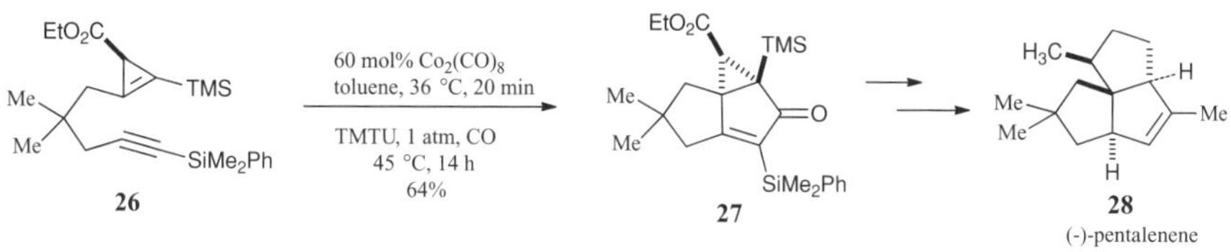

Scheme 8.4 *Pauson-Khand approach to (-)-pentalene **28**.*

8.4 The Tandem Pauson-Khand Reaction Directed Towards the Synthesis of Dicyclopentapentalenes

Dicyclopenta[*a,e*]pentalene **29** and Dicyclopenta[*a,f*]pentalene **30** are attractive synthetic targets for various reasons. Syntheses of the 14 π annulenes could provide important information in terms of bonding character, Hückel stabilization, aromaticity, and resonance energy. Annulenes **29** and **30** resemble nonbenzenoid hydrocarbons perturbed by centrally located sigma bonds. Questions of whether these annulenes are aromatic or exist as highly reactive olefins can only be answered by synthesis and characterization of these molecules. Although electronic delocalization may play a role in the preparation and isolation of 14 π cross-linked annulenes such as dicyclopenta[*a,e*]pentalene **29** and dicyclopenta[*a,f*]pentalene **30**, (Scheme 8.5), certainly the olefinic intermediates (cyclopentadienes) toward these polyquinenes are expected to be highly reactive.

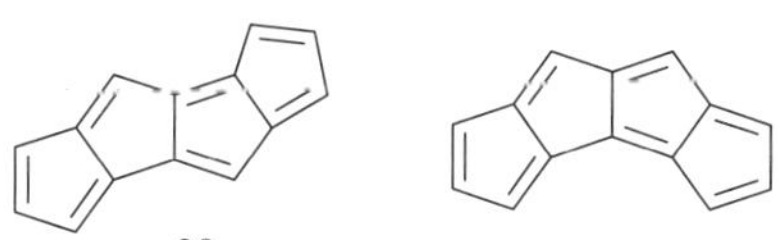

Scheme 8.5 *Dicyclopenta[a,e]pentalene **29** and dicyclopenta[a,f]pentalene **30**.*

The Weiss reaction[8] and the Pauson-Khand reaction[2] have been shown to be extremely useful methods for the preparation of bicycle[3.3.0]octane rings systems. The parent tetracyclic skeletons of annulenes **29** and **30** have been synthesized by Eaton,[12] McKervey,[13]

Kotha,[14] Mehta,[15] and Cook.[16] Recently, a new approach to these molecules has been designed and executed which employed a tandem Pauson-Khand reaction to regiospecifically generate the tetracyclic framework of cross linked annulenes **29** and **30**.[17] This work provided systems required for the construction of **29** and **30** in a higher oxidation state than those achieved previously.[17d]

The required bisalkene–bisalkynes **31** and **33** illustrated in Scheme 8.6 could be accessed on large scale.[17d] A variety of conditions for the Pauson-Khand reaction were employed while higher yields were realized with the molybdenum mediated tandem cyclization to provide tetracyclics **32** and **34**. The process was highly efficient with the generation of six carbon–carbon bonds in excess of 90% for each bond formed. The tetracylic ring systems **32** and **34** could also be synthesized rapidly on gram-scale; however, Van Ornum recently showed it was difficult to isolate a tetraene (a dicyclopentadiene) related to **32**.[18] Although related tetraenes had been observed by ^{1}HNMR spectroscopy,[18] attempts at isolation had resulted in polymerization or decomposition. Either the cyclopentadienes were too unstable or the acetonide began to decompose, consequently, this approach was modified to furnish intermediates in higher oxidation states using a new approach employing an allenic tandem Pauson-Khand reaction.[19]

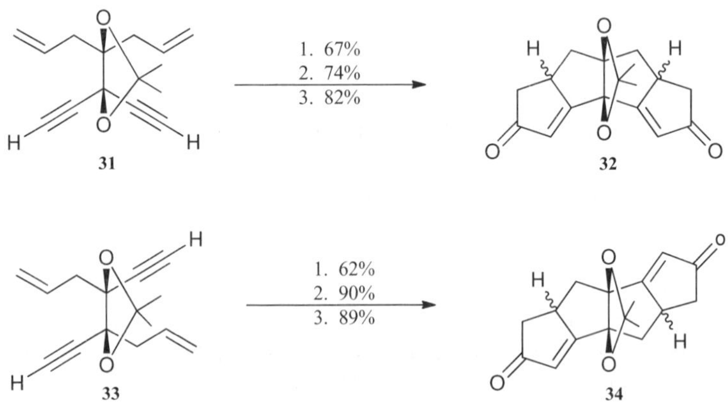

1. 3eq.$Co_2(CO)_8$, NMO, CH_2Cl_2
2. 0.2eq.$Co_2(CO)_8$, DME, CO, light (Q-Beam)
3. 2.4eq.$Mo(CO)_8$, 10 eq. DMSO, toluene, 100 °C, 4.5 hr

Scheme 8.6 *The tandem Pauson-Khand reaction.*

It was decided to stabilize the cyclopentadiene substituents with bulky groups, analogous to the previous approach of Hafner et al.[20] In addition, although the stereocenters in **32** and **34** would eventually be destroyed, it might be difficult to carry the mixture on in succeeding steps. Moreover, one might choose a precursor in a higher oxidation state closer in energy to the target 14 π cross-linked annulene. This would provide a tetracycle which would require fewer high energy steps to approach the target than previously envisaged.[17] Based upon a molybdenum hexacarbonyl mediated allenic Pauson-Khand reaction by Brummond[21] to provide cyclepentenones, a modified approach to provide the key diol is shown in Scheme 8.7.

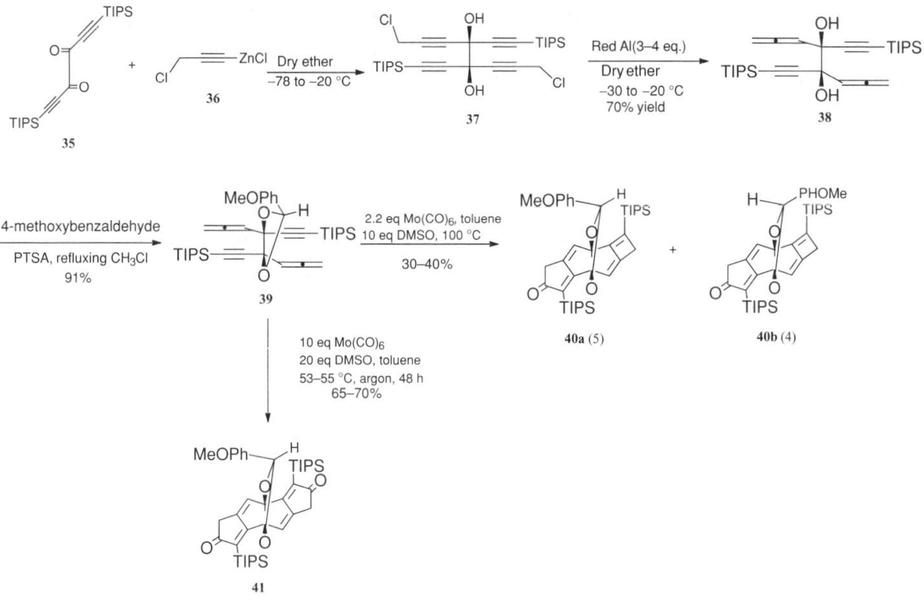

Scheme 8.7 *Studies on the allenic tandem Pauson-Khand reaction.*

The synthesis of the bisallene-bisalkyne **38** then took center stage. After a few attempts,[19] the desired (±)-diol **38** was obtained via the procedure illustrated in Scheme 8.7. The tetra-alkyne **37** was synthesized by treatment of the 1,2-diketone **35**[17e,22] with the alkynylide of propargyl chloride. The propargyl chloride **36** was stirred with butyllithium in diethyl ether at −78 °C to produce the alkynylide anion (the lithium anion survived only at low temperature).[23] Zinc proved to be the best choice as the metal counterpart of the alkynylide anion with respect to stereoselectivity and yield. A similar effect had been observed previously by the authors in another system.[17] When lithium was employed directly (M = Li) in this process, the yield was low and the undesired *meso* isomer predominated. When magnesium was employed, this reaction provided the tetrayne **37** in moderate yield with almost no stereoselectivity. Meso selectivity was also observed by Diederich et al.,[24] when trimethylsilyl acetylenyl magnesium was reacted with the same 1,2-diketone **35**.

However, when zinc was employed as the metal counterpart to stabilize the carbanion (produced by treating the lithium alkynylide of propargyl chloride with anhydrous $ZnCl_2$ at low temperature), the yield increased to over 80% and the desired dl diastereomer of **37** predominated in a ratio of 6:1 over the *meso* isomer. To date, the Felkin–Anh model[25] of addition can be employed to rationalize the observed stereoselectivity of the zinc-mediated process.

Lithium aluminum hydride has been reported in the literature[26] as the reagent of choice to convert propargyl chlorides into allenes. Unfortunately, when the tetrayne **37** was treated under the reported conditions,[26] a complex mixture of products was obtained with the desired bisallene-bisalkyne **38** present in very low yield. A number of other hydride sources were employed. Eventually, RedAl, $[(CH_3OCH_2CH_2O)_2AlH_2]Na$, was employed as the source of hydride under dilute conditions (−20 °C) to provide the desired bisallene-bisalkyne in 70% yield. Presumably, the red Al first reacted with the hydroxyl groups in **37** to give an intermediate in a rapid process. In a subsequent step, hydride was delivered intramolecularly to the sp carbon atom adjacent to the hydroxyl group through a five-membered-ring transition state. This process furnished the desired allene **38** and was presumably the rate-determining step. A similar regioselectivity was reported in the reductive opening of an epoxy alcohol.[27] It is noteworthy that dilute conditions were found to be optimal. At high concentrations of hydride, the nucleophile attacked the propargylic carbon atom in an S_N2 displacement of chloride and reduced the yield of allene **38**. It is also important to note that a different batch of Red Al would occasionally result in a lower yield (~40%) of **38**. If the process was maintained at low temperature (< −10 °C), then the byproducts isolated were monoallene and the starting **37**, which could be recycled to produce more of the desired bisallene **38**, if necessary. If one considers that two allenes were formed in the same one-pot process, then this was considered a pivotal reaction in this sequence. This was the first time, to these authors' knowledge, that Red Al was successfully employed for this transformation.[28]

A number of protecting groups were evaluated. The para-methoxybenzylidene acetal **39** could be formed more rapidly (4–5 h) by heating **38** with excess 4-methoxybenzaldehyde dimethyl acetal under the above conditions in greater than 90% yield and was chosen for the initial Pauson-Khand cyclization studies. Analogous to the conditions reported by Jeong[29] and Brummond et al.,[21] the bisallene **39** was heated in toluene to 100 °C in the presence of a slight excess of $Mo(CO)_6$ (2.2 equiv) and 10 equivalents of DMSO. In approximately 5 h, the starting **39** had been consumed and a new component of higher polarity was observed. The new component was isolated in approximately 40% yield and eventually found to be a mixture of diastereomers composed of the interesting [5.5.5.4] tetracycles **40a** and **40b** in a ratio of 5:4 (**40a**:**40b**). It was clear that the reaction conditions would have to be finely tuned to provide the desired [5.5.5.5] tetracycles. Finally, on the basis of an understanding of the reaction mechanism[19] modified reaction conditions were developed, which provided the desired [5.5.5.5] tetracycle **41** (Scheme 8.7).

In this modification, a large excess of $Mo(CO)_6$ was heated in toluene to form a saturated solution. The best temperature for this cycloaddition process had been narrowed to 53–55 °C. The saturated $Mo(CO)_6$ solution certainly provided a higher concentration of the Mo−alkyne complex, to promote the tandem Pauson-Khand reaction at the expense of the thermal [2+2] process. Again, the narrow temperature range (53 to 55 °C) was important to obtain the [5.5.5.5] system in tetracycle **41**, and suppress the thermally mediated [2+2] process. A lower temperature retarded the reaction rate, while higher temperatures effected more of the undesired [2+2] four-membered ring formation in tetracycles **40a** and **40b**.

Attention turned to removal of the protecting group to regenerate the vicinal diol. It was known that removal of the acetonide, para-methoxybenzylidene acetal or benzylidene acetal from the corresponding tetracyclic systems under mild conditions would be difficult, especially in the presence of the olefinic functionality. For this reason, the cyclic ortho ester acetal **42** was chosen for further study, since it would be easier to remove. Indeed, after a few attempts this protecting group was removed from dione **42** under mild conditions and in very high yield. As shown in Scheme 8.8, the cyclic ortho ester **42** was heated in aqueous

THF (THF:H_2O = 20:1 v/v) to a gentle reflux to provide the desired tetracyclic diol **43** in very high yield (> 95%).

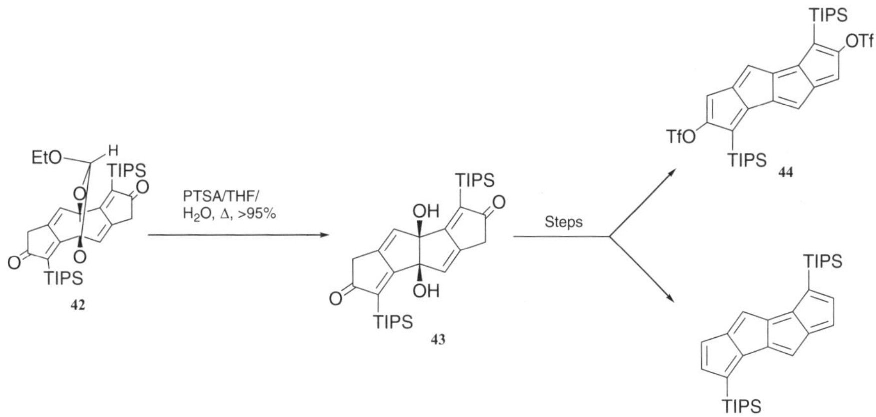

Scheme 8.8 *Final synthetic steps toward 14π annulenes **44** and **45**.*

Subsequent steps afforded the two 14π annulenes **44** and **45**. The two dicyclopenta[*a*,*e*]-pentalenes **44** and **45** underwent significant electronic delocalization, perhaps even aromaticity, as revealed by the planar nature of the X-ray structure of **44**. The tetracyclic rings in **44** assumed a flat geometry; the bond lengths of the tetracycle in **44** also fit well into the criteria for aromatic compounds. A comparison of the NMR and UV spectra of both **44** and **45** demonstrated that they both exhibited similar electronic properties; therefore, the purple colored 14 π cross-linked annulene **45** was planar as well as delocalized.

8.5 Enantioselective Total Synthesis of (-)-α-Kainic Acid

Isolated from the marine algae *Digenea simplex*[30] in 1953, (-)-α-kainic acid **54** possesses important biological activity exhibiting both anthelmintic and neurotransmitter activity[31]. Due to the continued level of interest in the therapeutic implications of (-)-α-kainic acid, several approaches via the Pauson-Khand reaction have been reported.[32] Hampered by low diastereoselectivity and the large number of steps, Helmchen set out to employ an enantioselective total synthesis of (-)-α-kainic acid[33]. This strategy established the necessary chiral centers incorporating an Ir-catalyzed allylic amination and a diastereoselective Pauson-Khand reaction.

Two complementary routes were developed which employed Ir-catalyzed reactions to provide the key enyne **47** of which the best route route is shown in Scheme 8.9. The starting carbonate **46** was reacted with the propargyl amine under the Ir-catalyzed conditions[34] to provide the corresponding 1,6-enyne in 99% enantiomeric excess. The standard Boc-protection under phase transfer conditions provided the Pauson-Khand precursors enyne **47**. The Pauson-Khand conditions which employed dicobaltoctacarbonyl in the presence of molecular sieves promoted the cyclization to furnish key enone **48** in 65% yield.

The desired diastereomer was obtained in a ratio of 88:12. The purified *trans*-**48** was obtained in 57% yield by chromatography. Subsequent reduction of alkene **48** via catalytic hydrogenation provided an epimeric mixture of ketones **49**. Bayer–Villager oxidation of **49** which employed *m*-CPBA gave the lactone **50** in 86% yield. The lactone ring opening was effected with $Ca(BH_4)_2$[35] in ethanol to give the corresponding diol which was reacted with TBDMSCl to selectively silylate the primary alcohol to give monol **51** in 92% yield. The methyl ketone **52** was then obtained from **51** under Ley–Griffith oxidation conditions[36] followed by transformation of **52** into the penultimate intermediate **53** with Tebbe's reagent.[37] Jones' oxidation of **53** furnished the dicarboxylic acids and this was followed by removal of the *N*-BOC protecting group under carefully controlled acidic conditions and ion exchange chromatography to provide (-)-α-kainic acid **54**. The overall yield from carbonate **46** was 12% utilizing the Ir-catalyzed allylic amination reaction and a highly diastereoselective intramolecular Pauson-Khand reaction.

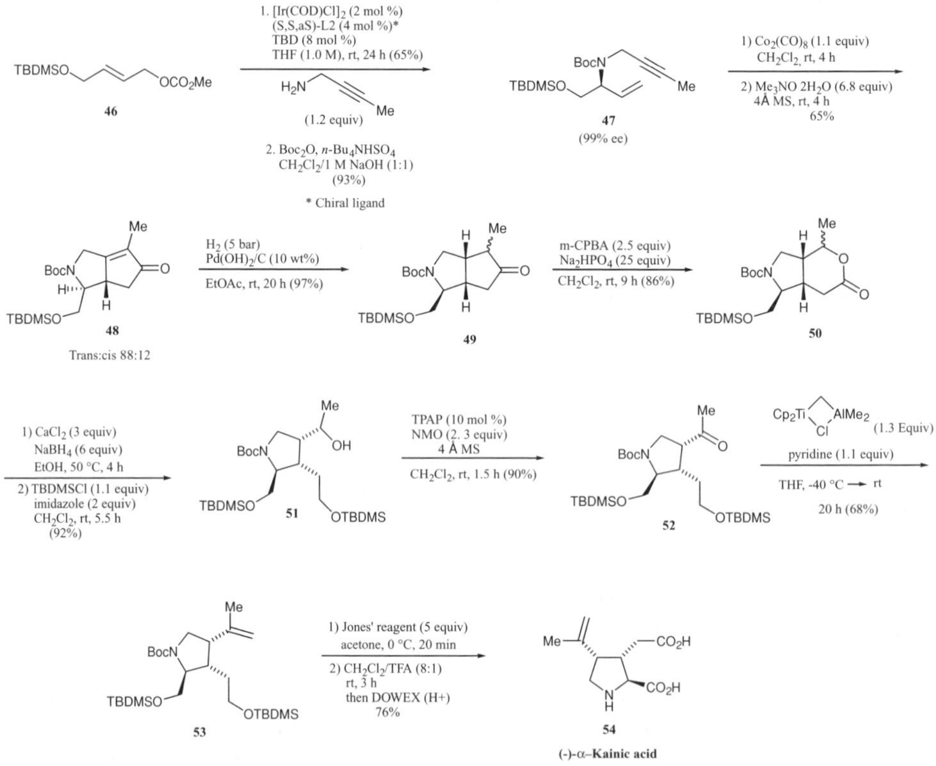

Scheme 8.9 *Enantioselective total synthesis of (-)-α-kainic acid.*

8.6 The Total Synthesis of Paecilomycine A

Paecilomycine A is a terpenoid-derived natural product isolated from *Isoria japonica*[38], Danishshefsky et al. utilized the Diels–Alder reaction with a Danishesky-diene to form the

A-ring and an intramolecular Pauson-Khand reaction to form the bicyclic enone sector.[39] As shown in Scheme 8.10, diene **55** underwent a cycloaddition reaction with dienophile **56** to provide adduct **57** in 86% yield. Subsequent transformations of the endo adduct **57** afforded the eneyne **58** appropriately set-up for the Pauson-Khand cyclization. The alkyne **58** was treated with dicobalt octacarbonyl in the presence of molecular sieves in toluene at room temperature for 2 hours after which was heated to 100 °C for 24 hours to afford a single stereoisomer **59** in 37% yield. Although the yield was lower than desired, the reaction was scaleable and the ketone was readily purified.

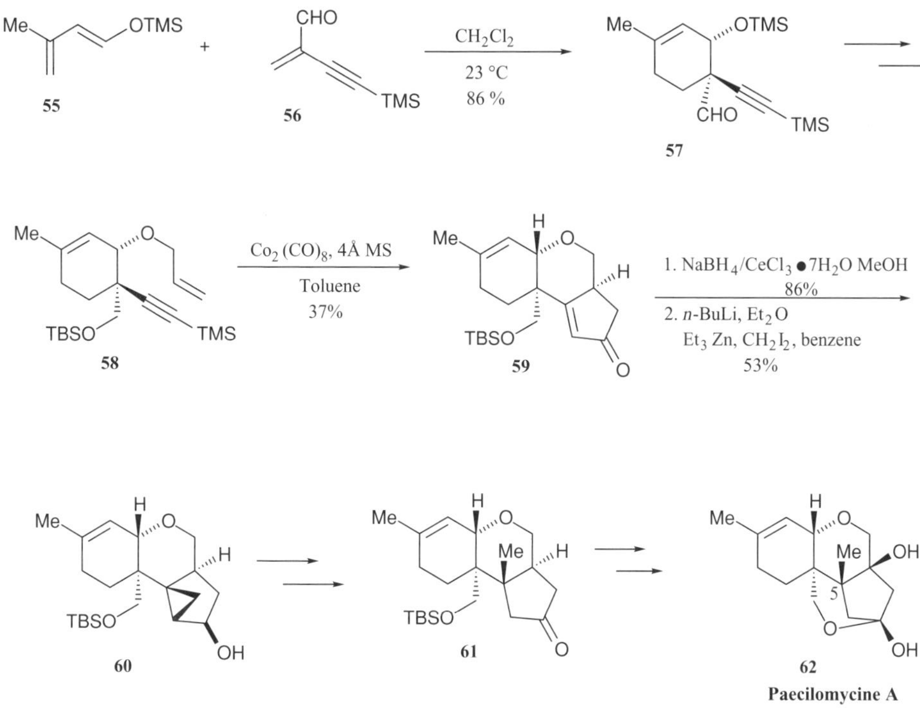

Scheme 8.10 *Total synthesis of paecilomycine A.*

The key tricyclic adduct **59** was reduced with sodium borohydride to the corresponding allylic alcohol. The alcohol was converted to cyclopropane **60** to provide the appropriate functionality for eventual installation of the β-methyl group at C(5) based on the earlier work of Corey et al.[40] which ultimately provided paecilomycine A **62**.

8.7 The Total Synthesis of (+)-Achalensolide

Bohlmann [41] isolated (+)-achalensolide **68** from the aerial parts of *Decachaeta thieleana* in 1983. Herz, in the same year, isolated **68** from *Stevia achalensis*, in Copina, Cordoba, Argentina.[42] Structurally, (+)-achalensolide possessed the bicycle[5.3.0]decane ring system which is typically present in the guaianolide skeletal system. Mukai et al. chose (+)-achelensolide as their target and employed Rh(I)-catalyzed allenic Pauson-Khand reactons, as shown in Scheme 8.11.[43]

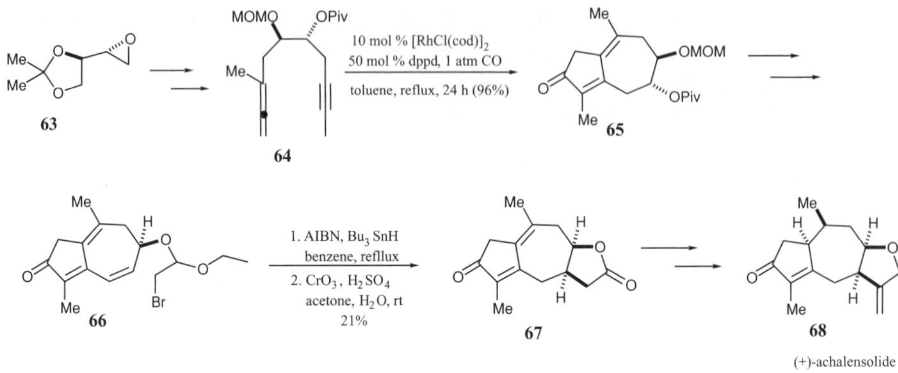

Scheme 8.11 *Total synthesis of (+)-achalensolide.*

The synthesis began with the known epoxide **63** derived from isoascorbic acid[44] from which the key allenyne intermediate **64** was synthesized in five steps. Several conditions were explored for the Rh(I)-catalyzed Pauson-Khand cyclization. The first conditions studied 10 mol% of $[RhCl(CO)_2]_2$ in refluxing toluene under an atmosphere of carbon monoxide to furnish only a 14% yield of bicyclic enone **65**. The yield was increased to 61% when $[RhCl(CO)dppp]_2$ was employed under the same conditions. Higher presences of carbon monoxide resulted in intractable mixtures. A third set of conditions were studied by preparing $[\{Rh(CO)(dppp)_2\}Cl]$ by *in situ* preparation of 10 mol% of $[RhCl(cod)]_2$ and 50 mol% 1,3-bis(diphenylphosphino)propane. When this mixture was heated in refluxing toluene under an atmosphere of carbon monoxide for 24 hours a 96% yield of the bicycle[5.3.0]decadienone **65** was obtained. Attempts to perform the Pauson-Khand reaction employing the Wilkinson catalyst $[(Ph_3P)RhCl]$ were unsuccessful.

With ample quantities of **65** in hand, the subsequent steps afforded bromoacetal **66** and this was followed by radical cyclization after which Jones oxidation gave tricyclic **67** in 15% yield from **65**. The four step synthesis from **65** which remained was carried out to provide the target **68** (+)-achalensolide. This completed the total synthesis of (+)-**68** from (-)-isoascorbic acid.

8.8 The Total Synthesis of (-)-Alstonerine

The alkaloid (-)-alstonerine **84** possesses structural features such as an indole ring and an azabridged bicyclic motif and has been the subject of three total syntheses and one formal synthesis.[45] In addition to the synthetic challenges, **84** may have cytoxic activity against human lung cancer cell lines.[46] Martin et al. have reported the use of the intramolecular Pauson-Khand reaction in the total synthesis of indole alkaloids and have recently directed their focus towards an enantioselective total synthesis of (-)-alstonerine[47].

The approach of Martin et al. centered on a series of model reactions of various piperidine or piperidone enynes to provide azabridged bicyclic cyclopentenones. As shown in Scheme 8.12, the appropriately functionalized series of piperidones related to **73** were synthesized from pyridine **69**. A number of Pauson-Khand promoters were studied with

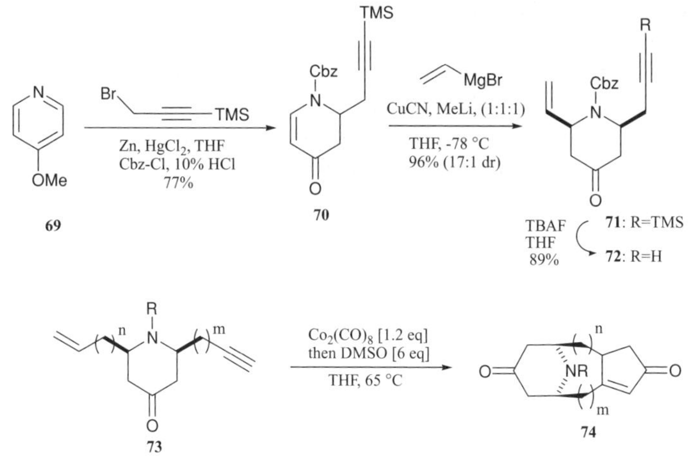

Entry	Substrate	Product	Yield (%)
1	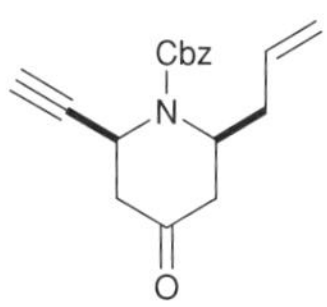 **75**	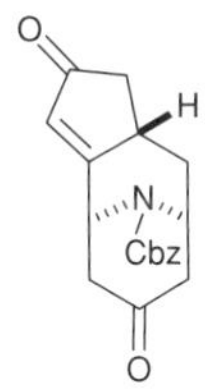 **76**	89
2	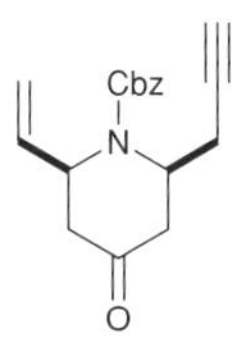 **77**	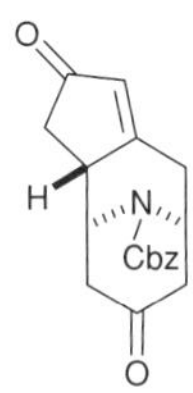 **78**	91
3	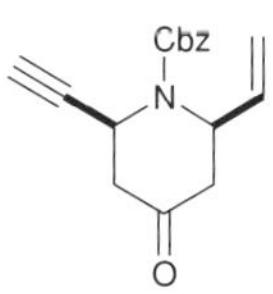 **79**	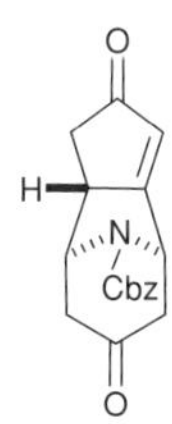 **80**	33 (3:1 dr)

Scheme 8.12 *Model studies of the Pauson-Khand reaction toward azabridged bicyclic enones.*

these enynes including NMO, BuSMe, and 4Å molecular sieves. The best conditions involved reacting enyne **75** with dicobal toctacarbonyl and this was followed by 6 equivalents of DMSO after which it was heated to 65 °C. The corresponding bicyclic enone **76** was obtained as a single diastereomer in 89% yield.

Martin et al. reported during the optimization of the cyclization conditions, that the handling and storage of Co_2CO_8 under argon led to the best results. Other catalytic variants were studied including cobalt and rhodium but starting enyne **75** was largely recovered. The optimized conditions were extended to the reaction of enynes **77** and **79** to provide yields of 91 and 33%, respectively. The positive results of the model studies supported the approach via the Pauson-Khand towards the synthesis of (-)-alstonerine. As shown in Scheme 8.13, the required starting enyne **81** had been previously synthesized by a four-step diastereoselective process from L-tryptophan.[48] The Pauson-Khand reaction was carried out with 1.2 equivalents of dicobalt octacarbonyl and 6 equivalents of DMSO at 65 °C in THF to give the diastereomer **82** in 92% yield. Enone **82** was not crystalline but was derivatized by BOC protection to give **83** followed by X-ray analysis to confirm the required stereochemistry necessary for eventual conversion to (-)-alstonerine **84**.

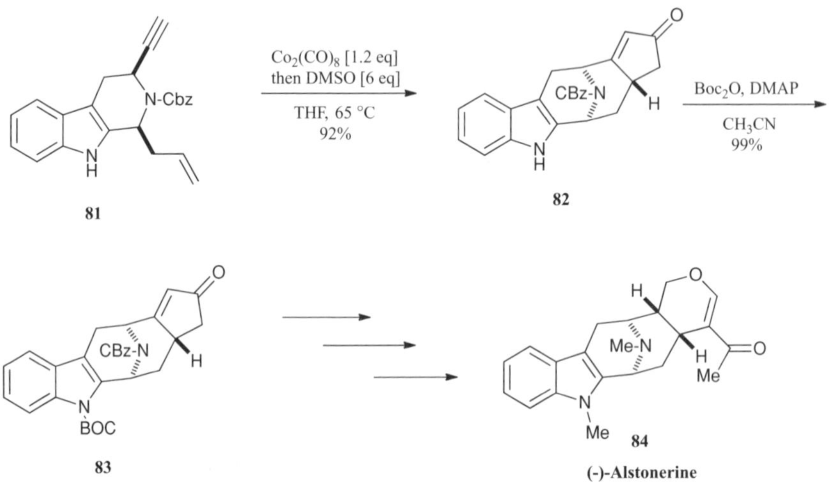

Scheme 8.13 *Intramolecular Pauson-Khand reaction towards (-)-alstonerine.*

8.9 The Total Synthesis of (±)-8α-Hydroxystreptazolone

The total synthesis of (±)-8α-hydroxystreptazolone **93** was recently reported by Nomura and Mukai[49] which employed the Pauson-Khand reaction with a 2-oxazolone derivative which served as an olefin moiety and amine equivalent. (±)-8α-Hydroxystreptazolone is an analog of streptazolin which has unique biological activity and has previously been synthesized by several groups.[50] The initial strategy of Mukai centered on the construction of 2-oxazolone-alkyne derivatives related to **85** in order to establish the appropriate Pauson-Khand cyclization for eventual synthesis of **93**. The standard conditions involved reaction of the alkyne **85** with dicobalt octacarbonyl in ether followed by the addition of the cyclization promoter, as shown in Table 8.1. The use of amine oxides as the oxidant generally provided

Table 8.1 *Pauson-Khand model studies of 2-oxazolone **85**.*

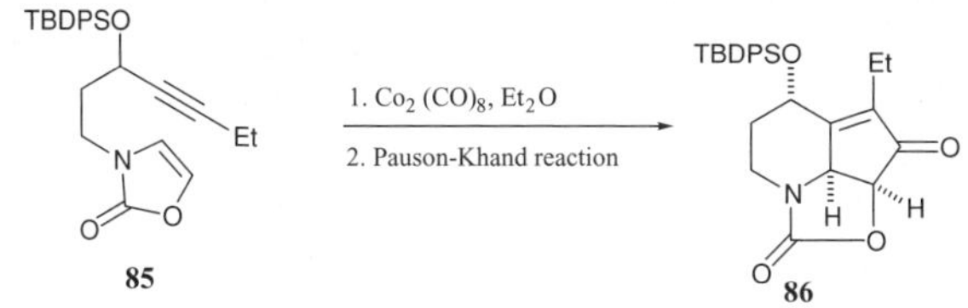

Entry	Promoter (eq)	Solvent	Time (h)	T (°C)	Yield (%)
1	heat	MeCN	75 min	75	trace
2	TMANO•H_2O	CH_2Cl_2	3 h	rt	37
3	TMANO•H2O	CH_2C1_2	5.5 h	reflux	55
4	NMO	CH_2Cl_2	20 h	rt	38
5	**TMANO/4Å MS**	**toluene**	**12 h**	**−10**	**60**
6	*i*-PrSMe	$C1(CH_2)_2C1$	45 min	reflux	16
7	$CyNH_2$	$C1(CH_2)_2C1$	30 min	reflux	11

satisfactory yields and conditions developed by Pérez-Castells[51] using TMANO and 4Å molecular sieves in toluene at a low temperature provided a 60% yield of tricyclic **86**. The conditions in entries 6 and 7 employing the conditions of Sugihara[52] provided lower yields of **86**, albeit they are useful in other systems.

Once effective conditions of the Pauson-Khand reaction were developed for the key oxazolone intermediates, the focus turned toward the total synthesis of the target molecule. As shown in Scheme 8.14, a mixture of diastereomers **89** was constructed by a two-step

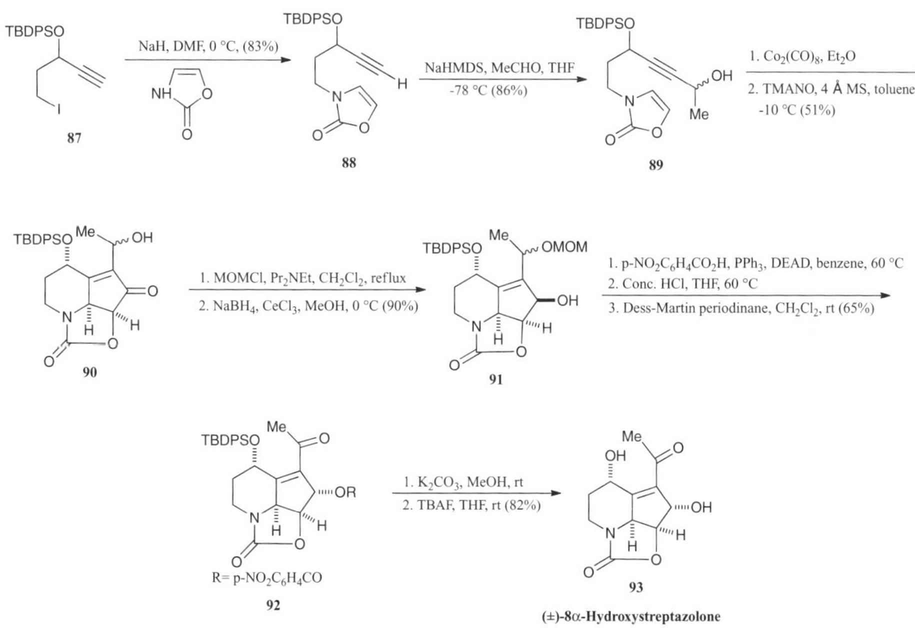

Scheme 8.14 *The total synthesis of (±)-8α-hydroxystreptazolone.*

sequence from starting alkyne **87** in high yield. The reaction of the diastereomeric mixture of yne-ols **89** with dicobalt octacarbonyl in ether afforded the alkyne complex followed by the addition of TMANO in the presence of 4Å molecular sieves in toluene at −10 °C. This process furnished the tricyclic enone **90** in 51% yield. The protection of the alcohol with MOMCl and reduction of the ketone resulted in alcohol **91** in 90% yield. Alcohol **91** was then reacted under Mitsunobu conditions and this was followed by regeneration of the secondary alcohol and oxidation using Dess–Martin conditions to provide the ketone **92** in 65% yield. The final steps to complete the total synthesis involved a two-step sequence to cleave the protecting groups to provide (±)-8α-hydroxystreptazolone **93** in 82% yield.

8.10 The Formal Total Synthesis of (±)-α- and β-Cedrene

The tricyclic sesquiterpenes α-cedrene **103** and β-cedrene **104** are natural products from the cedrene **102** family and possess the unique [5.3.1.0] tricylic skeleton. As a pioneer in the development of the Pauson-Khand reaction,[2] Pauson and co-workers set up to employ the cyclization reaction in efforts toward the formal total synthesis of cedrene analogs **103** and **104**[53]. As shown in Scheme 8.15, commercially available ketone **94** was taken

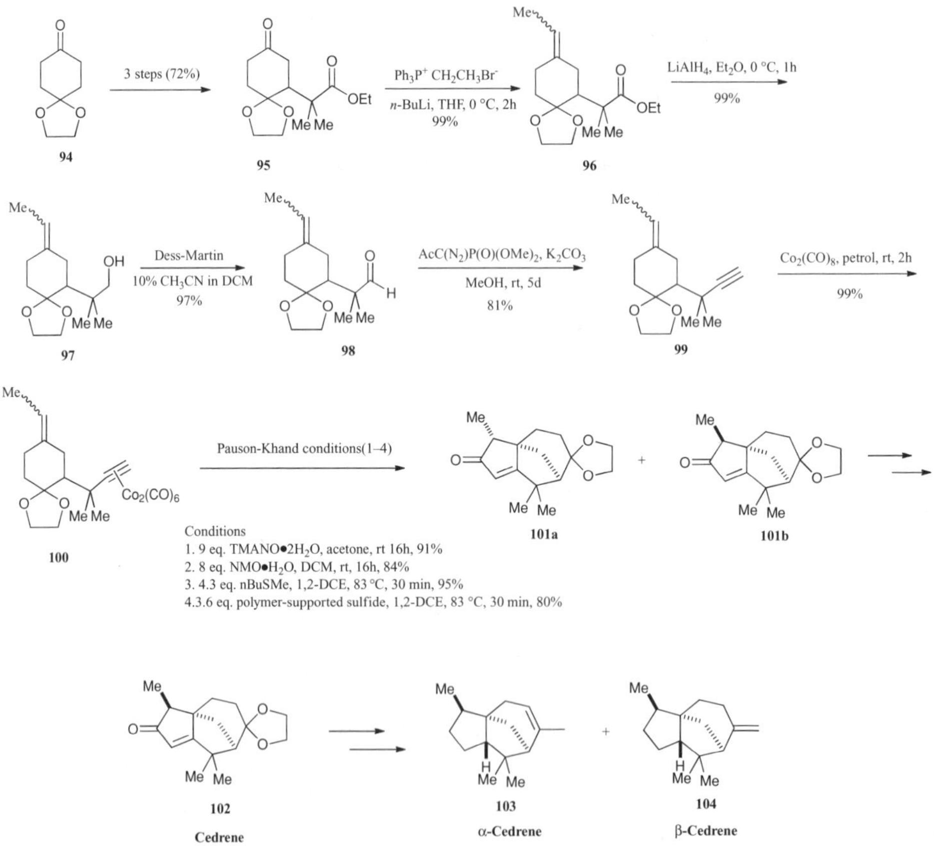

Scheme 8.15 The formal total synthesis of (±)-α- and β-cedrene.

through a three-step synthesis to give ester **95** in overall yield of 72%. The reaction of ketone **95** under standard Wittig conditions with the ethyltriphenylphosphonium bromide salt and *n*-butyl lithium provided the (Z) and (E) isomers of **96** in a 2:1 ratio, respectively. The subsequent reduction of ester **96** with lithium aluminum hydride provided the mixture of alcohols **97**, followed by oxidation to the corresponding aldehyde **98**. Addition of the alkyne group to **98** was achieved by employing acetyldiazomethylphosphonate under the Ohira–Bestmann conditions[54] to give alkyne **99** in 81% yield.[53] The reaction of the alkyne with dicobalt octacarbonyl in petrol provided the metallocycle **100** in nearly quantitative yield.

As shown in Scheme 8.15, a variety of conditions were employed including use of amine oxides,[6] polymer supported alkyl methyl sulfides,[55] and optimized conditions using Sugihara's procedure[52] to provide the desired enones **101a** and **101b** (2:1, respectively) in high yields. Both cyclizations at room temperature with amine oxides as well thermally mediated cyclizations proved successful. Both diastereomers were separated by silica gel chromatography with the minor enone **101b** possessing the required β-selectivity of the methyl group. Pauson was able to further improve the selectivity of the Wittig reaction which improved the selectivity of the Pauson-Khand intramolecular cyclization to provide **101a** and **101b** (1:2.4, respectively). Furthermore, epimerization conditions were developed by reacting **101a** with lithium hydroxide in aqueous THF to provide a mixture of **101a**:**101b** in a ratio of 1:9, respectively. With ample amounts of **101b** now in hand, further transformation to the target molecules **103** and **104** through cedrene **102** was carried out completing the formal total synthesis of (±)-α- and β-cedrene.

8.11 Additional Applications of the Pauson-Khand Reaction in Total Synthesis

Honda and Keneda[56] have used an intramolecular Pauson-Khand reaction to gain entry into the bicyclic ring system contained in (-)-incarvilline **109**. Because it is related to the biologically active parent compound incarvillateine,[57] (-)-incarvilline has been reported in studies on potential antinociceptive compounds related to incarvillateine.[58] The synthesis began with alcohol **105,** as shown in Table 8.2, from which the necessary enyne **106** was synthesized in a four-step sequence. The results of the study of the intramolecular Pauson-Khand cyclization are shown in Table 8.2. The alkyne **106** was first reacted with a slight excess of dicobaltoctacarbonyl in toluene under thermal conditions to afford the desired enone **107** and its diastereomer **108**. Optimization with various promoters including N-oxides and butyl methyl sulfides [59.52b] provided higher yields of **107** and **108**. The diastereomer selectivity was rationalized by steric repulsions through the cyclization intermediates leading to the intramolecular products. The subsequent steps completed the formal total synthesis of the monoterpene alkaloid (-)-incarvilline **109**.[57]

A related example of a Rh(I)-catalyzed [(3+2)+1] homologous Pauson-Khand cycloaddition has been reported by Yu[60] in the total synthesis of the furanoid sesquiterpene (±)-α-agarofuran **115**. Unique to this synthesis was the use of a 1-yne vinyl cyclopropane synthon and carbon monoxide to construct the key [4.4.0]bicyclic decalin system in **115**. The prior studies by Yu and co-workers[61] on intramolecular cyclizations using vinylcyclopropane synthons set the ground work for this total synthesis.

Table 8.2 *The Pauson-Khand cyclization in the total synthesis of (-)-incarvilline.*

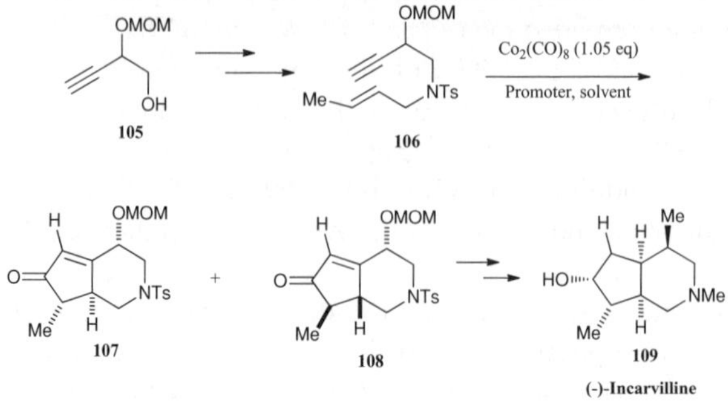

Entry	Promoter (eq)	Solvent	Atm	T (°C)	Time (h)	Yield (%)	
						107	**108**
1	none	toluene	Ar	110	2	54	7
2	NMO (10)	CH_2C1_2	Ar	rt	9	45	5
3	*n*-BuSMe (3.5)	DCE	Ar	83	24	58	7
4	*n*-BuSMe (3.5)	DCE	CO	83	2.5	66	7
5	*t*-BuSMe (3.5)	DCE	Ar	83	2.5	62	6
6	***t*-BuSMe (3.5)**	**DCE**	**CO**	**83**	2.5	**73**	**8**

In this vein, a four -tep process shown in Scheme 8.16 from cyclopropane ester **110** furnished the required vinylcyclopropane substrate **114** in good overall yield. Model studies on simpler vinylcyclo substrates with various rhodium catalysts were performed to establish the best conditions for construction of the bicyclic core of agarofuran. Reaction

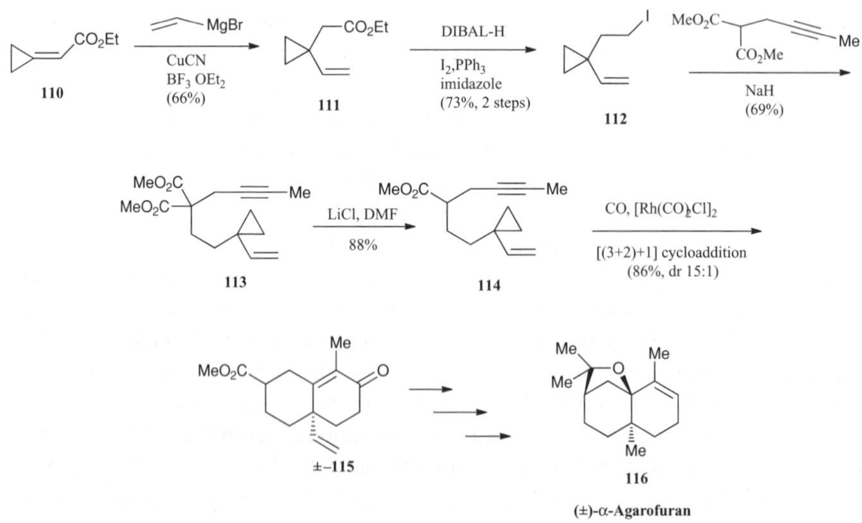

Scheme 8.16 *The Rh(I)-catalyzed Pauson-Khand reaction towards (±)-α-agarofuran.*

of vinylcyclopropane **114** under the conditions of the homo-Pauson-Khand reaction gave enone **115** in 86% yield with high diastereoselectivity. The structure of **115** was confirmed by X-ray analysis of the 2,4-dinitrophenylhydrazone derivative. The further steps provided (±)-α-agarofuran **116** to complete the total synthesis and validate the first synthetically practical homo-Pauson-Khand [(3+2)+1]cycloaddition employing vinylcyclopropane intermediates. This is an important advance in Pauson-Khand reaction.

Mukai and co-workers completed the first total syntheses of two tricyclic sesquiterpenes represented by **120** via a Rh(I)-catalyzed Pauson-Khand-related reaction.[62] As shown in Scheme 8.17, acetal **117**, derived from dimethyl D-tartrate, was converted into the appropriate ene-allenene **118**. The initial reaction conditions chosen, which employed 5 mol% $[RhCl(CO)_2]_2$, 5 atm CO in refluxing toluene did not provide the expected enone **118**. Several other catalysts and conditions failed to provide the desired results. Success was achieved by the reaction of allene **118** with a cationic rhodium catalyst under an atmosphere of carbon monoxide in refluxing toluene to provide the key bicycle[4.30]nonene **119** in 74% yield. The stereochemistry was confirmed by NOE experiments and rationalized by the formation of the expected rhodacycle in which the intermediate adopts the more stable chair conformation. Completion of the total synthesis relied upon the successful conversion of enone **119** into the stereoselective isopropylcyclopropane ring to provide both tricyclic sesquiterpenes represented by **120**.

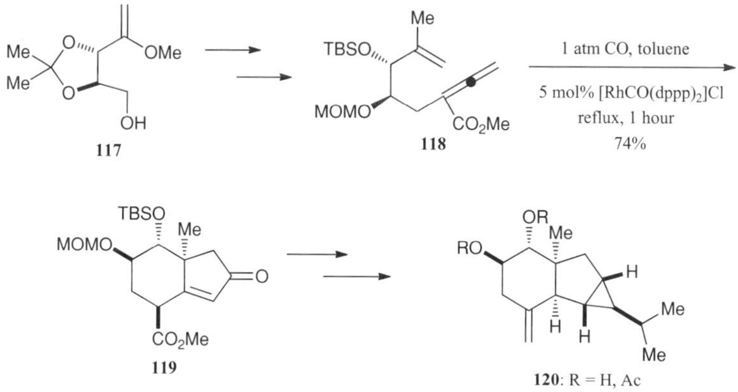

Scheme 8.17 The total synthesis of sesquiterpenes via an allenyl Pauson-Khand-type reaction.

Isocarbacyclin **124** is an analog of the vasodilator prostacyclin and has been previously prepared via an enantioselective total synthesis described by Saito.[63] Because of the recognition of the bicyclic core present in isocarbacyclin, a synthetic Pauson-Khand approach outlined in Scheme 8.18 was developed. L-Ascorbic acid **121** was converted via a series of steps to enyne **122** on multi-gram scale. The reaction of **122** was then studied under both stoichiometric and catalytic Pauson-Khand conditions.

Choice of the stoichiometric conditions by which alkyne **122** reacted with dicobalt octacarbonyl in CH_2Cl_2 (followed by heating to 65 °C in acetonitrile) furnished enone

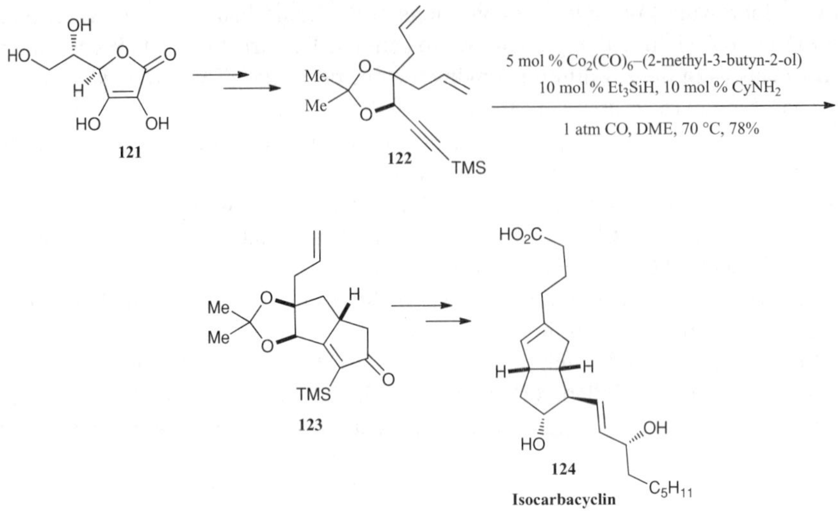

Scheme 8.18 *The diastereoselective total synthesis of isocarbacyclin.*

123 in 78% yield. The catalytic conditions employing the Livinghouse procedure,[64] as shown in Scheme 8.17, afforded the enone **123** in 76% yield based on a 67% conversion of **122**. Saito cited that although the catalytic process was successful, the reaction was slow and difficult to quantitatively convert monocycle **122** to tricycle **123**. A number of steps, including access to side chain analogs, were then carried out to complete the total synthesis of isocarbacyclin.

The unusual fused [5.8.5] ring system, the dicyclopenta[a,d]cyclooctene, commonly exists in the natural fusicoplagin A,[65] ophiobolins,[66] and cotylenol.[67] The employment of diene-diyne **125** from previously reported chemistry[17] to generate tetracycle **126** was studied in a variety of conditions via the tandem Pauson-reaction as shown in Scheme 8.19.[68] A variety of metal catalysts were screened in efforts to improve the diastereoselectivity of the key tetracyclic intermediate **126a**.

The best conditions obtained, to date, were found in entries 5 and 6 wherein the reaction was tuned to provide a higher diastereoselectivity by employing iridium and rhodium catalysts, respectively. All diastereomeric ratios were determined by integration of the NMR spectrum of the crude reaction material. With the tetracylic ketone **126** in hand, it was heated in an aqueous solution of THF in the presence of a catalytic amount of pTSA to provide diol **127** in 72% yield. The ring expansion strategy via an oxidative glycol cleavage with sodium periodate in a phosphate buffer provided a successful entry into the functionalized [5.8.5] ring systems in **128** and **129**.

8.12 Conclusions

For nearly 40 years, the Pauson-Khand reaction has been a powerful synthetic method for the construction of cyclopentenones and bicyclic enones. The examples highlighted in

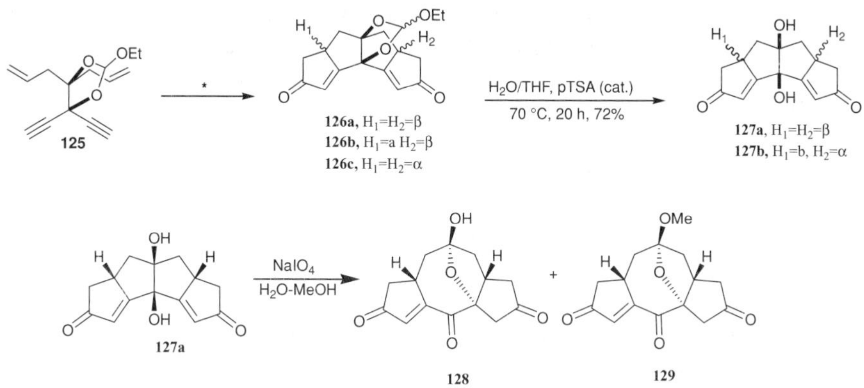

*Conditions for the tandem Pauson-Khand cyclization

1. $Co_2(CO)_8$ (25 mol%), DME, CO (1 atm), hν (Q-Beam), 60 °C, 48 h, 55%, **126a:126b:126c**: = 2:3:1, d.r (**126b**) = 1:1.2
2. $Co_2(CO)_8$ (25 mol%), DME, CO (1 atm), 60 °C, 36h, 65% **126a:126b:126c**: = 1.5:1:trace, d.r (**126b**) = 2:1
3. $Mo(CO)_6$ (240 mol%), DMSO (10 eq), toluene, Ar, 90–100 °C, 4–8 h, 72%, **126a:126b:126c** = 1:2.5:trace, d.r (**126b**) = 1.5:1
4. $[Rh(CO)_2Cl]_2$ (10 mol%), toluene, CO (1 atm), 100 °C, 28 h, 36%, **126a:126b:126c** = 1:1.2:trace, d.r (**126b**) = 5:3
5. $[Ir(COD)Cl]_2$ (20 mol%), Ph_3P (40 mol%), CO (1 atm), toluene, reflux, 24 h, 76%, **126a:126b** = 4:1 d.r (**126b**) = 5:1
6. **$[RhCl(CO)dppp]_2$ (20 mol%), CO (1 atm), CH_3CN, reflux, 24h, 68%, 126a:126b = 7:1, d.r (126b) = 10:1**

Scheme 8.19 *The Rh and Ir mediated Pauson-Khand reaction for simple entry into [5.8.5] ring systems.*

this chapter illustrate the fundamental role the process has played in the total synthesis of complex molecules of natural and computational interest.[69] Initially the process began by employing cobalt as the metal mediator, after which the scope of the reaction was expanded to include a variety of stoichiometric and catalytic conditions with different metal mediators to improve diastereoselectivity and yield. Due to the robustness of the process to tolerate a variety of functional groups, the use of the Pauson-Khand reaction will continue to make significant contributions in synthetic organic chemistry for years to come.

Acknowledgements

The authors would like to dedicate this book chapter in memory of Andrew J. Strom, pre-pharmacy student at Concordia University Wisconsin.

References

[1] (a) Nicolaou, K.C., Vourloumis, D., Winssinger, N., and Baran, P.S. (2000) The art and science of total synthesis at the dawn of the twenty-first century. *Angew. Chem. Int. Ed.*, **29**, 44–122; (b) Bertz, S.H. (1981) The first general index of molecular complexity. *J. Am. Chem. Soc.*, **103**, 3599–3601.

[2] (a) Khand, I.U., Knox, G.R, Pauson, P.L., and Watts, W.E. (1971) A cobalt induced cleavage reaction and a new series of arenecobalt carbonyl complexes. *J. Chem. Soc. Chem. Commun.*, 36a–36a; (b) Khand, I.U., Knox, G.R, Pauson, P.L., Watts, W.E and Foreman, M.I. (1973) Organocobalt complexes. Part II. Reaction of acetylenehexacarbonyldicobalt complexes, $(R^1C_2R^2)Co_2(CO)_6$, with norbornene and its derivatives. *J. Chem. Soc., Perkin Trans.* **1**, 977–981.

[3] Reviews of the Pauson-Khand reaction: (a) Pauson. P.L. (1985) The Khand reaction: A convenient and general route to a wide range of cyclopentenone derivatives. *Tetrahedron*, **41**, 5855–5860; (b) Pauson. P.L. (1987) *Organometallics in Organic Synthesis*, Springer, Berlin; (c) Schore, N.E. (1991) The Pauson-Khand cycloaddition reaction for synthesis of cyclopentenones. *Org. React.*, **40**, 1–90; (d) Pérez-Castells, J. (2006) Cascade reactions involving Pauson-Khand and related processes. *Top. Organomet. Chem.*, **19**, 207–257; (e) Brummond, K.M. and Kent, J.L. (2000) Recent advances in the Pauson-Khand reaction and related [2+2+1] cycloadditions. *Tetrahedron*, **56**, 3263–3283; (f) Urgoiti, J.B., Añorbe, L., Pérez-Serrano, L., Domínguez, G. and Pérez-Castells, J. (2004) The Pauson-Khand reaction, a powerful synthetic tool for the synthesis of complex molecules. *Chem. Soc. Rev.*, 33, 32–42.

[4] Enocki, N., Furusaki, A., Suehiro, K., Ishida, R., and Matsumoto, T. (1983) Epoxydictymene, a new diterpene from the brown alga. *Tetrahedron Lett.*, **24**, 4341–4342.

[5] (a) Jamison, T. F., Shambayati, S., Crowe, W. E. and Schreiber, S. L. (1994) Cobalt-mediated total synthesis of (+)-epoxydictymene. *J. Am. Chem. Soc.*, **116**, 5505–5506.

[6] Shambayati, S., Crowe, W. E. and Schreiber, S. L. (1990) N-oxide promoted Pauson-Khand cyclizations at room temperature. *Tetrahedron Lett.*, **31**, 5289–5292.

[7] Billington, D. C., Helps, I. M., Pauson, P. L., Thomson, W. and Willison, D. J. (1988) The effect of ultrasound and of phosphine and phosphine-oxides on the Khand reaction. *J. Organomet. Chem.*, **354**, 233–242.

[8] (a) Piers, E. and Karunaratne, V. (1984) Methylenecyclopentane annulation: a synthesis of the sesquiterpenoid (±)-pentalenene. *J. Chem. Soc. Chem. Commun.*, 959–960; (b) Piers, E. and Karunaratne, V. (1989) Bifunctional reagents in organic synthesis. Total syntheses of the sesquiterpenoids (±)-pentalenene and (±)-9-epi-pentalenene. *Can. J. Chem.*, **67**, 160–164.

[9] (a) Paquette, L. A. and Annis, G. D. (1982). Total synthesis of (±)-pentalenene. *J. Am. Chem. Soc.*, **104**, 4504–4506; (b) Mehta, G. and Rao, K. S. (1986) General transannulation approach to angular triquinanes. Total syntheses of (±)-pentalenene and (±)-epi-pentalenene. *J. Am. Chem. Soc.*, **108**, 8015–8021; (c) Ohfune, Y., Shirama, H. and Matsumoto, T. (1976) Formolysis of protoilludyl cation equivalents a new mode of rearrangement of bicyclo[4.2.0]oct-2-yl cation. *Tetrahedron Lett.* 2869–2872; (d) Crimmins, M. T. and DeLoach, J. A. (1986) Intramolecular photocycloadditions-cyclobutane fragmentation: total synthesis of (±)-pentalenene, (±)-pentalenic acid, and (±)-deoxypentalenic acid. *J. Am. Chem. Soc.*, **108**, 800–806; (e) Hua, D. H. (1986) Asymmetric total synthesis of (+)-pentalenene via chiral sulfinylallyl anions. Hydrolytic ring closure of enol thioether ketones. *J. Am.Chem. Soc.*, **108**, 3835–3835; (f) Pattenden, G. and Teague, S. J. (1987) Transannular cyclisation as a stratagem in synthesis. A total synthesis of (±)-pentalenene. *Tetrahedron*, **43**, 5637–5652.

[10] Schore, N. E. and Rowley E. G. (1988) Diastereofacialselectivity in intramolecular Pauson-Khand cycloaddition. Highly stereoselective synthesis of pentalenene. *J. Am. Chem. Soc.*, **110**, 5224–5225.

[11] Pallerla, M. K. and Fox, J. M. (2007) Enantioselective synthesis of (-)-pentalenene. *Org. Lett.*, **9**, 5625–5626.

[12] (a) Eaton, P. E., Müller, R. H., Carlson, G. R., Cullison, D. A., Cooper, G. F., Chou, T. C. and Krebs, E. P. (1977) Synthesis of peristylane and the acs-(Cs)-C14-tetraquinane, C14-pentaquinane, and norperistylane systems. *J. Am. Chem. Soc.*, **99**, 2751–2767; (b) Eaton, P. E., Srikrishna, A. and Uggeri, F. (1984) Cyclopentannulation of bicyclo[3.3.0]octane-3,7-dione. A more convenient synthesis of the [5]peristylane system. *J. Org. Chem.*, **49**, 1728–1732.

[13] McKervey, M. A., Vibuljan, P. and Ferguson, G.J. (1981) Synthesis and *X*-ray crystal structure of a (C_2)-C_{20}-hexaquinane derivative, a potential dodecahedrane precursor. *J. Chem. Soc. Chem. Commun.*, 912–914.

[14] Kotha, S., Brahmachary, E., Sivakumar, R., Joseph, A. and Sreenivasachary, N. (1997) A simple route to cyclopentane annulation. *Tetrahedron Lett.*, **38**, 4497–4500.

[15] Mehta, G. and Rao, K. S. (1985) Reductive carbon-carbon cleavage in caged systems. A new general synthesis of linearly fused cis-syn-cis-triquinanes. *J. Org. Chem.*, **50**, 5537–5543.

[16] Lannoye, G., Sambasivarao, K., Wehrli, S., Weiss, U. and Cook, J. M. (1988) General approach to the synthesis of polyquinenes via the Weiss reaction. 6. Progress toward the synthesis of dicyclopentapentalenes. *J. Org. Chem.*, **53**, 2327–2340.

[17] (a) Van Ornum, S. G. and Cook, J. M. (1996) Generation of four five-membered rings in a one pot process. Studies directed toward the synthesis of dicyclopenta[a,d]pentalene via the tandem Pauson-Khand reaction. *Tetrahedron Lett.*, **37**, 7185–7188; (b) Van Ornum, S. G. and Cook, J. M. (1997) Formation of six carbon-carbon bonds in a one pot process. Generation of the dicyclopenta[a,e]pentalene ring system *via* the tandem Pauson-Khand reaction. *Tetrahedron Lett.*, **38**, 3657–3658; (c) Van Ornum, S. G., Bruendl, M. M. and Cook, J. M. (1998) The livinghouse catalytic approach to the tandem Pauson-Khand reaction. Entry into the parent ring systems of dicyclopenta[*a,f*]pentalene and dicyclopenta[*a,e*]pentalene. *Tetrahedron Lett.*, **39**, 6649–6650; (d) Van Ornum, S. G., Bruendl, M. M., Cao, H., Reddy, M., Grubisha, D. S., Bennett, D. W. and Cook, J. M. (2000) Utility of the tandem Pauson-Khand reaction in the construction of tetracycles. *J. Org. Chem.*, **65**, 1957–1971; (e) Cao, H., Van Ornum, S. G. and Cook, J. M. (2000) The molybdenum-mediated tandem Pauson-Khand reaction. High yield entry into [5.5.5.5]tetracyclic systems. *Tetrahedron Lett.*, **41**, 5313–5316.

[18] Van Ornum, S. G. (1998) *Ph.D. Thesis*; Department of Chemistry; University of Wisconsin-Milwaukee, Milwaukee, Wi.

[19] Cao, H. (2003) *Ph.D. Thesis*; Department of Chemistry; University of Wisconsin-Milwaukee: Milwaukee.

[20] Stowasser, B. and Hafner, K. (1986) Synthesis of a dicyclopenta[*a, e*]pentalene by [6 + 2]-cycloaddition of 1,3-di-*tert*-butyl-5-vinylidenecyclopentadiene and aonsecutive 8π-electrocyclic reaction. *Angew. Chem., Int. Ed. Engl.*, **25**, 466–467.

[21] (a) Kent, J. L., Wan, H. and Brummond, K. M. (1995) A new allenic Pauson-Khand cycloaddition for the preparation of α-methylene cyclopentenones. *Tetrahedron Lett.*,

36, 2407–2410; (b) Brummond, K. M., Wan, H. and Kent, J. L. (1998) An intramolecular allenic [2 + 2 + 1] cycloaddition. *J. Org. Chem.*, **63**, 6535−6545.

[22] Faust, R. and Weber, C. (1997) One-step synthesis of dialkynyl-1,2-diones and their conversion to fused pyrazines bearing enediyne units. *Tetrahedron*, **53**, 14655–14670.

[23] Brandsma, L. and Verkrujisse, H. D. (1981) In *Synthesis of Acetylenes, Allenes and Cumulenes*; Elsevier, 24.

[24] Kammermeier, S., Tykwinski, R. R., Siemsen, P., Seiler, P. and Diederich, F. (1998) Synthesis of a highly strained permethylenated cycloocta-1,5-diyne derivative by acid-catalysed thermal rearrangement. *Chem. Commun.*, 1285–1286.

[25] (a) Mengel, A. and Reiser, O. (1998) Around and beyond Cram's rule. *Chem. Rev.*, **99**, 1191–1224. (b) Anh, N. T. (1980) Regio and Stereoselectivities in Some Nucleophilic Reactions. *Top. Curr. Chem.*, **88**, 145−162.

[26] Brandsma, L. and Verkrujisse, H. D. (1981) In *Synthesis of Acetylenes, Allenes and Cumulenes*; Elsevier, 155.

[27] Ma, P., Martin, V. S., Masamune, S., Sharpless, K. B. and Viti, S. M. (1982) Synthesis of saccharides and related polyhydroxylated natural products. 2. Simple deoxyalditols. *J. Org. Chem.*, **47**, 1378–1380.

[28] Cao, H., Flippen-Anderson, J. and Cook, J. M. (2003) The synthesis of a dicyclopenta[*a*,*e*]pentalene via a molybdenum hexacarbonyl-mediated tandem allenic Pauson-Khand reaction. *J. Am. Chem. Soc.*, **125**, 3230–3231.

[29] Jeong, N., Lee, S. J., Lee, B. Y. and Chung, Y. K. (1993) Molybdenum mediated preparation of cyclopentenones. *Tetrahedron Lett.*, **34**, 4027–4030.

[30] (a) Murakami, S., Takemoto, T., and Shimizu, Z. (1953) The effective principle of Digenea simplex Aq I. Separation of the effective fraction by liquid chromatography. *J. Pharm. Soc. Japan.*, **73**, 1026-1028; (b) Morimoto, H. (1955) Studies on the Active Components of *Digenea simplex* Ag. and Related Compounds. XXV. *J. Pharm. Soc. Japan.*,**75**, 937–40.

[31] (a) Watkins, J. C., Krogsgaard-Larsen, P. and Honoré, T. (1990) Structure-activity relationships in the development of excitatory ammo acid receptor agonists and competitive antagonists. *Trends Pharmacol. Sci.*, **11**, 25–33; (b) Cantrell, B. E., Zimmerman, D. M., Monn, J. A., Kamboj, R. K., Hoo, K. H., Tizzano, J. P., Pullar, I. A., Farell, L. N. and Bleakman, D. J. (1996) Synthesis of a series of aryl kainic acid analogs and evaluation in cells stably expressing the kainate receptor humGluR6. *J. Med. Chem.*, **39**, 3617–3624.

[32] (a) Yoo, S-E., Lee, S. H., Jeong, N., and Cho, I. (1991) A total synthesis of (-)-α-kainic acid by the Pauson-Khand reaction. *Tetrahedron Lett.*, **34**, 3435–3438; (b) Yoo, S-E., and Lee, S. H. (1991) A Total synthesis of (-)-alpha-kainic acid involving a Pauson-Khand reaction as the key step. *J. Org. Chem.*, **59**, 6968–6972; (c) Takano, S., Inomata, K., and Ogasawara, K. (1992). A new enantiospecific route to (-)-kainic acid *via* the intramolecular Pauson-Khand reaction. *J. Chem. Soc. Chem. Commun.*, 169–170.

[33] Farwick, A. and Helmchen, G. (2010) Enantioselective total synthesis of (-)-α-kainic acid. *Org. Lett.*, **12**, 1108–1111.

[34] Helmchen, G., Dahnz, A., Dübon, P., Schelwies, M., and Wiehofen, R. (2007) Iridium-catalysed asymmetricallylic substitutions. *Chem. Commun.* 675–691.

[35] Helmchen, G. and Prelog, V. (1972) Über atropisomere Biphenyl-Derivate mit einer Pseudoasymmetrieachse. *Helv. Chim. Acta.*, **55**, 2599–2611.

[36] Ley, S. V., Norman, J., Griffith, W. P. and Marsden, S. P. (1994) Tetrapropylammonium perruthenate, $Pr4N^+RuO_4^-$, TPAP: A catalytic oxidant for organic synthesis. *Synthesis*, 639–666.

[37] Tebbe, F. N., Parshall, G. W. and Reddy, G. S. (1978) Olefin homologation with titanium methylene compounds. *J. Am. Chem. Soc.*, **100**, 3611–3613.

[38] Kikuchi, H., Miyagawa, Y., Sahashi, Y., Inatomi, S, Haganuma, A., Nakahata, N. and Oshima, Y. (2004) Novel trichothecanes, paecilomycine A, B, and C, isolated from entomopathogenic fungus, *Paecilomyces tenuipes. Tetrahedron Lett.*, **45**, 6225–6228.

[39] Min, S-J., and Danishefsky, S. J. (2007) Total synthesis of paecilomycine A. *Angew. Chem Int. Ed.*, **46**, 2199–2202.

[40] Corey, E. J. and Virgil, S. C. (1990) Enantioselective total synthesis of a protosterol, 3.beta.,20-dihydroxyprotost-24-ene. *J. Am. Chem. Soc.*, **112**, 6429–6431.

[41] Castro, V., Ciccio, F., Alvardo, S., Bohlmann, F., Schmeda-Hirschmann, G. and Jakupovic, J. (1983) Decathieleanolid, ein dimeres guajanolid aus *Decachaeta thieleana Liebigs, Ann. Chem.*, 974–981.

[42] Oberti, J. C., Sosa, V. E., Herz, W., Prasad, J. S. and Goedken, V. L. (1983) Crystal structure and stereochemistry of achalensolide, a new guaianolide from Stevia achalensis. *J. Org. Chem.*, **48**, 4038–4043.

[43] Hirose, T., Miyakoshi, N. and Mukai, C. (2008) Total synthesis of (+)-achalensolide based on the Rh(I)-catalyzed allenic Pauson-Khand reaction. *J. Org. Chem.*, **73**, 1061–1066.

[44] Gravier-Pelletier, C., Dumas, J., Le Merrer, Y. L. and Depezay, J.-C. (1992) A General way, from L-Ascorbic and D-isoascorbic acids, to homochiral α-hydroxy, α,β-hydroxy, and α,β-epoxyaldehydes, useful building blocks for the synthesis of linear oxygenated acids metabolites. *Carbohydr. Chem.*, **11**, 969–998.

[45] (a) Tran, Y. S. and Kwon, O. (2005) An application of the phosphine-catalyzed [4 + 2] annulation in indole alkaloid synthesis: formal syntheses of ($\pm$)-alstonerine and ($\pm$)-macroline. *Org. Lett.*, **7**, 4289–4291; (b) Bi, Y., Zhang, L.-H., Hamaker, L. K. and Cook, J. M. (1994) Enantiospecific synthesis of (-)-alstonerine and (+)-macroline as well as a partial synthesis of (+)-villalstonine. *J. Am. Chem. Soc.*, **116**, 9027–9041; (c) Liao, X., Zhou, H., Wearing, X. Z., Ma, J. and Cook, J. M. (2005) The first regiospecific, enantiospecific yotal synthesis of 6-oxoalstophylline and an improved total synthesis of alstonerine and alstophylline as well as the bisindole alkaloid macralstonine. *Org. Lett.*, **7**, 3501–3504; (d) Liao, X., Zhou, H., Yu, J. and Cook, J. M. (2006) An improved total synthesis of (+)-macroline and alstonerine as well as the formal total synthesis of (−)-talcarpine and (−)-anhydromacrosalhine−methine. *J. Org. Chem.*, **71**, 8884-8890; (e) Cook, J. M., LeQuesne, P. W. and Elderfield, R. C. (1969) Alstonerine, a new indole alkaloid from *Alstonia muelleriana. J. Chem. Soc., Chem. Commun.*, 1306–1307.

[46] Keawpradub, N., Eno-Amooquaye, E., Burke, P. J. and Houghton, P. J. (1999) Cytotoxic activity of indole alkaloids from *alstonia macrophylla. Planta Med.*, **65**, 311–315.

[47] Miller, K. A., Shanahan, C. S. and Martin, S. F. (2008) The Pauson-Khand reaction as a new entry to the synthesis of bicyclic heterocycles: application to the enantioselective total synthesis of (-)-alstonerine. *Tetrahedron*, **64**, 6884–6900.

[48] Neipp, C. E. and Martin, S. F. (2003) Synthesis of bridged azabicyclic structures via ring-closing olefin metathesis. *J. Org. Chem.*, **68**, 8867–8878.

[49] (a) Numura, I. and Mukai, C. (2004) Studies on the total synthesis of steptazolin and its related natural products: first total synthesis of (±)-8α-hydroxystreptazolone. *J. Org. Chem.*, **69**, 1803–1812; (b) Numura, I. and Mukai, C. (2002) Total synthesis of (±)-8α-hydroxystreptazolone. *Organic Lett.*, **4**, 4301–4304.

[50] (a) Kozikowski, A. P. and Park, P. J. (1985) Total synthesis of streptazolin - an application of the aza-analogue of the ferrier rearrangement. *J. Am. Chem. Soc.*, **107**, 1763–1765; (b) Kozikowski, A. P. and Park, P. J. (1990) Synthesis of streptazolin: use of the aza-Ferrier reaction in conjunction with the INOC process to deliver a unique but sensitive natural product. *J. Org. Chem.*, **55**, 4668–4682; (c) Flann, C. J. and Overman, L. E. (1987) Enantioselective total synthesis of streptazolin. The tandem use of iminium ion vinylsilane cyclizations and intramolecular acylations. *J. Am. Chem. Soc.*, **109**, 6115–6118; (d) Yamada, H., Aoyagi, S. and S. Kibayashi, C. (1996) Stereoselective total synthesis of natural (+)-streptazolin via a palladium-catalyzed enyne bicyclization approach. *J. Am. Chem. Soc.*, **118**, 1054–1059; (e) Huang, S., and Commins, D. L. (2000) Total synthesis of (+)-streptazolin. *J. Chem. Soc., Chem. Commun.*, 569–570.

[51] Pérez-Serrano, L., Casarrubios, L., Domínguez, G. and Pérez-Castells, J. (1999) Pauson-Khand reaction induced by molecular sieves. *Org. Lett.*, **1**, 1187–1188.

[52] (a) Sugihara, T., Yamada, M., Ban, H., Yamaguchi, M. and Kaneko, C. (1997) Rate enhancement of the Pauson-Khand reaction by primary amines. *Angew. Chem., Int. Ed.*, **36**, 2801–2804; (b) Sugihari, T., Yamada, M. and Nishizawa, M. (1999) The intra- and intermolecular Pauson-Khand reaction promoted by alkyl methyl sulfides. *Synlett*, **6**, 771–773.

[53] Crawford, J. J., Kerr, W. J., McLaughlin, M., Morrison, A. J. Pauson, P. L. and Thurston, G. J. (2006) Use of a highly effective intramolecular Pauson-Khand cyclization for the formal total synthesis of (±)-α- and β-cedrene by preparation of cedrone. *Tetrahedron*, **62**, 11360–11370.

[54] (a) Ohira, S. (1989) Methanolysis of dimethyl (1-diazo-2-oxopropyl) phosphonate: generation of dimethyl (diazomethyl) phosphonate and reaction with carbonyl compounds. *Synth. Commun.*, **19**, 561–564; (b) Müller, S., Liepold, B., Roth, G. J. and Bestmann, H. J. (1996) An improved one-pot procedure for the synthesis of alkynes from aldehydes. *Synlett*, 521–522.

[55] Kerr, W. J., Lindsay, D. M., McLaughlin, M. and Pauson, P. L. (2000) A polymer-supported alkyl methyl sulfide as an efficient promoter of the Khand cyclisation reaction. *Chem. Commun.*, 1467–1468.

[56] Honda, T., and Kaneda, K. (2007) Diastereoselective formal synthesis of a monoterpene alkaloid, (-)-incarvalline. *J. Org. Chem.*, **72**, 6541–6547.

[57] Nakamura, M., Chi, Y.–M., Yan, W.–M., Nakasugi, Y., Yoshizawa, T., Irino, N., Hashimoto, F., Kinjo, J., Nohara, T. and Sakurda, S. (1999) Strong antinociceptive effect of incarvillateine, a novel monoterpene alkaloid from *incarvillea sinensis*. *J. Nat. Prod.*, **62**, 1293–1294.

[58] (a) Ichikawa, M., Takahashi, M., Aoyagi, S. and Kibayashi, C. (2004) Total synthesis of (−)-incarvilline, (+)-incarvine C, and (−)-incarvillateine. *J. Am. Chem. Soc.*, **126**, 16553–16558; (b) Ichikawa, M., Aoyagi, S. and Kibayashi, C. (2005) Total synthesis of (-)-incarvalline. *Tetrahedron Lett.*, **46**, 2327-2329; (c) Hong, B.–C., Gupta, A. K., Wu, M.–F., Liao, J.–H. and Lee, G.–H. (2003) Hetero [6+3] cycloaddition of fulvenes with *N*-alkylidene glycine esters: a facile synthesis of the delavayine and incarvillateine framework. *Org. Lett.*, **5**, 1689–1692.

[59] Sugihara, T., Yamaguchi, M. and Nishizawa, M. (1999) Activation of alkyne-dicobalt hexacarbonyl complexes by the Lewis bases a new trend of the Pauson-Khand reaction. *J. Synth. Org. Chem. Jpn.*, **57**, 158–169.

[60] Jiao, L., Zhou, L.–G. and Yu, Z.–X. (2010) Rh(I)-catalyzed [(3+2)+1] cycloaddition of 1-yne/ene-vinylcyclopropanes and CO: homologous Pauson-Khand reaction and the total synthesis of (±)-α-agarofuran. *Org. Lett.*, **12**, 2528–2531.

[61] (a) Jiao, L., Ye, S. and Yu, Z.-X. (2008) Rh(I)-catalyzed intramolecular [3 + 2] cycloaddition of *trans*-vinylcyclopropane-enes. *J. Am. Chem. Soc.*, **130**, 7178–7179; (b) Jiao, L., Lin, M. and Yu, Z.–X. (2010) Rh(I)-catalyzed intramolecular [3 + 2] cycloaddition reactions of 1-ene-, 1-yne- and 1-allene-vinylcyclopropanes. *Chem. Commun.*, 46, 1059–1061; (c) Li, Q., Jiang, G.–J., Jiao, L., and Yu, Z.–X. (2010) Reaction of α-ene-vinylcyclopropanes: type II intramolecular [5+2] cycloaddition or [3+2] cycloaddition? *Org. Lett.*, **12**, 1332–1335.

[62] Hayashi, Y., Miyakoshi, N., Kitagaki, S. and Mukai, C. (2008) Stereoselective total syntheses of uncommon sesquiterpenoids isolated from *Jatropha neopauciflora*. *Org. Lett.*, **10**, 2385–2388.

[63] Ishikawa, T., Ishii, H., Shimizu, K., Nakao, H., Urano, J., Kudo, T. and Saito, S. (2004) Diastereoselective total synthesis of isocarbacyclin from L-ascorbid acid. *J. Org. Chem.*, **69**, 8133–8135.

[64] Belanger, D. B. and Livinghouse, T. (1998) Hexacarbonyldicobalt-alkyne complexes as convenient $Co_2(CO)_8$ surrogates in the catalytic Pauson-Khand reaction. *Tetrahedron Lett.*, **39**, 7641–7644.

[65] Hasimoto, T., Tori, M., Taira, Z. and Asakawa, Y. (1985) New highly oxidized fusicoccane diterpenoids from the liverwort *plagiochila acanthophylla japonica*. *Tetrahedron Lett.*, **26**, 6473–6476.

[66] Nozoe, S., Morisaki, Tsuda, K., Iitaki, Y., Takahashi, T., Tamura, S., Ishibashi, K. and Shirasaka. M. (1968) The structure of ophiobolin, a C_{25} terpenoid having a novel skeleton. *J. Am. Chem. Soc.*, **87**, 4968–4970.

[67] Sassa, T., Takahama, A. and Shindo, T. (1975) The stereostructure of cotylenol, the aglycone of cotylenins leaf growth substances. *Agr. Biol. Chem.*, **39**, 1729–1734.

[68] Cao, H., Munda, S. R. and Cook, J. M. (2003) Studies on the Rh and Ir mediated tandem Pauson-Khand reaction. A new entry into the dicyclopenta[a,d]cyclooctene ring system. *Tetrahedron Lett.*, **44**, 6165–6168.

[69] Additional examples of the Pauson-Khand reaction in total synthesis: (a) Krafft, M.E., Cheung, Y.Y. and Abboud, K. A. (2001) Total synthesis of (±)-asteriscanolide. *J. Org. Chem.*, **66**, 7443–7448; (b) Ishizaki, M., Niimi, Y., Hoshino, O., Hara, H. and Takahashi, T. (2005) A formal total synthesis of Lycopodium alkaloid, (±)-magellanine, by using the intramolecular Pauson-Khand reaction. *Tetrahedron*, **61**, 4053–4065; (c) Nakayama, A., Kogure, N., Kitajima, Lycopodium Alkaloids, Lycoposerramine-C and

phlegmariurine-A. *Org. Lett.*, **11**, 5554–5557; (d) Bernardes, V., Kann, N., Riera, A., Moyano, A., Pericas, M.A. and Greene, A. E. (1995) Asymetric Pauson-Khand cyclization: a formal total synthesis of natural brefeldin A. *J. Org. Chem.*, **60**, 6670–6671; (e) Donkervoort, J.G., Gordon, A.R., Johnstone, C., Kerr, W.J. and Lange, U. (1996) Development of modified Pauson-Khand reactions with ethylene and utilisation in the total synthesis of (+)-taylorione. *Tetrahedron*, **52**, 7391–7420; (f) Fuyuhiko Inagaki, Masahiko Kinebuchi, Naoki Miyakoshi and Chisato Mukai (2010) Formal Synthesis of (+)-Nakadomarin A. *Org. Lett.*, **12**, 1800–1803.

9

Heterogeneous Catalytic Pauson-Khand Reaction

Young Keun Chung

The Pauson-Khand Reaction: Scope, Variations and Applications, First Edition. Edited by Ramon Rios Torres.

9.1 Introduction

Ever since the catalytic Pauson-Khand reaction (PKR) was discovered,[1] most studies have focused on its homogeneous reactions.[2] Knowledge about such reactions through extensive studies has provided significant help in developing new homogeneous catalysts. In contrast, the development of heterogeneous catalysts for PKR lags far behind, presumably due to difficulties in studying the reaction process. For a PKR, only limited types of heterogeneous reactions have been reported that include reactions between a liquid and a solid, which is the most common type of a heterogeneous PKR, and a handful of PKRs between an organic solvent and water (in biphasic conditions).[3] Reactions between a gas and a solid have not been reported.

While heterogeneous solid catalysts have inherent operational advantages that include ease of handling or separation for recycling, both of which are difficult in their homogeneous counterpart, the complexity of the reaction that arises mainly from the various irregular interactions between the stabilizer, solvent or substrate and the metal surface often hampers the design of a reaction. Thus, tailoring the heterogeneous catalysts to reach the desired selectivity is not easy, and further understanding regarding the reaction mechanisms of sufficient precision is still required although it is extremely difficult.

Distinguishing heterogeneous catalysis from homogeneous is an important issue and yet is not always obvious. For example, problems regarding the true homogeneity or real heterogeneity of catalytic systems used in the Pd(0)-catalyzed C–C cross-coupling reactions have often been a subject of debate[4] because they seem to be essential for understanding the reaction mechanism and function of Pd catalysts. For this reason, many useful methods for assessing heterogeneity and homogeneity in catalytic reactions have been proposed.[5] In relation to PKRs, however, this kind of problem has not arised yet and thus we will directly prompt discussion about the variety of heterogeneous catalysis of the PK and related reactions instead of starting with the general aspects of heterogeneity of them. For a general discussion of homogeneous/heterogeneous catalytic systems, please see other books and reviews.[6]

9.2 Development of Heterogeneous Catalysts for PKR

Endeavors to develop heterogeneous catalysts of the PKR have been channeled into two different approaches: one is the immobilization of homogeneous catalysts (e.g., cobalt carbonyls) on solid supports such as insoluble polymers[7] and dendrons[8] and the other is the use of metallic cobalt itself as a catalyst. The development of heterogenized homogeneous catalysts, which are intended to combine the advantages of homogeneous catalysts with those of heterogeneous ones, seems to be highly desirable. However, they suffer from many difficulties that include tedious and/or expensive syntheses of polymeric ligands, reduced catalytic activities and selectivities relative to the homogeneous catalyst, and excessive leaching of the catalytic species from the solid support. A catalytic system that overcomes all these problems has not yet been reported. The use of metallic cobalt(0) as a catalyst had been anticipated from previous uses of *in situ* generated Co(0) species to generate cobalt carbonyls.[9] For example, $Co(acac)_2$ with $NaBH_4$ acts as a catalytic system under mild conditions.[10] This study suggested that such a system could be catalytic providing it leads to *in situ* generation of low valency cobalt, which could then react with available CO to afford cobalt carbonyl species. Based on this fact, the first practical heterogeneous PKR catalyst of metallic cobalt was developed in 2000.[11] Since then, cobalt nanoparticles,[12] heterobimetallic cobalt and rhodium nanoparticles,[13] and useful variants of them[14] have been reported.

9.2.1 Polymer-Supported Catalytic Systems

The immobilization of homogeneous catalysts on solid supports and polymer-supported complexes has been actively studied in many catalytic systems.[15] However, only two cases have been reported for the PKR.[7,8] The first was reported by Gibson et al.[7] They prepared resin-bound phosphine-substitued cobalt carbonyl complexes via a reaction of polymer-bound triphenylphosphine with $Co_2(CO)_8$ in THF. The obtained resin was much more stable than bare $Co_2(CO)_8$. Using 5 mol% of catalyst precursor, they obtained products of an intramolecular PKR of terminal enynes in 50–60% yield (Scheme 9.1).

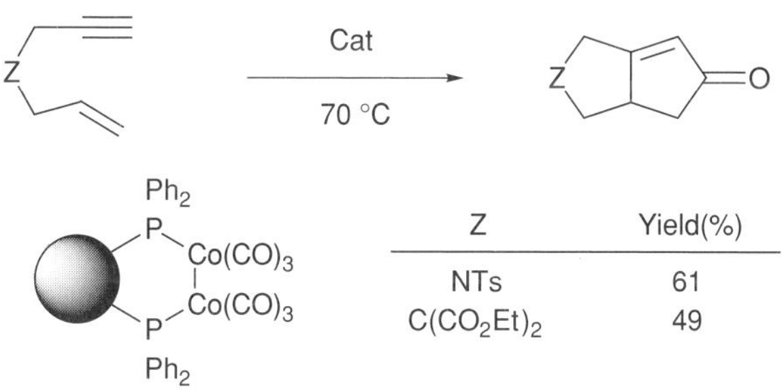

Z	Yield(%)
NTs	61
$C(CO_2Et)_2$	49

Scheme 9.1

They presented only two examples. When the catalytic system was recycled, the yield of allyl propargyl tosylamide dropped to 28%.[7b] According to their microanalysis of the resin after catalysis, up to 13% of the cobalt was lost during the first cycle. Nevertheless, the mild reaction conditions (70 °C and 1.05 bar CO) compared to those (130 °C and 20 atm CO) of metallic cobalt system[11] is noteworthy.

Portnoy et al. reported[8] the building of a catalyst on polyether dendron-functionalized polystyrene. They used Wang resin and three dendronized supports functionalized with 2- and 4-(diphenylphosphino)benzoic acid. Using cobalt complexes immobilized on the second- and third-generation dendron-functionalized polystyrene as catalysts, they studied an intramolecular PKR (Scheme 9.2).

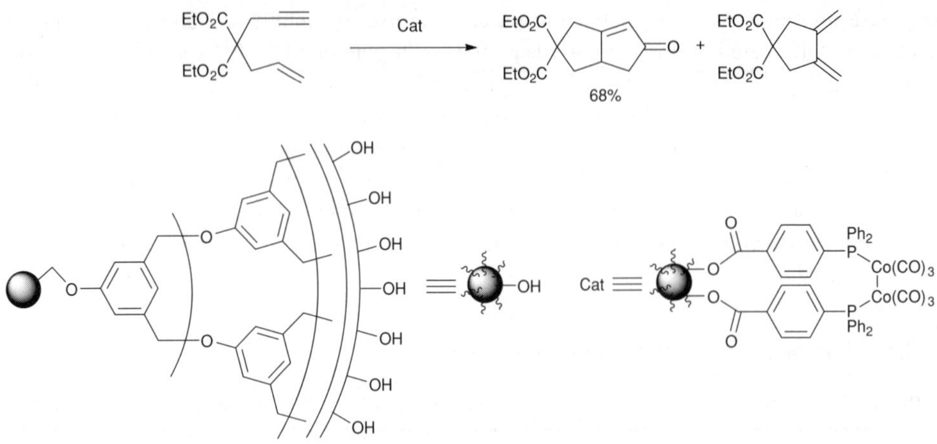

Scheme 9.2

In the catalytic reaction, along with the regular PKR product, an Alder-ene-type byproduct was obtained under all reaction conditions. They presented only one example (the best yield of the PKR product: 68%). Those catalytic cycles work under mild reaction conditions (5 mol% Co, 1.05 bar CO, THF, 70 °C, 24 h; 5 mol% of Co dimer, 1 atm CO, THF, 70 °C, 24 h) compared to those of the metallic cobalt system.

9.2.2 Bulk Cobalt as a Catalyst[11,17]

The first metallic cobalt catalyst for the PKR was prepared by decomposing $Co_2(CO)_8$ in a refluxing toluene solution or using the conventional impregnation method that adopts cobalt nitrate as a cobalt source with mesoporous silica such as SBA-15 and MCM-41.[11] The choice of mesoporous silica as a support was due to its large surface areas and porous structures.[16] The metallic cobalt immobilized on mesoporous silica was quite effective for the intramolecular PKR although harsh reaction conditions were necessary (Scheme 9.3).

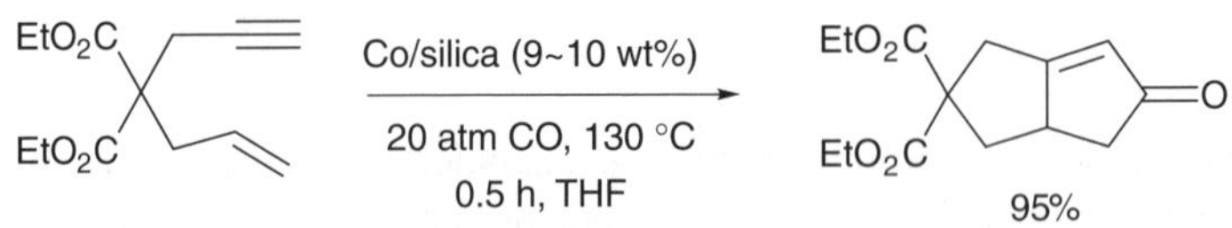

Scheme 9.3

To obtain good results, the cobalt loading should be adjusted to 9–10 wt% and the reaction temperature and the pressure of CO should be at least 130 °C and 20 atm, respectively. The catalyst was air-stable and reusable, and exhibited an excellent catalytic performance for many intramolecular PKRs (Table 9.1). The reaction gave good yields with mono-substituted olefins regardless of the substitution pattern of alkynes, while the internal alkyne substrate needed a relatively long reaction time. However, the catalytic system displayed

Table 9.1

Substrate	Time (h)	Product (%)	
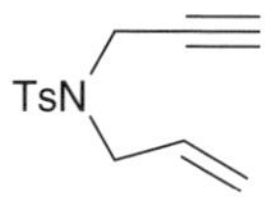	6.5	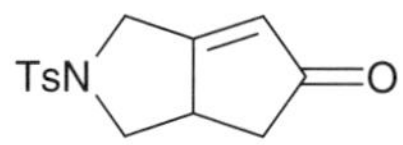	88
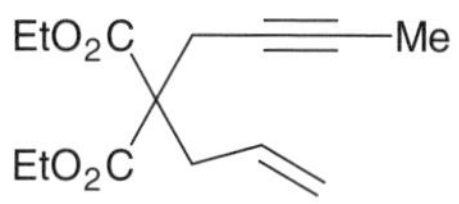	18	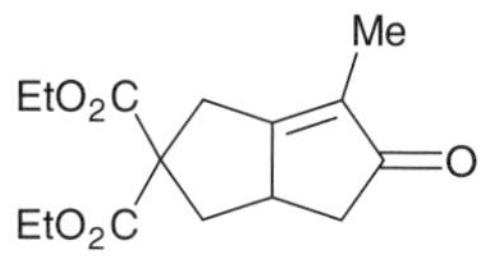	92
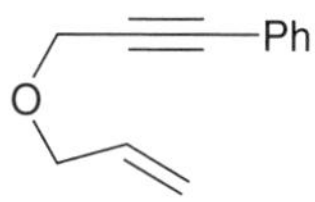	18	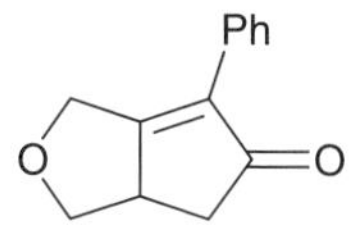	98
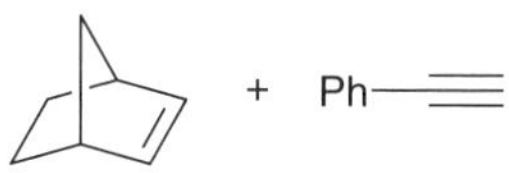	6.5	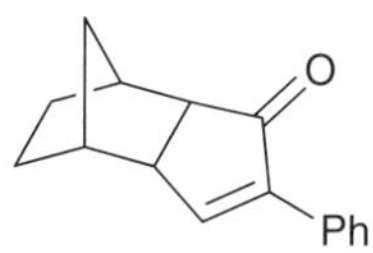	11
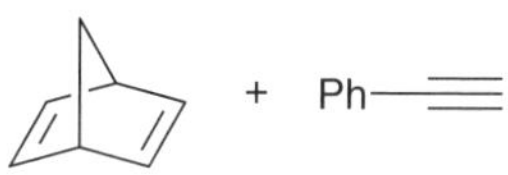	6.5	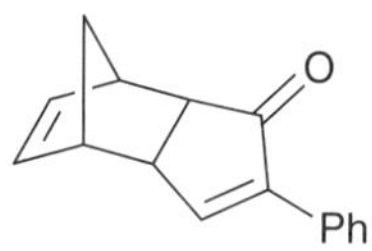	36

low activity in the intermolecular PKR. The bulk metallic cobalt on mesoporous silica was easily recovered and reused without loss of activity.

When the support was charcoal, the catalytic system (**Co/C**) exhibited an excellent catalytic performance for intra- and intermolecular PKRs (Table 9.2).[17] However, an *n*-alkyl halide substituted substrate was found to be inappropriate for the **Co/C** system. The optimum cobalt loading on charcoal was about 12 wt%. The catalytic system is stable and can be reused at least 10 times without losing any catalytic activity. The use of other carbon supports such as graphite and mesoporous carbon showed almost the same catalytic activity as that of cobalt on charcoal even though they have different surface areas and structures. The metallic cobalt showed ferromagnetism and was thus easily recovered by filtration or by using a magnet.

9.2.3 Raney Cobalt[14a]

Recently, skeletal metal catalysts[18] have attracted a lot of attention especially due to their microscopic spongy network of pores. One of the famous groups of skeletal catalysts is Raney catalysts. Recently, Leitner et al. reported[14a] on the use of the commercially available Raney cobalt in catalytic intra- and intermolecular PKRs. According to their report, their reaction conditions (the substrate/Co = 17.0, 23 bar CO, 130 °C; the substrate/Co ratio = 6.0, 35 bar CO, 130 °C) were quite similar to those of bulk metallic cobalt on charcoal,

Table 9.2

Substrate	Product			Yield (%)
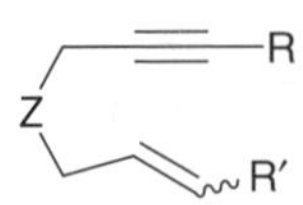	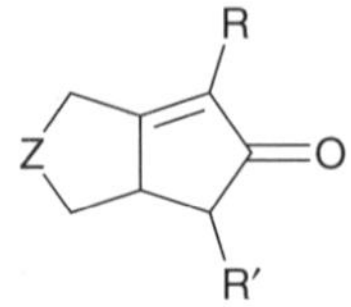			
	Z	R	R′	
	$C(CO_2Et)_2$	Me	H	98
	O	Ph	H	98
	NTs	Me	H	98
	O	H	Ph	93
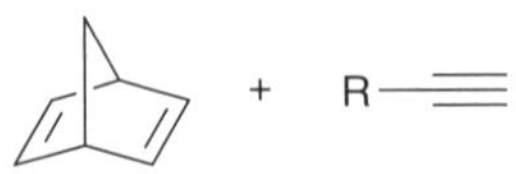	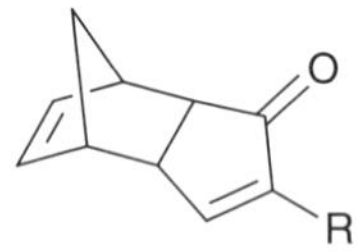			
		R		
		Ph		98
		n-Pent		98
		Cl		22

although the average size of Raney cobalt seems to be in the range of nanoparticles.[19] The catalytic system was quite effective for the intramolecular PKR of carbon-tethered enynes (Table 9.3). However, rather moderate yields (52–56%) were observed for the intermolecular reaction of norbornene with 1-pentyne, 1-octyne or 4-pentyn-1-ol. Raney cobalt is commercially available, cheap, easy to handle and nontoxic. Moreover, the catalytic system can be efficiently separated and reused by filtration and by exploiting its ferromagnetic properties.

9.2.4 Mesoporous Organized Zirconium Oxide Silica Powders as Catalyst[21]

The mesoporous silicas have been considered fascinating materials for many technological applications due to their porous and morphological characteristics.[20] Sanchez et al. reported[21] the use of mesoporous mixed zirconia-silica powders (**ZS20$_c$**) in the PKR. Under the optimized reaction conditions (4 mol% zirconium at 130 °C under 30 bar of CO for 48 h in toluene), they studied a reaction between cyclohexene and alkynes such as phenylacetylene, 1-octyne, 4-octyne and methyl butyndioate (Scheme 9.4). The conversion rates for terminal alkynes were low, 20% for phenylacetylene and 12% for 1-octyne. In the cases of 4-octyne and methyl butynedioate, no conversion was observed. One notable thing is that the acetylenic moiety was located in β position of the carbonyl group in the reaction product. Overall, the observed regioselectivity is rather unusual and thus interesting, but

Table 9.3

Substrate	Product	Yield (%)
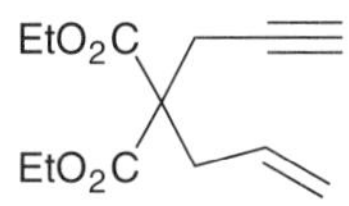	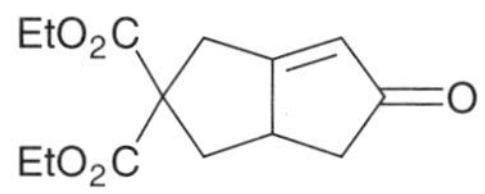	91.5
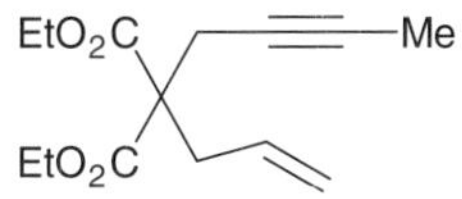	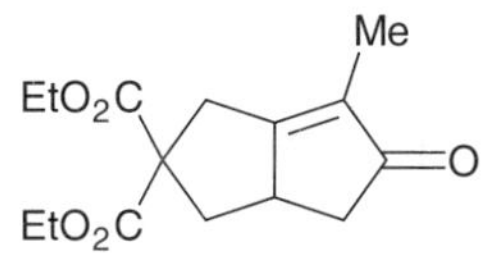	94
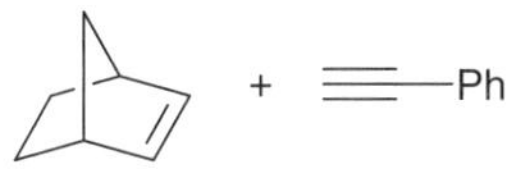	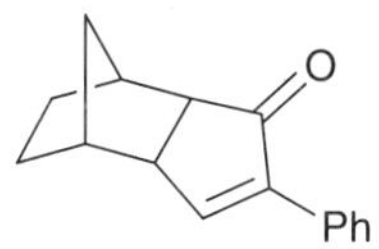	86
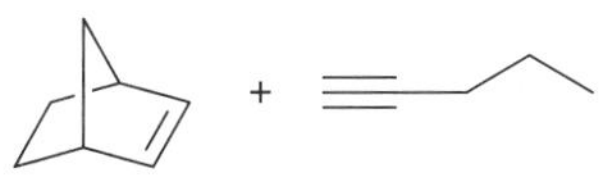	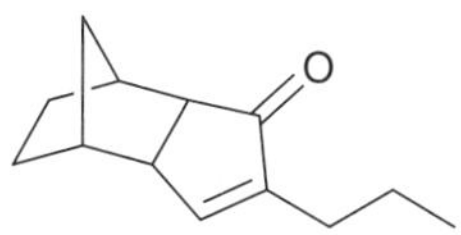	56
	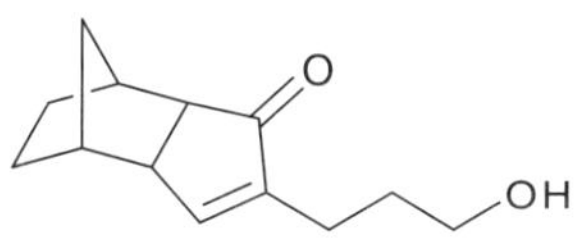	52.5

the observed yields were poor and the scope of the alkyne substrate was too narrow to be used in practical synthesis.

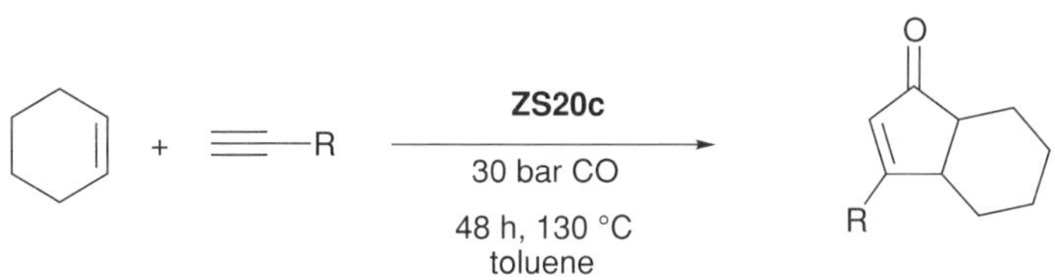

R	Yield (%)
Ph	20
n-C_6H_{13}	12

Scheme 9.4

9.2.5 Mesoporous Graphitic Carbon as a Promoter[22]

Antonietti et al.[22] studied the use of mesoporous graphitic carbon nitride (**mpg-C_3N_4**) (Figure 9.1) as a new class of heterogeneous metal-free catalysts for the activation of

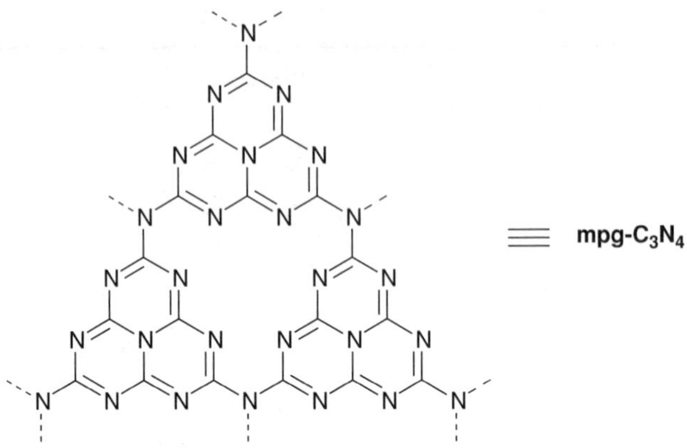

Figure 9.1 [25] Reproduced by permission of the Royal Society of Chemistry.

CO_2. They discovered that phenol, biphenyl and CO could be generated by the reaction of benzene and CO_2 in the presence of **mpg-C_3N_4**. The generated CO was quenched in a PK-type reaction (Scheme 9.5). When benzene, 1-hexene and dimethylacetylene-1,2-dicarboxylate (DMAD) were reacted in the presence of **mpg-C_3N_4** at 150 °C under 10 bar of CO_2 for 2 d, 70% of the PK product and 30% of a Diels–Alder reaction product between DMAD and biphenyl were observed. Although the experimental details have not been reported, it is reasonable to say that the reaction is promoted.

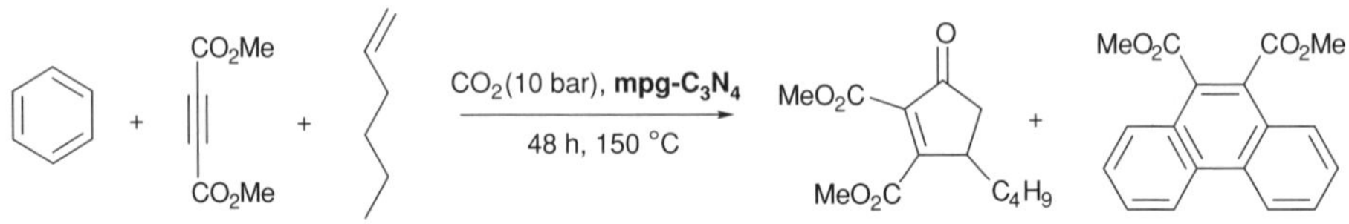

Scheme 9.5

9.3 Transition Metal Nanoparticle Catalyst

When bulk metallic cobalt was used as a catalyst in the PKR, the catalytic system was effective but it still required harsh reaction conditions, presumably due to the low surface-to-volume ratio of cobalt. However, if cobalt nanoparticles were used as catalysts, the surface-to-volume ratio of cobalt and the number of active sites dramatically increased because smaller particles have more edges and kinks than larger particles.[23] Small nanoparticles provide a tremendous driving force for diffusion especially at elevated temperatures. In addition, sometimes an inexplicable synergistic effect is also expected.[24] For these reasons researchers have dedicated themselves to using transition metal nanoparticles as catalysts. Since the first cobalt nanoparticle was used as a catalyst in the PKR, heterobimetallic nanoparticle systems such as Co/Ru,[25] Co/Rh[13,26] and Co/Pd[27] have been developed. We will discuss each topic briefly.

9.3.1 Colloidal Cobalt Nanoparticles

The cobalt nanoparticles prepared by the thermal decomposition of $Co_2(CO)_8$ were used in the PKR.[12] The intramolecular PKR of an enyne has been studied under optimized reaction conditions (Co NPs (25 mg, 45 wt% cobalt), enyne (0.42 mmol), 130 °C, THF, 5 atm of CO, 12 h) (Scheme 9.6).

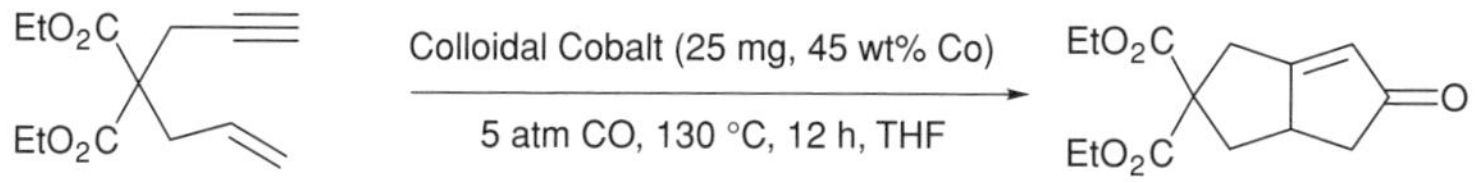

Temp (°C)	CO(atm)	Yield(%)
130	5	97
110	5	26
130	3	45
130	1	0

Scheme 9.6

The cobalt nanoparticle catalyst is more active than other types of heterogeneous catalysts such as metallic cobalt supported on mesoporous silica or charcoal.[11,17] The catalyst maintained its high activity even after five cycles of recycling and reuse. A TEM image (Figure 9.2b) of the catalyst after running four reaction cycles showed that the particle size of the cobalt nanoparticles was unchanged and confirmed that no particle agglomeration was taking place.

9.3.2 Cobalt Nanoparticles on Charcoal

To combine the merits of conventional heterogeneous catalysts with those of cobalt nanoparticles, such as the high catalytic activity, cobalt nanoparticles was immobilized on a support. The first choice of catalyst support in the developmental stage for liquid-phase reactions was charcoal because of its confirmed applicability in wide range.[28] Cobalt nanoparticles on charcoal (**CNC**) were prepared by the thermal reduction of dicobalt octacarbonyl.[29] **CNC** is quite stable even in air for several months.

9.3.2.1 CNC-Catalyzed PKR

For the normal intramolecular PKRs, the **CNC** system gave almost the same result as the cobalt nanoparticles (Scheme 9.7). The catalyst maintained its high activity even after being

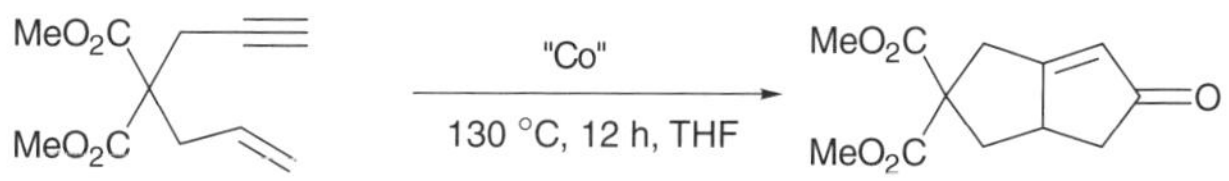

Co NP 97%
Co NP on charcoal 98%

Scheme 9.7

Table 9.4

Substrate	Product	Yield (%)
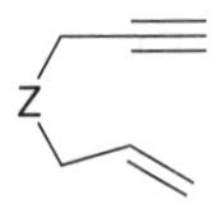	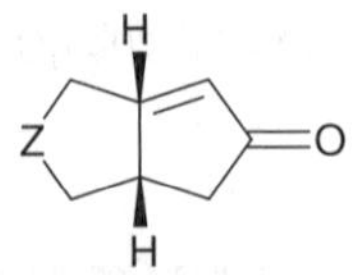	
	Z	
	$C(CO_2Et)_2$	98
	NBoc	90
	NTs	94
	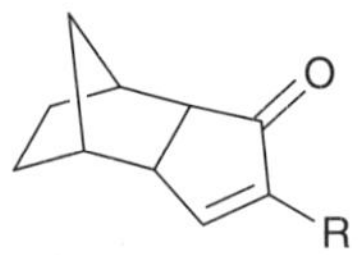	
	R	
	Ph	54
	n-Bu	43

recycled five times. However, for internal enyne and heteroatom-bridged enynes a higher CO pressure of 10 atm is required. The **CNC** is also effective for intermolecular PKR under 15 atm of CO.

9.3.2.2 *CNC-catalyzed one-pot reductive PKR*

The sequential reaction of PKR and hydrogenation is a useful method for the synthesis of natural products.[30] The known reductive PKRs, however, have used stoichiometric or substoichiometric amounts of cobalt complexes.[30] The use **CNC** as a bifunctional catalyst in the reductive PKR in a one-pot reaction was thus a meaningful. Under the optimum conditions (**CNC** (0.1 g, 12 wt%), substrate (0.48 mmol), THF, 130 °C, 5 atm CO, 5 atm H_2, 18 h), the desired products were successfully synthesized by **CNC** through the intramolecular reductive PKRs, but the yields of the intermolecular version were moderately low (Scheme 9.8 and Table 9.4).

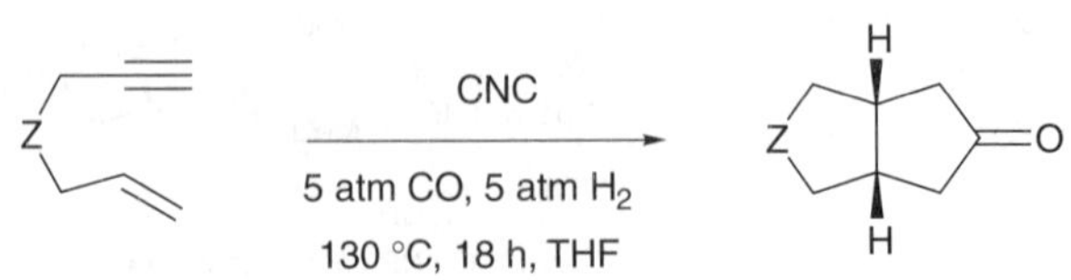

Scheme 9.8

The reductive PKR provides a straightforward approach to the bicyclic and tricyclic ketone frames which appear as core skeletons in many natural syntheses, such as those of the linear and angularly fused triquinane sesquiterpenes (Scheme 9.9).[32]

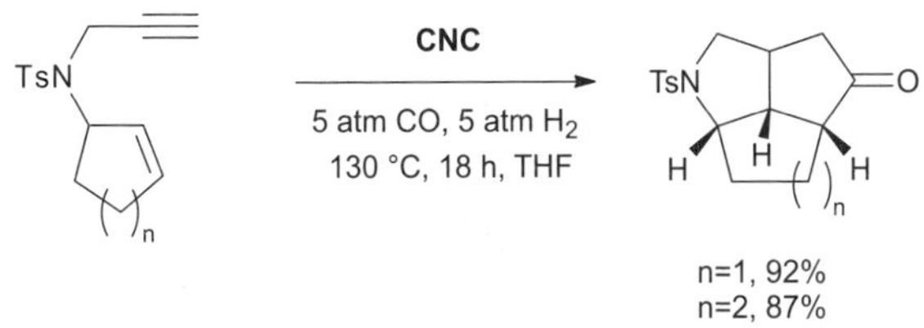

Scheme 9.9

9.3.3 Poly(ethylene glycol)-Stabilized Cobalt Nanoparticles[14b]

The use of insoluble polymeric materials as a support in heterogeneous catalysis often involves problems that include lowered reactivities, extended reaction times, diffusion-limited reactivity and reagent leaching. To alleviate these problems, Leitner et al. reported[14b] the use of PEG-stabilized cobalt nanoparticles as readily available, highly active and partly recyclable catalysts for the PKR (Scheme 9.10).

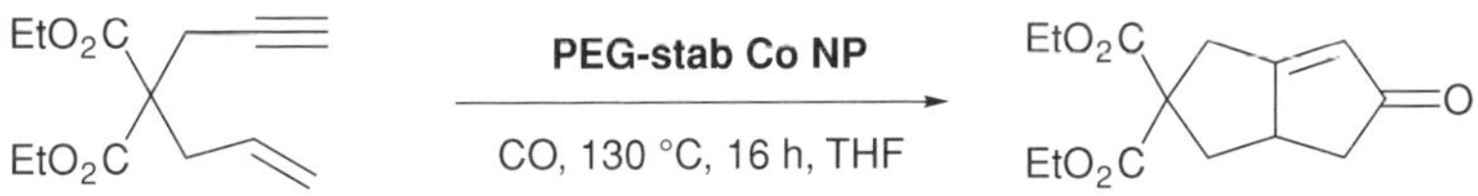

Co (mol%)	Press. (bar)	Conv. (%)	Select. (%)
3	23	>99	98
3	10	35	90
50	10	97	98
50	5	85	82

Scheme 9.10

As expected, the catalytic system was quite active for intramolecular PKRs of carbon-tethered enynes and for intermolecular PKRs of norbornene with phenylacetylene, 4-pentyn-1-ol, 1-pentyne and 1-octyne (Table 9.5). At high cobalt loading (50 mol%), the pressure of CO can be lowered 5–10 bar at 130 °C, but at low cobalt loading (3 mol%) suitable for the practical application, a high CO pressure (23–35 bar) was inevitable. Most reactions were thus carried out at 130 °C under high CO pressure and a reaction time of 16 h. Contrary to their expectations, PEG-stabilized cobalt nanoparticles were only partly recyclable due to the deactivation of the catalytic system resulting from the noticeable leaching (up to 0.74% per cycle based on the contamination in the product) and changes in the particle morphology.

Table 9.5

Substrate	Product	Yield (%)
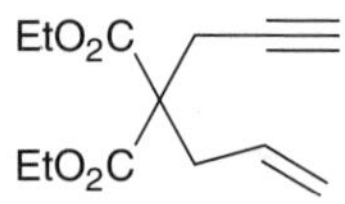	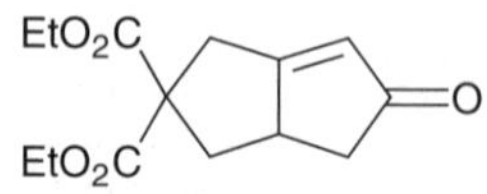	89
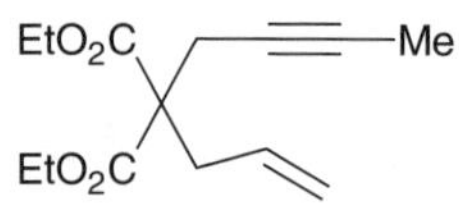	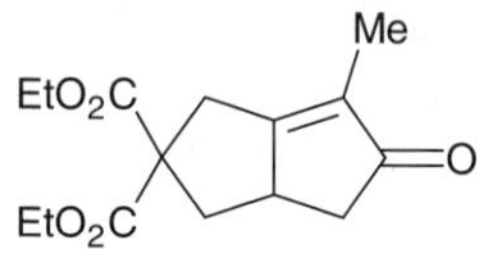	90
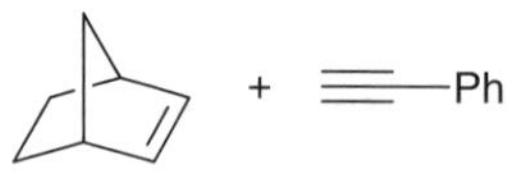	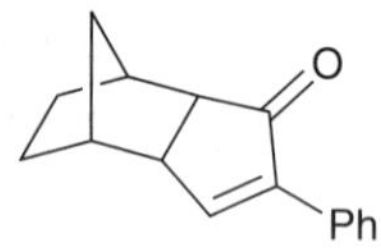	88
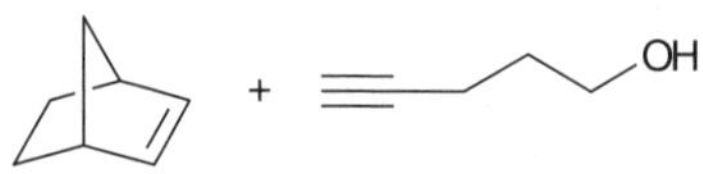	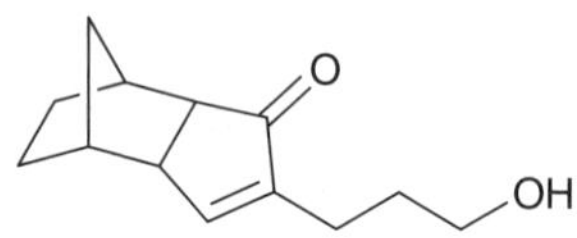	82

9.4 Bimetallic Nanoparticle Catalysts

A representative structure of bimetallic nanoparticle catalyst systems displays a core/shell, cluster-in-cluster or alloy (Figure 9.2). Besides the original function that can act as a catalyst in two consecutive reactions, these systems may provide a way to enhance the activity relative to the monometallic catalysts.

9.4.1 Immobilized Heterobimetallic Ru/Co Nanoparticle[25]

Heterobimetallic ruthenium/cobalt nanoparticles immobilized on charcoal (**RuCNC**) were developed as catalysts in the PKR using pyridylmethyl formate as a substitute of carbon monoxide (Scheme 9.11).[25] Among organic substitutes screened in the PKR,[33] pyridylmethyl formate was of choice. When either ruthenium carbonyl or colloidal cobalt nanoparticles were used as a catalyst in the presence of pyridylmethyl formate, no reaction product was observed.

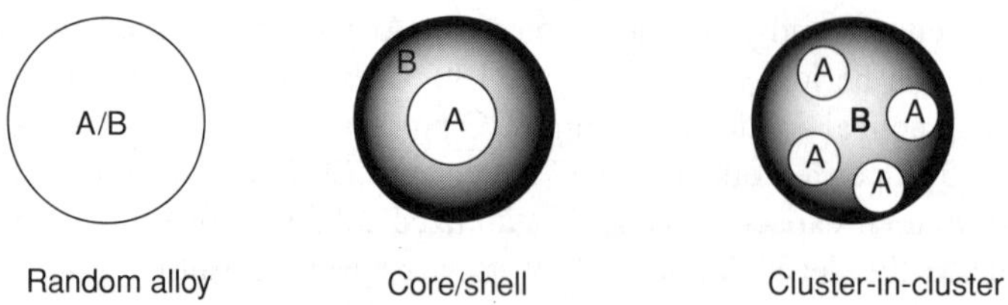

Figure 9.2 [36a] Reproduced by permission of EDP Sciences.

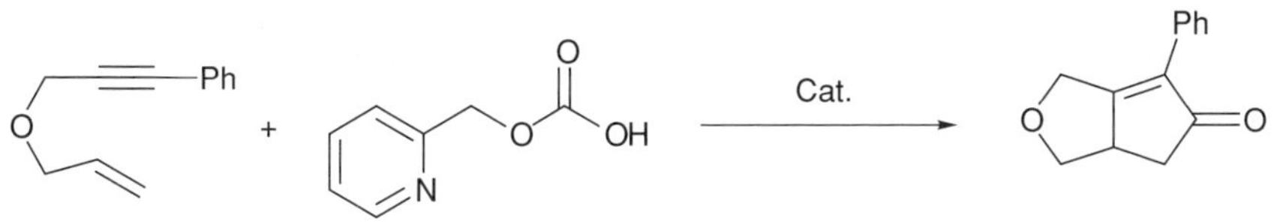

Cat.	Yield (%)
$Ru_3(CO)_{12}$/**CNC**	98
$Ru_3(CO)_{12}$	-
CNC	-
Ru CNC	98

Scheme 9.11

A TEM image (Figure 9.3) reveals the cluster-in-cluster structure, with mean sizes of 5 and 18 nm for ruthenium and cobalt, respectively.

Under the optimized reaction conditions (1.5 equiv 2-pyridylmethyl formate, THF, 12 h, and 130 °C), the **RuCNC** was a highly effective catalyst for intra- and intermolecular Pauson-Khand-type reactions (Table 9.6). The catalyst system could be re-used five times without loss of activity. Elemental analysis (by ICP-AES) of the reaction mixture after completion of the reaction showed that less than 0.1 ppm of the ruthenium and cobalt species had bled. The reusability of the catalyst and the environmentally friendly nature of the process are especially attractive.

9.4.2 Heterobimetallic Co/Rh Nanoparticles[13]

Enhancement of the catalytic activity of heterobimetallic (Co/Rh) nanoparticles in the PKR compared to that of mono-metallic ones (Co or Rh) was reported by Chung's group.[13] The choice of cobalt and rhodium metals as ingredients for the heterobimetallic nanoparticles was made given the established performance of each constituent in the PKR.[34,35]

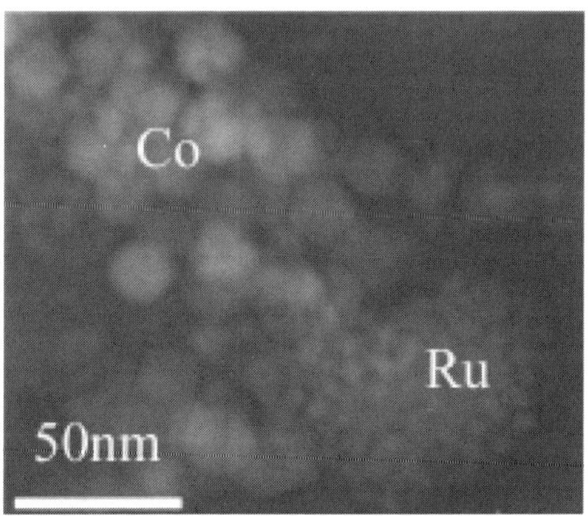

Figure 9.3 *HR-REM of **RuCNC**. Reprinted with permission from [27]. Copyright 2002 American Chemical Society.*

Table 9.6

Substrate	Product	Yield (%)
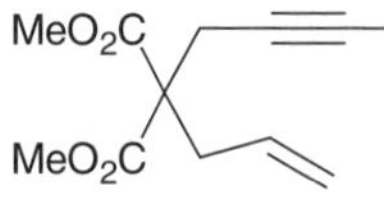	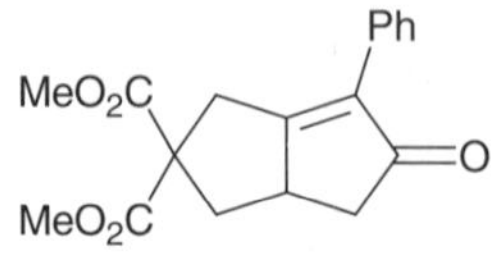	93
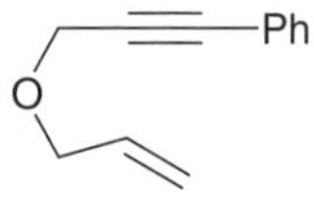	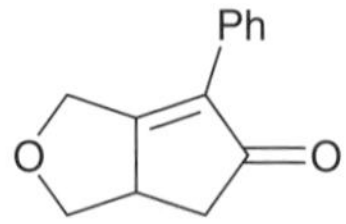	97
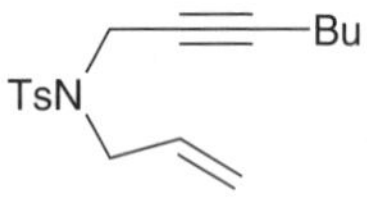	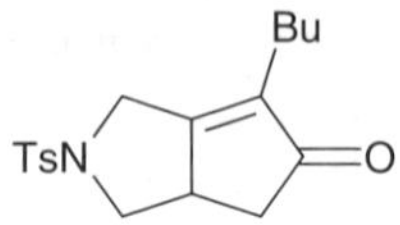	89
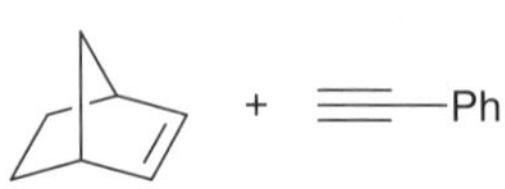	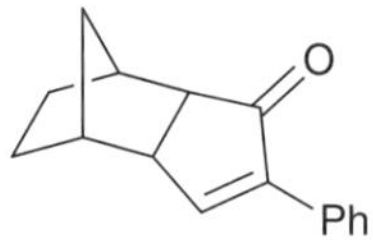	90
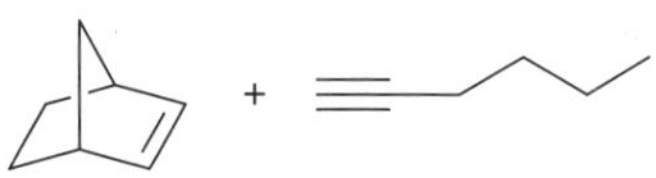	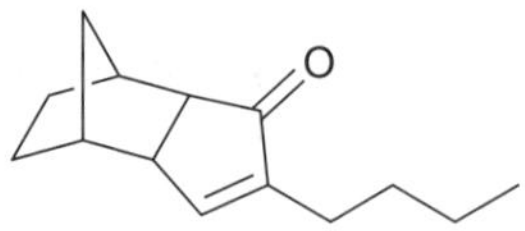	93

Cobalt-rhodium heterobimetallic nanoparticles were synthesized from $Co_nRh_{4-n}(CO)_{12}$ and immobilized on charcoal. The method used in this study gave fixed stoichiometric Co/Rh bimetallic nanoparticles.

Several years ago, Casanove reported[36] on the structure and chemical order in Co–Rh (Co_xRh_{1-x}) nanoparticles. Their synthesis was based on the co-decomposition under dihydrogen (3 bar) of [$Co(\eta^3$-$C_8H_{13})(^4\eta$-$C_8H_{12})$] and [Rh(acetylacetate)(η^4-C_8H_{12})], in a solvent containing either a polymer or a ligand. Based on their study, a marked segregation of cobalt atoms at the surface was proposed (Figure 9.4).

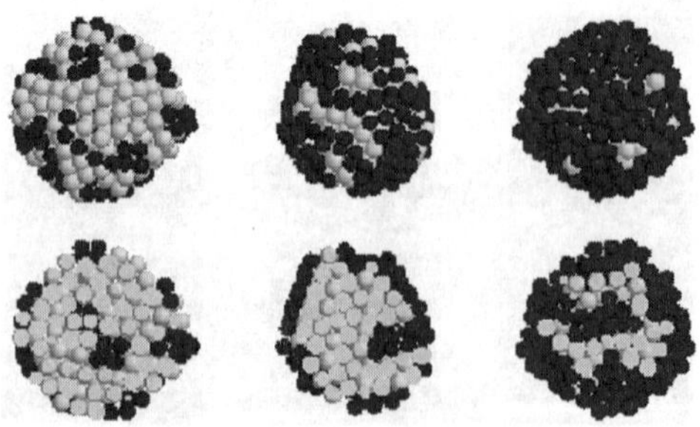

Figure 9.4 *420-atom clusters after MCM simulated. Black balls: Co; White balls: Rh. From Left to right: $Co_{0.25}Rh_{0.75}$, $Co_{0.5}Rh_{0.5}$, and $Co_{0.75}Rh_{0.25}$. Top: general view; bottom: cross-sections [from ref. 36].*

9.4.2.1 *Co_2Rh_2-catalyzed PKR under 1 atm of CO*

The use of heterobimetallic Co/Rh nanoparticles as catalysts in the intramolecular PKR of allyl propargyl ether was studied (Scheme 9.12).

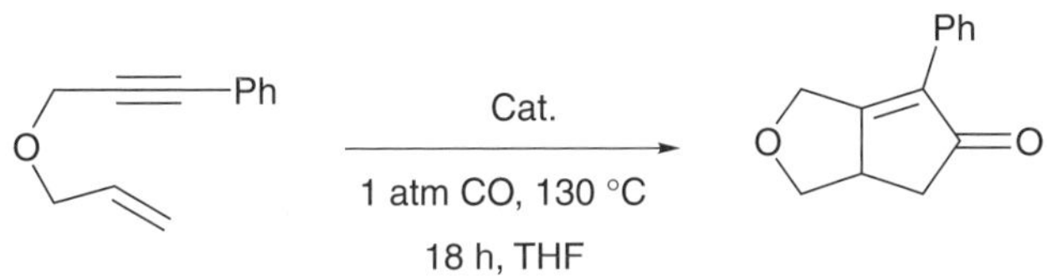

Cat.	Yield (%)
Co_2Rh_2	87
Co_3Rh	65
Co NP	0
Rh NP	23
A mixture of Co and Rh NP	12

Scheme 9.12

Under 1 atm CO, no reaction was observed with **CNC**, and 23% of the reaction product was obtained with rhodium nanoparticles on charcoal. When **Co_3Rh** nanoparticles (derived from $Co_3Rh(CO)_{12}$) were used, the expected product was obtained in 65% yield. The use of **Co_2Rh_2** nanoparticles (derived from $Co_2Rh_2(CO)_{12}$) as a catalyst gave 87% of the reaction product. Interestingly, when a mixture of colloidal cobalt and rhodium nanoparticles was used as a catalyst under the same reaction conditions, only 12% of the product was obtained. This study showed that the catalytic activities towards specific products can often be tuned by changing the composition of heterobimetallic nanoparticle catalysts. Using **Co_2Rh_2** as a catalyst, the optimized reaction conditions were established as 1 atm of CO, 130 °C, THF, and 18 h. The catalytic system maintained its high level of activity even after being recycled five times.

The catalytic system was effective in the inter- and intramolecular PKR (Table 9.7). All the substrates including heteroatom-tethered enynes gave high yields (81–92%). Even in the intermolecular PKRs, moderate yields (59–68%) were obtained. This result is remarkable because many useful rhodium catalysts are generally less effective for the intermolecular PKR[34] and cannot be reused. Thus, the **Co_2Rh_2** catalytic system seems to overcome the disadvantages of homogeneous rhodium catalysts.

9.4.2.2 *Co_2Rh_2-catalyzed Pauson-Khand-type reaction in the presence of aldehydes instead of carbon monoxide*[13]

Owing to the recent attention given to green chemistry, in many cases instead of carbon monoxide aldehyde,[37] normally prepared from the corresponding alcohol, was used as the substitute for carbon monoxide. Very recently, Chung[38] reported the use of an alcohol as a source of carbon monoxide in the presence of $[Rh(CO)Cl(dppp)]_2$ (dppp = bis(diphenylphosphino)propane).

Table 9.7

Substrate	Product	Yield (%)
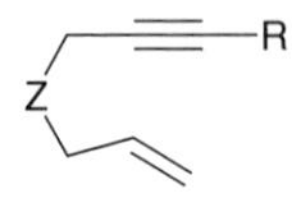	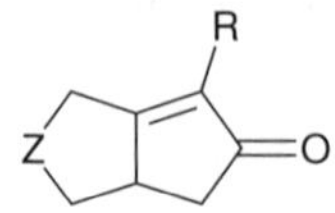	

Z	R	Yield (%)
$C(CO_2Et)_2$	Bu	82
$C(CO_2Et)_2$	Ph	92
O	Bu	89
O	Ph	87
NTs	Me	81
NTs	Bu	91

Substrate	Product	Yield (%)
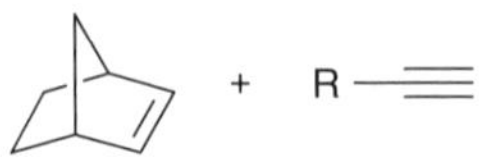	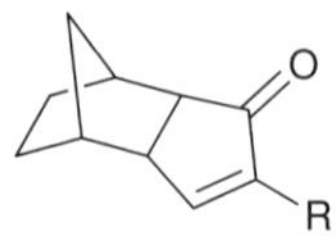	

R	Yield (%)
Bu	60
Ph	63
TMS	65
cyclohexyl	68

Substrate	Product	Yield (%)
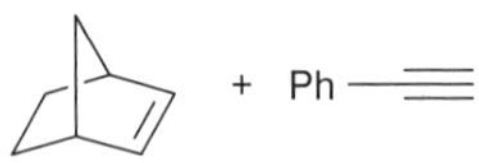	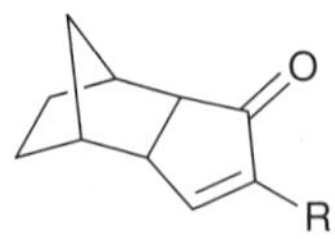	59

Using $\mathbf{Co_2Rh_2}$ nanoparticles as the catalyst, aldehydes proved to be effective, but 2-pyridylmethyl formate was not (Scheme 9.13). Among the aldehydes, α,β-unsaturated aldehydes such as cinnamaldehyde, crotonaldehyde and acrolein gave good results. The reusability test of the catalytic system showed that the catalytic activity was retained for at least five runs.

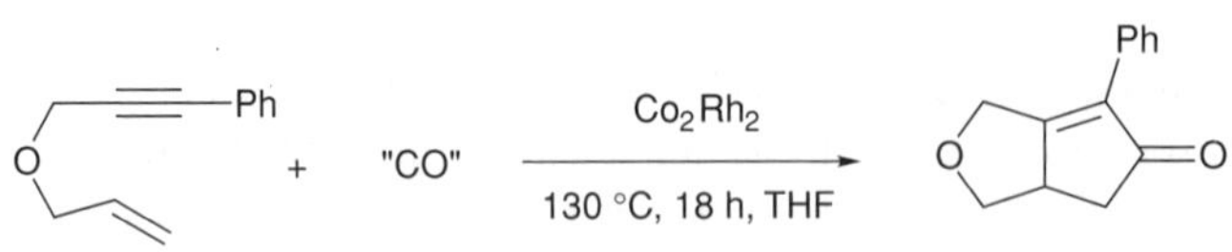

"CO".	Yield (%)
Acrolein	87
Crotonaldehyde	93
Cinnamaldehyde	91
Benzaldehyde	43
1-naphthaldehyde	35
2-pyridylmethyl formate	0

Scheme 9.13

Table 9.8

Substrate	Product		Yield (%)
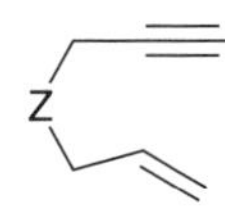	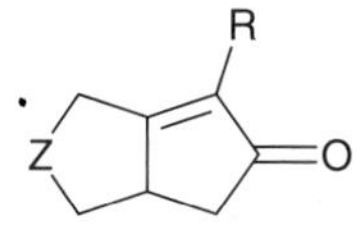		
	Z	R	
	$C(CO_2Et)_2$	Ph	97
	O	Bu	85
	NTs	Me	87
	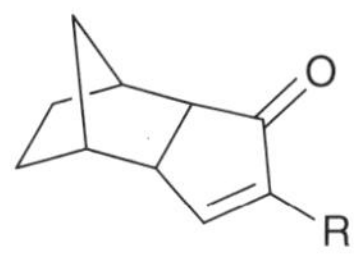		
	R		
	Bu		64
	Ph		65
	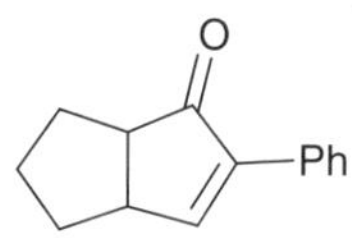		49
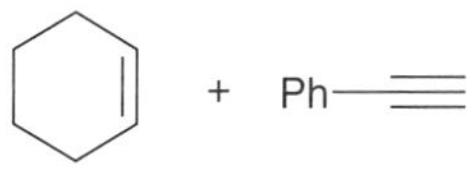	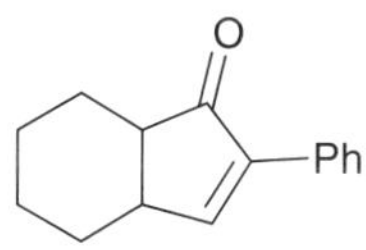		17

The screening result of various enyne substrates for the intramolecular PKR under optimized reaction conditions (**Co_2Rh_2**/C (0.05 g), enyne (0.96 mmol), crotonaldehyde (2.4 mmol), THF, 130 °C, and 18 h) is shown in Table 9.8. The catalytic system is effective for giving high yields of the intramolecular PKR products. No additives were necessary to enhance or promote the catalytic reaction. Moreover, multi-gram quantities of cyclopentenones were obtained without any difficulties. The catalytic system is also effective in the intermolecular PK-type reaction although the yields were not high, partly due to the trimerization of alkynes. For example, in a reaction of phenylacetylene with norbornene, the PKR product and trimerization product were obtained in 64% and 23% yields, respectively.

*9.4.2.3 **Co_2Rh_2**-catalyzed asymmetric catalytic Pauson-Khand-type reaction*[13]

Compared to the huge advances seen in homogeneous asymmetric catalysis, most of the transition metal nanoparticle-catalyzed reactions have been limited to non-asymmetric reactions[39] until the advent of some notable reports.[15d,40]

A practical and realistic asymmetric PKR catalyzed by heterogeneous catalysts was reported by Chung's group.[13] They studied a $\mathbf{Co_2Rh_2}$-catalyzed asymmetric PKR in the presence of a chiral diphsosphine, which included (*R*)-BINAP, (*R*)-tol-BINAP, (*R*,*R*)-NORPHOS, (*R*)-DIOP and (*S*,*S*)-BDPP (Scheme 9.14). Based on the ee and % yield, (*S*,*S*)-BDPP was choice. Using crotonaldehyde as a CO source, a highly eantioselective reaction was observed under the optimized reaction conditions (enyne (0.96 mmol), crotonaldehyde (2.4 mmol), (*S*,*S*)-BDPP (0.11 mmol), $\mathbf{Co_2Rh_2}$ on charcoal (0.05 g, total weight of metals 12.2 mg), THF (5 mL), 18 h, 130 °C).

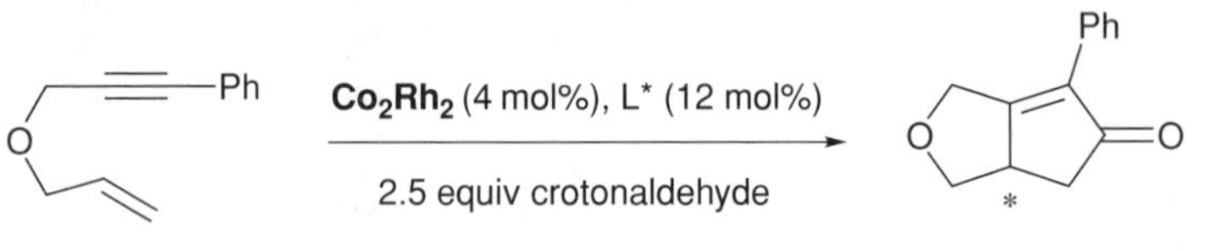

"L*"	Yield (%)	ee (%)
(*R*)-BINAP	93	79
(*R*)-tol-BINAP	91	73
(*R*,*R*)-NORPHOS	83	-
(*R*)-DIOP	90	34
(*S*,*S*)-BDPP	91	84

Scheme 9.14

Various enyne substrates for the intramolecular PKR were tested under the optimized reaction conditions (Table 9.9). The yields were high (75–94%) and the ee values were moderate to high (51–87%). In some cases, the observed ee values were comparable to those obtained in homogeneous catalysis using rhodium catalysts derived from $[RhCl(CO)_2]_2$/(*S*)-BINAP and AgOTf.[41] According to their study on the reusability of the catalytic system, the yields and ee values were retained over at least five runs. However, the chiral diphosphine

Table 9.9

Substrate		Product	
Z	R	Yield (%)	ee (%)
O	Ph	91	84
O	Bu	75	62
NTs	Ph	94	85
$C(CO_2Et)_2$	Me	93	87
$C(CO_2Et)_2$	Ph	94	51

becomes insufficient for recyclization and thus should be supplemented. One of the advantages of this catalytic system is that (*S,S*)-BDPP can be recovered in high yield (up to 87%) from the reaction mixture after the completion of the run and the recovered (*S,S*)-BDPP can be used in further reactions.

9.4.2.4 *Co_2Rh_2-catalyzed PK-type reaction using olefinic aldehydes as an alkene and CO source*[44]

From previous studies,[42,43] the use of α,β-unsaturated aldehyde as the source of alkene and CO could be envisioned. However, the use of an α,β-unsaturated aldehyde as the source of alkene and CO in the PKR was first realized by Chung's group.[44] They studied a reaction of an alkyne with α,β–unsaturated aldehydes such as acrolein, crotonaldehyde, and cinnamylaldehyde in the presence of **Co_2Rh_2** (Scheme 9.15). A PK-type reaction product was obtained in reasonable to high yields (Scheme 9.16). It is somewhat surprising that α,β-unsaturated aldehydes, known as poor substrates for the PKR, can be used as a source of CO and alkene. The reaction was highly regioselective with respect to alkenes and alkynes. Treatment of the unsymmetric internal alkyne afforded one isomer as the sole product.

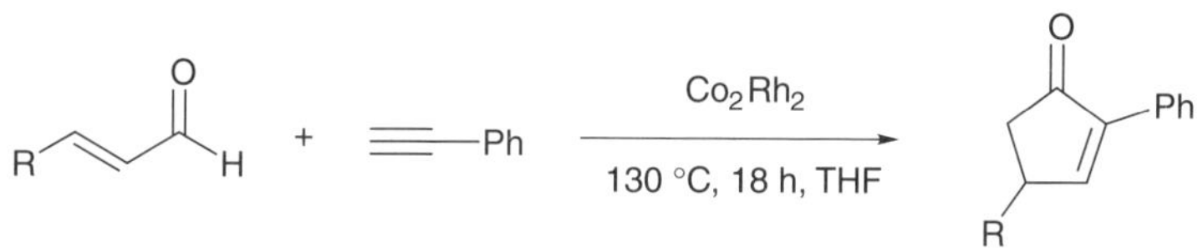

R	Yield (%)
H	60
Me	77
Ph	75

Scheme 9.15

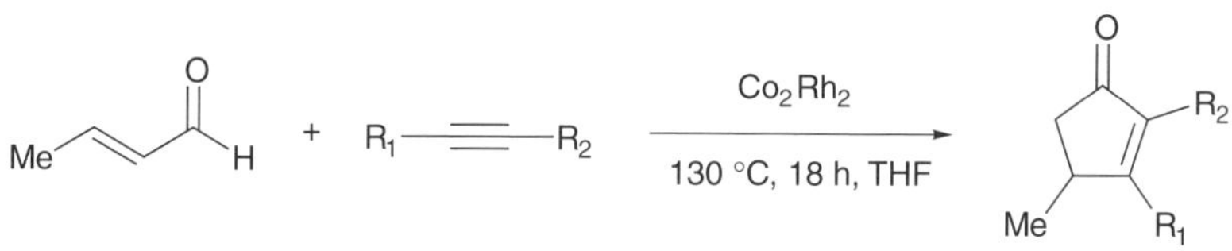

R_1	R_2	Yield (%)
H	Ph	77
H	Bu	70
Ph	Ph	71

Scheme 9.16

According to the reusability test of the catalytic system for the reaction between cinnamic aldehyde and phenylacetylene, the catalytic activity of **Co_2Rh_2** nanoparticles was retained for at least five runs.

9.4.2.5 Co_2Rh_2-catalyzed allenic PKR[48]

The involvement of allenes in the substrate list has expanded the synthetic utility of the PKR. The first intramolecular allenic PKR was reported by Narasaka[45] in the presence of $Fe(CO)_4(NMe_3)$ under photoirradiative conditions. Shortly thereafter, Brummond[46] developed an allenic variant of the PK-type cycloaddition which provides access to molecules possessing 4-alkylidene or α–methylene cyclopentenones, depending upon the reactant structure and reaction conditions.[47]

The first nanoparticle-catalyzed allenic PKR was reported by Chung et al. using $\mathbf{Co_2Rh_2}$ as a catalyst.[48] For reference purposes, they also studied other transition-metal nanoparticles including Rh, Co and a mixture of Co and Rh nanoparticles as a catalyst under the same reaction conditions. The best yield (85%) was obtained with $\mathbf{Co_2Rh_2}$, under the 2 atm of CO at 110 °C for 6 h (Scheme 9.17). The cyclization occurs selectively with the distal double bond of the allene, to yield bicyclo[4.3.0]nonenone. The catalyst maintained its high activity even after being recycled six times.

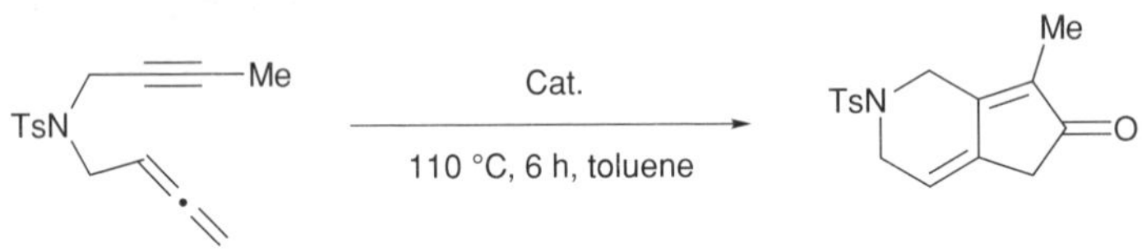

Cat.	Yield (%)
Co_2Rh_2 (10 mol%)	85
Co (10 mol%)	29
Rh (10 mol%)	trace
Co (5 mol%) + Rh (5 mol%)	21

Scheme 9.17

The catalytic system was also effective for various allenyne substrates (Table 9.10). However, the introduction of a substituent(s) to the allene terminus was detrimental. It seemed that the steric effect of the substituents on the allene in the $\mathbf{Co_2Rh_2}$-catalyzed

Table 9.10

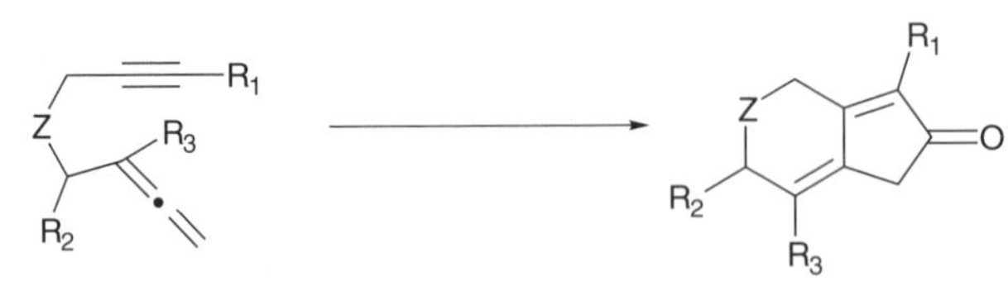

Substrate				Product
Z	R_1	R_2	R_3	Yield (%)
NTs	Me	H	H	85
NTs	H	H	H	28
NTs	Me	Me	H	70
NTs	Me	H	Ph	74
$C(CO_2Et)_2$	Ph	H	H	73
O	Me	H	H	59

[2+2+1] cycloaddition reactions might be larger than that in the homogeneous reactions. Thus, the reaction was highly dependent upon the substituent(s) on the allene moiety.

The reaction was extended to the catalytic formation of bicyclo[5.3.0]dec-1,7-dien-9-ones in acceptable yields (Scheme 9.18). Complete chemoselectivity was observed in the formal [2+2+1]-cycloaddition, leading to the exclusive construction of the bicyclo[5.3.0]dec-1,7-dien-9-one framework in acceptable yields compared to those in previous works using a homogeneous catalyst.[42]

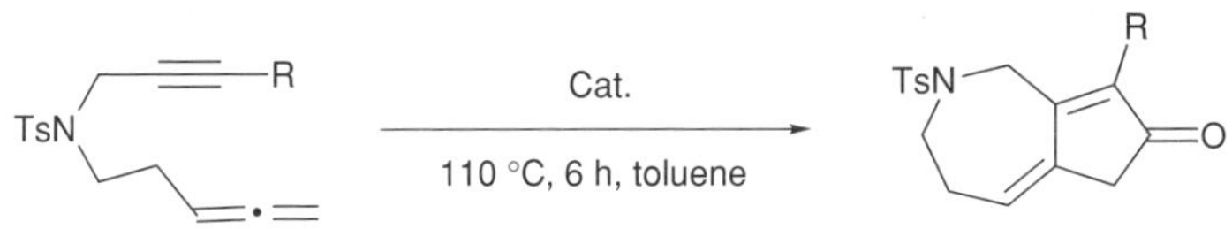

R	Yield (%)
H	64
Me	73
Ph	72

Scheme 9.18

9.4.2.6 *Co_2Rh_2-catalyzed PK-type reaction of allenes*[51]

Recently, the chemistry of allenes has attracted a lot of attention, presumably due to their high reactivity. However, although many transition metal-catalyzed cycloaddition of allenes have been disclosed,[49,50] there was no report on the use of allenes as substrates in the intermolecular PK-type reaction until 2008. The first use of allenes in inter- and intramolecular PK-type reactions was reported by Chung[51] using **Co_2Rh_2** as a catalyst. They studied the catalytic formation of cyclopentenones via a [2+2+1] cycloaddition of two allenes and carbon monoxide.

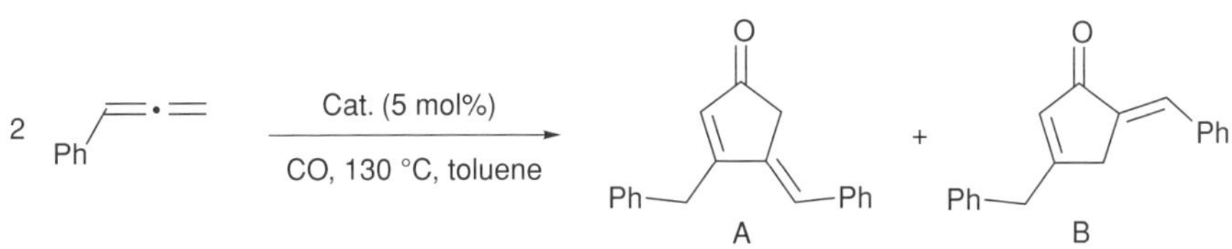

Cat.	CO (atm)	Time (h)	Yield(A / B) (%)
Co_2Rh_2	1	18	45/8
Co_2Rh_2	2	18	59/3
Co_2Rh_2	2	12	62/10
Co_2Rh_2	2	6	68/12
$[Rh(CO)_2Cl]_2$	2	6	32/25

Scheme 9.19

Using phenylallene as a substrate, they screened the reaction conditions including the reaction temperature, the reaction time and the CO pressure (Scheme 9.19). Two carbonylated cyclic enones were isolated in various ratios and yields depending upon the reaction conditions. These two cyclic enones are easily separated by chromatography and show characteristic ^{1}H NMR patterns. The allenes formally serve both as an excellent alkene- and alkyne-type moiety in the PK-type [2+2+1] cycloaddition with carbon monoxide. The optimum reaction conditions were established as follows: 1.0 mmol allene, 5 mol% $\mathbf{Co_2Rh_2}$ catalyst, 5 mL toluene, 130 °C, CO (2 atm), and 6 h. For reference purposes, they also studied the same reaction in the presence of $[Rh(CO)_2Cl]_2$, which showed a lower yield (57%). They tested other allenes under the optimized reaction conditions (Scheme 9.20). The overall yields from carbonylative cycloaddition reactions were relatively high (76–92%). Thus this process provides a rapid and atom-economical method for the synthesis of a variety of cyclopentenones with an exocyclic double bond. This study widens the scope of a substrate in the PK-type reaction.

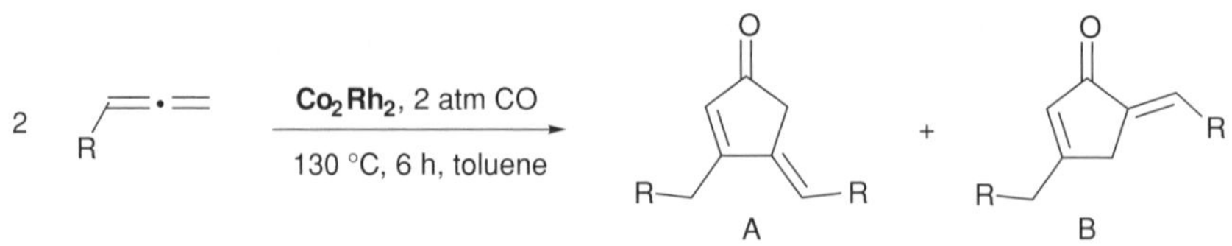

R	Yield(A / B) (%)
4-tolyl	70/14
1-naphthyl	72/20
4-anisyl	61/0
4-acetylphenyl	64/20
hexyl	52/7

Scheme 9.20

9.4.2.7 Co_2Rh_2-catalyzed PK-type reaction of bisallenes[51]

Compared to the advances in the chemistry of allenes and related compounds, those of bisallenes lags far behind,[52] presumably because of their high instability and the difficulty of their preparation.

Chung reported[51] the first use of bisallenes in a catalytic PK-type cycloaddition in the presence of carbon monoxide using $\mathbf{Co_2Rh_2}$ as a catalyst (Scheme 9.21).

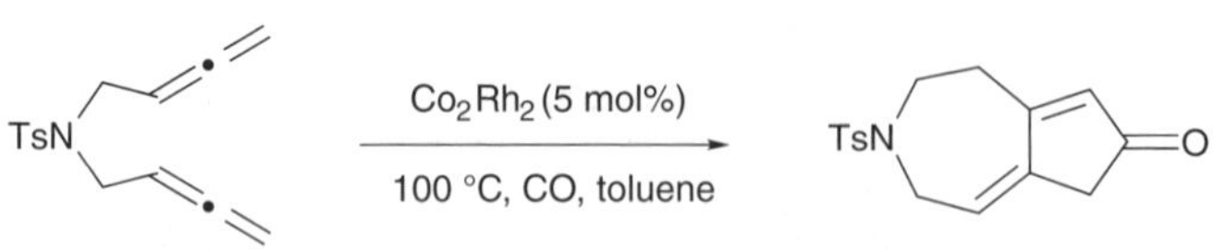

CO (atm)	Time (h)	Yield (%)
1	12	33
2	12	66
3	4	70
2	4	68

Scheme 9.21

Table 9.11

Substrate	Product			Yield (%)
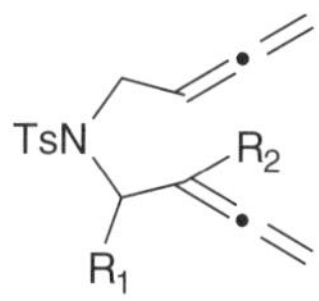	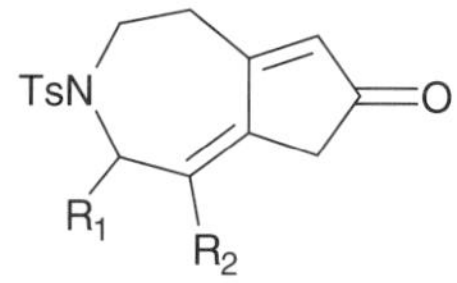			
		R_1	R_2	
		Me	H	67
		H	Ph	74
		H	Me	53
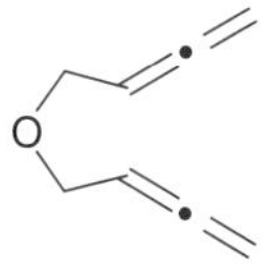	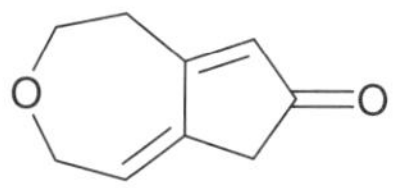			45

A carbonylated bicyclic enone was isolated in reasonable to high yields depending upon the reaction conditions. The bisallenes serve both as an excellent alkene- and alkyne-type moiety in the PK-type [2+2+1] cycloaddition with carbon monoxide.

They investigated the carbonylative cycloaddition of other bisallenes under optimized reaction conditions (0.3 mmol allene, 5 mol% $\mathbf{Co_2Rh_2}$ catalyst, 2 atm CO, 3 mL toluene, 100 °C, and 4 h) (Table 9.11). The corresponding intramolecular [2+2+1] cocyclization product was isolated in reasonable to high yield. The high degree of regiocontrol is noteworthy.

9.5 Sequential Action of Two Different Catalysts in One-Pot Reactions

Although significant breakthroughs regarding the development of tandem reactions have been achieved recently, those in the field of PKR are relatively rare. The first demonstration was reported by Jeong's group in 2000.[53] They showed the multiple C–C bond forming transformations using two different Pd and Rh complexes as catalysts (Scheme 9.22).

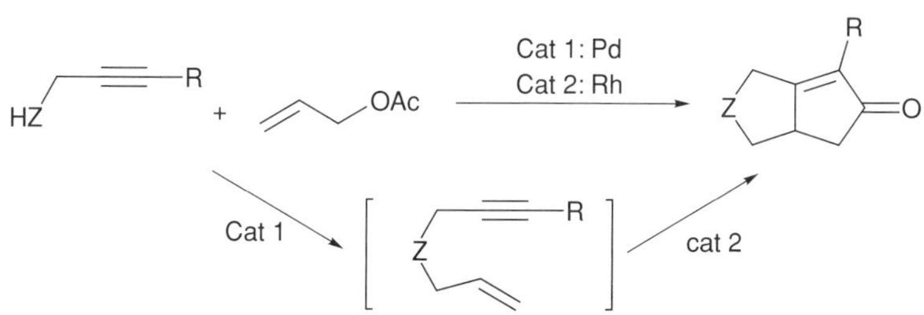

Scheme 9.22

One year later, Evans reported[54] a single rhodium-catalyst to facilitate both transformations, the allylic alkylation and PK annulation reaction, in a tandem sequence, just by the controlling the reaction temperature to modulate the catalytic activity.

9.5.1 Chiral Pd(II) Complex and Bulk Cobalt on Charcoal-Catalyzed Tandem Asymmetric Allylic Alkylation and PK Annulation Reactions[55]

A tandem reaction related to the PKR catalyzed by a combination of homogeneous and heterogeneous catalysts was studied by Chung's group (Scheme 9.23).[55] They used a combination of chiral Pd(II) complex and bulk cobalt metal on charcoal as the catalyst system. This is a rare example of the use of a combination of homogeneous and heterogeneous catalysts in a tandem reaction.[56] The $[(\eta^3\text{-allyl})PdCl]_2$ catalyzed an asymmetric allylic

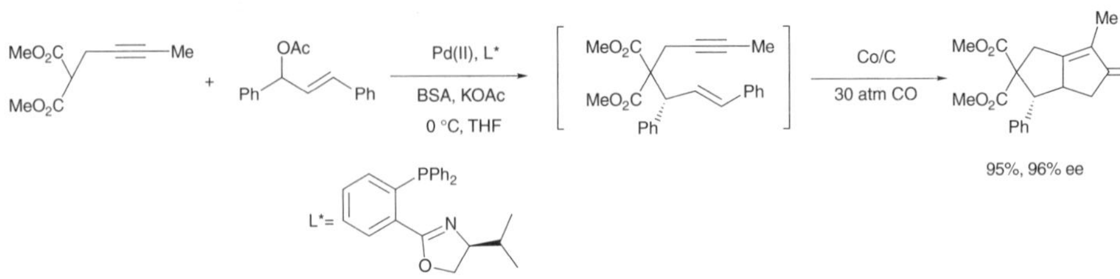

Scheme 9.23

alkylation to generate enantioselective enynes and the heterogeneous Co/C carried out a catalytic PKR of the *in situ* generated enantiomerically enriched enynes. Chiral diphosphines, phosphinooxazoline and *(R,R)*-NORPHOS were screened as an ancillary ligand and phosphinooxazoline was the choice. High yield up to 95% and high *ee* value up to 96% were observed. The enantioselectivity of this tandem reaction was absolutely dependent on the Pd-catalyzed asymmetric allylic alkylation. The success of this tandem catalytic reaction is primarily due to the use of Co/C. When $Co_2(CO)_8$ instead of Co/C was used as a catalyst, only 2% of the desired product was isolated. They attributed the generation of acetic acid in the palladium-catalyzed allylic alkylation step for the background of the low yield. The usefulness of the combined catalyst is highlighted by the synthesis of various tricyclic enones (Scheme 9.24). *trans*-1,2-Bis[2-(diphenylphosphino)benzamido]cyclohexane[57] was used as a chiral ligand. High yields (90–97%) were observed, but the ee values (31–94% ee) were highly sensitive to the ring size.This study shows that the tandem action of a homogeneous chiral Pd(II) catalyst and a heterogeneous Co/C catalyst leads to a highly enantioselective two-step one-pot PK-type reaction.

9.5.2 Pd(II) Complex and Cobalt Nanoparticle-Catalyzed Three-Step One-Pot Synthesis of Fenestranes[60]

The fenestranes are a unique class of hydrocarbons. Major advances in the synthesis of fenestranes were achieved by the application of transition metal-induced transformations.[58,59] Chung reported the use of cobalt nanoparticles in combination with a palladium complex to develop an efficient synthetic procedure both reducing the number of steps and increasing

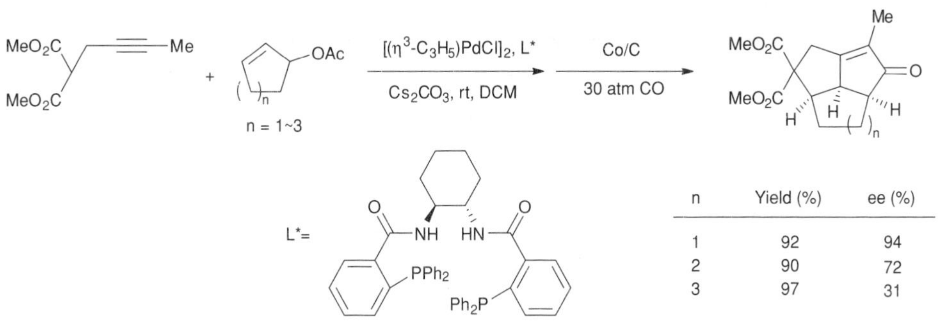

n	Yield (%)	ee (%)
1	92	94
2	90	72
3	97	31

Scheme 9.24

the yield (Scheme 9.25).[60] This was the three-step one-pot synthesis of [5.5.5.5]fenestranes from the readily available 1,6-enynes and alkynes as the starting material. Two different catalysts, cobalt nanoparticles and allyl palladium chloride, were capable of promoting three sequential but independent catalytic cycles in the desired chronological order, thereby allowing a new molecular queuing process to take place. This is another example of a combination of homogeneous and heterogeneous catalysts in a tandem reaction. The intriguing interplay of sequential reactions gives the highly funtionalized [5.5.5.5]fenestranes in high yields. It should be noted, however, that the steric effect of the substituent(s) on the enyne substrate and alkyne diester is an important factor allowing the Pd-catalyzed allylic alkylation reaction to proceed. The bulk cobalt on charcoal instead of cobalt nanoparticles was not effective in the third step. Cobalt nanoparticles were tolerant of virtually all the reaction conditions and quite effective for both the first and third reactions.

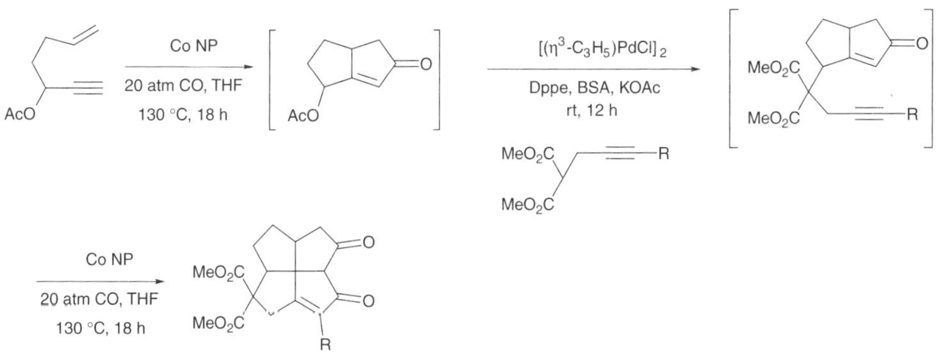

R	Yield (%)
Me	84
Ph	74

Scheme 9.25

This study might open a number of doors for PKRs and related organic reactions that could be effectively catalyzed by a combination of homogeneous and heterogeneous catalysts.

Table 9.12

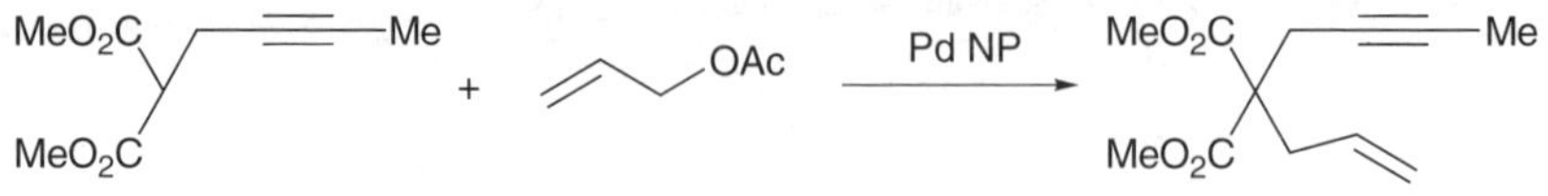

Solvent	Base	Time (h)	Yield (%)
THF	NaH	10	88
THF	NaH	6	89
THF	K_2CO_3	6	N.R.
THF	BSA	6	23
THF	$NaHCO_3$	6	78
DMF	NaH	6	87
CH_2Cl_2	NaH	6	60

9.5.3 Palladium and Cobalt Nanoparticles-Catalyzed Tandem Allylic Alkylation and PKRs[63]

Most of the known combinations of bimetallic nanoparticles have been used as catalysts for specific single reactions utilizing the synergistic effect between the different metal species;[61] the combination of bimetallic nanoparticles for a sequential reaction of two specific reactions corresponding to two different metals is relatively rare.[62]

The Chung group prepared palladium and cobalt nanoparticles immobilized on silica (**PCNS**) and used them in the sequential allylic alkylation and PKRs (Scheme 9.26).[27]

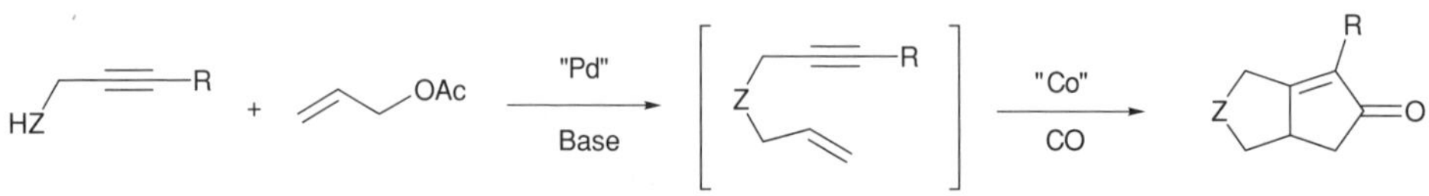

Scheme 9.26

They first prepared palladium nanoparticles immobilized on silica (1 wt % of Pd) by the reaction of $Pd(OAc)_2$ with colloidal silica in a solvent mixture of ethanol and toluene (v/v, 6:94) and used them for the study of the allylic alkylation (Table 9.12). This was the first palladium nanoparticle catalyzed-allylic alkylation.

Then the palladium and cobalt nanoparticles immobilized on silica (**PdCNS**) (14.5 wt% of Co and 1 wt% of Pd) were prepared by heating Pd on silica with colloidal cobalt in ethanol at reflux (Figure 9.5). The HRTEM revealed that the bimetallic nanoparticles are in a cluster-in-cluster structure.

Using **PdCNS** as a catalyst, they studied the sequential allylic alkylation and PKR (Scheme 9.27). The **PdCNS** was very efficient but the reaction was highly sensitive to the

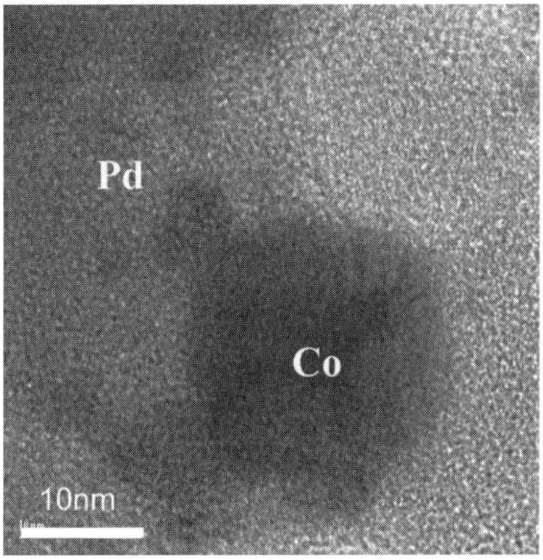

Figure 9.5 HR-TEM of **PdCNS**.

CO pressure and reaction temperature. The best isolated yield was 88% at 130 °C under a CO pressure of 10 atm.

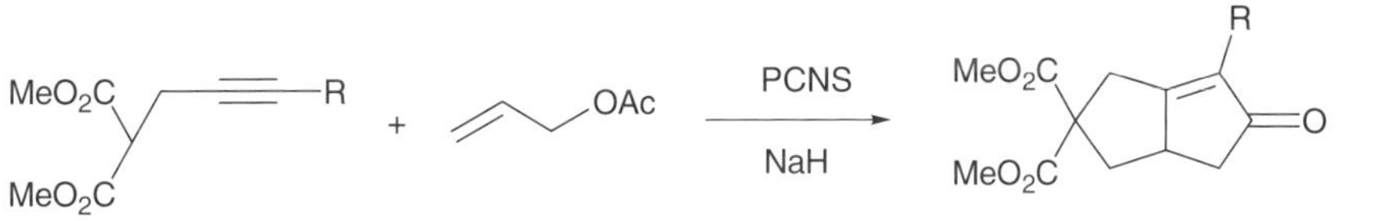

Scheme 9.27

The catalytic reaction was nonetheless effective with substituted enynes and allylic acetate (Table 9.13). The heteroatom-tethered substrates were screened to generate the corresponding products in high yields. Especially when cyclic allyl acetate was used, a tricyclic enone was obtained in 73% yield.

9.6 Conclusion

This chapter focuses on the development of heterogeneous catalytic PKRs. Many novel catalysts have been developed for the PKRs. Heterogeneous catalytic systems for the PKR have now reached a significant level of maturity. In particular, the utilization of heterobimetallic cobalt-rhodium nanoparticles as catalysts has showed enhanced activity and widened the scope of substrates in the PKR compared to those of monometallic nanoparticles, and has uncovered several new novel reactions. In addition, a combination of two different metal nanoparticles, such as cobalt/palladium and cobalt/ruthenium, shows considerable potential for future research into the development of combined catalysts, which

Table 9.13

Substrate	Product	Yield (%)
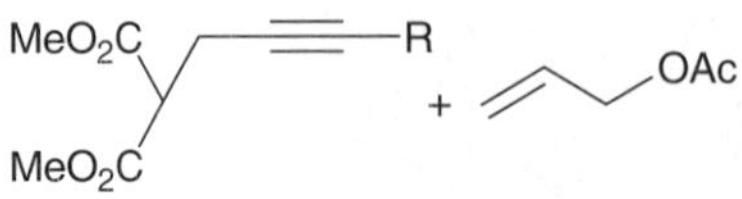	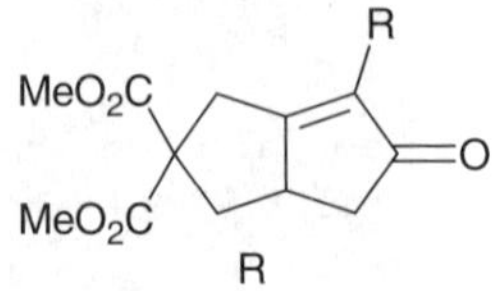	
	R	
	Me	88
	Bu	82
	TIPS	84
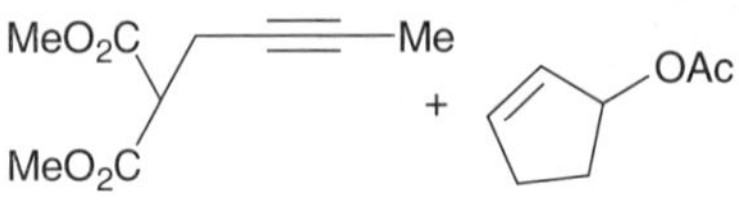	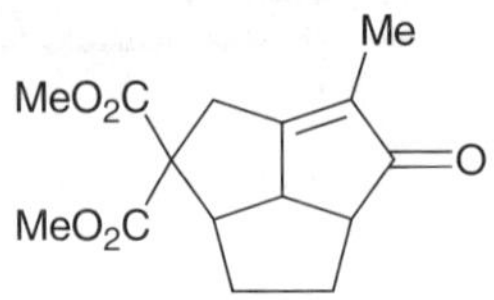	73
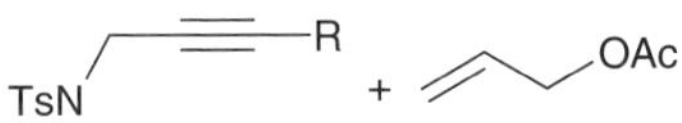	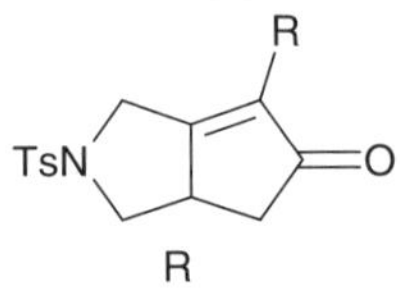	
	R	
	Bu	68
	Ph	75

can be applied to an *in situ* generation of highly reactive intermediates and may lead to new activities and other possibilities. Moreover, the combination of a homogeneous catalyst and a heterogeneous catalyst, for example a palladium complex and cobalt metal, may be a very useful catalytic system and provide new reaction protocols that allow important building blocks in previously unknown fashions. Recently, there has been much progress in catalytic versions of the asymmetric PKRs. However, fine-tuning of the chiral ligands to increase both product and optical yield is still needed. Overall, the use of heterogeneous catalysts in the PKRs and related reactions is still a burgeoning area.

References

[1] (a) I. U. Khand, G. R. Knox, P. L. Pauson, W. E. Watts and M. I. Foreman, Organocobalt complexes. Part II. Reaction of acetylenehexacarbonyl dicobalt complexes $(R^1C_2R^2)Co_2(CO)_6$, with norbornene and its derivatives, *J. Chem. Soc. Perkin Trans.* **1** 977 (1973); (b) V. Rautenstrauch, P. Megard, J. Conesa and W. Kuster, 2-Pentylcyclopent—2-en-1-one by catalytic Pauson-Khand reaction, *Angew. Chem. Int. Ed. Engl.* **29**, 1413 (1990)

[2] (a) N. Jeong, S. H. Hwang, Y. S. Lee and Y. K. Chung, Catalytic conversion of the intramolecular Pauson-Khand reaction, *J. Am. Chem. Soc.* **116**, 3159 (1994); (b)

B. Y. Lee, Y. K. Chung, N. Jeong, Y. S. Lee and S. H. Hwang, (Indenyl)cobalt(I)-catalyzed cocyclization of alkyne, alkene, and carbon monoxide to cyclopentenones, *J. Am. Chem. Soc.* **116**, 8793 (1994); (c) F. A. Hicks and S. L. Buchwald, Highly enantioselective catalytic Pauson-Khand type formation of bicyclic cyclopentenones, *J. Am. Chem. Soc.* **118**, 11688 (1996); (d) N. Jeong, H. H. Hwang, Y. W. Lee and J. S. Lim, Catalytic Pauson-Khand reaction in super critical fluids, *J. Am. Chem. Soc.* **119**, 10549 (1997); (e) J. W. Kim and Y. K. Chung, Pauson-Khand reaction catalyzed by $Co_4(CO)_{12}$, *Synthesis* **142** (1998).

[3] L. Chen and C. J. Li, Catalyzed reactions of alkynes in water, *Adv. Synth. Catal.* **348**, 1459 (2006).

[4] A. M. Trzeciak and J. J. Ziółkowski, Monomolecular, nanosized and heterogenized palladium catalysts for the Heck reaction, *Coord. Chem. Rev.* **251**, 1281 (2007).

[5] M. Lamblin, L. Nassar-Hardy, J.-C. Hierso, E. Fouquet and F.-X. Felpin, Recyclable heterogeneous palladium catalysts in pure water: Sustainable developments in Suzuki, Heck, Sonogashira and Tsuji-Trost reactions, *Adv. Synth. Catal.* **352**, 33 (2010).

[6] (a) J. M. Thomas and W. J. Thomas, *Principles and Practice of Heterogeneous Catalysis*, VCH, Weinheim, 1997; (b) G. Ertl, H. Knötzinger, F. Schüth, and J. Weitkamp (Eds.), *Handbook of Heterogeneous Catalysis,* 2nd edn, Wiley-VCH, Weinheim, 2008; (c) D. J. Cole-Hamilton, R. P. Tooze, Eds., *Catalyst Separation, Recovery and Recycling: Chemistry and Process Design*, Springer: Berlin, 2006; (d) G. Schmid, Large clusters and colloids. Metals in the embryonic state, *Chem. Rev.* **92**, 1709 (1992); (e) L. N. Lewis, Chemical catalysis by colloids and clusters, *Chem. Rev.* **93**, 2693 (1993); (f) R. M. Crooks, M. Zhao, L. Sun, V. Chechik and L. K. Yeung, Dendrimer-encapsulated metal nanoparticles: synthesis, characterization, and applications to catalysis, *Acc. Chem. Res.* **34**, 181 (2001); (g) J. Schulz, A. Roucoux and H. Patin, Reduced transition metal colloids: A novel family of reusable catalysts?, *Chem. Rev.* **102**, 3757 (2002); (h) F. Fernández-García, A. Martinez-Arias, J. C. Hanson and J. A. Rodriguez, Nanostructured oxides in chemistry: characterization and properties, *Chem. Rev.* **104**, 4063 (2004); (i) D. Astruc, F. Lu and J. R. Aranzaes, Nanoparticles as recyclable catalysts: The frontier between homogeneous and heterogeneous catalysis, *Angew. Chem., Int. Ed.* **44**, 7852 (2005); (j) A.-H. Lu, E. L. Salabas and F. Schüth, Magnetic nanoparticles: synthesis, protection, functionalization, and application, *Angew. Chem., Int. Ed.* **46**, 1222 (2007); (k) M. R. Buchmeiser, Polymer-supported well-defined metathesis catalysts, *Chem. Rev.* **109**, 303 (2009); (l) Z. Wang, G. Chen and K. Ding, Self-supported catalysts, *Chem. Rev.* **109**, 322 (2009); (m) A. F. Trindade, P. M. P. Gois and C. A. M. Alfonso, Recyclable stereoselective catalysts, *Chem. Rev.* **109**, 418 (2009); (o) A. Schatz, O. Reiser and W. J. Stark, Nanoparticles as semi-heterogeneous catalyst supports, *Chem. Eur. J.* **16**, 8950 (2010); (p) H.-J. Freund, Model studies in heterogeneous catalysis, *Chem. Eur. J.* **16**, 9384 (2010).

[7] (a) A. C. Comley, S. E. Gibson and N. J. Hales, Polymer-supported cobalt carbonyl complexes as novel solid-phase catalysts of the Pauson-Khand reaction, *Chem. Commun.* 305 (2000). (b) A. C. Comley, S. E. Gibson, N. J. Hales, C. Johnstone and A. Stevenazzi, The application of polymer-bound carbonylcobalt(0) species in linker chemistry and catalysis, *Org. Biomol. Chem.* **1**, 1959 (2003).

[8] A. Dahan and M. Portnoy, Dendritic effect in polymer-supported catalysis of the intramolecular Pasuon-Khand reaction, *Chem. Commun.* 2700 (2002).

[9] (a) A. Devasagayaraj and M. Periasamy, A simple convenient synthesis of alkyne-$Co_2(CO)_6$ complexes and their utilization in the Pauson-Khand cyclopentenone synthesis, *Tetrahedron Lett.* **30**, 595 (1989); (b) A. Devasagayaraj, S. A. Rao and M. Periasamy, Simple syntheses of $Co_2(CO)_8$ and $NaCo(CO)_4$ reagents by the reduction of $CoCl_2$ (or $CoBr_2$) under carbon monoxide at atmospheric pressure, *J. Organomet. Chem.* **403**, 387 (1991); (c) M. Periasamy, M. R. Reddy and A. Devasagayaraj, Simple, convenient methods for the preparation of alkyne-$Co_2(CO)_6$ complexes from $CoBr_2$ for application in Pauson-Khand cyclopentenone synthesis, *Tetrahedron* **50**, 6955 (1994); (d) M. L. N. Rao and M. Periasamy, Reductive Pauson-Khand reaction using $(RC{\equiv}CR')Co_2(CO)_6/CF_3COOH$ system, *J. Organomet. Chem.* **532**, 143 (1997); (e) T. Rajesh and M. Periasamy, Carbonylative cyclization of alkynes using cobalt carbonyl species prepared via reduction of $CoBr_2$ with Zn under a carbon monoxide atmosphere, *Organometallics* **18**, 5709 (1999); (f) T. Rajesh and M. Periasamy, A sub-stoichiometric version of the Pauson-Khand reaction, *Tetrahedron Lett.* **40**, 817 (1999).

[10] (a) N. Y. Lee, Y. K. Chung, Synthesis of cyclopentenones: The new catalytic cocyclization reaction of alkyne, alkene, and carbon monoxide employing catalytic $Co(acac)_2$ and $NaBH_4$, *Tetrahedron Lett.* **37**, 3145 (1996); (b) Y. K. Chung, Transition metal alkyne complexes: the Pauson-Khand reaction, *Coord. Chem. Rev.* **188**, 297 (1999).

[11] S.-W. Kim, S. U. Son, S. I. Lee, T. Hyeon, Y. K. Chung, Cobalt on mesoporous silica: The first heterogeneous Pauson-Khand catalyst, *J. Am. Chem. Soc.* **122**, 1550 (2000).

[12] S.-W. Kim, S. U. Son, Y. K. Chung, T. Hyeon, Colloidal cobalt nanoparticles: a highly active and reusable Pauson-Khand catalyst, *Chem. Commun.* 2212 (2001).

[13] K. H. Park, S. U. Son, Y. K. Chung, Immobilized Co/Rh heterobimetallic nanoparticle-catalyzed Pauson-Khand-type reaction, *Adv. Synth. Catal.* **347**, 854 (2005).

[14] (a) J.-L. Muller, A. Rickers and W. Leitner, Raney Cobalt: an effective and recyclable catalysts for the Pauson-Khand reaction, *Adv. Synth. Catal.* **349**, 287 (2007); (b) J.-L. Muller, J. Klankermayer and W. Leitner, Poly(ethylene glycol) stabilized Co nanoparticles as highly active and selective catalysts for the Pauson-Khand reaction, *Chem. Commun* 1939 (2007).

[15] For reviews on immobilized catalysts, see: (a) D. E. de Vos, I. F. Vankelecom, P. A. Jacobs, Eds. *Chiral Catalyst Immobilization and Recycling*, Wiley-VCH, Weinheim, 2000; (b) A. Kirschning, Ed. Immobilized Catalysts: Solid Phases, Immobilization and Applications in *Topics in Current Chemistry*, Springer, Berlin, 2004, 242, 1; (c) D. R., Buchmeiser, Ed. *Polymeric Materials in Organic Synthesis and Catalysis*, Wiley-VCH, Weinheim, 2006; (d) K. Ding, Y. Uozumi, Eds. *Handbook of Asymmetric Heterogeneous Catalysis*, Wiley-VCH, Weinheim, 2008; (e) M. Heitbaum, F. Glorius and I. Escher, Asymmetric heterogeneous catalysis, *Angew. Chem. Int. Ed.* **45**, 4732 (2006).

[16] (a) C. T. Kresge, M. E. Leonowicz, W. J. Roth, J. C. Vartuli and J. S. Beck, Ordered mesoporous molecular sieves synthesized by a liquid-crystal template mechanism, *Nature* **359**, 710 (1992); (b) J. S. Beck, J. C. Vartuli, W. J. Roth, M. E. Leonowicz, C. T. Kresge, K. D. Schmitt, C. T. W. Chu, D. H. Olson, E. W. Sheppard, S. B. McCullen, J. B. Higgins and J. L. Schlenker, A new family of mesoporous molecular

sieves prepared with liquid crystal templates, *J. Am. Chem. Soc.* **114**, 10834 (1992); (c) D. Zhao, J. Feng, Q. Huo, N. Melosh, G. H. Fredrickson, B. F. Chmelka and G. D. Stucky, Triblock Copolymer Syntheses of Mesoporous Silica with Periodic 50 to 300 Angstrom Pores, *Science* **279**, 548 (1998); (d) P. Yang, D. Zhao, D. I. Maroglese, B. F. Chmelka and G. D. Stucky, Generalized syntheses of large-pore mesoporous metal oxides with semicrystalline frameworks, *Nature* **396**, 152 (1998); (e) A. Corma, From microporous to mesoporous molecular sieve materials and their use in catalysis, *Chem. Rev.* **97**, 2373 (1997).

[17] S. U. Son, S.-I. Lee and Y. K. Chung, Cobalt on charcoal: A convenient and inexpensive heterogeneous Pauson-Khand catalyst, *Angew. Chem., Int. Ed.* **39**, 4158 (2000).

[18] B. Cornils, W. A. Herrmann, R. Schlögl and C.-H. Wong (Eds). *Catalysis from A to Z*, Wiley-VCH, Weinheim, 2000, p 531.

[19] 21. For example, Raney® 2700 cobalt has a BET Surface area of 12 m^2/g and a particle size range of 20-50 nm. References: B. H. Gross, R. C. Mebane, D. L. Armstrong, *Appl. Catal. A: Gen.* **219**, 281-289 (2001); Raney® Technical Manual, 4th Edition, W. R. Grace Company, Chattanoogo, Davison, TN 1996, p 21 and 23.

[20] L. F. Giraldo, B. L. López, L. Pérez, S. Urrego, L. Sierra and M. Mesa, Mesoporous silica applications, *Macromol. Symp.* **258**, 129 (2007).

[21] F. Goettmann, P. Le Floch and C. Sanchez, Highly regioselective terminal hydroformylation and Pauson-Khand reaction catalysed by mesoporous organized zirconium oxide based powders, *Chem. Commun.* 180 (2006).

[22] F. Goettmann, A. Thomas and M. Antonietti, Metal-free activation of CO_2 by mesoporous graphitic carbon nitride, *Angew. Chem., Int. Ed.* **46**, 2717 (2007).

[23] A. Z. Moshfegh, Nanoparticle catalysts, *J. Phys. D: Appl. Phys.* **42**, 233001 (30 pp) (2009).

[24] (a) J. Huang, T. Jiang, B. X. Han, W. Z. Wu, Z. M. Liu, Z. L. Xie and J. L. Zhang, A novel method to immobilize Ru nanoparticles on SBA-15 firmly by ionic liquid and hydrogenation of arene, *Catal. Lett.* **103**, 59 (2005); (b) J. Huang, T. Jiang, H. Gao, B. Han, Z. Liu, W. Wu, Y. Chang and G. Zhao, Pd nanoparticles immobilized on molecular sieves by ionic liquids: heterogeneous catalysts for solvent-free hydrogenation, *Angew. Chem., Int. Ed.* **43**, 1397 (2004); (c) G. Vijayaraghvan and K. J. Stevenson, Synergistic assembly of dendrimer-templated platinum catalysts on nitrogen-doped carbon nanotube electrodes for oxygen reduction, *Langmuir* **23**, 5279 (2007); (d) M. A. Gelesky, C. W. Scheeren, L. Foppa, F. A. Pavan, S. L. P. Dias and J. Dupont, Metal nanoparticle/ionic liquid/cellulose: new catalytically active membrane materials for hydrogenation reactions, *Biomacromolecules* **10**, 1888 (2009).

[25] K. H. Park, S. U. Son and Y. K. Chung, Immobilized heterobimetallic Ru/Co nanoparticle-catalyzed Pauson-Khand-type reactions in the presence of pyridylmethyl formate, *Chem. Commun.* 1898 (2003).

[26] J. H. Park and Y. K. Chung, Cobalt-rhodium heterobimetallic nanoparticle-catalyzed reactions, *Dalton Trans.* 2369 (2008).

[27] K. H. Park, S. U. Son and Y. K. Chung, Sequential actions of palladium and cobalt nanoparticles immobilized on silica: one-pot synthesis of bicyclic enones by catalytic allylic alkylation and Pauson-Khand reaction, *Org. Lett.* **4**, 4361 (2002).

[28] D. M. de la Torre and M. M. Guijarro, Covalent bonds on activated carbon, *Eur. J. Org. Chem.* **27**, 5147 (2010).

[29] S. U. Son, K. H. Park and Y. K. Chung, Cobalt nanoparticles on charcoal: a versatile catalyst in the Pauson-Khand reaction, hydrogenation, and the reductive Pauson-Khand reaction, *Org. Lett.* **4**, 3983 (2002).

[30] Reviews: (a) P. Rylander, *Catalytic Hydrogenation in Organic Syntheses*, Academic Press, New York, 1979; (b) M. Hudlicky, *Reductions in Organic Chemistry*, Wiley, New York, 1984; (c) R. L. Augustine, The stereochemistry of hydrogenation of α,β-unsaturated ketones, *Adv. Catal.* **25**, 56 (1976); (d) P. Evans, Y. Kavanagh and M. O'Brien, Stereocontrolled preparation of bicyclic alkaloid analogues: an approach towards the kinabalurine skeleton, *Tetrahedron* **65**, 8259 (2009); (e) A. Farwick and G. Helmchen, Enantioselective total synthesis of (-)-α-kainic acid, *Org. Lett.* **12**, 1108 (2010).

[31] (a) M. E. Kraft, L. V. R. Boñaga, J. A. Wright and C. Hirosawa, Cobalt Carbonyl-Mediated Carbocyclizations of Enynes: Generation of Bicyclooctanones or Monocyclic Alkenes, *J. Org. Chem.* **67**, 1233 (2002); (b) D. P. Becker and D. L. Flynn, Synthesis of N-BOC-3-azabicyclo[3.3.0]octan-7-one via reductive Pauson-Khand cyclization and subsequent conversion to a novel diazatricyclic ring system, *Tetrahedron* **49**, 5047 (1993).

[32] (a) N. E. Schore and E. G. Rowley, Diastereofacialselectivity in intramolecular Pauson-Khand cycloaddition: highly stereoselective synthesis of pentalenene, *J. Am. Chem. Soc.* **110**, 5224 (1988); (b) J. Cassayre and S. Z. Zard, A short synthesis of (-)-dendrobine. Some observations on the nickel mediated radical cyclisation and on the Pauson-Khand reaction, *J. Organomet. Chem.* **624**, 316 (2001); (c) J. Cassayre and S. Z. Zard, A short synthesis of (-)-dendrobine, *J. Am. Chem. Soc.* **121**, 6072 (1999); (d) C. Mukai, M. Kobayashi, I. J. Kim and M. Hanaoka, Total synthesis of (±)-ceratopicanol, *Tetrahedron* **58**, 5225 (2002); (e) M. E. Krafft, G. M. Kyne, C. Hirosawa, P. Schmidt, K. A. Abboud and N. L'Helias, An approach to triquinane synthesis using the Pauson-Khand reaction, *Tetrahedron* **62**, 11782 (2006).

[33] (a) T. Morimoto and K. Kakiuchi, Evolution of carbonylation catalysis: no need for carbon monoxide, *Angew. Chem. Int. Ed.* **43**, 5580 (2004); (b) T. Morimoto, M. Fujioka, K. Fuji, K. Tsutsumi and K. Kakiuchi, Rhodium-catalyzed CO gas-free carbonylative cyclization using aldehydes, *Pure Appl. Chem.* **80**, 1079 (2008).

[34] I. U. Khand, G. R. Knox, P. L. Pauson and W. E. Watts, A cobalt induced cleavage reaction and a new series of arenecobalt carbonyl complexes, *J. Chem. Soc., Chem. Commun.* **36** (1971).

[35] T. Kobayashi, Y. Koga and K. Narasaka, The rhodium-catalyzed Pauson-Khand reaction, *J. Organomet. Chem.* **624**, 73 (2001).

[36] (a) M. C. Fromen, J. Morillo, M. J. Casanove and P. Lecante, Structure and chemical order in Co-Rh nanoparticles, *Europhys. Lett.* **73**, 885 (2006); (b) M. C. Fromen, P. Lecante, M. J. Casanove, P. B. Guillemaud, D. Zitoun, C. Amiens, B. Chaudret, M. Respaud and R. E. Benfield, Structural study of bimetallic Co_xRh_{1-x} nanoparticles: size and composition effects, *Phys. Rev. B* **69**, 235416 (2004).

[37] (a) T. Morimoto, K. Fuji, K. Tsutsumi and K. Kakiuchi, CO-transfer carbonylation reactions: a catalytic Pauson-Khand-type reaction of enynes with aldehydes as source of carbon monoxide, *J. Am. Chem. Soc.* **124**, 3806 (2002); (b) T. Shibata, N. Toshida and K. Takagi, Rhodium complex-catalyzed Pauson-Khand-type reaction with aldehydes as a CO source, *J. Org. Chem.* **67**, 7446 (2002); (c) H.W. Lee, L. N. Lee, A. S.

C. Chan and F. Y. Kwong, Microwave-assisted rhodium-complex-catalyzed cascade decarbonylation and asymmetric Pauson-Khand-type cyclizations, *Eur. J. Org. Chem.* 3403 (2008).

[38] J. H. Park, Y. Cho and Y. K. Chung, Rhodium-catalyzed Pauson-Khand-type reaction using alcohol as a source of carbon monoxide, *Angew. Chem., Int. Ed.* **49**, 5138 (2010).

[39] S. Roy and M. A. Pericàs, Functionalized nanoparticles as catalysts for enantioselective processes, *Org. Biomol. Chem.* **7**, 2669 (2009).

[40] (a) S. Jansat, M. Gomez, K. Philippot, G. Muller, E. Guiu, C. Claver, S. Castillon and B. Chaudert, A case for enantioselctive allylic alkylation catalyzed by palladium nanoparticles, *J. Am. Chem. Soc.* **126**, 1592 (2004); (b) E. Schmidt, A. Vargas, T. Mallat and A. Biaker, Shape-selective enantioselective hydrogenation of Pt nanoparticles, *J. Am. Chem. Soc.* **131**, 12358 (2009); (c) M. L. Kantam, S. Laha, J. Yadav, P. R. Likhar, B. Sreedhar and B. M. Choudary, Asymmetric hydrosilylation of prochiral ketones catalyzed by nanocrystalline copper(II) oxide, *Adv. Synth. Catal.* **349**, 1797 (2007).

[41] N. Jeong, B. K. Sung and Y. K. Choi, Rhodium(I)-catalyzed asymmetric intramolecular Pauson-Khand-type reaction, *J. Am. Chem. Soc.* **122**, 6771 (2000).

[42] K. M. Brummond, H. Chen, K. D. Fisher, A. D. Kerekes, B. Rickards, P. C. Sill and S. J. Geib, An allenic Pauson-Khand-type reaction: a reversal in π-bond selectivity and the formation of seven-membered rings, *Org. Lett.* **4**, 1931 (2002).

[43] For reviews, see: (a) K. M. Brummond and J. E. DeForrest, Synthesizing allenes today, *Synthesis* 795 (2007); (b) S. Ma, Some typical advances in the synthetic applications of allenes, *Chem. Rev.* **105**, 2829 (2005); (c) L. K. Sydnes, Allenes from cyclopropanes and their use in organic synthesis, *Chem. Rev.* **103**, 1133 (2003).

[44] K. H. Park, S. U. Son and Y. K. Chung, A Pauson-Khand-type reaction between alkynes and olefinic aldehydes catalyzed by rhodium/cobalt heterobimetallic nanoparticles: an olefinic aldehyde as an olefin and CO source, *Org. Lett.* **6**, 1183 (2004).

[45] (a) K. Narasaka and T. Shibata, Conversion of 1-(ω-alkynyl)-1,2-propadienyl sulfides to bicyclic dienones by the use of iron carbonyl complex, *Chem. Lett.* 315 (1994); (b) T. Shibata, Y. Koga and K. Narasaka, Intra- and intermolecular allene-alkyne coupling reactions by the use of $Fe(CO)_4(NMe_3)$, *Bull. Chem. Soc. Jpn.* **68**, 911 (1995).

[46] (a) K. M. Brummond, H. Wan and J. L. Kent, An intramolecular allenic [2+2+1] cycloaddition, *J. Org. Chem.* **63**, 6535 (1998); (b) K. M. Brummond, The allenic Pauson-Khand cycloaddition. Dependence in π–bond selectivity on substrate structure, *Tetrahedron Lett.* **39**, 931 (1998); (c) J. L. Kent, H. Wan and K. M. Brummond, A new allenic Pauson-Khand cycloaddition for the preparation of α–methylene cyclopentenones, *Tetrahedron Lett.* **36**, 2407 (1995).

[47] B. Alcaide and P. Almendros, The allenic Pauson-Khand reactions in synthesis, *Eur. J. Org. Chem.* 3377 (2004).

[48] J. H. Park, S. Y. Kim, S. M. Kim, S. I. Lee and Y. K. Chung, Allenic Pauson-Khand reaction catalyzed by cobalt-rhodium heterobimetallic nanoparticles, *Synlett* 453 (2007).

[49] For reviews, see: (a) R. W. Bates and V. Satcharoen, Nucleophilic transition metal based cyclization of allenes, *Chem. Soc. Rev.* **31**, 12 (2002); (b) R. Zimmer, C. U. Dinesh, E. Nandanan and F. A. Khan, Palladium-catalyzed reactions of allenes, *Chem. Rev.* **100**, 3067 (2000). For selected recent examples of metal-catalyzed reactions with allenes, see: (c) H. A. Wegner, A. de Meijere and P. A. Wender, Transition

metal-catalyzed intermolecular [5+2] and [5+2+1] cycloadditions of allenes and vinylcyclopropanes, *J. Am. Chem. Soc.* **127**, 6530 (2005). (d) J. Barluenga, R. Vicente, P. Barrio, L. A. López and M. Tomás, Metal-controlled selective [3+2] cyclization reactions of alkenyl Fischer carbene complexes and allenes, *J. Am. Chem. Soc.* **126**, 5974 (2004).

[50] For review, see: (a) A. S. K. Hashmi, How and selective transition metal catalyzed reactions of allenes, *Angew. Chem., Int. Ed.* **39**, 3590 (2000). For recent papers, see: (b) M. Kajitani, I. Kamiya, A. Nomoto, N. Kihara and A. Ogawa, Transition-metal-catalyzed carbonylation of allenes with carbon monoxide and thiols, *Tetrahedron* **62**, 6355 (2006). (c) D. Regás, J. M. Ruiz, M. M. Afonso and J. A. Palenzuela, Hetero Diels-Alder reaction of vinyl allenes and aldehydes. An experimental and computational study, *J. Org. Chem.* **71**, 9153 (2006). (d) P. A. Wender, M. P. Croatt and N. M. Deschamps, Metal-catalyzed [2+2+1] cycloaddition of 1,3-dienes, allenes, and CO, *Angew. Chem., Int. Ed.* **45**, 2459 (2006).

[51] J. H. Park, E. Kim, H.-M. Kim, S. Y. Choi and Y. K. Chung, Cobalt/rhodium heterobimetallic nanoparticle-catalyzed carbonylative [2+2+1] cycloaddition of allenes and bisallenes to Pauson-Khand-type reaction products, *Chem. Commun.* 2388 (2008).

[52] (a) Y. Abe, K. Kuramoto, M. Ehara, H. Nakatsuji, M. Suginome, M. Murakami and Y. Ito, A mechanism for the palladium-catalyzed regioselective silaboration of allene: a theoretical study, *Organometallics* **27**, 1736 (2008); (b) Y.Nakao, Y. Hirata, M. Tanaka and T. Hiyama, Nickel/BPh_3-catalyzed alkynylcyanation of alkynes and 1,2-dienes: an efficient route to highly functionalized conjugated enynes, *Angew. Chem. Int. Ed.* **47**, 385 (2008); (c) M. Aoki, S. Izumi, M. Kaneko, K. Ukai, J. Takaya and N. Iwasawa, Ni(0)-promoted hydroxycarboxylation of 1,2-dienes by reaction with CO_2 and O_2, *Org. Lett.* **9**, 1251 (2007); (d) T. Ohmura, H. Taniguchi and M. Suginome, Palladium-catalyzed asymmetric silaboration of allenes, *J. Am. Chem. Soc.* **128**, 13682 (2006); (e) Y. Nakao, Y. Hirata and T. Hiyama, Cyanoesterification of 1,2-dienes: synthesis and transformations of highly functionalized α–cyanomethylacrylate esters, *J. Am. Chem. Soc.* **128**, 7420 (2006).

[53] N. Jeong, S. D. Seo and J. Y. Shin, One-pot preparation of bicyclopentenones from propargyl malonates (and propargylsulfonamides) and allylic acetates by a tandem action of catalysts, *J. Am. Chem. Soc.* **122**, 10220 (2000).

[54] P. A. Evans and J. E. Robinson, Regio- and diastereoselective tandem rhodium-catalyzed allylic alkylation/Pauson-Khand annulations reactions, *J. Am. Chem. Soc.* **123**, 4609 (2001).

[55] S. U. Son, K. H. Park, H. Seo, Y. K. Chung and S.-G. Lee, Catalytic asymmetric synthesis of cyclopentenones from propargyl malonates and allylic acetate by successive action of homogeneous palladium(II) and cobalt on charcoal catalysts in one-pot reaction, *Chem. Commun.* 2440 (2001).

[56] (a) M. Trincado, H. Grutzmacher, F. Vizza and C. Bianchini, Domino rhodium/palladium-catalyzed dehydrogenation reactions of alcohols to acids by hydrogen transfer to inactivated alkenes, *Chem. Eur. J.* **16**, 2751 (2010); (b) P. Barbaro, C. Bianchini, V. Dal Santo, A. Meli, S. Moneti, R. Psaro, A. Scaffidi, L. Sordelli and F. Vizza, Hydrogenation of arenes over silica-supported catalysts that combine a grafted rhodium complex and palladium nanoparticles: evidence for substrate activation on $Rh_{single\text{-}site}$-Pd_{metal} moieties, *J. Am. Chem. Soc.* **128**, 7065 (2006); (c) P. Barbaro, C.

Bianchini, V. Dal Santo, A. Meli, S. Moneti, C. Pirovano, R. Psaro, L. Sordelli and F. Vizza, Benzene hydrogenation by silica-supported catalysts made of palladium nanoparticles and electrostatically immobilized single site, *Organometallics* **27**, 2809 (2008); (d) D. T. Thompson, Combined homogeneous and heterogeneous catalysts, *Platinum Metals Rev.* **43**, 114 (1999); (e) H. Gao and R. J. Angelici, Combination catalysts consisting of a homogeneous catalyst tethered to a silica-supported palladium heterogeneous catalyst: arene hydrogenation, *J. Am. Chem. Soc.* **119**, 6937 (1997).

[57] (a) B. M. Trost and R. C. Bunt, Asymmetric induction in allyl alkylations of 3-(acyoxy)cycloalkenes, *J. Am. Chem. Soc.* **116**, 4089 (1994); (b) B. M. Trost and R. C. Bunt, On ligand design for catalytic outer-sphere reactions: a simple asymmetric vinylglycinol, *Angew. Chem., Int. Ed. Engl.* **35**, 99 (1996).

[58] (a) R. Keese, R. Guidettigrept and B. Herzog, Synthesis of [5.5.5.5]fenestranes by Pd-catalyzed carbonylation-cyclisation, *Tetrahedron Lett.* **33**, 1207 (1992); (b) D. H. Kim, S. U. Son, Y. K. Chung and S.-G. Lee, Catalytic one-pot synthesis of [5.5.5.6]fenestrane systems via dicobalt octacarbonyl-catalyzed tandem cycloaddition of dienediynes, *Chem. Commun.* 56 (2002).

[59] (a) A. van der Waals and R. Keese, Synthesis of a [5.5.5.5]fenestrenedione *via* tandem Pauson-Khand tetracyclisation, J. Chem. Soc., *Chem. Commun.* 570 (1992); (b) M. Thommen, A. L. Veretenov, R. Guidetti-Grept and R. Keese, The tandem Pauson-Khand reaction, *Helv. Chim. Acta* **79**, 461 (1996); (c) M. Thommen, P. Gerber and R. Keese, Synthesis of a [5.5.5.5]fenestrane via Pauson-Khand reaction, *Chimia* **45**, 21 (1991).

[60] S. U. Son, K. H. Park and Y. K. Chung, Sequential actions of cobalt nanoparticles and palladium(II) catalysts: three-step one-pot synthesis of fenestranes from an enyne and an alkyne diester, *J. Am. Chem. Soc.* **124**, 6838 (2002).

[61] (a) S. Hermans, R. Raja, J.M. Thomas, B.F.G. Johnson, G. Sankar and D. Gleeson, Solvent-free, low-temperature, selective hydrogenation of polyenes using a bimetallic nanoparticle Ru-Sn catalyst, *Angew. Chem., Int. Ed.* **40**, 1211 (2001); (b) F. Coloma, J. Llorca, N. Homs, P.R. de la Piscina, F. Rodriguez-Reinoso and A.Sepulveda-Escribano, Crotonaldehyde hydrogenation over alumina- and silica-supported Pt–Sn catalysts of different composition. *In situ* DRIFT study, *Phys. Chem. Chem. Phys.* **2**, 3063 (2000); (c) D.S. Shephard, T. Maschmeyer, G. Sankar, J.M. Thomas, D. Ozkaya, B.F.G. Johnson, R. Raja, R.D. Oldroyd and R.G. Bell, Preparation, characterisation and performance of encapsulated copper–ruthenium bimetallic catalysts derived from molecular cluster carbonyl precursors, *Chem. Eur. J.* **4**, 1214 (1998).

[62] For sequential reactions directed by core/shell catalytic reactors, see: (a) Y. Wei, S. Soh, M. M. Apodaca, J. Kim and B. A. Grzybowski, Sequential reactions directed by core/shell catalytic reactors, *Small* **6**, 857 (2010); for a bimetallic heterogeneous catalyst, see: (b) C. Thiot, M. Schmutz, A. Wagner and C. Mioskowski, A one-pot synthesis of (*E*)-disubstituted alkenes by a bimetallic [Rh–Pd]-catalyzed hydrosilylation/Hiyama cross-coupling sequence, *Chem. Eur. J.* **13**, 8971 (2007).

10

Other Transition Metal-Mediated Cyclizations Leading to Cyclopentenones

Ramon Rios Torres and Jan Vesely

Every day you may make progress. Every step may be fruitful. Yet there will stretch out before you an ever-lengthening, ever-ascending, ever-improving path. You know you will never get to the end of the journey. But this, so far from discouraging, only adds to the joy and glory of the climb."

—-Sir Winston Churchill

The Pauson-Khand Reaction: Scope, Variations and Applications, First Edition. Edited by Ramon Rios Torres.

10.1 Introduction and Background

The cyclopentenone ring structure is a constituent of a broad range of natural products and compounds that are of biological or pharmaceutical interest (Scheme 10.1). Cyclopentenones have attracted considerable attention as constituents of prostaglandins.[1,2,3] Prostaglandins play an important role in the human body, display significant anti-tumor activity, and have also been reported to inhibit the influenza virus and human immunodeficiency virus (HIV).[4] For example, Punaglandin (**2**), a halogen-containing prostaglandin, reportedly exhibits strong anti-neoplastic activity.[5]

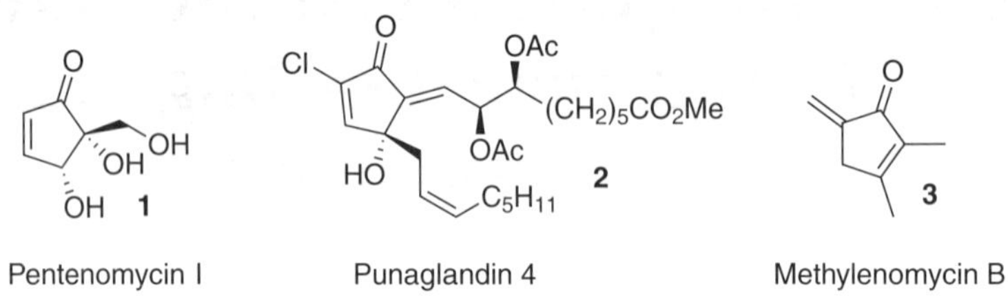

Scheme 10.1 *Compounds with a cyclopentadienone in their structure.*

Moreover, cyclopentenones are useful building blocks for organic synthesis owing to the inherent reactivity of the α,β-unsaturated carbonyl moiety.[6] There are numerous methods for the synthesis of cyclopentenone derivatives. One of the most widely used approaches is the Pauson-Khand reaction already discussed in this book. However, there are numerous other methods for the synthesis of cyclopentenone derivatives. This chapter focuses on the synthesis of cyclopentenones via transition metal catalysis; however, a brief review of some other approaches is first presented.

One classical approach to the synthesis of cyclopentenones without the use of transition metals is via intramolecular aldol reactions.[7] A beautiful example of the synthesis of a cyclopentenone via an intramolecular aldol reaction is the synthesis of cyclopentenone **5**, a key intermediate in the synthesis of (±)-hirsutene reported by List (Scheme 10.2).[8]

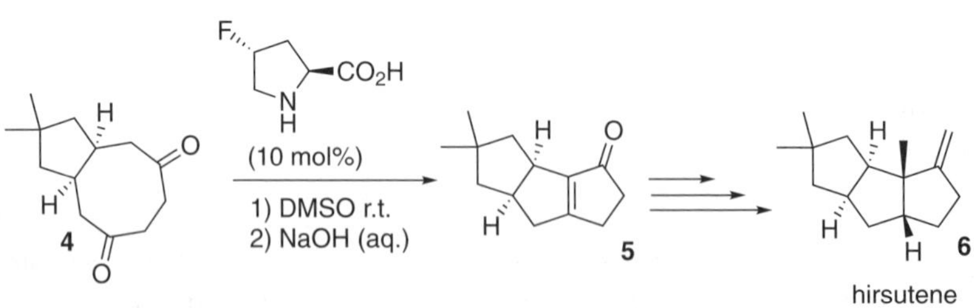

Scheme 10.2 *Synthesis of cyclopentenones via an intramolecular aldol reaction.*

Intramolecular Wittig type reactions[9] are an alternative option for the synthesis of cyclopentenones as illustrated by the preparation of **8**, which is a key intermediate in the synthesis of prostanoid Punaglandin (**2**) (Scheme 10.3).[10]

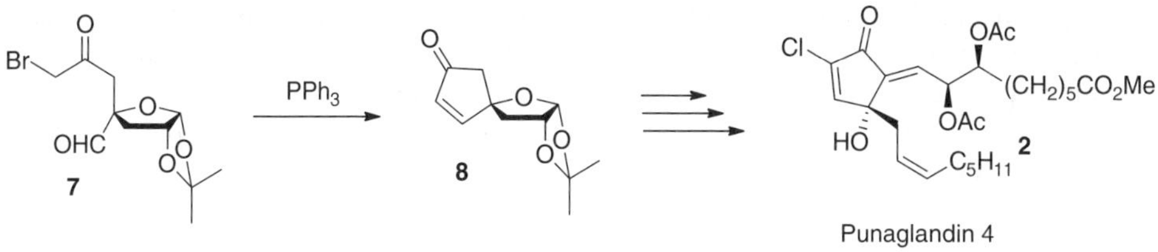

Scheme 10.3 *Synthesis of cyclopentenones via an intramolecular Wittig reaction.*

With the advent of organocatalysis, the Nazarov reaction[11] has emerged as a common method for the synthesis of cyclopentenones. The Nazarov reaction is an acid-induced cyclization of allyl vinyl and divinyl ketones that yields substituted cyclopentenones. An elegant example of this reaction was described by Rueping and co-workers (Scheme 10.4).[122]

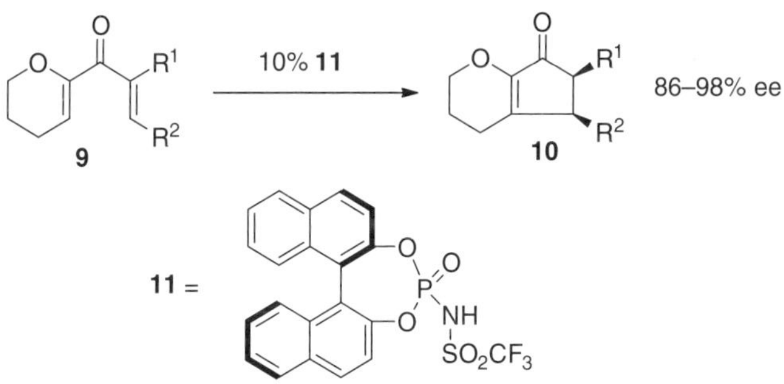

Scheme 10.4 *Nazarov reaction reported by Rueping.*

The final example in this category of reactions is the preparation of cyclopentenones by intramolecular 1,5-C-H insertions.[13] Alkylidene carbenes undergo regio- and stereoselective intramolecular C-H insertion reactions affording cyclopentenes. Alkylidene carbenes can be generated from alkynes under flash vacuum pyrolysis conditions by a reversible 1,2-shift. This strategy has been used to prepare a key intermediate in the synthesis of (±)-isocomene, as illustrated in Scheme 10.5.[14]

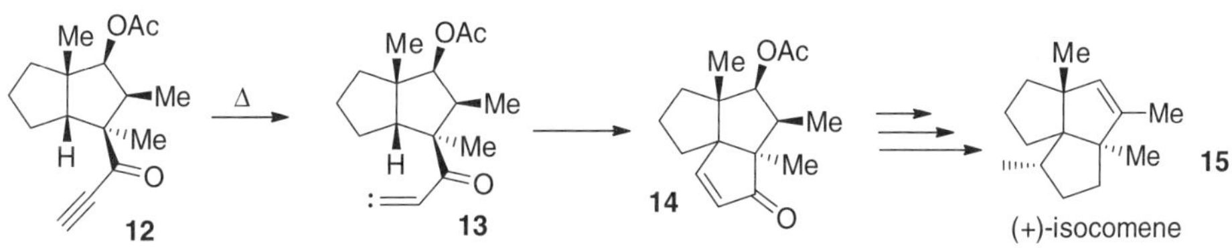

Scheme 10.5 *Synthesis of cyclopentenones via 1,5-C-H insertion.*

In spite of all the efforts expended in the quest for new non-metallic methodologies for the synthesis of cyclopentenones, transition metal-mediated strategies remain the most common approach. Apart from the Pauson-Khand reaction and related transformations, numerous other approaches have also emerged through the years. These approaches are discussed in this chapter.

10.2 [4+1] Strategies for the Synthesis of Cyclopentenones

The synthesis of cyclopentenones via [4+1] strategies can be achieved by means of various methodologies. The use of 1,3-butadienes represents one such example. Iron tricarbonyl 1,3-diene complexes can react with aluminum halides, resulting in decomplexation of the iron complex, to afford the corresponding cyclopentenones in good to excellent yields.[15] These reactions are normally carried out at atmospheric pressure and at room temperature and are complete in short reaction times (usually in minutes). In 1992, Frank-Neumann and co-workers reported a nice example of this reaction.[16] They used tricarbonyl iron complexes of 1,1,3 and 1,1,2,3-substituted 1,3-dienes that were decomplexed under mild conditions by aluminum halides to furnish the corresponding conjugated cyclopentenones. Depending on the substitution pattern, spirocyclic and/or cyclocondensed cyclopentenones can be obtained. One important feature of the reaction is its stereospecificity, which allows the building of optically active compounds from complexes with optically active ligands. For example, the chiral diquinane **18** has been synthesized from the chiral diene **16**. The corresponding cycloadduct was obtained as a 9:1 mixture of diastereomers, indicating that the cyclocarbonylation is a stereoselective process (Scheme 10.6). This study revealed that the reaction proceeds via an acyl π-allyl complex, whose formation is promoted by the coordination of aluminum bromide with a carbon monoxide ligand.

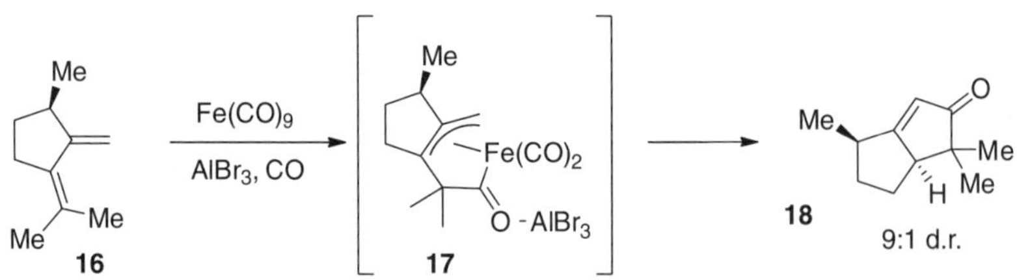

Scheme 10.6 *Synthesis of cyclopentenones via an iron catalyzed [4+1] cyclocarbonylation.*

The palladium-catalyzed carbonylative cyclization of 1-iodo-1,4-alkadienes, reported by Negishi and co-workers, is another alternative for the synthesis of cyclopentenones.[17,18] For successful conversion to cyclopentenones, the reaction has to be carried out in the presence of alcohols and under a carbon monoxide atmosphere (40–100 atm). These conditions lead to the obtention of monocyclic cyclopentenones (Scheme 10.7a). When the reaction is carried out in the absence of alcohol, the 1-iodo-1,4-alkadienes are carbonylated to α-methylene-cyclopentenones (Scheme 10.7b).

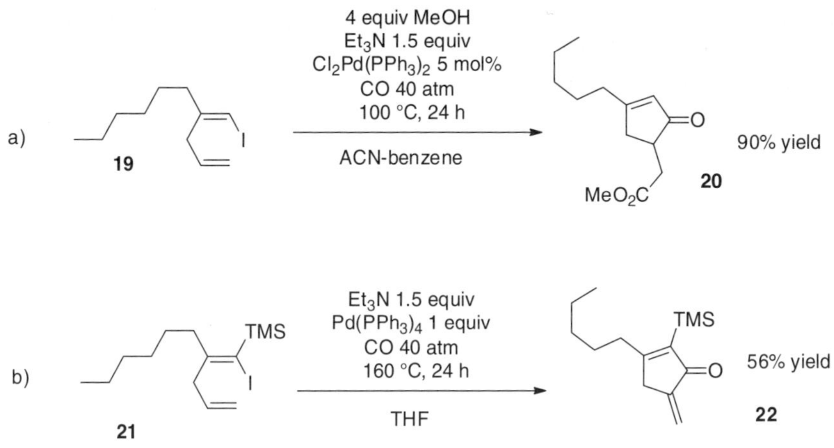

Scheme 10.7 *Palladium catalyzed [4+1] cyclocarbonylation.*

The same research group described a similar reaction with 2-iodophenyl derivatives that led to the synthesis of indenones (Scheme 10.8).[19]

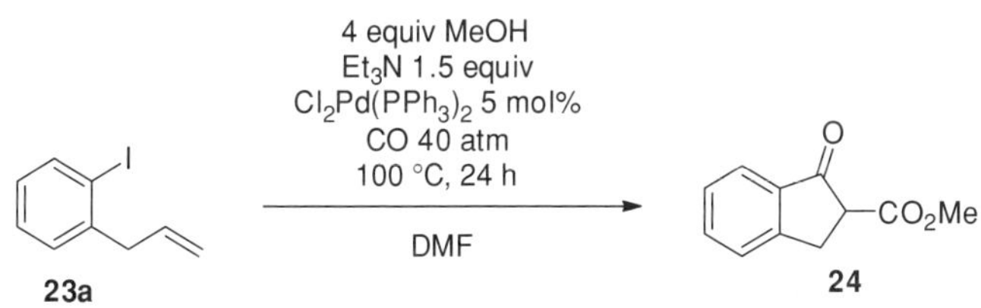

Scheme 10.8 *Synthesis of indenones via a Pd-catalyzed [4+1] cyclocarbonylation.*

In 2003, Larock and co-workers reported the use of dienyl triflates, bromides and iodides for the synthesis of cyclopentenones via a palladium-catalyzed reaction.[20] The reaction proceeded optimally with the use of 10 mol% of $Pd(OAc)_2$, 2 equiv. of pyridine, 1 equiv. of $n\text{-}Bu_4NCl$, 1 atm. of CO, a reaction temperature of 100 °C and DMF as a solvent. This methodology is particularly efficient for substrates that contain a terminal olefin, such as **23**. The reaction renders the final compounds in excellent yields as shown in Scheme 10.9. Moreover, it appears that the nature of X (bromine, iodine or triflate) has little influence on the outcome of the reaction.

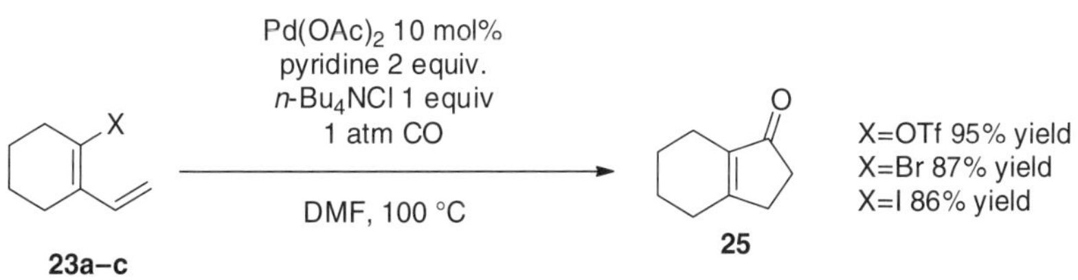

Scheme 10.9 *Palladium catalyzed [4+1] cyclocarbonylation reported by Larock.*

The commonly accepted mechanism (Scheme 10.10) for the preceding reaction is as follows: first the reduction of $Pd(OAc)_2$ occurs to render the active Pd(0). Next, an oxidative addition of the aryl iodide to Pd(0) takes place followed by coordination and insertion of carbon monoxide to produce an acylpalladium intermediate. The next step is an acylpalladation of the neighboring carbon–carbon bond followed by reversible palladium β-hydride elimination and re-addition to form a palladium enolate. Finally, protonation by H_2O affords the indanone and regenerates the Pd(II) salt, which is once again reduced to Pd(0) by CO as shown in Scheme 10.10.

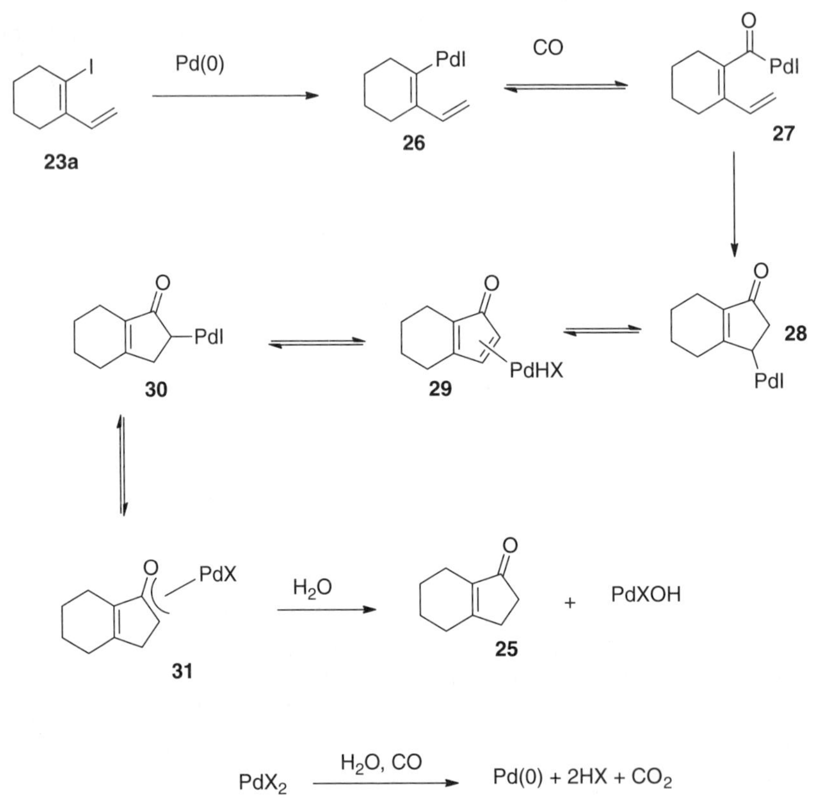

Scheme 10.10 *Mechanism for the palladium-catalyzed synthesis of cylopentenones.*

Bertus and Szymoniak developed a novel method for the synthesis of cyclopentenones via a titanium mediated [4+1] assembly of 1,3-dienes.[21] They studied the reaction of various dienes with nitriles using titanium catalysis. The reaction is strongly dependent on the temperature: when extremely low temperatures (−60 °C) were used for the addition of the nitrile, the only product obtained was cyclopropylamine **35**, whereas at higher temperatures (−20 °C) the reaction renders cyclopentenyl amine **36**, as depicted in Scheme 10.11.

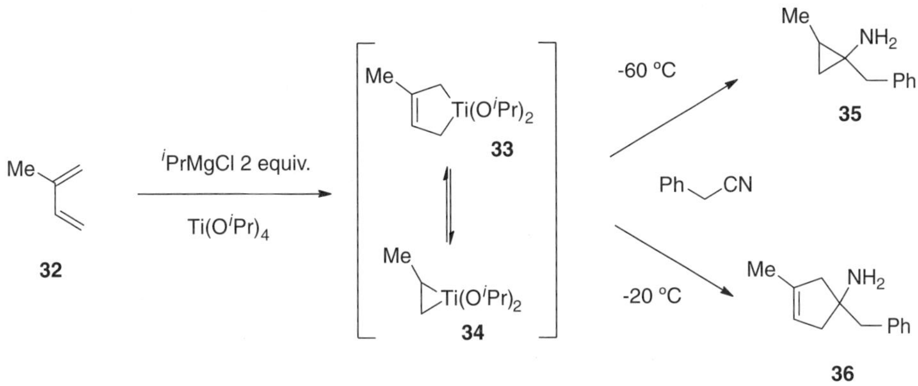

Scheme 10.11 *Titanium-catalyzed [4+1] reaction.*

This reaction offers a new route for accessing 3-substituted-2-cyclopentenones when 2-trimethylsilyloxybutadiene **37** is used as the starting diene, using the previously established conditions, as shown in Scheme 10.12.

39a R=Ph 75% yield
39b R=Bn 69% yield
39c R=^{n}C$_9$H$_{19}$ 62% yield
39d R=(CH$_2$)$_3$Cl 54% yield

Scheme 10.12 *Titanium catalyzed [4+1] synthesis of cyclopentenones.*

Another example of a [4+1] annulation was developed by Moses in 2006,[22] which employs stereoselective annulation reactions with silyl vinylketenes derived from Fisher carbene complexes. As depicted in Scheme 10.13, the Fisher carbene complexes first react thermically with triisopropyl (TIPS) or *tert*-butyldimethylsilyl (TBS) substituted alkynes to form the silyl vinylketenes. Notably, silyl vinylketenes are viable precursors to cyclopentenones or other cycloadducts whereas vinylketenes are highly unstable and exhibit a preference to undergo dimerization and/or [2+2] cycloaddition reactions. This difference in behavior is due to the presence of the silyl substituent that provides significant stabilization to the ketene moiety and suppresses the dimerization pathway, thereby facilitating its participation in [4+1] cycloaddition reactions with carbenoid reagents. In the subsequent step, the silyl vinylketene reacts with a carbenoid reagent such as dimethylsulfonium methylide (as previously reported by Danheiser).[23] The reaction tolerates dimethylsulfonium methylide and diazomethane reagents as carbenoids rendering the final cyclopentenones in good yields.

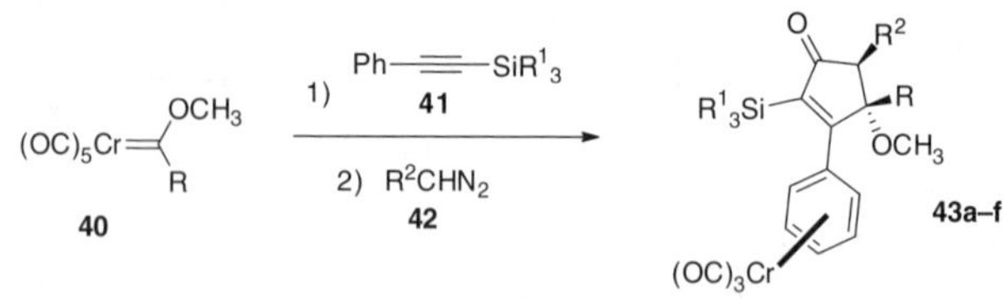

43a R=Me R^1=TIPS R^2=H 84% yield
43b R=n-Bu R^1=TIPS R^2=H 83% yield
43c R=Ph R^1=TIPS R^2=H 72% yield
43d R=Ph R^1=TBS R^2=H 70% yield
43e R=Ph R^1=TIPS R^2=Ph 73% yield
43f R=Ph R^1=TBS R^2=Ph 76% yield

Scheme 10.13 *Synthesis of cyclopentenones reported by Moses.*

A more detailed mechanism for the addition of the carbenoid to the silyl vinylketene is outlined in Scheme 10.14. Initially, there is a stereoselective addition of the carbenoid reagent to the ketene to provide the (Z)-enolate. Subsequent ionization of the leaving group generates the 2-oxidopentadienyl cation **46** as the sterically favored intermediate. The last step is a concerted 4π electrocyclic ring closure that preserves the original alkene geometry and provides the observed diastereomer of the cyclopentenone. It should be noted that in previous studies, Danheiser and co-workers[23] suggested a different mechanism that cannot be discounted. However, kinetic experiments are in concordance with the mechanism depicted in Scheme 10.14.

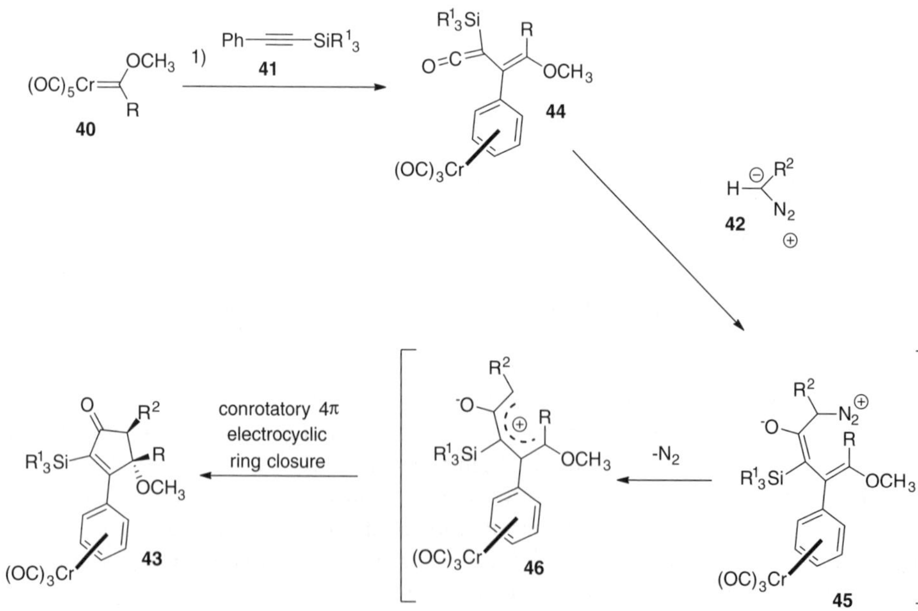

Scheme 10.14 *Mechanism for the [4+1] synthesis of cyclopentenones reported by Moses.*

10.3 [3+2] Strategies for the Synthesis of Cyclopentenones

10.3.1 Iron Promoted Reactions

The use of [3+2] cycloadditions is another commonly employed strategy for the synthesis of cyclopentenones using organometallic compounds. Specifically, the iron carbonyl-promoted cyclocouplings between secondary dibromoketones and enamines provides a synthetic route to α,α-dialkylated cyclopentenones (Scheme 10.15). The ketone is converted to the iron enolate **49** by diiron nonacarbonyl, and undergoes cycloaddition with nucleophilic alkenes such as **50**, affording the cyclopentenones. The use of morpholino-enamines leads to products that rapidly eliminate morpholine, providing cyclopentenones in a single one-pot procedure.[24–27]

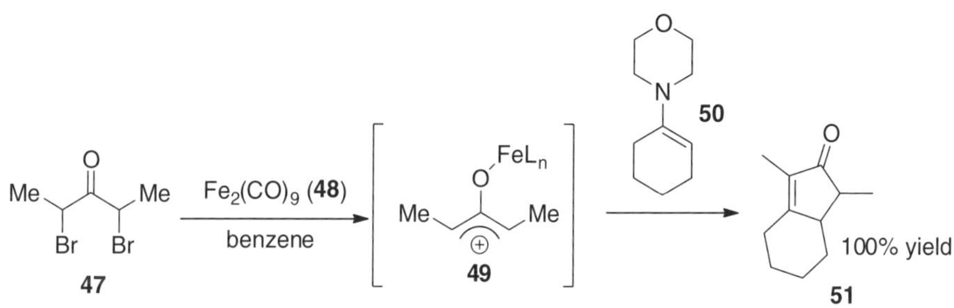

Scheme 10.15 Iron catalyzed synthesis of cyclopentenones.

Another important consideration for this reaction is that enamines of both aldehydes and ketones can be used, and bicyclic cyclopentenones can be easily obtained using cyclic enamines as a starting material. However, one important limitation of this methodology is, in fact, the impossibility of preparing α,β-unsubstituted cyclopentenones.

A fascinating variation of this method employs iron complexes such as **52**, with ketenes such as diphenyl ketene **53** (Scheme 10.16).[28,29] The generated iron complex can then be easily cleaved using a variety of oxidative and non-oxidative procedures.

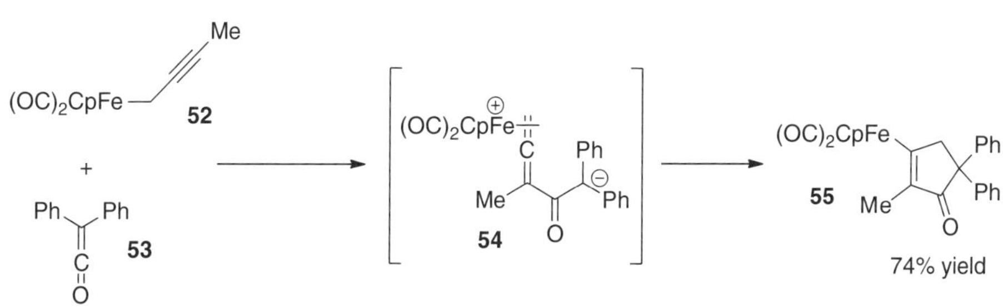

Scheme 10.16 Synthesis of cyclopentenones from ketenes.

10.3.2 Synthesis of Cyclopentenones via [3+2] Cycloadditions of Fisher Alkenyl Complexes

One important strategy for the synthesis of cyclopentenones is based on the [3+2] cycloaddition between chromium-alkenyl complexes and alkynes. The reaction was developed by Barluenga and co-workers using a Rh(I) complex as catalyst.[30] Initially, chromium methoxycarbene complexes react with rhodium(I) complexes to afford rhodium(I) alkenyl(methoxy)carbine complexes in high yields. Remarkably, the simultaneous transfer of two ligands (carbine and CO) represents a rare process. Next, a metalla-Diels–Alder reaction takes place to render a rhodium cyclohexadiene species that produces the desired cyclopentadiene. The regioselectivity of the reaction is strongly dependent on the nature of the alkyne. In the case of terminal alkynes the regiochemistry is simply dictated by steric discrimination, the larger alkyne substituent being placed away from the rhodium. In the case of internal electron-poor alkynes, the steric differentiation is not as pronounced and the electronic factors, which favor the interaction between the electrophilic metal center and the nucleophilic C-alkyne, dominate leading to cycloadducts with opposite regiochemistry, as illustrated in Scheme 10.17.

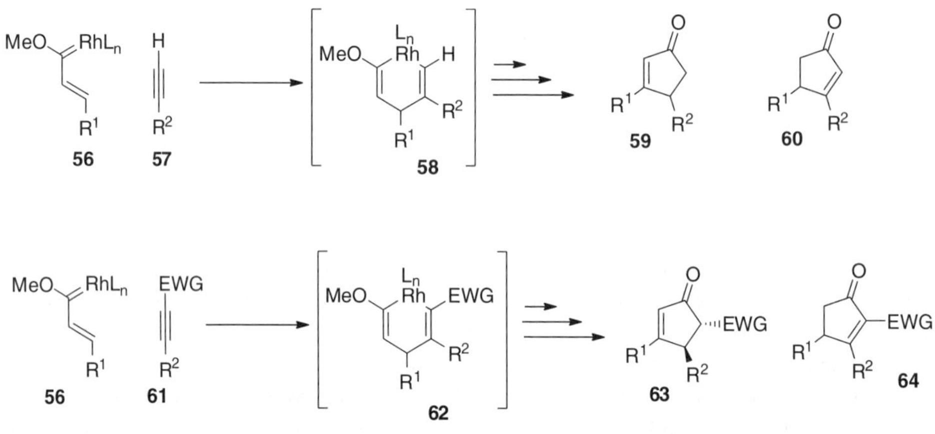

Scheme 10.17 Regioselectivity in the Rhodium catalyzed [3+2] cycloaddition of Fisher alkenyl complexes.

It should be noted that most of these types of cyclopentenones are not easily accessible by the intermolecular Pauson-Khand reaction. As shown in Scheme 10.18, the rhodium-catalyzed [3+2] cycloaddition affords the corresponding cyclopentenones in good yields and with total regioselectivity.

An alternative [3+2] strategy for the synthesis of cyclopentenones is the nickel-catalyzed cyclization of alkenyl Fisher carbene complexes and internal alkynes.[31] As was highlighted in the foregoing discussion, alkenyl Fisher carbenes react with terminal alkynes or electron-poor alkynes under Rh(I) catalysis to render cyclopentenones. In order to circumvent the limitations of that method, Barluenga and co-workers developed a powerful methodology

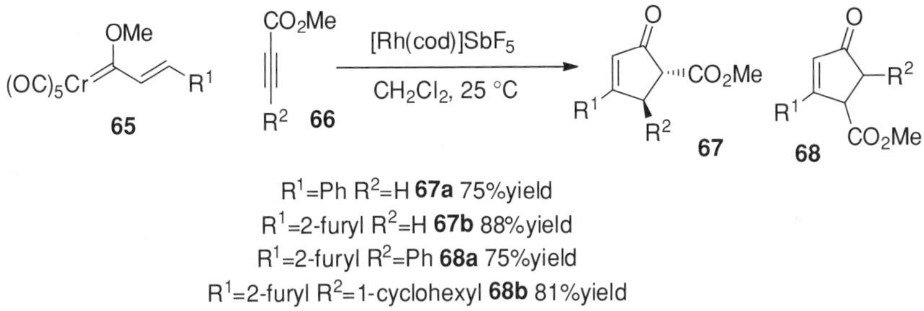

Scheme 10.18 *Rhodium catalyzed [3+2] cycloaddition.*

using Nickel(0) complexes as catalysts. The reaction tolerates unactivated alkynes, boron or tin substituted alkynes and electron rich alkynes furnishing the final cyclopentenones with moderate to good yields (43–75%) and with total regioselectivity as shown in Scheme 10.19.

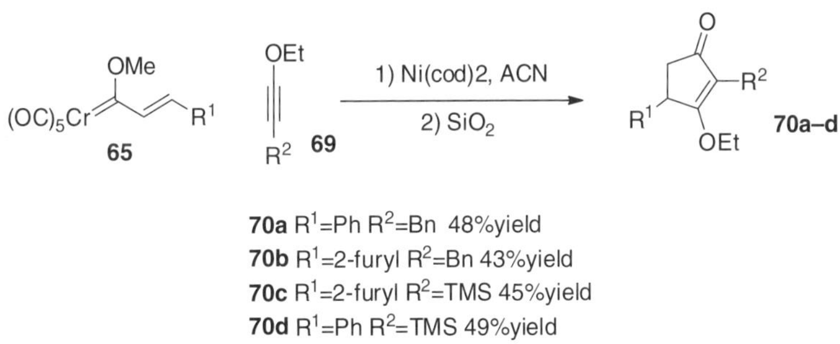

Scheme 10.19 *Nickel catalyzed [3+2] cycloaddition.*

The proposed mechanism (Scheme 10.20) involves Cr-Ni exchange, followed by regioselective insertion of a disubstituted alkyne unit into the nickel carbine complex leading to carbine **71**; this species undergoes ring closure rendering the corresponding cyclopentadiene **74**, possibly via the nickelacyclohexadiene species **73**. The origin of the regioselectivity lies in the alkyne insertion step and is consistent with the trend found in the insertion of these types of alkynes into Fischer complexes of chromium. Remarkably, when terminal alkynes were used a [3+2+2] cyclization took place affording the corresponding 7-membered carbocycle.[322]

10.4 Nickel(0) and Palladium(0) Synthesis of Cyclopentenones

A transformation reaction using Ni and Pd complexes, which is similar to the Pauson-Khand reaction, has also been reported. The main difference between the techniques is that the nickel and palladium route requires the use of allyl halides and alkynes to furnish the

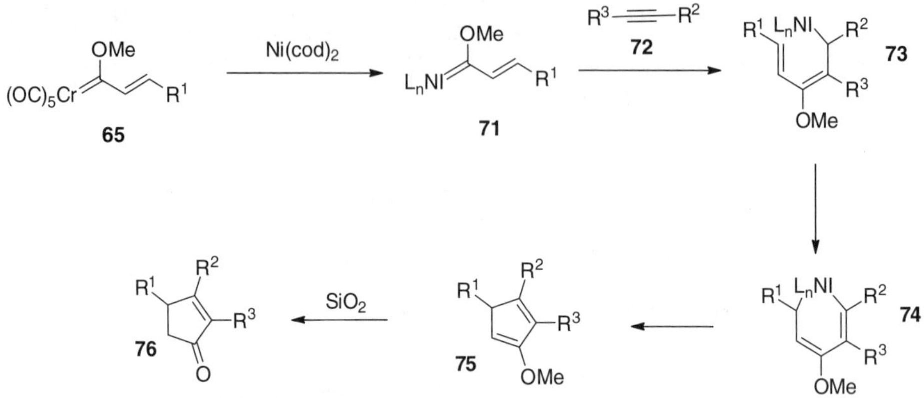

Scheme 10.20 *Proposed mechanism for the Nickel catalyzed [3+2] cycloaddition.*

desired cyclopentenones. In this methodology (Scheme 10.21), nickel and palladium act as the transition metal promoters and the double bond of the allyl halide serves as the alkene component of a formal [2+2+1] cycloaddition. Although several examples of these processes under mild conditions can be found in the literature, this reaction has attracted much less interest than the Pauson-Khand reaction due to the fact that most of those cyclizations demand the use of stoichiometric amounts of the hazardous $Ni(CO)_4$ complex.[33–37]

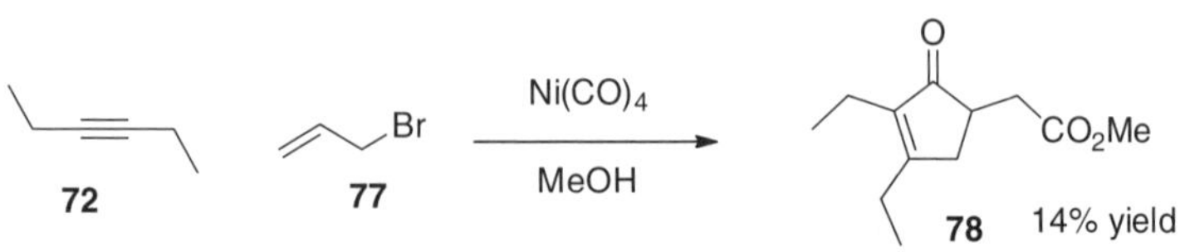

Scheme 10.21 *Ni(0) catalyzed synthesis of cyclopentenones.*

From the mechanistic point of view, the reaction proceeds via a well-defined sequence of steps, supported by mechanistic studies (Scheme 10.22).[38] The first step corresponds to the formation of the allyl nickel or allyl palladium complexes, generated by oxidative addition of Ni(0) or Pd(0) into the allyl halide. Next, the allyl complex undergoes alkyne insertion and concurrent carbonylation to render intermediate **81**. A 5-exo-trig cyclization followed by a second carbonylation then furnishes the complex **82**, which evolves into the final compound via methanol cleavage.

One important feature of the Ni(0)-catalyzed reaction is its high regioselectivity as a result of the polarization of the triple bond. The π-allyl nickel complex acts as an electrophile and adds to the most negative part of the alkyne. Importantly, the steric factors do not seem to affect the mode of addition. The functional group tolerance in the alkyne is high and this may be exploited to improve the yield and regioselectivity of the reaction. The intermolecular counterpart of this reaction has been particularly useful for the synthesis of bicyclo[3.3.0]octenine derivatives,[39] and the monocyclic antibiotic methylomycin B has

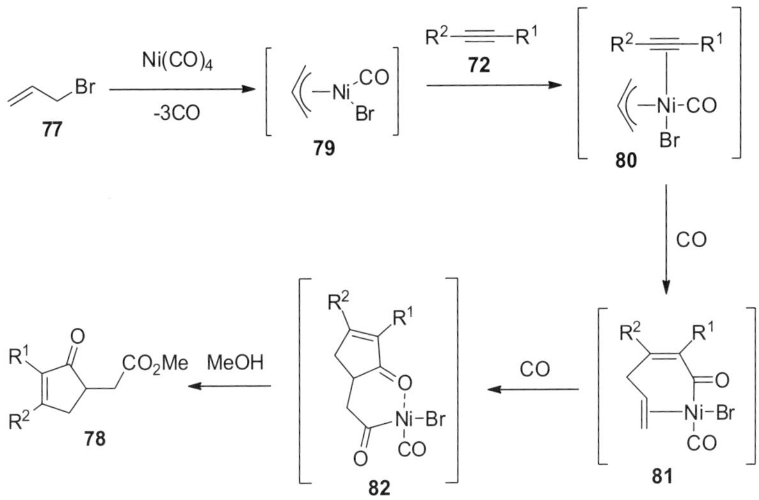

Scheme 10.22 *Proposed mechanism.*

been prepared in just four steps via this reaction using allyl bromide and 2-butynol as starting materials.[40] Remarkably, this cyclization is a useful methodology for the one-step synthesis of several ring-fused systems and spiro cyclopentenones (Scheme 10.23) in good yields and with total stereocontrol, when cyclic allylic systems and alkynes are used as starting materials.[41–43]

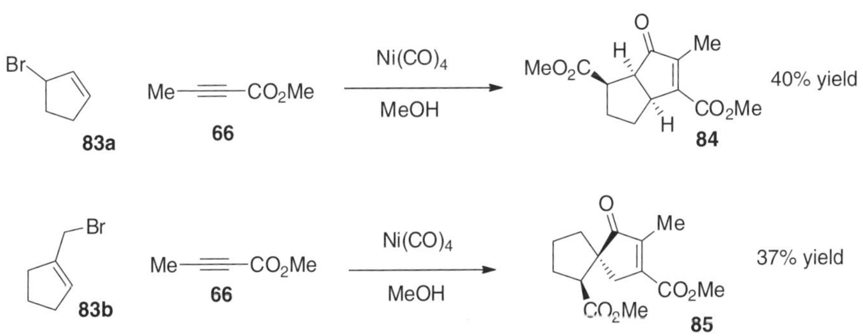

Scheme 10.23 *Synthesis of fused and spiro cyclopentenones.*

In another example, depicted in Scheme 10.24, the chiral acetylenic sulfoxide **86** reacts with the cyclic allyl halide **87**, to furnish cyclopentenone **88** in low yields and with excellent diastereoselectivity.[44,45]

An interesting variation of the Ni(0)-catalyzed reaction has been reported using isocyanides (as isoelectronic counterparts to carbon monoxide) to synthesize 1-amino-2-cyclopentenes, which may be hydrolyzed to the corresponding cyclopentenones (Scheme 10.25).[46,47]

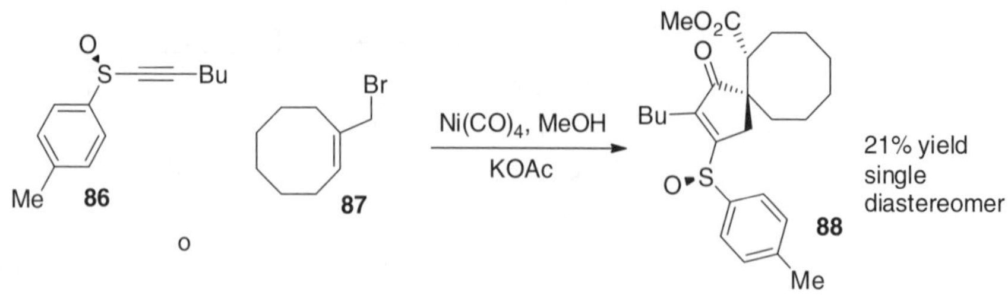

Scheme 10.24 *Diastereoselective Ni(0) catalyzed synthesis of cyclopentenones.*

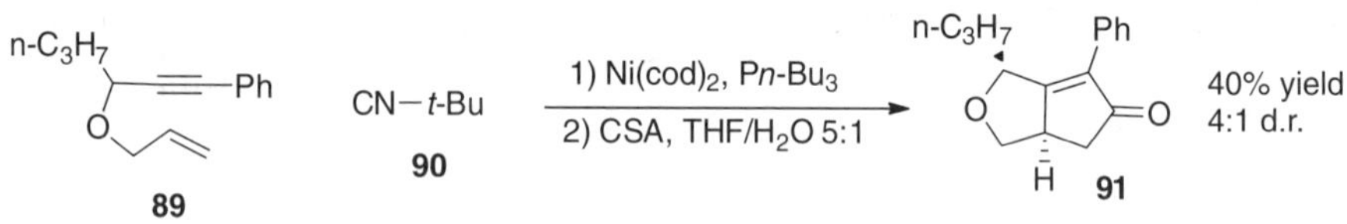

Scheme 10.25 *Synthesis of cyclopentenones avoiding the use of CO.*

Since 1998, Mackenzie's π-allyl nickel complexes (generated from $Ni(cod)_2$, enals, and trimethylsilylchloride), have been used in the synthesis of annulated cyclopentenones.[48–52] In this variant (Scheme 10.26), the silyloxy substituted nickel π-allyl complex reacts with the alkyne under a carbon monoxide atmosphere and then methanol is added to furnish the cyclopentenone.

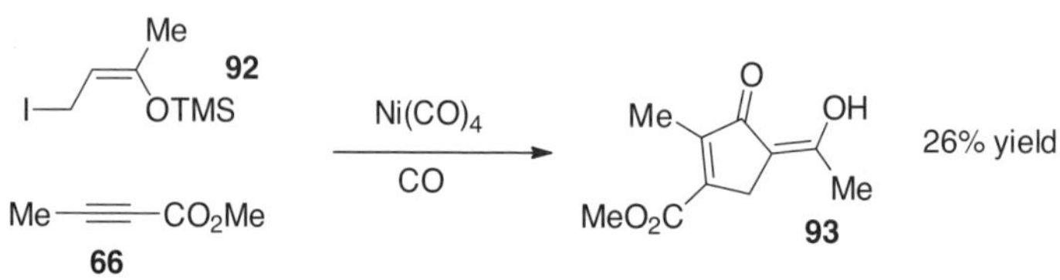

Scheme 10.26 *McKenzies's π allyl complex in the synthesis of cyclopentenones.*

Interestingly, the synthesis of annulated cyclopentenones via an intramolecular palladium- or nickel-catalyzed metallo-ene-type cyclization/carbonylation, using carbon monoxide at atmospheric pressure has been reported by Oppolzer (Scheme 10.27). The reaction allows the use of various nickel or palladium sources, such as $Ni(CO)_3PPh_3$ or $Ni(COD)_2$/1,4-bis(diphenylphosphino)butane or Pd(dibenzylidene-acetone)$_2$/PPh_3, with enynes bearing an allylic halide or acetate moiety, furnishing mono- or bicyclic products. The formation of cyclopentenones or diene derivatives is highly dependent on the choice of metals and ligands and on the substitution pattern of the substrate. The mechanism is similar to that described in Scheme 10.22, affording the cyclopentenones or diene derivatives in good yields.[53–55] An intriguing utilization of this methodology is its application to the synthesis of natural products such as (±)-hirsutene.[55–58]

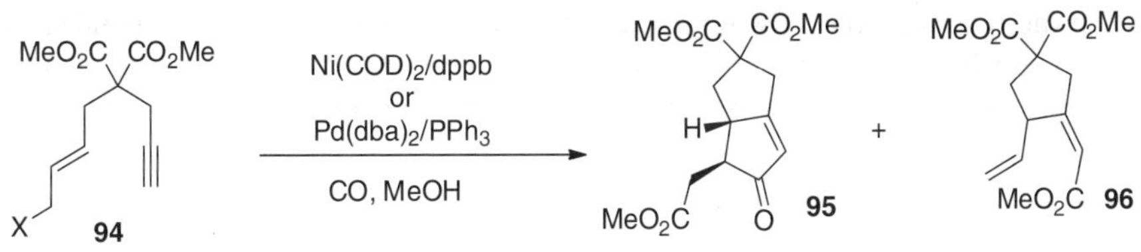

Scheme 10.27 *Ni(0) or Pd(0) synthesis of fused cyclopentenones.*

Another interesting variation of this reaction is the synthesis of α-methylenecyclopentenones fused to five-membered rings via a palladium-catalyzed carbonylative cyclization of enynes (Scheme 10.28).[59]

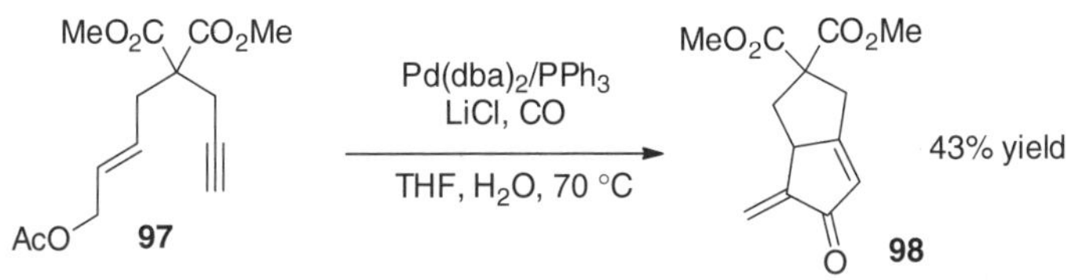

Scheme 10.28 *Synthesis of α-methylenecyclopentenones.*

10.5 Metal Carbine Strategies for the Synthesis of Cyclopentenones

10.5.1 Cyclopropylcarbene Chromium Complexes

The reactions of cyclopropylcarbene chromium complexes with alkynes have attracted significant attention from several research groups. For example, Herndon and co-workers reported that cyclopropylchromium carbine **99** reacts with diphenylacetylene **72a** affording a mixture of cyclopentenones **100** and **101** with a ratio of 1:20,[60–64] as depicted in Scheme 10.29. The major product of the reaction is the thermodynamically stable cyclopentenone **101**, in which the *trans* stereochemistry of the original alkyne substituents is retained in the product. In the commonly accepted mechanism for this reaction, the products are believed to be formed via "*in situ*" reduction of the cyclopentadienone moiety to the corresponding cyclopentenone derivatives under the reaction conditions.[64]

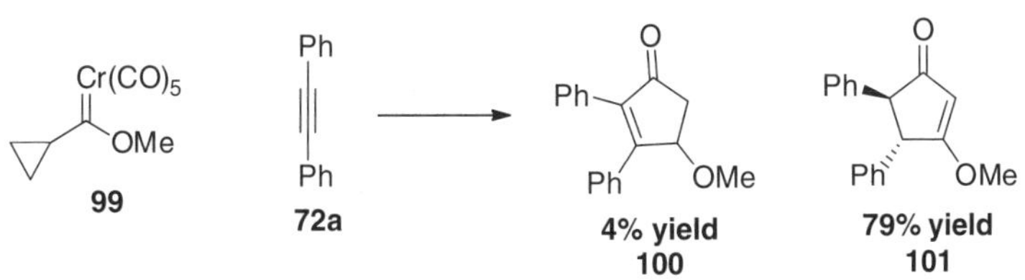

Scheme 10.29 *Cylopentenone synthesis from,cyclopropyl chromium complexes reported by Herndon.*

When the reaction proceeds intramolecularly via an alkyne substituent bonded to the carbine moiety by means of the alkoxy substituent, the reaction furnishes a cyclopentenone-fused oxygen heterocycle as shown in Scheme 10.30. This represents a nice extension of this methodology.[65]

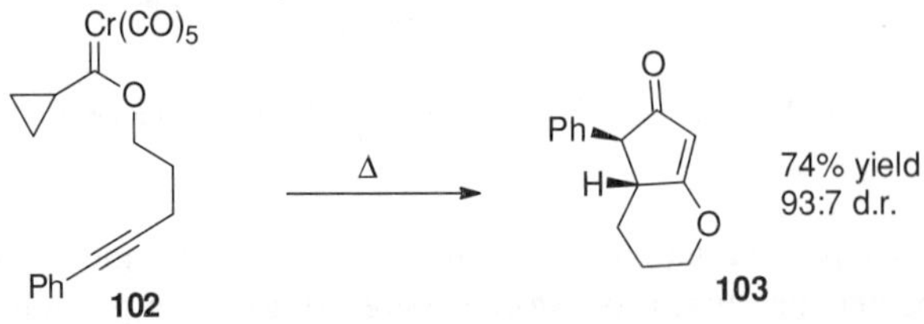

Scheme 10.30 *Intramolecular cylization of cyclopropyl chromium complexes.*

In the proposed mechanism (Scheme 10.31), the alkyne inserts into the chromium complex with formation of a metallacyclobutene **105**, which is converted into the vinylcarbene complex **106**. This mechanism is similar to mechanism described for the Dötz reaction.[66] Final conversion into a cyclopentadienone then occurs via a 1,5-alkyl shift, followed by CO insertion, conversion to the pentadienyl complex, ring closure and fragmentation with ethylene loss. Subsequent reduction with Cr(0) in water renders the cyclopentenone.

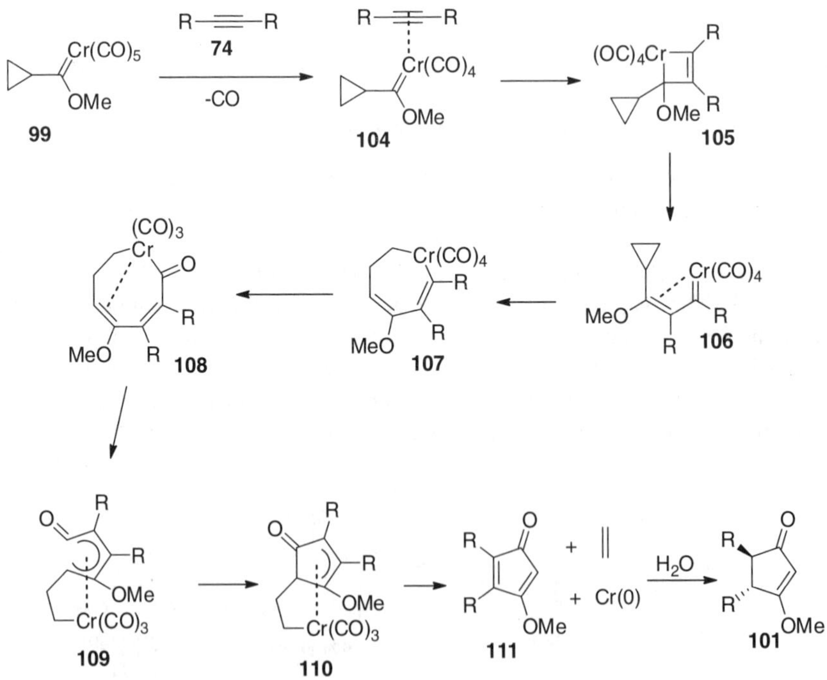

Scheme 10.31 *Proposed mechanism for cyclopentenone synthesis by cyclization of cyclopropyl chromium complexes.*

On the other hand, the coupling of propargyl alcohol derivatives with cyclopropylcarbene chromium complexes produces alkylidene cyclopentenones.[67]

10.5.2 Rhodium(II)-Catalyzed Strategies

Another interesting route to the synthesis of cyclopentenones is the rhodium-catalyzed intramolecular cyclization of α-diazo ketones bearing alkyne units. This methodology gives access to a plethora of substituted cyclopentenones.

The proposed mechanism proceeds via decomposition of the α-diazoketone **112** to generate a rhodium-stabilized carbenoid intermediate **113** as shown in Scheme 10.32. Next, addition of the rhodium carbenoid onto the alkyne π-bond generates a vinyl carbenoid intermediate **114**, in which the carbene-like character has been transferred to the more remote carbon of the alkyne. This intermediate may react further in either an intra- or intermolecular fashion to afford novel products. However, several studies from Padwa laboratories suggested that the mechanism is highly dependent on the solvent employed.[68,69]

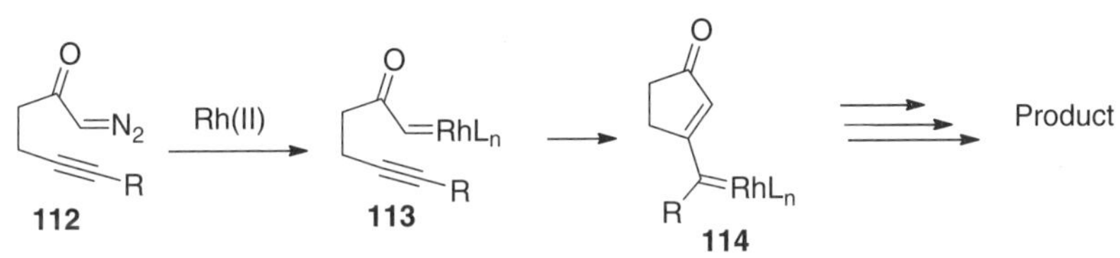

Scheme 10.32 *Rh (II) catalyzed synthesis of cyclopentenones.*

In one example of the rhodium-catalyzed intramolecular cyclization, the α-diazido ketone **115** undergoes insertion of the internal alkyne, followed by an external alkyne insertion to render cyclopentenone **116**, along with several byproducts. Ensuing treatment of **116** with catalytic amounts of a rhodium(II) complex leads to clean isomerization giving **117** as a single isomer in good yields (Scheme 10.33).[70–72]

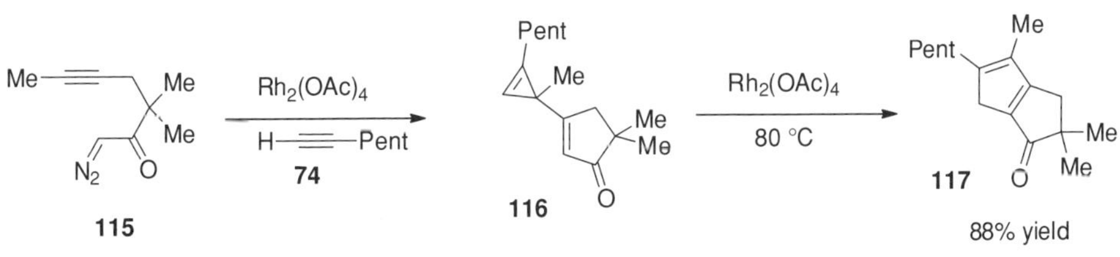

Scheme 10.33 *Intramolecular Rh(II) catalyzed synthesis of fused cyclopentenones.*

Another methodology for the synthesis of fused cyclopentenones is illustrated in Scheme 10.34. This reaction proceeds via the rhodium(II)-catalyzed diazo decomposition of a mixture of α,β-unsaturated carbonyl compounds **118** and **119**.[73] It should be noted that $Rh_2(acam)_4$ has to be used as the source of Rh in this reaction given that when $Rh_2(OAc)_4$ was used, the reaction only afforded complex mixtures.

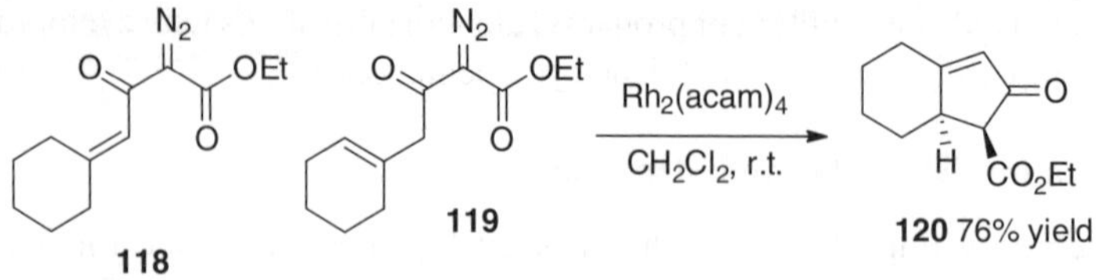

Scheme 10.34 *Rh(II)-catalyzed synthesis of cyclopentenones from α,β-unsaturated carbonyl compounds.*

10.6 Other Methodologies

The selective monodecarbonylative coupling of cyclobutenediones with alkenes is yet another method employed in the synthesis of cyclopentenones. Kondo and co-workers discovered the ruthenium-catalyzed reconstructive synthesis of cyclopentenones by an unusual coupling reaction of cyclobutenediones with alkenes, involving a C–C bond cleavage.[74] The suggested mechanism (Scheme 10.35) for the reaction is as follows: first, the oxidative addition of cyclobutenedione **121** into an active ruthenium complex occurs selectively at the C2–C3 bond under the direction of an alcoxy substituent to generate a ruthenacyclopentenedione intermediate. It is important to use the appropriate carbon monoxide pressure (3 atm) in order to control the selective mono-decarbonylation of the ruthenacyclopentenedione leading to a ruthenecyclobutenone intermediate while concomitantly suppressing the complete decarbonylation leading to an alcoxyalkyne and CO. Next, stereoselective cis-carboruthenation of 2-norbenene **124** and reductive elimination with retention of stereochemistry renders the corresponding exo-cyclopentenone **127**.

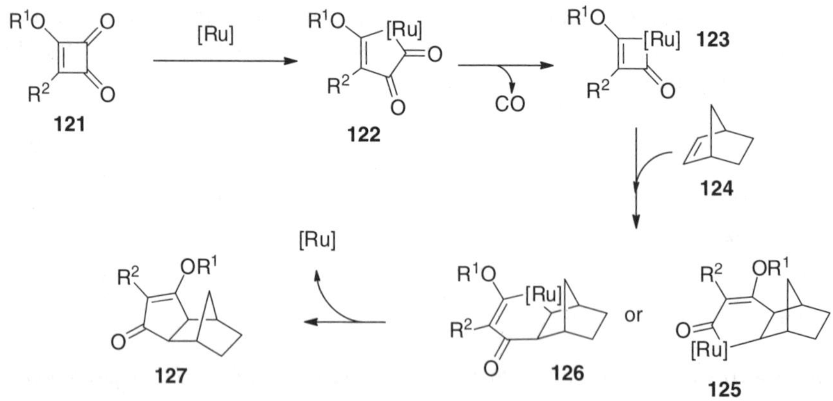

Scheme 10.35 *Proposed mechanism for the Ru-catalyzed monodecarbobylation of cyclobutadienones.*

This methodology affords the final cyclopentenones in moderate to good yields and with total control of stereoselectivity as shown in Scheme 10.36.

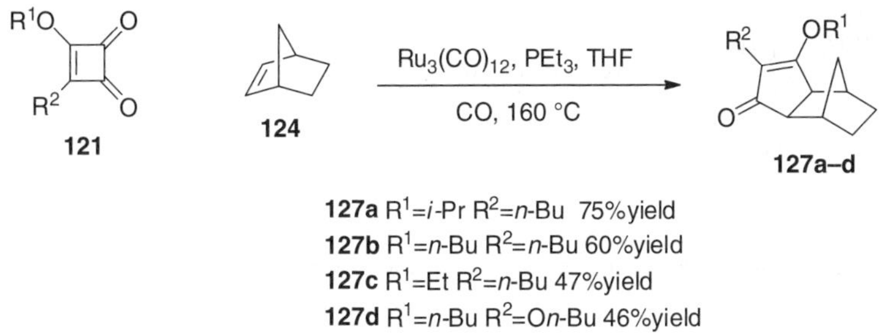

Scheme 10.36 *Ru catalyzed synthesis of cyclopentenones.*

Cyclopentenones may also be synthesized by the rhodium-catalyzed hydroacylation of alkynals.

It has been reported that Rhodium(I) complexes catalyze the intramolecular hydroacylation of a variety of 4-alkynals **128** to generate cyclopentenones **131**.[75,76] The accepted mechanism indicates a *trans* addition of the rhodium hydride to the alkyne to generate the six-membered rhodium metallacyclohexene **130**. Finally, reductive elimination of complex **130** renders the desired cyclopentenone and regenerates the rhodium(I) catalyst as illustrated in Scheme 10.37.

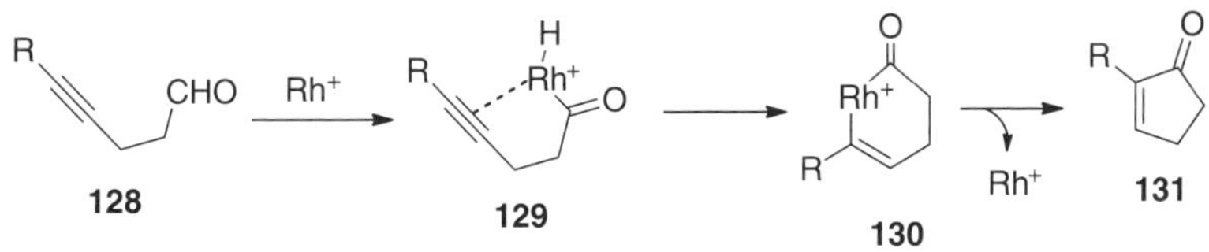

Scheme 10.37 *Proposed mechanism for the Rh catalyzed hydroacylation.*

The reaction is simply catalyzed by $[Rh(dppe)]_2(BF_4)_2$ affording the final cyclopentenones in good yields as shown in Scheme 10.38.

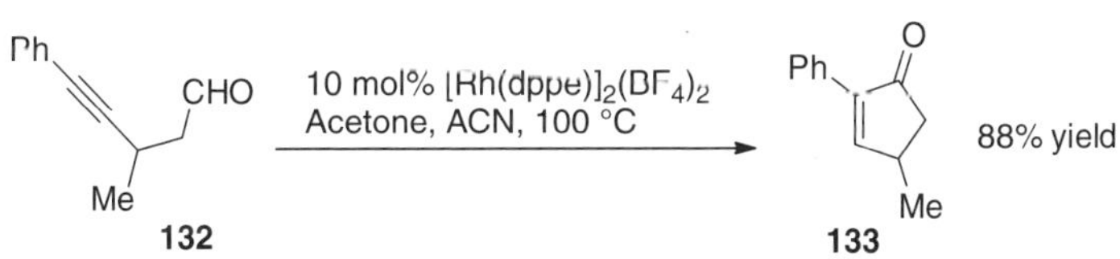

Scheme 10.38 *Rh catalyzed hydroacylation.*

The scope of the reaction was subsequently expanded by the use of 5-alkynals. Cyclopentenones were obtained via a tandem hydroacylation/double bond migration that takes place at elevated temperatures. This methodology gives access to α,β-disubstituted cyclopentenones such as dihydrojasmone (Scheme 10.39).[77]

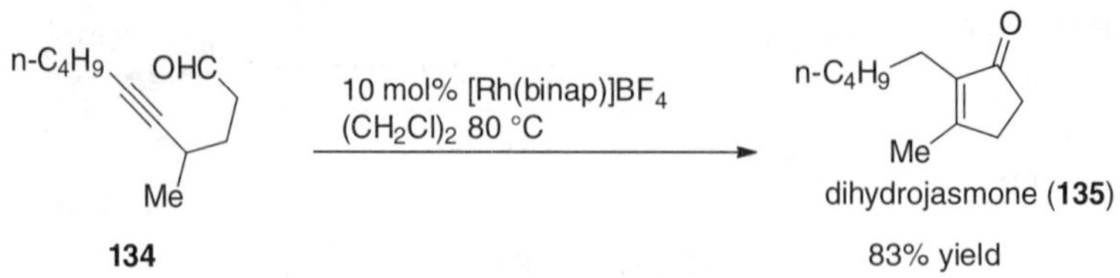

Scheme 10.39 *Tandem hydroacylation/bond migration reaction.*

A different approach to the synthesis of cyclopentenones exploits the ring expansion of vinyl cyclobutanols. Vinyl cyclobutanols, in the presence of palladium(II) catalysts and 2 equivalents of benzoquinone render the desired cyclopentenones. The mechanism is illustrated in Scheme 10.40, with **137** as the key intermediate. Through this methodology 5,5-, 5,6-, and 5,7-ring systems are readily accessible.[78]

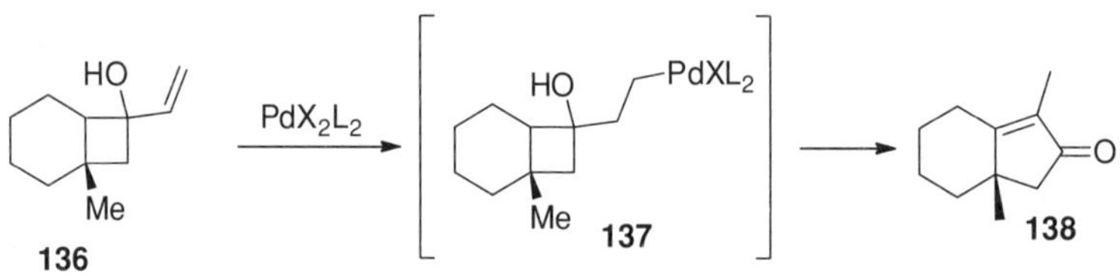

Scheme 10.40 *Vinyl cyclobutanols ring expansion of vinyl cyclobutanols catalyzed by Pd.*

A similar approach involves the ring expansion of three-membered spiropentane rings that takes place under an atmosphere of carbon monoxide, and is catalyzed by Rh(I) complexes. The reaction generates the cyclopentenones with good yields and regioselectivities.[79]

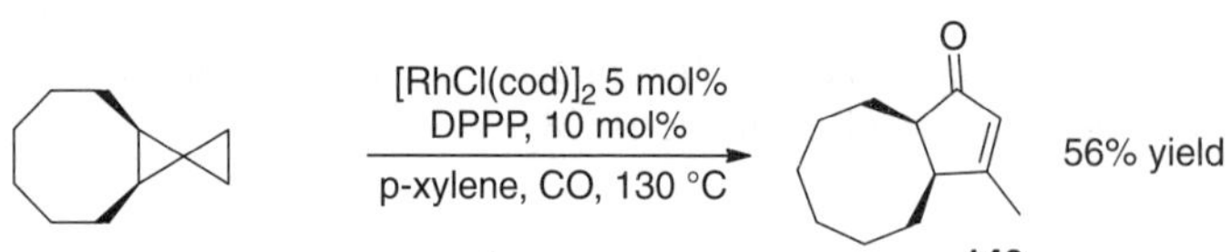

Scheme 10.41 *Spiropentanes ring expansion catalyzed by Pd.*

The formation of cyclopentenone **146** from spiropentane **141** can be explained by the pathway depicted in Scheme 10.41. Initially, Rh(I) oxidatively adds to the C4–C5 bond of the corresponding spiropentane to form spirocyclic rhodiacyclobutane **142**. Insertion of carbon monoxide generates rhodacyclopentanone **143**. Selective migration of the methylene carbon (C2) onto rhodium through a β-carbon elimination then converts the spirocyclic skeleton into the six-membered rhodiacycle **144**. Finally, reductive elimination furnishes 3-methylene cyclopentanone **145**, which isomerizes to the cyclopentenone **146** under the reaction conditions (Scheme 10.42).

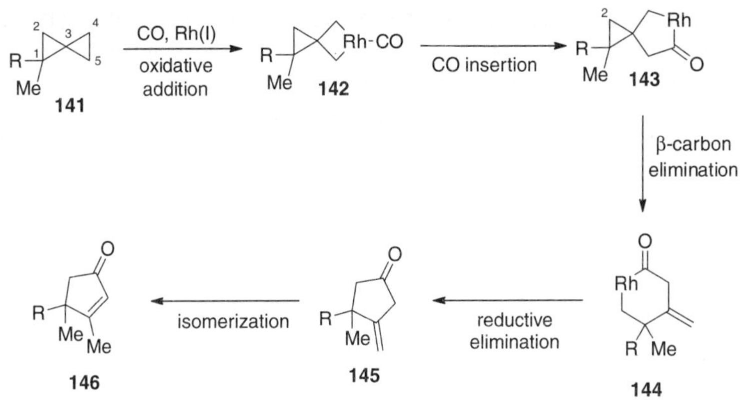

Scheme 10.42 *Proposed mechanism for the Rh-catalyzed carbonylation of spiropentanes.*

The Nazarov reaction employing organometallic reagents is another interesting approach to the synthesis of cyclopentenones.[11] The Nazarov reaction is formally a cyclization of divinyl ketones under the influence of very strong acids. In the organometallic version of this reaction, a metallic Lewis acid such as Cu and Pd, is used instead of a strong Brønsted acid (Scheme 10.43).

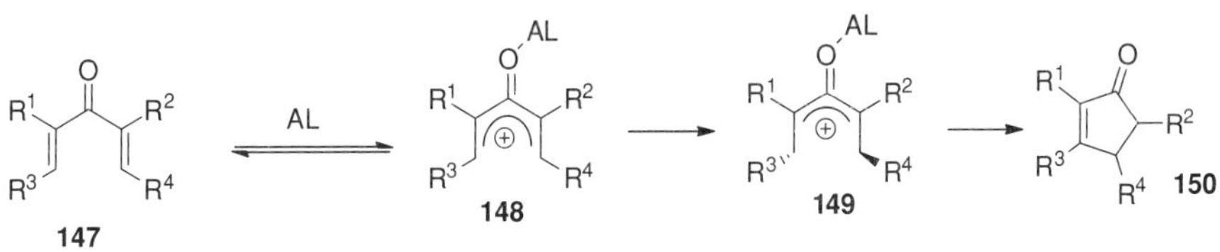

Scheme 10.43 *Nazarov reaction.*

Although the Nazarov reaction is generally promoted by one or more equivalents of a Lewis acid ($SnCl_4$, $TiCl_4$, $FeCl_3$ or $AlCl_3$), the Nazarov reaction catalyzed by transition metal salts has recently received significant attention. For example, the first palladium-catalyzed Nazarov cyclization of divinyl ketones was developed by Tius in 2003.[80] The reaction takes place under mild conditions and leads, in the case of α-alcoxy dienones, to 2-hydroxycyclopentenones in good yields (Scheme 10.44).

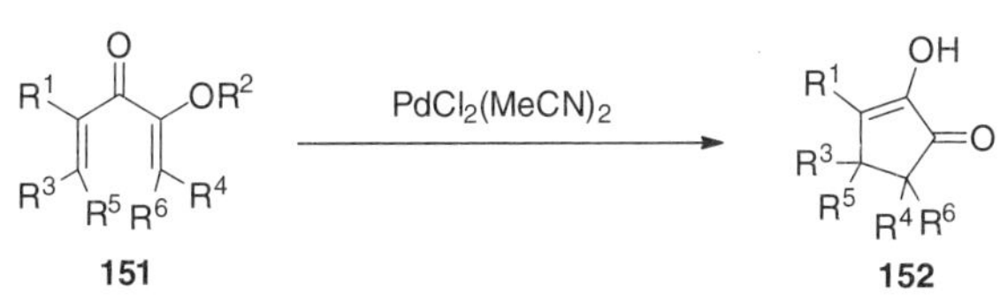

152a R^1=Me R^2=Et R^3=Ph R^4=R^5=R^6=H 91% yield
152b R^1=R^3=Me R^2=Et R^4=R^5=R^6=H 70% yield
152c R^1=R^5=Me R^2=Et R^5=Ph R^4=R^6=H 92% yield

Scheme 10.44 *Nazarov reaction catalyzed by Pd(II).*

The mechanism of the palladium-catalyzed Nazarov reaction is as follows: the reaction is initiated by the complexation of a palladium salt with the double bond of the divinyl ketone. Once formed, the π complex **154** undergoes intramolecular attack to produce a palladium enolate **155**. Next, hydrolysis with loss of ethanol renders **156** and HCl. Decomposition of the palladium enolate by the strong acid regenerates the Pd(II) catalyst and leads to **157** (Scheme 10.45).

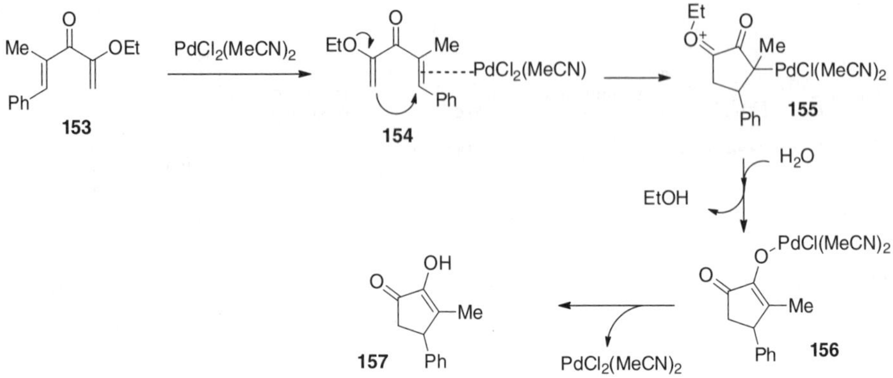

Scheme 10.45 *Proposed mechanism for the Pd(II) catalyzed Nazarov reaction.*

This example represents the only Nazarov reaction reported thus far, in which the transition metal catalyst does not act as a simple Lewis acid. The existing literature contains a large number of examples of catalytic Nazarov reactions promoted by Lewis acids. The use of transition metals as simple Lewis acids is beyond the scope of this chapter and we only highlight one very recent example of a Nazarov reaction catalyzed by chiral copper salts, reported recently by Tang and co-workers.[81] Chiral tris(oxazoline) copper(II) catalysts were found to promote the reaction of divinylketones rendering the final cyclopentenones in excellent yields and enantioselectivities as shown in Scheme 10.46.

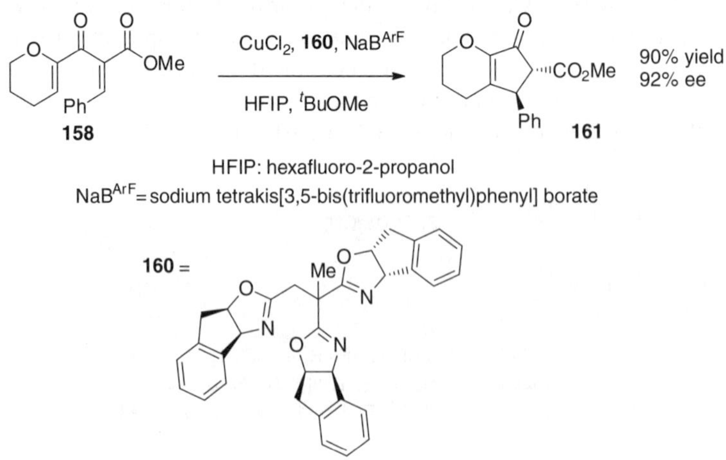

Scheme 10.46 *Copper-catayzed Nazarov reaction.*

The Au(I)-catalyzed hydrative rearrangement of 1,1-diethylcarbinol acetates is an intriguing methodology employed in the synthesis of cyclopentenones.[82] This reaction is strongly dependent on the reaction conditions as shown in Scheme 10.47.

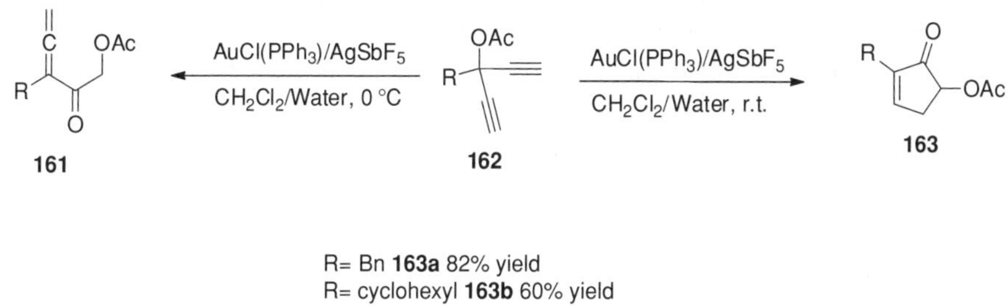

Scheme 10.47 *Au(I)-catalyzed hydrative rearrangement of 1,1-diethylcarbynol acetates.*

The Au(I)-catalyzed hydrative rearrangement reaction begins with the formation of **164** via an Au catalyzed [3,3]-sigmatropic shift of the acetate group (Scheme 10.48). Due to the extraordinary alkynophilicity of gold, Au coordinates with the triple bond of **164**, which induces oxacyclization/1,3-dioxolium ion generation via an internal nucleophilic attack by the carbonyl oxygen of the acetoxy group at the C-2 position of the allene to furnish **165**. Next, a molecule of water is expected to be involved at this stage to form an intermediate **166**, which is converted into **167** by simple deprotodemetalation. Interestingly, Au(I) further activates the allene functionality, which induces the sequential 1,3-dioxole ring opening and the 5-endo-dig carbocyclization to furnish the gold-coordinated cyclopentenone **168**. The latter, upon protodemetalation, renders the cyclopentenone **163** as illustrated in Scheme 10.48.

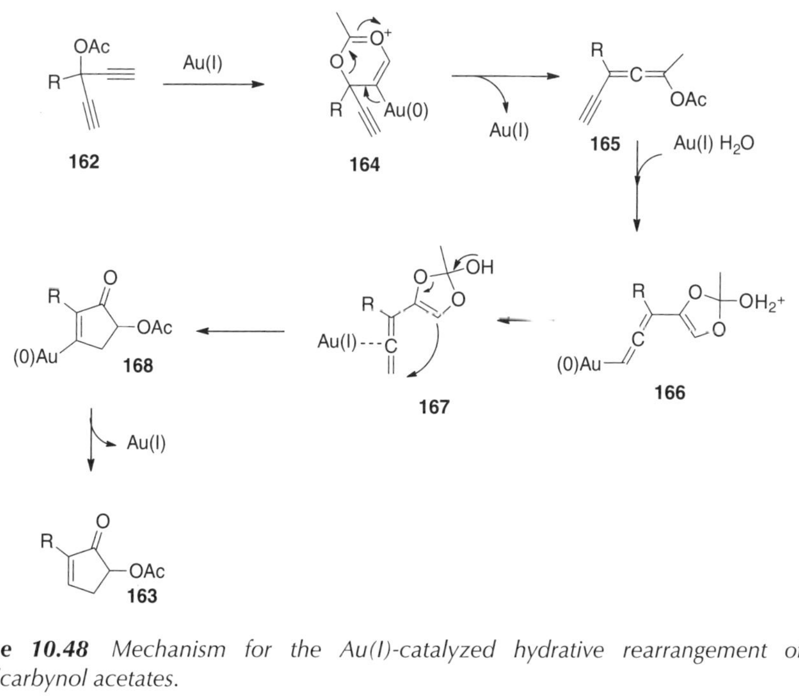

Scheme 10.48 *Mechanism for the Au(I)-catalyzed hydrative rearrangement of 1,1-diethylcarbynol acetates.*

Rhee and co-workers described the use of Au(I) catalysts for the synthesis of highly substituted 2-cyclopentenones from 5-siloxypent-3-en-1-ynes.[83] The mechanism of the reaction could be easily rationalized by an initial siloxycyclization of **167** followed by ionization of a C–O bond to generate a carbocation intermediate **169**. Intramolecular carbocyclization and the subsequent elimination of cationic gold(I) via a cyclic intermediate **170** with the assistance of iPrOH produces the highly substituted cyclopentenone (Scheme 10.49).

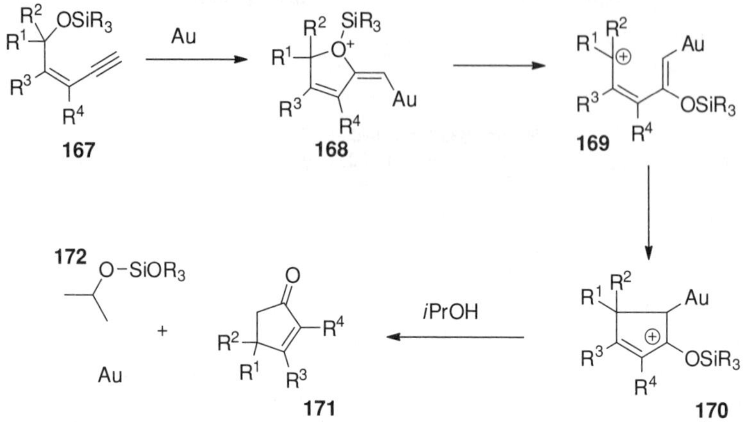

Scheme 10.49 *Mechanism for the Au-catalyzed synthesis of cyclopentenones.*

The reaction gives access to densely substituted cyclopentenones possessing a quaternary carbon center, in good yields as shown in Scheme 10.50.

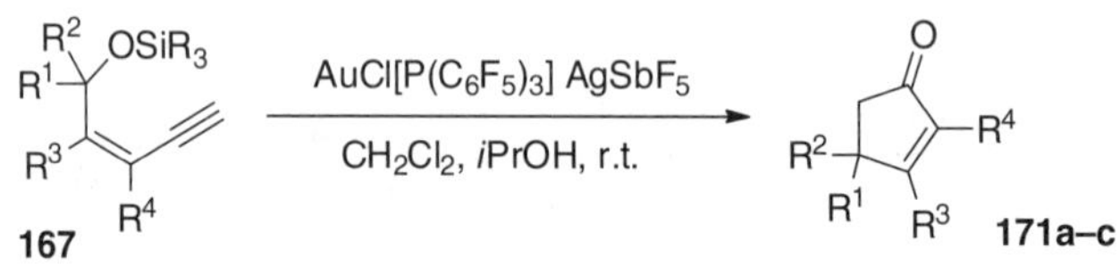

171a R^1=Ph R^2=Me R^3=H R^4=*n*C8H17 97% yield
171b R^1=Ph R^2=Et R^3=H R^4=*n*C8H17 91%yield
171c R^1=allyl R^2=Me R^3=R^4=-(CH2)3- 91%yield

Scheme 10.50 *Cylopentanone synthesis reported by Rhee.*

10.7 Conclusions

In this chapter, we presented a variety of transition-metal mediated alternatives to the Pauson-Khand reaction for the synthesis of cyclopentenones. For example, the long-known nickel-mediated carbonylation reaction represents a very attractive methodology and offers a great degree of regioselectivity in the intramolecular version. Unfortunately, however, nickel tetracarbonyl, which is required in stoichiometric amounts for the reaction to occur, is very hazardous to handle and the scope of the reaction is very limited. In contrast, Pd catalysis or alkenylchromium carbene compounds represent powerful alternatives to the venerable Pauson-Khand reaction, complementing and improving it.

An important advantage of the nickel-catalyzed transformations is the use of aldehydes as carbon monoxide surrogates, in which case the only limitation of the reaction is the nature of the reactant itself: the need to use 4-pentynal or 5-hexynal reduces the applicability of this methodology. Another interesting alternative is the use of Au catalysis for the isomerization of alkynes and intramolecular cyclization. Gold catalysis is, nowadays, one of the most exciting fields in organic catalysis and new developments are expected. However, it is clear that circumventing the use of highly toxic carbon monoxide is the alternative of choice and recent developments in this area are very promising.

Experimental

Typical experimental Procedure for the nickel catalyzed carbonylative cycloaddition of allyl halides and acetylenes reported by Moretó et al.[38]

A solution of nickel(0) tetracarbonyl (40 mmols) in dry acetone (5 mL) was added dropwise at 15 °C into a solution of the alkyne (20 mmols) and allyl chloride (40 mmols) in acetone-methanol (40:5 mL) placed in a thermostated raction flask, equipped with magnetic stirrer, dropping funnel, thermometer, gas inlet, and mercury valve, which had been previously purged with Argon, and the reaction mixture was kept for 15 h at 14 °C. Then, the temperature was raised to 40 °C and after one additional hour a stream of nitrogen was passed through the mixture for 5 h, keeping the temperature at 40 °C, to remove any unreacted nickel(0) tetracarbonyl and most of the solvent. The remaining solvent was removed under vacuum and the crude reaction product was treated with water (40 mL) and repeatedly extracted with dichloromethane. The extract was dried over magnesium sulfate and after removal of the solvent, flash chromatography of the residue afforded the cyclopentenone.

Typical experimental procedure for the reaction of Fischer carbene complexes with alkynes reported by Moser.[22]

A solution of carbene complex (2.0 mmols) and TIPS-substituted phenyl acetylene (2.40 mmols) in benzene (20 mL) was degassed using several freeze–pump–thaw cycles. The solution was then heated at reflux for 12 h. Upon cooling to room temperature, the solvents were removed under reduced pressure and the residue purified via flash chromatography.

Typical experimental procedure for the rearrangenment of 1,1-diethynilcarbinol-acetates reported by Oh.[80]

In a small flask, 3 mol% of $AuCl(PPh_3)$ and SbF_6 was taken in 1.0 mL of wet dichloromethane at 0 °C. It was then allowed to reach room temperature (25 °C) followed by addition of 1,1-diethylcarbinolacetate dissolved in the minimum volume of wet dichloromethane. The resulting mixture was stirred for 1–2 h at room temperature and monitored periodically by TLC. Upon completion, the solvent was subjected to flash column chromatography to afford the pure cyclopentenones.

References and Notes

[1] For an excellent and inspirational review about transition metal catalyzed synthesis of cyclopentenones see: S. E. Gibson, S. E. Lewis, N. Mainolfi, "Transition metal-mediated routes to cyclopentenones" *J. Org. Chem.*, **689**, 3873–3890, (2004).

[2] H.-W. Frühauf, "Metal-Assisted Cycloaddition Reactions in Organotransition Metal Chemistry" *Chem. Rev.*, **97**, 523–596, (1997).
[3] R. A. Ellison, "Methods for the synthesis of 3-oxocyclopentenes" *Synthesis*, 397–412, (1973).
[4] R. Noyori, H. Kozano, M. Mori, R. Hirata, Y. Shiga, T. Kokura, M. Suzuki, "Unnatural prostaglandins of biochemical and physiological significance" *Pure Apll. Chem.*, **66**, 1999–2005, (1994).
[5] R. Noyori, M. Suzuki, "Organic synthesis of prostaglandins: advancing biology" *Science*, **259**, 44–45, (1993).
[6] A. Cordova (Ed.), *Enantioselective Conjugate Addition*, Wiley, (2010).
[7] E. M. Carreira, in: E. N. Jacobsen, A. Pfaltz, H. Yamamoto (Eds.), *Comprehensive Asymmetric Catalysis*, vol III, Springer, Heidelberg, p. 997 (Chapter 29), (1999).
[8] C. L. Chandler, B. List, "Catalytic, asymmetric transannular aldolizations: total synthesis of (+)-hirsutene" *J. Am. Chem. Soc.* **130**, 6737–6739, (2008).
[9] K. B. Becker, "Cycloalkenes by intramolecular Wittig reaction" *Tetrahedron*, **36**, 1717–1745, (1980).
[10] C. Kuhn, J.-C. Florent, "A carbohydrate approach to 4-hydroxy-2-cyclopentenone moiety of antitumor prostanoid punaglandin IV via alkylation of ester urinate" *Tetrahedron Lett.*, **39**, 4247–4250, (1998).
[11] For an exceptional review about Nazarov reaction see: H. Pellissier "Recent developments in the Nazarov process" *Tetrahedron.*, **61**, 6479–6517, (2005).
[12] M. Rueping, W. Ieawsuwan, A. P. Antonchick, B. J. Nachsteim, "Chiral Bronsted acids in the catalytic asymmetric Nazarov Cyclization- The first enantioselective organocatalytic electrocyclic reaction", *Angew. Chem. Int. Ed.*, **46**, 2097–2100, (2007).
[13] W. Kirmse, "Alkenylidenes in organic synthesis" *Angew. Chem. Int. Ed.*, **36**, 1164–1170, (1997).
[14] G. G. G. Manzardo, M. Karpf, A. S. Dreiding, "Formal total synthesis of (+−)-isocomene by application of the alpha -alkynone cyclization" *Helv. Chem. Acta*, **69**, 659–669, (1986).
[15] M. Franck-Neumann, E. L. Michelotti, R. Simler, J.-M. Vernier, "Cyclocarbonylation of acyclic 1,3-dienes via their tricarbonyl iron complexes: cyclopenten-2-ones and dicarbonyl cyclopentadienyl iron halides" *Tetrahedron Lett.*, **33**, 7361–7364, (1992).
[16] M. Franck-Neumann, J.-M. Vernier, "Cyclocarbonylation of 1,1,3-trisubstituted 1,3-dienes via their tricarbonyl iron complexes. Synthesis of cyclocondensed and (or) spirocyclic (optically active) cyclopenten-2-ones" *Tetrahedron Lett.*, **33**, 7365–7368, (1992).
[17] E. Negishi, S. Ma, J. Amanfu, C. Coperet, J. A. Miller, J. M. Tour, "Palladium-Catalyzed Cyclization of 1-Iodo-Substituted 1,4-, 1,5-, and 1,6-Dienes as Well as of 5-Iodo-1,5-dienes in the Presence of Carbon Monoxide" *J. Am. Chem. Soc.*, **118**, 5919–5931 (1996).
[18] J. M. Tour, E. Negishi, "Metal promoted cyclization. 9. Controlled and catalytic acylpalladation. A novel route to cyclopentenone and cyclohexenone derivatives" *J. Am. Chem. Soc.*, **107**, 8289–8291, (1985).

[19] E. Negishi, C. Copéret, S. Ma, T. Mita, T. Sugihara, J. M. Tour, "Palladium-Catalyzed Carbonylative Cyclization of 1-Iodo-2-alkenylbenzenes". *J. Am. Chem. Soc.*, **118** (25), 5904–5918, (1996).

[20] S. V. Gagnier, R. C. Larock, "Palladium-Catalyzed Carbonylative Cyclization of Unsaturated Aryl Iodides and Dienyl Triflates, Iodides, and Bromides to Indanones and 2-Cyclopentenones" *J. Am. Chem. Soc.*, **125**, 4804–4807, (2003).

[21] C. Laroche, P. Bertus, J. Szymoniak, "Titanium-mediated [4+1] assembly of 1,3-dienes and nitriles: formation of 3-cyclopentenyl amines and cyclopentenones" *Chem. Commun.*, 3030–3035, (2005).

[22] W. H. Moser, L. A. Feltes, L. Sun, M. W. Giese, R. W. Farrell, "Stereoselective [4+1] annulation reactions with silyl vinylketenes derived from Fischer carbene complexes" *J. Org. Chem.*, **71**, 6542–6546, (2006).

[23] J. L. Loebach, D. M. Bennet, R. L. Danheiser, "The Reaction of (Trialkylsilyl)vinylketenes with Carbenoid Reagents: A New [4+1] Annulation Route to Cyclopentenones" *J. Am. Chem. Soc.*, **120**, 9690–9691, (1998).

[24] R. Noyori, "Organic syntheses via the polybromo ketone-iron carbonyl reaction" *Acc. Chem. Res.*, **12**, 61–66, (1979).

[25] Y. Hayakawa, K. Yokoyama, R. Noyori, "Novel cyclopentenone synthesis via the iron carbonyl aided cyclocoupling between alpha, alpha '-dibromo ketones and enamines" *J. Am. Chem. Soc.*, **100**, 1799–1806, (1978).

[26] R. Noyori, Y. Hayakawa, M. Funakura, H. Takaya, "Carbon-carbon bond formations promoted by transition metal carbonyls. III. Mechanistic aspects of the reaction of alpha,alpha 1-dibromo ketones and iron carbonyl. Reductive rearrangements of dibromo ketones" *J. Am. Chem. Soc.*, **94**, 7202–7204, (1972).

[27] R. Noyori, Y. Hayakawa, H. Takaya, S. Murai, R. Kobayashi, N. Sonoda, "Carbon-carbon bond formation promoted by transition metal carbonyls. 16. Reaction of alpha alpha '-dibromo ketones and iron carbonyls. Mechanistic aspects" *J. Am. Chem. Soc.*, **100**, 1759–1765, (1978).

[28] H. L. Stokes, L. Ni, J. A. Belot, M. E. Welker, "Reactions of transition-metal h1-allyl and propargyl complexes with ketenes and some new nonoxidative transition-metal-carbon bond cleaving reactions of the ketene cycloadducts which yield cyclopentenones" *J. Organomet. Chem.*, **487**, 95–104, (1995).

[29] L. Ni, J. A. Belot, M. E. Welker, "Nonoxidative transition-metal-carbon bond cleaving reactions and their application in a transition-metal mediated cyclopentenone synthesis" *Tetrahedron Lett.*, **33**, 177–180, (1992).

[30] J. Barluenga, R. Vicente, L. A. Lopez, E. Rubio, M. Tomas, C. Alvarez-Rua "New Fischer carbene complexes of Rhodium(I): Preparation and 2-cyclopentenone ring synthesis by annelation to alkynes" *J. Am. Chem. Soc.*, **126**, 470–471, (2004).

[31] J. Barluenga, P. Barrio, L. Riesgo, L. A. Lopez, M. Tomas, "A general and regioselective synthesis of cyclopentenone derivatives through Nickel(0)-mediated [3+2] cyclization of alkenyl Fischer carbene complexes and internal alkynes" *J. Am. Chem. Soc.*, **129**, 14422–14426, (2007).

[32] J. Barluenga, P. Barrio, L. A. Lopez, M. Tomas, S. Garcia-Granda, C. Alvarez-Rua, "Nickel(0)-Mediated [3+2+2] and [2+2+2+1] Cyclization Reactions of Chromium Fischer Carbene Complexes and Alkynes" *Angew. Chem. Int. Ed.*, **42**, 3008–3011, (2003).

[33] G. P. Chiusoli, "Catalysis of olefin and carbon monoxide insertion reactions" *Acc. Chem. Res.*, **6**, 422. (1973).
[34] L. Cassar, G. P. Chiusoli, M. Foa, "Stereospecific character of cyclizing carbonylation of allyl halides" *Tetrahedron Lett.*, **8**, 285–287, (1967).
[35] L. Cassar, G. P. Chiusoli, "New types of carbonylation with insertion of acetylene" *Tetrahedron Lett.*, **6**, 3295–3299, (1965).
[36] F. Camps, J. M. Moreto, L. Pages, "Influence of functional substitution of allyl halides on their nickel tetracarbonyl promoted carbonylative cycloaddition with acetylenes" *Tetrahedron*, **15**, 3147–3162, (1992).
[37] F. Camps, J. Coll, J. M. Moreto, J. Torras, "Studies for control of the nickel tetracarbonyl promoted carbonylative cycloaddition allyl halides and acetylenes" *J. Org. Chem.*, **54**, 1969–1978, (1989).
[38] A. Llebaria, J. M. Moreto, "The [$Ni(CO)_4$]-promoted carbonylative cycloaddition of alkynes and allyl halides for the synthesis of 2-cyclopentenone derivatives" *J. Organomet. Chem.*, **451**, 1–2, (1993).
[39] F. Camps, J. Coll, J. M. Moreto, J. Torras, "Stereoselective synthesis of bicyclo[3.3.0]oct-1-en-3-one derivatives" *Tetrahedron Lett.*, **28**, 4745–4748, (1987).
[40] F. Camps, J. Coll, J. M. Moreto, J. Torras, "A total synthesis of methylenomycin B" *Tetrahedron Lett.*, **26**, 6397–6398, (1985).
[41] L. Pages, A. Llebaria, F. Camps, E. Molins, C. Miravitlles, J. M. Moreto, "Ring-fused and spiro cyclopentenones by $Ni(CO)_4$-promoted intermolecular carbonylative cycloaddition of acetylenes with 3-halo- and 1-(halomethyl)cycloalkenes" *J. Am. Chem. Soc.*, **114**, 10449–10461, (1992).
[42] F. Camps, A. Llebaria, J. M. Moreto, L. Pages, "Fused bicyclic systems through a nickel mediated intermolecular cyclopentenone annulations" *Tetrahedron Lett.*, **33**, 109–112, (1992).
[43] F. Camps, A. Llebaria, J. M. Moreto, L. Pages, "New practical synthesis of spirocyclopentenones" *Tetrahedron Lett.*, **33**, 113–114, (1992).
[44] J. M. Villar, A. Delgado, A. Llebaria, J. M. Moretó, "An approach to homochiral fused and spiro cyclopentenones" *Tetrahedron: Asymmetry*, **6**, 665–668, (1995).
[45] J. M. Villar, A. Delgado, A. Llebaria, J. M. Moretó, "Asymmetric approaches to cyclopentenones in the Ni(0)-promoted cyclocarbonylation reaction of allyl halides and acetylenes" *Tetrahedron*, **52**, 10525–10546. (1996).
[46] K. Tamao, K. Kobayashi, Y. Ito, "Nickel(0)-promoted cyclization of enzymes with isocyanides: a new route to polycyclic cyclopentenone skeletons" *J. Am. Chem. Soc.*, **110**, 1286–1288, (1988).
[47] K. Tamao, K. Kobayashi, Y. Ito, "Nickel(0)-mediated intramolecular cyclizations of enynes, dienynes, bis-dienes, and diynes" *Synlett*, 539–546, (1992).
[48] G. Garcia-Gomez, J. M. Moreto, "Intramolecular Ni-Mediated Cyclizations with α,ω-Dienals" *Eur. J. Org. Chem.*, 1359–1369, (2001).
[49] G. Garcia-Gomez, J. M. Moreto, "Cyclopentenones from Carbonylative Cycloaddition of Mackenzie's Allyl Nickel Complexes and Acetylenes" *J. Am. Chem. Soc.*, **121**, 878–880, (1999).
[50] G. Garcia-Gomez, J. M. Moreto, "Change of the reaction pattern by methodological variations in a multicomponent assembly promoted by Ni complexes" *Chem. Eur. J.*, **7**, 1503–1509, (2001).

[51] G. G. Gomez, X. Camps, A. J. Cayuela, J. M. Moreto, "Ni(COD)$_2$: an efficient promoter in the construction of 5-cyclopent-2-enone derivatives from allyl halides, alkynes and carbon monoxide" *Inorg. Chim. Acta*, **296**, 94–102, (1999).
[52] F. Vilaseca, L. Pages, J. M. Villar, A. Llebaria, A. Delgado, J. M. Moreto, "Preference for beta -H elimination in the termination of the Ni-promoted carbonylative cycloaddition of 2-haloethylidenecycloalkanes and alkynes" *J. Organomet. Chem.*, **551**, 107–115, (1998).
[53] W. Oppolzer, "Intramolecular, Stoichiometric (Li, Mg, Zn) and Catalytic (Ni, Pd, Pt) Metallo-Ene Reactions in Organic Synthesis [New Synthetic Methods (75)]" *Angew. Chem. Int. Ed.*, **28**, 38–52, (1989).
[54] W. Oppolzer, T. H. Keller, M. Bedoya-Zurita, C. Stone, "Diastereocontrolled nickel(0)- and palladium(0)-catalyzed "metallo-ene-type" cyclizations/ carbonylations" *Tetrahedron. Lett.*, **30**, 5883–5886, (1989).
[55] W. Oppolzer, M. Bedoya-Zurita, C. Y. Switze, "Nickel-catalyzed intramolecular allylation of alkenes and alkynes coupled with beta -elimination or carbonylation" *Tetrahedron Lett.*, **29**, 6433–6436, (1988).
[56] W. Oppolzer, C. Robyr, "Synthesis of (+-)-hirsutene by a catalytic allylpalladium-alkyne cyclization/carbonylation cascade" *Tetrahedron*, **50**, 415–424, (1994).
[57] W. Oppolzer, J.-Z. Xu, C. Stone, "Preparation of a key intermediate for the synthesis of the (+-)-pentalenolactone E methyl ester by catalytic 'palladium-ene cyclization'/methoxycarbonylation" *Hel. Chim. Acta*, **74**, 465–468, (1991).
[58] W. Oppolzer, "Regio- and stereocontrolled catalytic palladium- and nickel 'ene-type' cyclizations" *Pure Appl. Chem.*, **62**, 1941–1948, (1990).
[59] N. C. Ihle, C. H. Heathcock, "Palladium-catalyzed intramolecular alkyne-carbon monoxide-alkene insertion cascade for synthesis of alpha –methylenecyclopentenones" *J. Org. Chem.*, **58**, 560–563, (1993).
[60] J. W. Herndon, P. P. Patel, "Effect of solvent polarity extremes on the coupling of cyclopropylcarbene-chromium complexes and alkynes: synthesis of beta -alkoxycyclopentadienones, 2-cyclopentene-1,3-diones, and cis 4,5-disubstituted cyclopentenones" *Tetrahedron Lett.*, **38**, 59–62, (1997).
[61] J. W. Herndon, S. U. Tumer, W. F. K. Schnatter, "Cyclopentenones from the reaction of alkynes with cyclopropylcarbenechromium complexes" *J. Am. Chem. Soc.*, **110**, 3334–3335, (1988).
[62] J. W. Herndon, S. U. Tumer, "Cyclopentadienones in the reaction of alkynes with cyclopropylcarbene-chromium complexes" *Tetrahedron Lett.*, **30**, 295–296, (1989).
[63] J. W. Herndon, D. K. Hill, L. A. McMullen, "Conversion of 2-alkenyl cyclopropylcarbene-chromium complexes to 5-alkenyl-2-cyclopentenones: a stereospecific three- to five-carbon ring expansion reaction" *Tetrahedron Lett.*, **36**, 5687–5690, (1995).
[64] S. U. Tumer, J. W. Herndon, L. McMullen, "Reaction of alkynes with cyclo propylcarbene-chromium complexes: a versatile [4+2+1-2] cycloaddition reaction for the construction of cyclopentenones" *J. Am. Chem. Soc.*, **114**, 8394–8404, (1992).
[65] J. W. Herndon, J. J. Matasi, "Synthesis of cyclopentane-fused oxygen heterocycles from the intramolecular reaction of alkynes with cyclopropylcarbene-chromium complexes" *J. Org. Chem.*, **55** (3), 786–788, (1990).

[66] B. A. Anderson, W. D. Wulff, A. L. Rheingold, "Characterization of the first .eta.4-vinylketene metal complex from the reaction of a Group 6 Fischer carbene complex and an alkyne" *J.Am. Chem. Soc.*, **112**, 8615–8617, (1990).

[67] J. W. Herndon, J. Zhu, D. Sampedro, "Synthesis of alkylidenecyclopentenones and dialkylidenecyclopentenones via the coupling of propargylic alcohol and 2-alkyne-1,4-diol derivatives with cyclopropylcarbene-chromium complexes" *Tetrahedron*, **56** (28), 4985–4993, (2000).

[68] A. Padwa, D. J. Austin, S. L. Xu, "Rhodium(II) catalyzed cyclizations of α-diazo substituted alkynes. A new mode of reaction" *Tetrahedron Lett.*, **32**, 4103–4106, (1991).

[69] A. Padwa, U. Chiacchio, D. J.Fairfax, J.M. Kassir, A. Litrico, M. A. Semones, S. L. Xu, "A comparative study of the decomposition of o-alkynyl-substituted aryl diazo ketones. Synthesis of polysubstituted β-naphthols via arylketene intermediates" *J. Org. Chem.*, **58**, 6429–6437, (1993).

[70] A. Padwa, "Rhodium(II) mediated cyclizations of diazo alkynyl ketones" *J. Organomet. Chem.*, **617-618**, 3–16, (2001).

[71] A. Padwa, D. J. Austin, S.L. Xu, "Control of chemoselectivity in the rhodium(II)-catalyzed alkyne insertion reaction of alpha -diazo ketones" *J. Org. Chem.*, **57** (5), 1330–1331, (1992).

[72] T. R. Hoye, C. J. Dinsmore, "Double (internal/external) alkyne insertion reactions of alpha -diazo ketones" *Tetrahedron Lett.*, **32** (31), 3755–3758, (1991).

[73] G. Deng, X. Tian, J. Wang, "A novel approach to bicyclic fused cyclopentenone derivatives" *Tetrahedron Lett.*, **44** (3), 587–589, (2002).

[74] T. Kondo, A. Nakamura, T. Okada, N. Suzuki, K. Wada, T. Mitsudo, "Ruthenium-Catalyzed Reconstructive Synthesis of Cyclopentenones by Unusual Coupling of Cyclobutenediones with Alkenes Involving Carbon-Carbon Bond Cleavage" *J. Am. Chem. Soc.*, **122**, 6319–6320, (2000).

[75] K. Tanaka, G. C. Fu, "A versatile new method for the synthesis of cyclopentenones via an unusual rhodium-catalyzed intramolecular trans hydroacylation of an alkyne" *J. Am. Chem. Soc.*, **123** (46), 11492–11493, (2001).

[76] K. Tanaka, G. C. Fu, "Enantioselective Synthesis of Cyclopentenones via Rhodium-Catalyzed Kinetic Resolution and Desymmetrization of 4-Alkynals" *J. Am. Chem. Soc.*, **124** (35), 10296–10297, (2002).

[77] K. Takeishi, K. Sugishima, K. Sasaki, K. Tanaka, "Rhodium catalyzed intramolecular hydroacylation of 5- and 6-alkynals: convenient synthesis of α-alkylidenecycloalkanones and cycloalkenones" *Chem. Eur. J.*, **10**, 5681–5688, (2004).

[78] G. R. Clark, S. Thiensathit, "A palladium-catalyzed rearrangement of 1-vinyl-1-cyclobutanols" *Tetrahedron Lett.*, **26** (21), 2503–2506, (1985).

[79] T. Matsuda, T. Tsuboi, M. Murakami, "Rhodium-catalyzed carbonylation of spiropentanes" *J. Am. Chem. Soc.*, **129**, 12596–12597, (2007).

[80] C. Bee, E. Leclerc, M. A. Tius, "The Palladium(II)-Catalyzed Nazarov Reaction" *Org. Lett.*, **5**, 4927–4930, (2003).

[81] P. Cao, C. Deng, Y.-Y. Zhou, X.-L. Sun, J.-C. Zheng, Z. Xie, Y. Tang, "Asymmetric Nazarov reaction catalyzed by chiral trisoxazoline/copper(II)" *Angew. Chem. Int. Ed.*, **49**, 4463–4465, (2010).

[82] C. H. Oh, S. Karmakar, "Au(I)-catalyzed hydrative rearrangement of 1,1-diethylcarbinol acetates to functionalized cyclopentenones and allenones" *J. Org. Chem.*, **74**, 370–374, (2009).

[83] S. E. An, J. Jeong, B. Baskar, J. Lee, J. Seo, Y. H. Rhee, "Gold(I)-catalyzed synthesis of highly substituted 2-cyclopentenones from 5-Siloxypent-3-en-1-ynes" *Chem. Eur. J.*, **15**, 11837–11841, (2009).

Index

The Pauson-Khand Reaction: Scope, Variations and Applications, First Edition. Edited by Ramon Rios Torres.